Analysis of Turbulent Flows with Computer Programs

THIRD EDITION

Tuncer Cebeci

Formerly Distinguished Technical Fellow,
Boeing Company, Long Beach, California
horizonpublishing.net

ELSEVIER

AMSTERDAM • BOSTON • HEIDELBERG • LONDON
NEW YORK • OXFORD • PARIS • SAN DIEGO
SAN FRANCISCO • SINGAPORE • SYDNEY • TOKYO

Butterworth-Heinemann is an Imprint of Elsevier

BH

Butterworth-Heinemann is an imprint of Elsevier
The Boulevard, Langford Lane, Kidlington, Oxford OX5 1GB, UK
225 Wyman Street, Waltham, MA 02451, USA

First edition 1974
Second edition 2004
Third edition 2013

British Library Cataloguing in Publication Data
A catalogue record for this book is available from the British Library

Library of Congress Cataloging-in-Publication Data
A catalog record for this book is availabe from the Library of Congress

ISBN: 978-0-08-098335-6

13 14 15 16 17 10 9 8 7 6 5 4 3 2 1

Dedication

This book is dedicated to the memory of my beloved wife, Sylvia Holt Cebeci, my best friend of many years. She will always be with me, in my heart, in my memories and in all the loving ways she touched my life each day.

I will always remember her note to me on my birthday in 2011 prior to her death on May 4, 2012:

Dear TC,

The twilight times bring you automatic admission to a special club; to share the game of special 50's, 60's music; remembering your friends and death of a special friend; being a historian simply because you live long enough. You and I have been so fortunate to share a golden life everyday.

Love, Sylvia

Contents

Preface to the Third Edition

The first edition of this book, *Analysis of Turbulent Boundary Layers*, was written in the period between 1970 and early 1974 when the subject of turbulence was in its early stages and that of turbulence modeling in its infancy. The subject had advanced considerably over the years with greater emphasis on the use of numerical methods and an increasing requirement and ability to calculate turbulent two- and three-dimensional flows with and without separation. The tools for experimentation were still the traditional Pitot tube and hot wire-anemometer so that the range of flows that could be examined was limited and computational methods still included integral methods and a small range of procedures based on the numerical solution of boundary layer equations and designed to match the limited range of measured conditions. There have been tremendous advances in experimental techniques with the development of non-intrusive optical methods such as laser-Doppler, phase-Doppler and particle-image velocimetry, all for the measurement of velocity and related quantities and of a wide range of methods for the measurement of scalars. These advances have allowed an equivalent expansion in the range of flows that have been investigated and also in the way in which they could be examined and interpreted. Similarly, the use of numerical methods to solve time-averaged forms of the Navier-Stokes equations, sometimes interactively with the inviscid-flow equations, has expanded, even more so with the rise and sometimes fall of Companies that wished to promote and sell particular computer codes. The result of these developments has been an enormous expansion of the literature and has provided a great deal of information beyond that which was available when the first edition was written. Thus, the topics of the first edition needed to be re-examined in the light of new experiments and calculations, and the ability of calculation methods to predict a wide range of practical flows, including those with separation, to be reassessed.

The second edition, entitled *Analysis of Turbulent Flows*, undertook the necessary reappraisal, reformulation and expansion, and evaluated the calculation methods more extensively but also within the limitations of two-dimensional equations largely because this made explanations easier and the book of acceptable size. In addition, it was written to meet the needs of graduate students as well as engineers and so included homework problems that were more sensibly formulated within the constraints of two independent variables. References to more complex flows, and particularly those with separation, were provided and the relative merits of various turbulence models considered.

The third edition, entitled *Analysis of Turbulent Flows with Computer Programs*, keeps the structure of the first and second editions the same. It expands the solution of the boundary-layer equations with transport-equation turbulence models, considers the solution of the boundary-layer equations with flow separation and provides computer programs for calculating attached and separating flows with several turbulence models.

The second edition and the contents of this new edition should be viewed in the context of new developments such as those associated with large-eddy simulations (LES) and direct numerical solutions (DNS) of the Navier-Stokes equations. LES existed in 1976 as part of the effort to represent meteorological flows and has been rediscovered recently as part of the recognition of the approximate nature of solutions of time-averaged equations as considered here. There is no doubt that LES has a place in the spectrum of methods applied to the prediction of turbulent flows but we should not expect a panacea since it too involves approximations within the numerical method, the filter between time-dependent and time-average solutions and small-scale modeling. DNS approach also has imperfections and mainly associated with the computational expense which implies compromises between accuracy and complexity or, more usually, restriction to simple boundary conditions and low Reynolds numbers. It is likely that practical aerodynamic calculations with and without separation will continue to make use of solutions of the inviscid-flow equations and some reduced forms of the Navier-Stokes equations for many years, and this book is aimed mainly at this approach.

The first and second editions were written with help from many colleagues. AMO Smith was an enthusiastic catalyst and ideas were discussed with him over the years. Many colleagues and friends from Boeing, the former Douglas Aircraft Company and the McDonnell-Douglas Company, have contributed by discussion and advice and included K. C. Chang and J. P. Shao. Similarly, Peter Bradshaw, the late Herb Keller of Cal Tech and the late Jim Whitelaw of Imperial College have helped in countless ways.

Indian Wells

Tuncer Cebeci

Computer Programs Available from horizonpublishing.net

1. Integral Methods.
2. Differential Method with CS Model for two-dimensional flows with and without heat transfer and infinite swept-wing flows.
3. Hess-Smith Panel Method with and without viscous effects.
4. Zonal Method for k-ε Model and solution of k-ε Model equations with and without wall functions.
5. Differential Method for SA Model and for a Plane Jet.
6. Differential Method for inverse and interactive boundary-layer flows with CS Model.

Chapter 1

Introduction

Chapter Outline Head

1.1 Introductory Remarks

Turbulence in viscous flows is described by the Navier–Stokes equations, perfected by Stokes in 1845, and now soluble by Direct Numerical Simulation (DNS). However, computing capacity restricts solutions to simple boundary conditions and

Analysis of Turbulent Flows with Computer Programs. http://dx.doi.org/10.1016/B978-0-08-098335-6.00001-X

moderate Reynolds numbers and calculations for complex geometries are very costly. Thus, there is need for simplified, and therefore approximate, calculations for most engineering problems. It is instructive to go back some eighty years to remarks made by Prandtl [1] who began an important lecture as follows:

> *What I am about to say on the phenomena of turbulent flows is still far from conclusive. It concerns, rather, the first steps in a new path which I hope will be followed by many others.*
> *The researches on the problem of turbulence which have been carried on at Göttingen for about five years have unfortunately left the hope of a thorough understanding of turbulent flow very small. The photographs and kineto-graphic pictures have shown us only how hopelessly complicated this flow is …*

Prandtl spoke at a time when numerical calculations made use of primitive devices – slide rules and mechanical desk calculators. We are no longer "hopeless" because DNS provides us with complete details of simple turbulent flows, while experiments have advanced with the help of new techniques including non-obtrusive laser-Doppler and particle-image velocimetry. Also, developments in large-eddy simulation (LES) are also likely to be helpful although this method also involves approximations, both in the filter separating the large (low-wave-number) eddies and the small 'sub-grid-scale' eddies, and in the semi-empirical models for the latter.

Even LES is currently too expensive for routine use in engineering, and a common procedure is to adopt the decomposition first introduced by Reynolds for incompressible flows in which the turbulent motion is assumed to comprise the sum of mean (usually time-averaged) and fluctuating parts, the latter covering the whole range of eddy sizes. When introduced into the Navier–Stokes equations in terms of dependent variables the time-averaged equations provide a basis for assumptions for turbulent diffusion terms and, therefore, for attacking mean-flow problems. The resulting equations and their reduced forms contain additional terms, known as the Reynolds stresses and representing turbulent diffusion, so that there are more unknowns than equations. A similar situation arises in transfer of heat and other scalar quantities. In order to proceed further, additional equations for these unknown quantities, or assumptions about the relationship between the unknown quantities and the mean-flow variables, are required. This is referred to as the "closure" problem of turbulence modeling.

The subject of turbulence modeling has advanced considerably in the last seventy years, corresponding roughly to the increasing availability of powerful digital computers. The process started with 'algebraic' formulations (for example, algebraic formulas for eddy viscosity) and progressed towards methods in which partial differential equations for the transport of turbulence quantities (eddy viscosity, or the Reynolds stresses themselves) are solved simultaneously with reduced forms of the

Navier–Stokes equations. At the same time numerical methods have been developed to solve forms of the conservation equations which are more general than the two-dimensional boundary layer equations considered at the Stanford Conference of 1968.

The first edition of this book was written in the period from 1968 to 1973 and was confined to algebraic models for two-dimensional boundary layers. Transport models were in their infancy and were discussed without serious application or evaluation. There were no similar books at that time. This situation has changed and there are several books to which the reader can refer. Books on turbulence include those of Tennekes and Lumley [2], Lesieur [3], Durbin and Petterson [5]. Among those on turbulence models the most comprehensive is probably that of Wilcox [6].

The second edition of this book had greater emphasis on modern numerical methods for boundary-layer equations than the first edition and considered turbulence models from advanced algebraic to transport equations but with more emphasis on engineering approaches. The present edition extends this subject to encompass separated flows within the framework of interactive boundary layer theory.

This chapter provides some of the terminology used in subsequent chapters, provides examples of turbulent flows and their complexity, and introduces some important turbulent-flow characteristics.

1.2 Turbulence – Miscellaneous Remarks

We start this chapter by addressing the question "What is turbulence?" In the 25th Wilbur Wright Memorial Lecture entitled "Turbulence," von Kármán [7] defined turbulence by quoting G. I. Taylor as follows:

Turbulence is an irregular motion which in general makes its appearance in fluids, gaseous or liquid, when they flow past solid surfaces or even when neighboring streams of the same fluid flow past or over one another.

That definition is acceptable but is not completely satisfactory. Many irregular flows cannot be considered turbulent. To be turbulent, they must have certain stationary statistical properties analogous to those of fluids when considered on the molecular scale. Hinze [8] recognizes the deficiency in von Kármán's definition and proposes the following:

Turbulent fluid motion is an irregular condition of flow in which the various quantities show a random variation with time and space coordinates, so that statistically distinct average values can be discerned.

In addition turbulence has a wide range of wave lengths. The three statements taken together define the subject adequately.

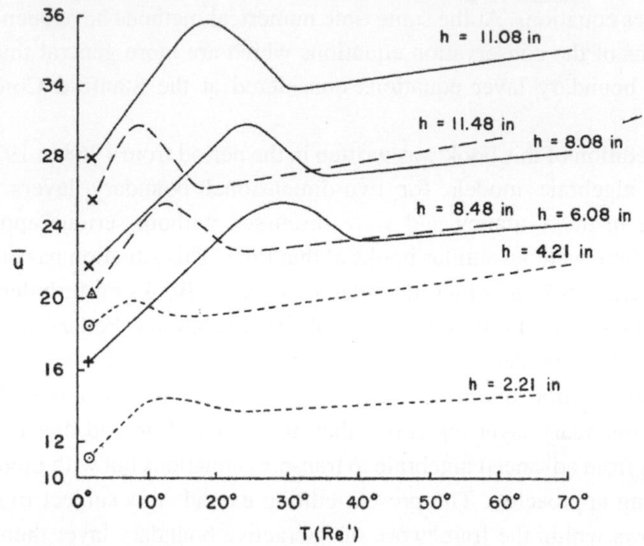

Fig. 1.1 Relation between $\bar{u}$, (expressed in Rhineland inches per second) and the temperature (expressed in degrees Reaumur) for various pipe diameters and heads h (in Rhineland inches), after tests by G. Hagen. — 0.281 cm diam.; – – – 0.405 cm diam.; - - - 0.596 cm diam. [10].

What were probably the first observations of turbulent flow in a scientific sense were described by Hagen [9]. He was studying flow of water through round tubes and observed two distinct kinds of flow, which are now known as laminar (or Hagen-Poiseuille) and turbulent. If the flow was laminar as it left the tube, it looked clear like glass; if turbulent, it appeared opaque and frosty. The two kinds of flow can be generated readily by many household faucets. Fifteen years later, in 1854, he published a second paper showing that viscosity as well as velocity influenced the boundary between the two flow regimes. In his work he observed the mean* velocity $\bar{u}$ in the tube to be a function of both head and water temperature. (Of course, temperature uniquely determines viscosity.) His results are shown in Fig. 1.1 for several tube diameters. The plot contains implicit variations of $\bar{u}$, r_0, and ν, the velocity, the tube radius, and the kinematic viscosity, respectively. This form of presentation displays no orderliness in the data. About thirty years later, Reynolds [11] introduced the parameter $R_r \equiv \bar{u}r_0/\nu$ an example of what is now known as the Reynolds number (with velocity and length scales depending on the problem). It collapsed Hagen's data into nearly a single curve. The new parameter together with

*For now, let "mean" denote an average with respect to time, over a time long compared with the lowest frequencies of the turbulent fluctuations. In Section 2.3 we will give more details of this and other kinds of averaging.

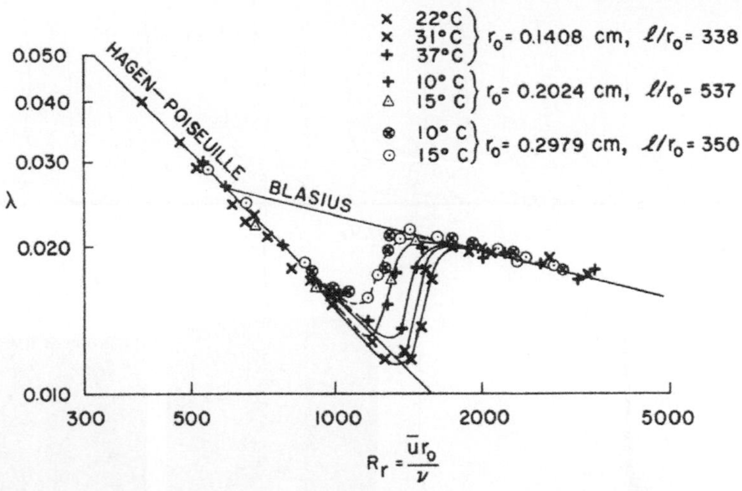

Fig. 1.2 Pressure-drop coefficient vs Reynolds number (Hagen's tests of Fig. 1.1 replotted; squarely cut-off entrance) [10].

the dimensionless friction factor λ, defined such that the pressure drop $\Delta p = \lambda(\varrho\bar{u}^2/2)\,(l/r_0)$, transforms the plot of Fig. 1.1 to that of Fig. 1.2. The quantity l is tube length; the other quantities have the usual meaning. Thus was born the parameter, Reynolds number. The term "turbulent flow" was not used in those earlier studies; the adjective then used was "sinuous" because the path of fluid particles in turbulent flow was observed to be sinusoidal or irregular. The term "turbulent flow" was introduced by Lord Kelvin in 1887.

In the definition of turbulence, it is stated that the flow is irregular. The extreme degree of irregularity is illustrated in Fig. 1.3. If a fine wire is placed transversely in flowing water and given a very short pulse of electric current, electrolysis occurs and the water is marked by minute bubbles of hydrogen that are shed from the length of the wire, provided that the polarity is correct. These bubbles flow along with the stream and mark it. In simple rectilinear flow, the displacement is $\Delta x = u\Delta t$, or, more generally, since u, v, and w motion can occur, $\Delta \mathbf{r} = \int_0^t \mathbf{v}\, d\tau$, where $\mathbf{r}$ is the displacement vector, $\mathbf{v}$ the velocity vector, and t and τ time. Hence the displacement is proportional to the velocity, provided that the times are not too long. The sequence of profiles in Fig. 1.3a was obtained by this hydrogen-bubble technique. All are for the same point in a boundary-layer flow, but at different instants. The variation from instant to instant is dramatic. Figure 1.3b, the result of superposition, shows the time-average displacement for the 17 profiles, and Fig. 1.3c shows the conventional theoretical shape. The average shape remains steady in time, and it is this steadiness of statistical values that makes analysis possible. But Fig. 1.3 shows strikingly

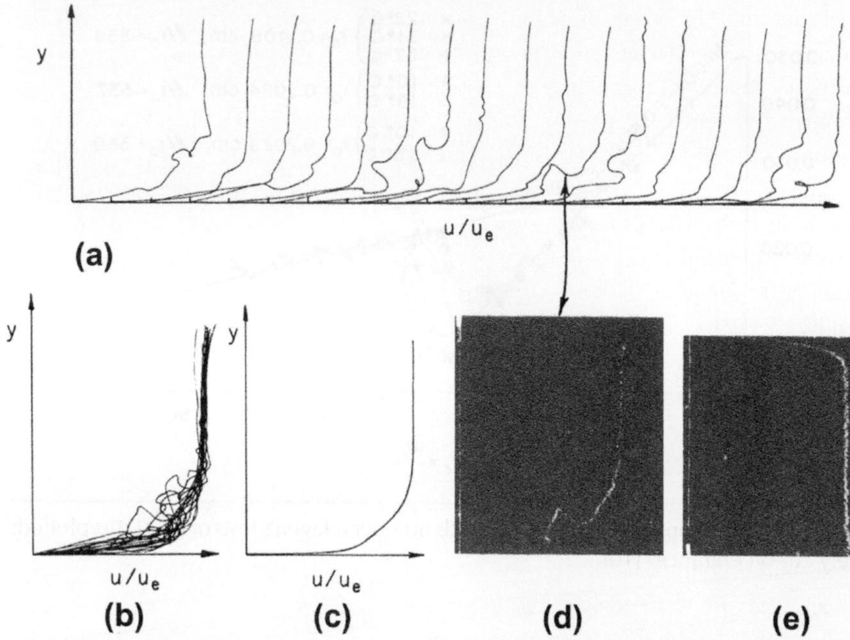

Fig. 1.3 Instantaneous turbulent boundary-layer profiles according to the hydrogen-bubble technique. Measurements were made at $R_x \approx 10^5$ on a flat plate 5 ft aft of leading edge. The boundary layer was tripped. (a) A set of profiles, all obtained at the same position from 17 runs. (b) The same set superimposed. (c) A standard mean profile at the same R_x. (d) Photograph of one of the hydrogen-bubble profiles. (e) A laminar profile on the opposite side of the plate.

that the flow is anything but steady; it is certainly not even a small-perturbation type of flow.

The Reynolds-number parameter has a number of interpretations, but the most fundamental one is that it is a measure of the ratio of inertial forces to viscous forces. It is well known that inertial forces are proportional to ϱV^2. Viscous forces are proportional to terms of the type $\mu \partial u / \partial y$, or approximately to $\mu V/l$, for a given geometry. The ratio of these quantities is

$$\varrho V^2/(\mu V/l) = \varrho Vl/\mu \equiv R_l, \tag{1.2.1}$$

which is a Reynolds number. Whenever a characteristic Reynolds number R_l is high, turbulent flow is likely to occur. In the tube tests of Fig. 1.2, the flow is laminar for all conditions where R_r is below about 1000, and it is turbulent for all conditions where R_r is greater than about 2000. Between those values of R_r is the transition region. Accurate prediction of the transition region is a complicated and essentially unsolved problem.

One fact that is often of some assistance in predicting transition will be mentioned here. Numerous experiments in tube flow with a variety of entrance conditions or degrees of turbulence of the entering flow exist. Preston [12] notes from this information that it seems impossible to obtain fully turbulent flow in a tube at Reynolds numbers R_r less than about 1300 to 2000. His observation is confirmed by the data of Fig. 1.2. Then by considering the similarity of the wall flow for both tube and plate he transfers this observation to low-speed flat-plate flow and concludes that turbulent flow cannot exist below a boundary-layer Reynolds number $R_\theta \equiv u_e\theta/\nu$ of about 320, where u_e is the edge velocity and θ is the momentum thickness defined by

$$\theta = \int_0^\infty \frac{u}{u_e}\left(1 - \frac{u}{u_e}\right) dy \tag{1.2.2}$$

If the laminar boundary layer were to grow naturally from the beginning of the flat plate, the x Reynolds number, $R_x = u_e x/\nu$, would be about 230,000 for $R_\theta = 320$. However, under conditions of very low turbulence in an acoustically treated wind tunnel, an x Reynolds number of 5,000,000 can be reached [13]. Hence, it has been demonstrated that there is a spread ratio of more than 20:1 in which the flow may be either laminar or turbulent. Preston's observation is of importance when turbulent boundary layers are induced by using some sort of roughness to trip the laminar layer, as in wind-tunnel testing. If the model scale is small, R_θ at the trip may be less than 320. Then the trip must be abnormally large – large enough to bring R_θ up to 320. Fortunately, however, the Reynolds number is often so great that there is no problem.

1.3 The Ubiquity of Turbulence

The following series of figures are some examples of turbulent flow that show its ubiquitous character. The eddies and billowing can be clearly seen in the cumulus cloud of Fig. 1.4. Figure 1.5 shows turbulent mixing of two different gases, smoke and air. Even at stellar magnitudes turbulence seems to occur (Fig. 1.6). Turbulent motion can occur at all speeds and under all sorts of conditions: in water at $M \approx 0$, in hypersonic flow, in channels, in rocket nozzles, or on or near external surfaces such as airfoils. Figure 1.7 shows the turbulence in a different way. It shows the wake of a small circular cylinder in a towing tank, made visible by aluminum powder. Although the wake is too close to the cylinder to produce fully developed turbulence, the erratic path lines do indicate turbulence and its wonderful complexity. Figure 1.8, taken at a ballistic range, reveals a turbulent wake at hypersonic speeds.

Fig. 1.4 Turbulent motion in a cumulus cloud.

Fig. 1.5 Turbulent motion in a smoke trail generated to indicate wind direction for landing tests.

1.4 The Continuum Hypothesis

The Navier–Stokes equations and their reduced forms leading to Euler (Chapter 2) and boundary-layer (Chapter 3) equations are derived by considering flow and forces about an element of infinitesimal size, with the flow treated as a continuum.

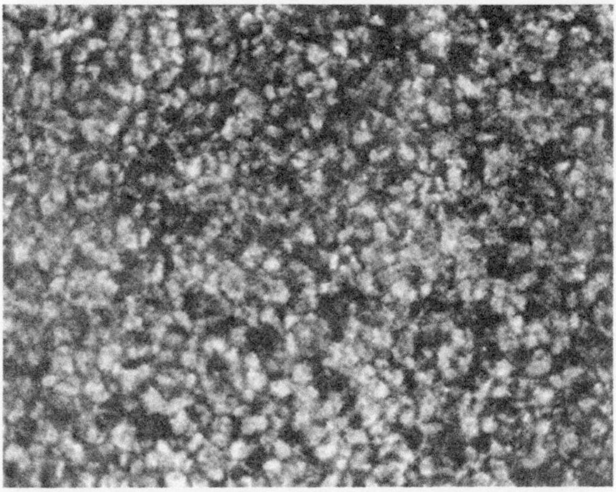

Fig. 1.6 Solar granulations – a highly magnified section of the sun's surface. This appears to be a random flow, a form of turbulence. The pattern changes continuously. It becomes entirely different after about ten minutes. Photo courtesy of Hale Observatories.

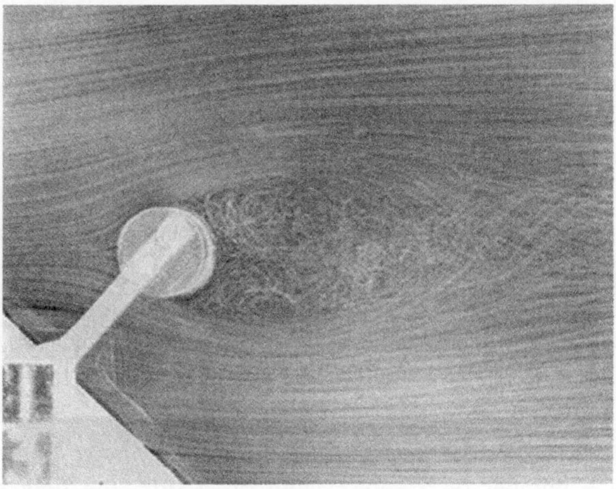

Fig. 1.7 The turbulent motion in the wake of a circular cylinder in water. Motion is made visible by aluminum powder.

Although turbulent eddies may be very small, they are by no means infinitesimal. How well does the assumption of continuity apply?

Avogadro's number states that there are 6.025×10^{23} molecules in a gram molecular weight of gas, which at standard temperature (0°C) and pressure (760 Torr) occupies 22,414 cm^3, which means 2.7×10^{19} molecules/cm^3. Hence

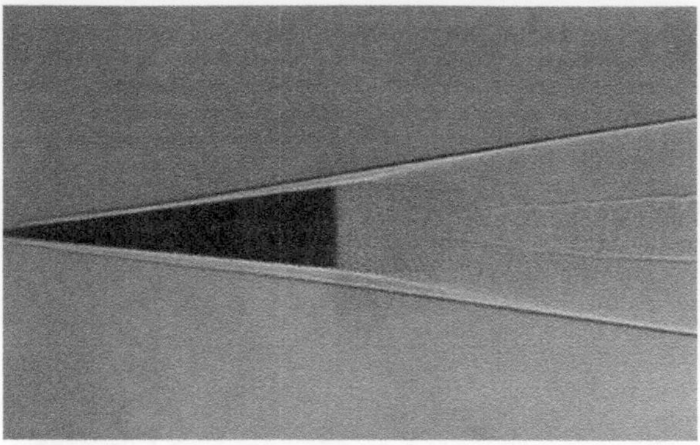

Fig. 1.8 Typical turbulent wake of a 6.3°-half-angle projectile ($M = 10.6$, $R_l = 10.7 \times 10^6$). Photo courtesy of Naval Ordnance Laboratory.

a cube whose edge is only 0.001 cm would contain 2.7×10^{10} molecules. At these standard conditions, the mean free path for gases such as air is approximately 10^{-5} cm, which is significantly smaller than the edge of the 0.001-cm cube. The total number of collisions γ per second in a cubic centimeter is $\gamma = \nu\, \bar{c}\, /2\lambda$, where ν is the number of molecules in a cubic centimeter, $\bar{c}$ is the mean velocity (for air roughly 5×10^4 cm/sec), and λ is 10^{-5} cm. For these representative numbers, $\gamma = 6.75 \times 10^{28}$ collisions/sec cm^3, and the collision frequency for a molecule is 5×10^9/sec or in a 0.001-cm cube the number is 6.75×10^{19} collisions/sec. Hence, under standard conditions, even very small eddies should obey the laws of continuum mechanics, and because the number of collisions per second is so great, reaction or readjustment times should be very small. Also, it appears that since both the number of molecules and the number of collisions are so great, the continuum hypothesis will hold even for moderately rarefied gas flows.

What is the size of the smallest eddies? What is termed the microscale is generally considered to be a measure of the average value of the smallest eddies. The microscale will be described in Sections 1.5 and 1.6. In Section 1.11 a value is given for a rather large-scale flow. The value is 0.05 in. or about 1 mm. Hence, with respect to such a number or a cube 1 mm on a side, the flow surely acts as a continuum.

In studying the final process of dissipation, Kolmogorov [14] deduced a still smaller length scale as well as a velocity scale. They are

$$\eta = (\nu^3/\varepsilon)^{1/4}, \qquad \upsilon = (\nu\varepsilon)^{1/4}, \tag{1.4.1}$$

where ε is a measure of the rate of dissipation of energy due to turbulence (see Section 3.5). Observe that the Reynolds number $\eta\upsilon/\nu$ formed from those two quantities is unity. A relationship between the Kolmogorov length scale η and the mean free path λ can be obtained by writing the definition of kinematic viscosity, namely, $\nu = 0.499\, \bar{c}\, \lambda$. Making use of that relationship, from Eq. (1.4.1) we can write

$$\eta/\lambda = \bar{c}/2v. \tag{1.4.2}$$

A representative value of ε is given in Fig. 4.6b in dimensionless form as $(\varepsilon\delta/u_\tau^3)$. Let us use the value 20. For the tests, u_τ/u_e was about $(0.0015)^{1/2} = 0.04$. For these test conditions, it follows that ε is approximately

$$\varepsilon \approx 12 \times 10^{-4} u_e^3/\delta. \tag{1.4.3}$$

With Eq. (1.4.3), we can write Eq. (1.4.2) as

$$\eta/\lambda \approx 3\mathrm{R}_\delta^{1/4}/(u_e/\bar{c}), \tag{1.4.4}$$

where $\mathrm{R}_\delta = u_e\delta/\nu$. Now a turbulent boundary layer has a thickness roughly equal to ten times the momentum thickness. Hence, with the value 320 presented in Section 1.1, the minimum value of $\mathrm{R}\delta$ is about 3000. Then according to Eq. (1.4.4) η is small, but it too is substantially larger than the mean free path. Note that Mach number has effectively been brought in by the term $u_e/\bar{c}$. Accordingly, only if the Mach number becomes quite large will any question arise as to the continuum hypothesis.

1.5 Measures of Turbulence – Intensity

Figure 1.9 shows the evolution of turbulence at a particular point on a 108-in.-chord plate as the tunnel speed, and hence chord Reynolds number, was increased. In the sequence, the transition position was moved relative to the hot wire by changing tunnel speed which is often a more convenient method than moving the hot wire through the transition region while holding tunnel speed constant. Until the last or fully-developed turbulent trace is reached, it is questionable that the fluctuations meet Hinze's requirement (Section 1.2) for discernment of statistically distinct average values. Certainly the traces are not long enough to indicate any statistical regularity. But the last trace seems to indicate that the fully turbulent state has arrived. Some features visible to the eye are: (1) average value of the velocity fluctuations; (2) range of magnitudes, or distribution, of these fluctuations, and (3) some sort of frequency or wave length, or distribution thereof (thousands of oscillations in 0.1 second or just a few?); (4) the shortest wave length; (5) the average wave length, etc. A number of useful measures have been developed, and it is our purpose here to acquaint the reader with a few of the more important ones.

Consider the bottom trace in Fig. 1.9 and work out a time average by taking samples periodically without bias, say at every 0.1 or 0.01 sec, or even more

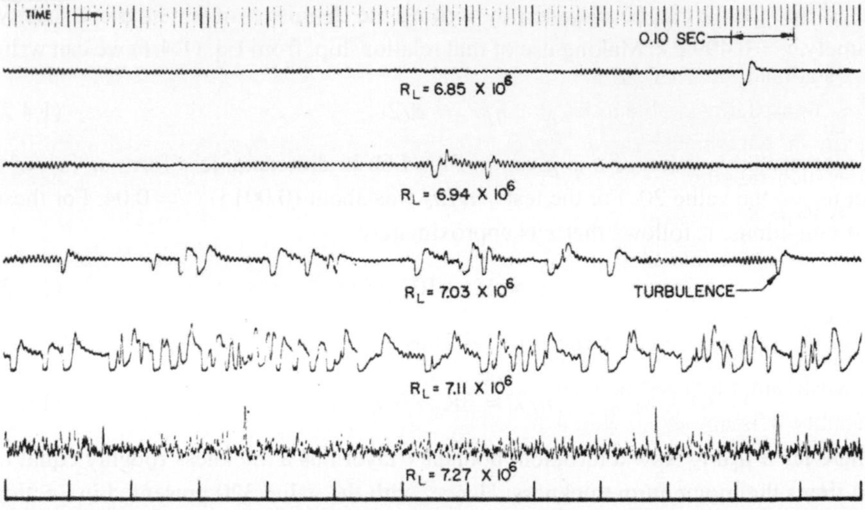

Fig. 1.9 Hot-wire records showing growth of laminar oscillations, their breakdown into turbulent spots, and development of fully turbulent flow. An 0.008-in.-diam. trip wire is located 8 in. behind the leading edge. The hot wire is located 56 in. behind the leading edge, 0.020 in. off the surface. Traces are u fluctuations. Gain is the same for all traces. Data from Smith and Clutter [15].

frequently. Or use random sampling, which is also a suitable method. Then, if fluctuations in u are being sampled,

$$u_{\text{mean}} \equiv \bar{u} = \lim_{x \to \infty} (1/n) \sum_{i=1}^{n} u_i. \tag{1.5.1}$$

Let the individual fluctuations about $\bar{u}$ be $u'_1 = u_1 - \bar{u}$, $u'_2 = u_2 - \bar{u}$, etc. By the definition of a mean, the average value of u' will be zero; that is,

$$\overline{u'} = \lim_{n \to \infty} (1/n) \sum_{i=1}^{n} u'_i = 0. \tag{1.5.2}$$

However, the mean of the squares of the fluctuating components is not equal to zero, since all are positive. For the u component, the mean square is in fact

$$\overline{(u')^2} = \lim_{n \to \infty} (1/n) \sum_{i=1}^{n} u_i'^2. \tag{1.5.3}$$

The root-mean-square of this quantity, that is the measure of the magnitude of the velocity fluctuations about the mean value, is called the intensity of turbulence. It is often expressed as the relative intensity by the three quantities

$$(\overline{u'^2})^{1/2}/\bar{u}, \qquad (\overline{v'^2})^{1/2}/\bar{u}, \qquad (\overline{w'^2})^{1/2}/\bar{u}, \tag{1.5.4}$$

A "stationary" turbulent flow is characterized by a constant mean velocity $\bar{u}$ (with $\bar{v} = \bar{w} = 0$ for suitable axes) and constant values of $\overline{u'^2}, \ \overline{v'^2}, \ \overline{w'^2}$. The true velocity at any instant is never known, but at least certain average properties can be specified. One such measure is the relative level of turbulence σ in a stream whose average velocity is $\bar{u}$:

$$\sigma = (1/\bar{u}) \ \left[(\overline{u'^2} + \overline{v'^2} + \overline{w'^2})/3\right]^{1/2}. \tag{1.5.5}$$

If the turbulence is isotropic, $\overline{u'^2} = \overline{v'^2} = \overline{w'^2}$. Isotropic turbulence can be developed in a wind tunnel by placing a uniform grid across the duct. A few mesh lengths downstream, the flow becomes essentially isotropic in its turbulence properties. The quantity σ is about 1.0% in a poor wind tunnel, 0.2% in a good general purpose tunnel, and as low as 0.01–0.02% in a well-designed low-turbulence tunnel.

The quantity σ is directly related to the kinetic energy of the turbulence, as will now be shown. Consider a flow whose mean velocity is $\bar{u}$, that is, $\bar{v} = \bar{w} = 0$. Its instantaneous velocity can be represented by

$$V = (\bar{u} + u')\mathbf{i} + v'\mathbf{j} + w'\mathbf{k}.$$

The instantaneous kinetic energy per unit mass is

$$\frac{1}{2}[(\bar{u} + u')^2 + v'^2 + w'^2],$$

and the mean kinetic energy per unit mass is

$$\frac{1}{2}\bar{u}^2.$$

To get the kinetic energy of the turbulence, we subtract the mean kinetic energy from the instantaneous kinetic energy and obtain

$$\frac{1}{2}(2u'\,\bar{u} + q^2)$$

where $q^2 = u'_j\,u'_j$ with $j = 1, 2, 3$. The mean kinetic energy of the turbulence per unit mass, k, can be obtained by taking the mean of the above expression. This gives

$$k = \frac{1}{2}\overline{q^2}. \tag{1.5.6}$$

With the relation given by Eq. (1.5.5), we can also write Eq. (1.5.6) as

$$k = \frac{3}{2}\bar{u}^2\sigma^2. \tag{1.5.7}$$

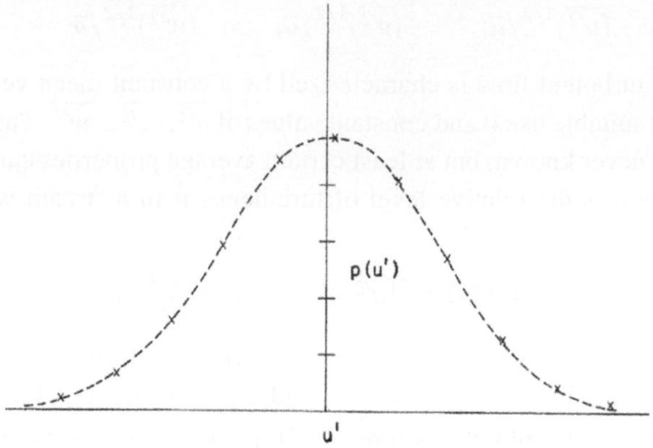

Fig. 1.10 Probability density function for the occurrence of various magnitudes of velocity fluctuations u' in a turbulent flow generated by a wire grid. Measurement was made 16 mesh widths downstream. Mesh Reynolds number $\bar{u}M/\nu = 9600$. Crosses represent measurements; dashed line represents a Gaussian or normal distribution [16].

Until now, we have considered only the mean intensity of the fluctuations. Let us now consider the distribution of the velocity fluctuations. Are the velocity fluctuations all about the same, or are some large and some small? The lowest trace in Fig. 1.9 shows a considerable variation. At least in homogeneous turbulence, for which the question has been studied in some detail, the distribution is nearly Gaussian. A typical result is shown in Fig. 1.10.

Even in two-dimensional mean flows, the turbulent fluctuations are three dimensional. That should be fairly evident from the appearance of turbulent water flow, cloud motion, smoke flow, etc. Figure 1.11 shows some typical measurements made in a thick two-dimensional boundary layer. The three fluctuating components differ of course, but not greatly. Observe that the fluctuations reach as much as 10% of the base velocity $\bar{u}$, which is consistent with the indications of Fig. 1.3.

1.6 Measures of Turbulence – Scale

The oscilloscope trace of a hot wire placed in a stream flowing at 100 mph would surely show a far more gradual fluctuation if the average eddy were 3 ft in diameter than it would if the average eddy were ½ in. in diameter. Hence, both scale and magnitude are parameters. For a stationary random-time series such as this is presumed to be, a statistical method has been developed to establish well-defined scales. Consider a stationary time series as in the sketch of Fig. 1.12, which could be the kind supplied by the experiment just mentioned. Suppose one reads base values at

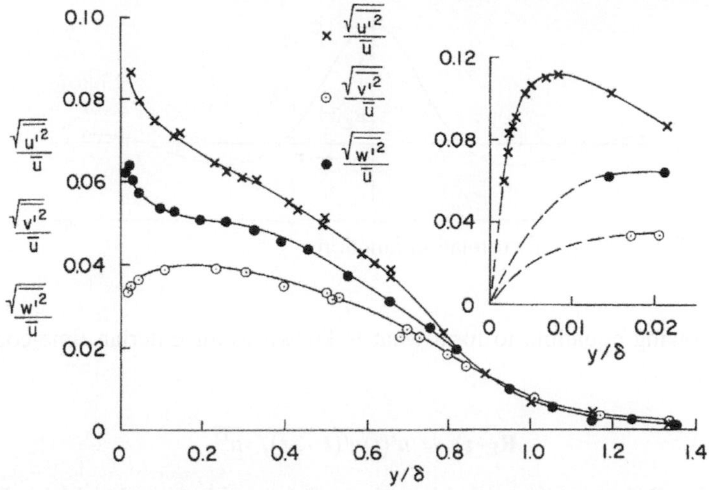

Fig. 1.11 Relative turbulence intensities in the flow along a smooth flat plate. The inset shows values very near the plate [17].

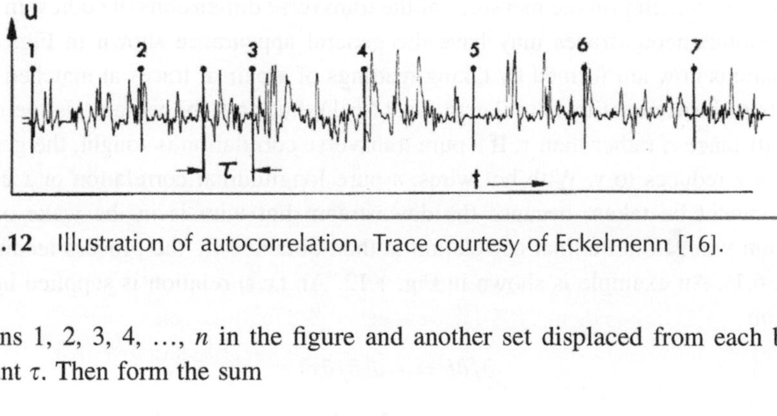

Fig. 1.12 Illustration of autocorrelation. Trace courtesy of Eckelmenn [16].

stations 1, 2, 3, 4, ..., n in the figure and another set displaced from each by an amount τ. Then form the sum

$$\overline{u'(t)u'(t-\tau)} = \lim_{n\to\infty} (1/n) \sum_{i=1}^{n} u_i'(t)u_i'(t-\tau). \tag{1.6.1}$$

This equation represents an autocorrelation function, since it is a function of the offset τ. In problems such as those considered in this book, the quantity defined by Eq. (1.5.1) converges to a unique function of τ that is independent of t for any particular steady turbulent flow. Because the correlation is all within a single trace, it is called an autocorrelation function. If $\tau \to 0$, the function becomes $\overline{u'(t)u'(t)}$ or $\overline{u'^2}$. This quantity can be conveniently introduced for normalizing purposes, as

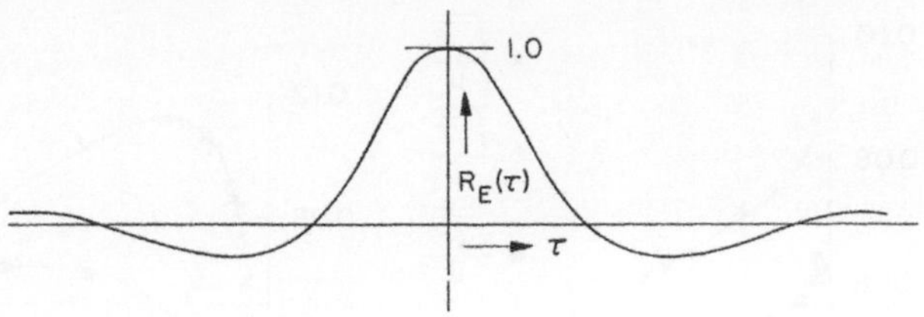

Fig. 1.13 Plot of a typical correlation function.

in the following equation, to form what is known as an Eulerian time-correlation coefficient:

$$\mathrm{R_E}(\tau) = \overline{u'(t)u'(t-\tau)} / \overline{u'^2}. \tag{1.6.2}$$

The function $\mathrm{R_E}(\tau)$ may have a wide variety of shapes; the one sketched in Fig. 1.13 is typical.

The correlation shown in Fig. 1.13 was obtained from a trace produced by a single instrument. In wind-tunnel tests two hot wires are offen placed abreast of each other with the distance between them varied in order to obtain transverse correlations, which provide measures of the transverse dimensions of eddies. In such cases, simultaneous traces may have the general appearance shown in Fig. 1.14. Correlations now are formed by taking readings of a pair of traces at matched time instants to form quantities similar to Eq. (1.6.1), but now the variable is the separation distance r, rather than τ. If a pure transverse correlation is sought, the general distance r reduces to y. With hot wires, a pure longitudinal correlation or x correlation cannot be taken, because the downstream hot wire is in the wake of the upstream wire. Longitudinal correlation is then obtained by the process leading to Eq. (1.6.1). An example is shown in Fig. 1.12. An (x, t) relation is supplied by the equation

$$\partial/\partial t = -\bar{u}(\partial/\partial x), \tag{1.6.3}$$

which is known as Taylor's hypothesis. The hypothesis simply assumes that the fluctuations are too weak to induce any significant motion of their own, so that disturbances are convected along at the mean stream velocity. It is quite accurate so long as the level of turbulence is low, for example, less than 1%. It is not exact, and has appreciable errors at high levels of turbulence [8].

In homogeneous turbulence, a transverse correlation coefficient appears as in Fig. 1.15, although the coefficient does not always become negative. The curvature at the vertex is determined by the smallest eddies. Hence, a measure of the smallest eddies is provided by the intercept of the osculating parabola. Theory shows that the correlation begins as a quadratic function; a linear term would of course destroy the

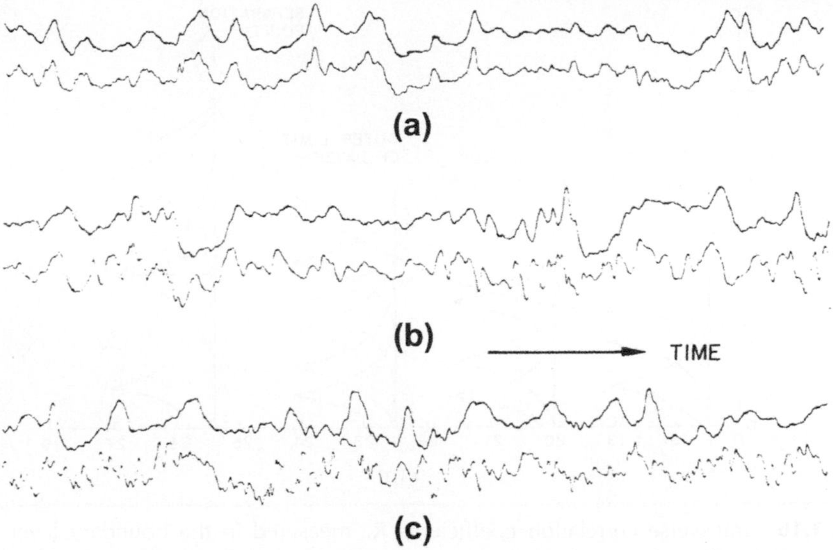

Fig. 1.14 Oscillograms illustrating transverse correlation: (a) nearly perfect correlation, two wires very close together; (b) moderate correlation, two wires a moderate distance apart; (c) very low correlation, two wires far apart [18].

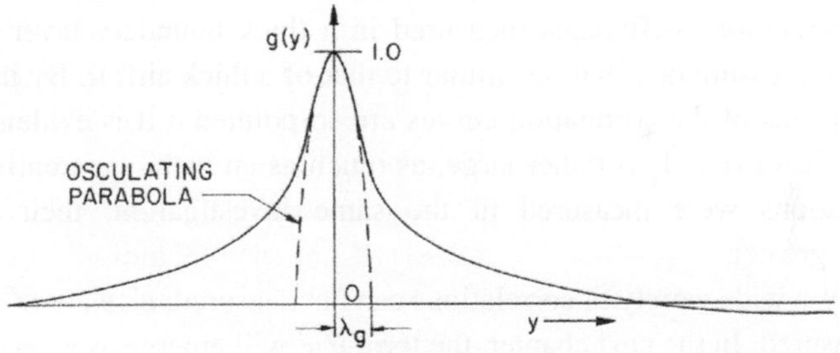

Fig. 1.15 Typical transverse correlation function. The microscale λ_g is the intercept of the osculating parabola. $1/\lambda_g^2 = -\frac{1}{2}(d^2g/dy^2)_0$.

required symmetry. The relation shown in the figure is readily derived by means of a Taylor's series for $g(y)$ that ends with the y^2 term. The length λ_g is known as the *microscale of the turbulence*. Since the value of g at $y = 0$ is normalized to unity, a second convenient measure is the area under half the curve, Λ_g; that is,

$$\Lambda_g = \int_0^\infty g(y)\ dy. \tag{1.6.4}$$

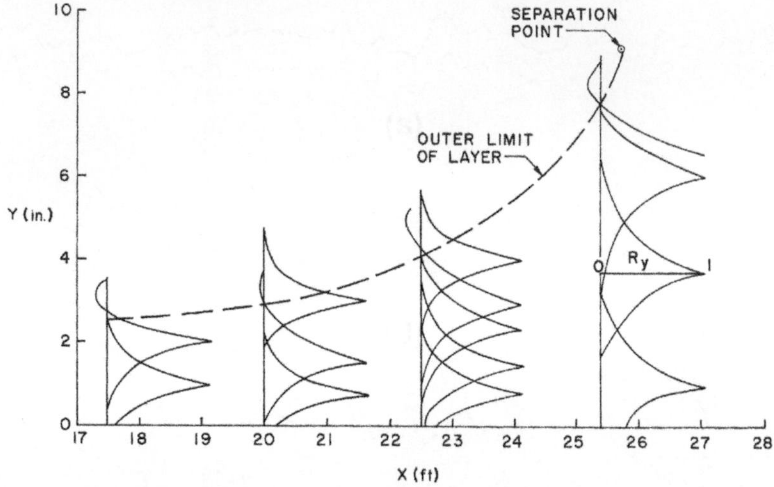

Fig. 1.16 Transverse correlation coefficients R_y measured in the boundary layer of a large airfoil-like body. At the $17\frac{1}{2}$ -ft station the edge velocity was 160 ft/sec. $R_y = \overline{u_1' u_2'}, (\overline{u_1'})^{1/2} (\overline{u_1'})^{1/2}$ [19].

That area is a well-defined measure of the approximate size of the largest eddies. For obvious reasons, it is called the *integral scale*. Figure 1.16 shows a number of transverse correlation coefficients measured in a thick boundary layer on a large body having a pressure distribution similar to that of a thick airfoil. By inspection – because the peaks of the correlation curves are so pointed – it is evident that λ_g is rather small. However, Λ_g is rather large, as much as an inch, apparently. Longitudinal correlations were measured in the same investigation; their scales are considerably greater.

Obviously, a wide variety of correlations can be measured; u', v', or w' may be the quantity measured. In the next chapter, the term $\overline{v'u'}$ will emerge as a very important quantity, which when multiplied by $-\varrho$ is known as a Reynolds shear-stress term. It is a correlation between two velocities at a point. It could be computed from oscilloscope traces, but two hot wires arranged in the form of an x can yield instantaneous u', v' directly. If v' were not related to u', the correlation would be zero. Actually, it is physically related to u'. The transverse correlations just discussed are known as *double correlations*. Correlations involving three or more measurements can be made; they are of importance in attempts to develop further the statistical theory of turbulence.

The rates of change of $(\overline{u'^2})^{1/2}/\bar{u}$ and λ_g downstream of a grid are of interest both from a practical standpoint and from the standpoint of the general theory of turbulence. Figure 1.17 shows typical results downstream of a grid at both large and small

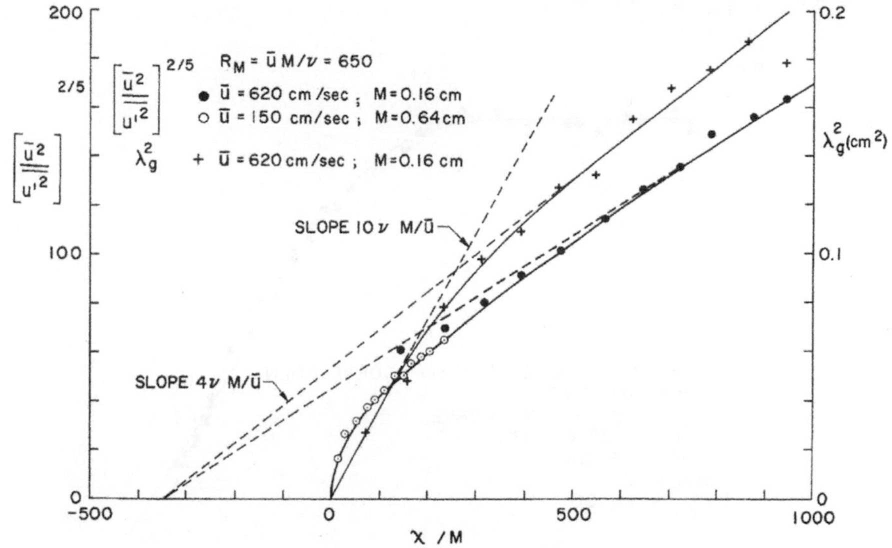

Fig. 1.17 Variations of $(\overline{u'^2}/\bar{u}^2)$ and λ_g downstream of a grid in a wind tunnel [20].

distances. At a short distance downstream of the grid, λ_g^2 has a slope $10\nu M/\bar{u}$; at large distances, the slope is $4\nu M/\bar{u}$, where M is the mesh size. In the initial stages of decay, $\bar{u}^2/\overline{u'^2}$ varies as $t^{5/2}$. A parameter that often arises is the Reynolds number of turbulence $R_\lambda = (\overline{u'^2})^{1/2}\lambda_g/\nu$. In initial stages of decay, $(\overline{u'^2})^{1/2} \sim t^{-1/2}$ and $\lambda_g \sim t^{1/2}$. Hence R_λ remains constant in this region because the t terms cancel. Since the data of Fig. 1.17 pertain to isotropic turbulence, the figure also provides information on the decay of the kinetic energy of turbulence.

1.7 Measures of Turbulence – The Energy Spectrum

Since turbulence has fluctuations in three directions, any complete study of the energy spectrum must necessarily involve a three-dimensional spectrum, or more specifically a correlation tensor involving nine spectrum functions. But our purpose here is only to introduce the general concept of a spectrum, and so we shall confine our discussion to the one-dimensional case. Just as with light, where the different colors (wave lengths or frequencies) may have different degrees of brightness, so may the signal in a turbulent flow have different strengths for different frequencies. For instance, the low-frequency portion of a u' trace might have little energy and the high-frequency portion much, or vice versa. The spectrum of turbulence relates the energy content to the frequency. Consider the band of frequencies between n and $n + dn$. Then define E_1 such that $E_1(n)\,dn$ is the contribution to $\overline{u'^2}$ of the

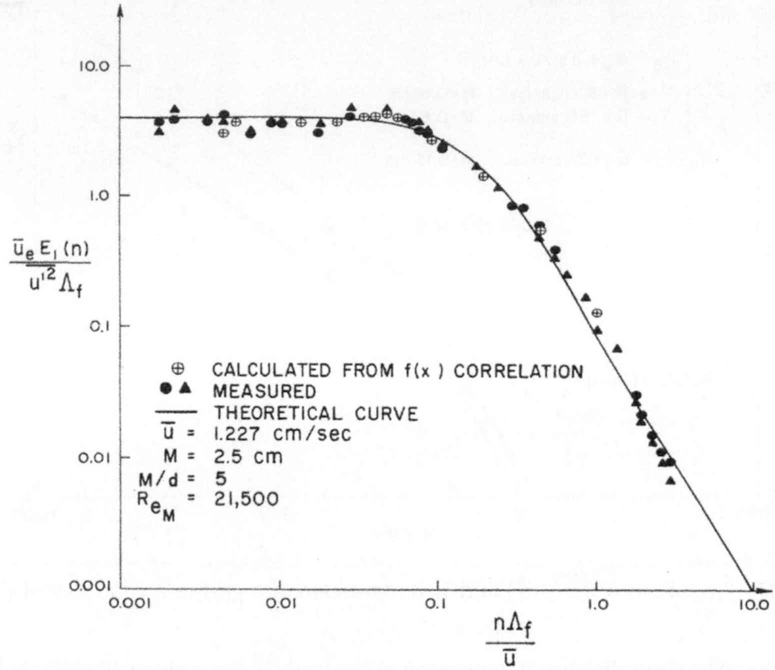

Fig. 1.18 The spectral function $E_1(n)$ for u' fluctuations downstream of a grid [21].

frequencies in this band. Obviously, the mean-square fluctuation covering all frequencies is

$$\overline{u'^2} = \int_0^\infty E_1(n)\ dn, \tag{1.7.1}$$

where E_1 is the spectral distribution function for the $\overline{u'^2}$ component. A typical form of the function for homogeneous grid turbulence is shown in a normalized form in Fig. 1.18. Spectral analysis is readily performed on digitized signals.

Different shapes of the curve will obviously indicate different distributions of the energy as a function of frequency. We shall now discuss some of the properties of a spectral function and show the reason for the normalized coordinates used in Fig. 1.18. If $f(x)$ is the longitudinal correlation function, that is, the companion of $g(y)$ of Fig. 1.15, it can be shown by the theory of Fourier transforms [8] that

$$f(x) = (1/\overline{u'^2}) \int_0^\infty E_1(n)\ \cos\,(2\pi nx/\bar{u})\ dn. \tag{1.7.2}$$

Also, by the inverse transform relation,

$$E_1(n) = 4\overline{u'^2}/\bar{u} \int_0^\infty f(x) \cos(2\pi n x/\bar{u})\, dx. \tag{1.7.3}$$

The primary derivation is from a time series, but here time is replaced by means of Taylor's hypothesis, that is, $t = x/\bar{u}$. Equations (1.7.2) and (1.7.3) show that the correlation function can be derived from spectral measurements, and vice versa. Now if we let $n \to 0$ in Eq. (1.7.3) we have

$$\lim_{n\to 0} \left[(\bar{u}/4\overline{u'^2})\, E_1(n) \right] = \int_0^\infty f(x)\, dx \equiv \Lambda_f, \tag{1.7.4}$$

which is analogous to Eq. (1.6.4); that is, the integral scale is just the product of the quantity $\bar{u}/4\,\overline{u'^2}$ and the limit of $E_1(n)$ as $n \to 0$. The microscale can be derived from the Fourier transform for $f(x)$, Eq. (1.7.2). By definition (see Fig. 1.15) the microscale λ_f is

$$1/\lambda_f^2 = -\frac{1}{2}(\partial^2 f/\partial x^2)_{x=0}. \tag{1.7.5}$$

Double differentiation of Eq. (1.7.2) yields

$$1/\lambda_f^2 = (2\pi^2/\bar{u}^2 \overline{u'^2}) \int_0^\infty n^2\, E_1(n)\, dn. \tag{1.7.6}$$

Because of the weighting factor n^2 in Eq. (1.7.6), it is evident that the microscale is chiefly determined by the higher frequencies.

An equation defining energy dissipation in isotropic turbulence is

$$d\overline{u'^2}/dt = -10\nu\overline{u'^2}/\lambda_f^2. \tag{1.7.7}$$

Since the higher frequencies generally determine the microscale, they generally determine the dissipation rate. The form of Eq. (1.7.4), together with the fact that Λ_f is one of the results, now explains the normalized coordinates of Fig. 1.18.

Power spectral information such as that just discussed is more than just a method of presenting data. In the analysis of linear oscillating systems subject to random forcing functions, the solution requires knowledge of the forcing function. The output is the mean square of the response [22].

Figure 1.18, which is typical of spectral measurements in a turbulent flow, shows that the energy is distributed over a very wide range of frequencies. One process that importantly contributes to the effect is vortex stretching. A turbulent flow, particularly shear layers, is a large array of small vortices. Not only because of their own interaction but also because of the mean velocity distribution in the boundary layer, vortices may find themselves convected into the regions of higher velocity. If so, they become stretched and their vorticity is intensified. Hence, more vorticity can be generated at higher wave number. Thus there is such a complicated interaction that vorticity of a wide variety of scales and strengths is generated. Batchelor [23] has a good introductory discussion of this phenomenon, together with the descriptive equations.

1.8 Measures of Turbulence – Intermittency

A laminar flow velocity profile asymptotes into the surrounding flow rapidly but continuously. In fact, the disturbance due to a laminar flow such as a boundary layer decays at least as fast as exp $(-ky^2)$, where k is near unity. Hence, although it decays rapidly, the boundary layer has no distinct edge. The situation is quite different in turbulent flows. There is a distinct edge, although it wanders around in random fashion. Clouds show the effect well. A cumulus cloud is just a well-marked turbulent flow on a giant scale. The line of demarcation between clear sky and cloud, which shows as visible turbulent eddies, is quite sharp. There is no gradual fading into the clear blue sky. Figure 1.4 is a picture of such a cloud, showing the sharp but irregular boundaries. The ambient air can be thought of as being contaminated by adjacent turbulence. The phenomenon is evident in any markedly turbulent flow – clouds, smokestack plumes, exhaust steam, dust storms, muddy water in clear water, etc.

Figure 1.19 shows the same basic phenomenon, in this case due to the wake of a bullet. If the wake were in motion as in a wind tunnel, it is clear that a hot wire or another sensor would be either entirely in the turbulence or out of it; and, judging by the appearance of the wake, the fraction of time the hot wire sees turbulence is a statistical function of the distance from the center of the wake. The fraction is called γ, the intermittency, a term introduced by Corrsin and Kistler [24]. Entirely outside a turbulent flow $\gamma = 0$, and entirely inside $\gamma = 1$.

Corrsin and Kistler appears to have first noticed the effect in 1943, during studies of a heated jet. Two important early specific studies of the phenomenon as it occurs in ordinary turbulent flows of air were conducted by Corrsin and Kistler [24] and Klebanoff [17]. Klebanoff made such measurements for a boundary-layer flow on a flat plate. He found that the intermittency was accurately described by the following equation, where δ is the mean thickness of the boundary layer:

$$\gamma = \frac{1}{2}(1 - \text{erf}\zeta), \qquad \text{where} \qquad \zeta = 5[(y/\delta) - 0.78]. \tag{1.8.1}$$

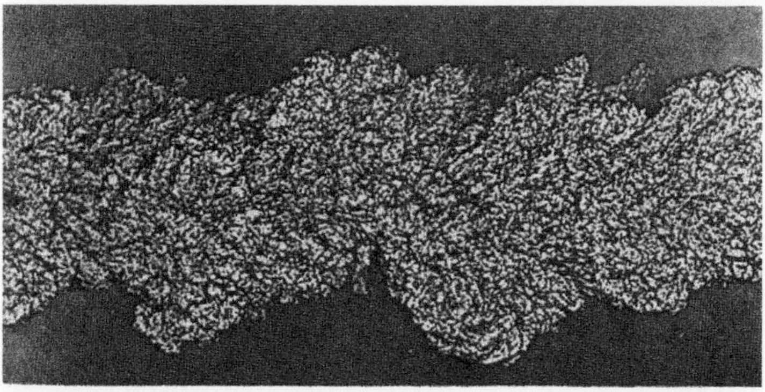

Fig. 1.19 Turbulent wake of a bullet, showing sharp but irregular boundary. The photo is a shadowgraph, which tends to accentuate the small-scale structure [24].

Hence the interface has a Gaussian probability distribution whose cumulative distribution is just Eq. (1.8.1). An appreciable portion of the flow, according to these measurements, is turbulent to a distance well beyond the mean edge of the boundary layer, in fact to $y/\delta \approx 1.20$. Also, an appreciable fraction is nonturbulent as far into the boundary layer as $y/\delta \approx 0.4$.

There is a relative velocity between the ragged edge and the main irrotational stream. Hence the flow can be viewed as a flow past a very rough surface. It is natural to ask what the effect is of this ragged randomly fluctuating boundary upon the exterior irrotational flow. Phillips [25], who studied the fluctuating velocity field induced by the turbulence in the main irrotational stream by means of a simplified model, found that the energy of the fluctuations decays asymptotically as the inverse fourth power of the distance from a representative mean plane. Experiments confirm the result.

1.9 The Diffusive Nature of Turbulence

On the molecular scale, motion of the molecules – hence diffusion – is quite a random process. An important reason is that a normal gas is such that the mean free path is far greater than the molecular diameter. On the scale of turbulence, the process is not nearly so random. Nevertheless, it may be helpful to indicate some of the gross features of a random motion, which is what a diffusion process amounts to.

A method starting from first principles is to consider a very general motion in three dimensions, where at first no assumption is made as to uniformity of steps. The motion is assumed to proceed in steps, which is certainly a correct assumption on the molecular scale. Each step may have any direction and any length. In Cartesian coordinates, each step has the components Δx_i, Δy_i, Δz_i, but for brevity we shall leave out the Δ's. Then after n steps the total distance traveled in the x, y, and z directions is

$$x = \sum_0^n x_i, \quad y = \sum_0^n y_i, \quad z = \sum_0^n z_i, \tag{1.9.1}$$

Obviously, the square of the total distance traveled after n steps is

$$r_n^2 = \left(\sum_0^n x_i\right)^2 + \left(\sum_0^n y_i\right)^2 + \left(\sum_0^n z_i\right)^2. \tag{1.9.2}$$

Now consider in detail the first, of x term, which may be written $x_0 + x_1 + x_2 + x_3 + \cdots$. Its square is

$$\begin{aligned}&(x_0 + x_1 + x_2 + x_3 + \cdots)^2\\ &\quad(x_0^2 + x_1^2 + x_2^2 + x_3^2 + \cdots) + 2x_0(x_1 + x_2 + x_3 + \cdots)\\ &\quad+2x_1(x_2 + x_3 + \cdots) + 2x_2(x_3 + \cdots) + \cdots.\end{aligned} \tag{1.9.3}$$

The expressions for displacements y and z are similar. The first term on the right is a series of squares and hence always positive, but the remaining terms all contain simple sums of displacements. If the number of steps is great and if the motion has a high degree of randomness, there will be nearly as many negative steps as positive. The cross-product quantities therefore become negligible in comparison with the first term. Therefore, as n becomes large

$$r_n^2 = \lim_{n\to\infty} \sum_0^n x_i^2 + y_i^2 + z_i^2. \tag{1.9.4}$$

This is a fully general result quite independent of step length. It states that the square of the distance traveled is equal to the sum of the squares of the displacements in the three coordinate directions. In any random motion, as in molecular motion, the ith path between collisions has a total length l_i which is exactly

$$l_i^2 = x_i^2 + y_i^2 + z_i^2. \tag{1.9.5}$$

Hence Eq. (1.9.4) can be written more compactly, but with the same generality, as

$$r_n^2 = \lim_{n\to\infty} \sum_0^n l_i^2. \tag{1.9.6}$$

If all paths are of equal length l, we can write

$$r_n^2 = \lim_{n\to\infty} nl^2 \quad \text{or} \quad r_n = l(n)^{1/2}. \tag{1.9.7}$$

If motion is at a uniform mean speed $\bar{c}$ as in molecular motion, the number of collisions or steps can be eliminated by the relation

$$\bar{c}t = nl, \quad (1.9.8)$$

giving

$$r(t) = (\bar{c}lt)^{1/2}. \quad (1.9.9)$$

But according to kinetic-gas theory, $\bar{c}\, l \approx 2\nu$, where l now is the molecular mean free path and ν the kinematic viscosity. Therefore,

$$r(t) \approx (2\nu t)^{1/2}. \quad (1.9.10)$$

That is, the mean distance reached by some kind of random-motion process is proportional to $\nu^{1/2}$ and $t^{1/2}$. The product $(\nu t)^{1/2}$ is fundamental to all diffusion processes. If ν is large, diffusion will be much greater than when ν is small.

Although the relations just derived are properly applicable only to molecular motion in gases, they still exhibit some of the gross behavior of turbulence, especially the high diffusivity. The development assumed that there was negligible correlation between the successive steps. However, when eddies are large, there must be some correlation at first, if the steps l are small. In fact, at the very beginning, before any changes of path occur, we obviously can write, starting from time $t = 0$,

$$r(t) = \bar{c}t. \quad (1.9.11)$$

Compare that with Eq. (1.9.9). The relations together show that a random motion where scales are large starts out as a linear function of time, but after correlation is lost, it becomes a square-root function of time and velocity. Our discussion considers only the very beginning of a random process and the final fully developed phases. Expansions of the type given in Eq. (1.9.3) bring in the notion of correlation. In an important paper on the subject of diffusion by continuous movements, Taylor [26] presented a method for analyzing the complete problem instead of just its limits. Correlation functions are a key feature of the analysis.

In turbulent flow, the process of transfer of momentum and other quantities is sufficiently similar to the molecular process to suggest the use of fictitious or eddy viscosity. It is interesting to compare values. For gases on the molecular scale, as was mentioned earlier, $\nu = 0.499\, \bar{c}l$. For turbulent flow in the outer parts of the boundary layer, a formula that gives good results is $\varepsilon_{\mathrm{m}} = 0.0168 u_e \delta^*$, where δ^* is the displacement thickness. Typical values are shown in Table 1.1.

The effective viscosity in the example is 400 times the true viscosity. Hence the diffusion rate [Eq. (1.9.10)] is 20 times as great. The primary reason for the large difference is the great difference in characteristic length – the mean free path. In the

TABLE 1.1 Comparison of typical molecular and turbulent viscosities

Flow	Characteristic velocity (cm/sec)	Characteristic length (cm)	Kinematic viscosity (cm^2/sec)
Molecular (laminar) ($\nu = 0.499\,\bar{c}l$)	$\bar{c} = 54{,}000$ (0°C)	9.4×10^{-6}	0.25
Turbulent[a] ($\varepsilon_m = 0.0168 u_e \delta^*$)	$\bar{c} = 6000$ (200 ft/sec)	1	100

[a]The quantities u_e and δ^* prove to be successful reference quantities, but the effective velocity and length are about 1/30 and 1/2 as much, respectively.

table the ratio is about 10^5. When allowance for the effective length is made, the ratio is still greater than 10^4. The large ratio is due to the fact that eddies are being dealt with, rather than molecules. In many cases, the characteristic velocities are not substantially different. In the table they are, but at 2000 ft/sec, a moderate supersonic velocity, $\bar{c}$ would exceed the molecular value of $\bar{c}$. Since a boundary layer at full scale may have a displacement thickness much larger than 1 cm, it is apparent that the two types of viscosity can easily differ by a thousand-fold. The strong diffusiveness helps turbulent flows to withstand much stronger adverse pressure gradients than laminar flows without separation.

The rise of smoke from a cigarette in quiet air has certain similarities to the random walk just discussed. The smoke first rises as a slender filament with very little diffusion, because the flow is laminar. Then transition takes place and the diffusion is greatly increased. If any one element of smoke is traced, it can be seen that it wanders back and forth as it rises, in much the random way just visualized. The strong difference in diffusion rate can be put to good use as an indicator of transition. A filament of gas injected into a laminar boundary layer will not diffuse much, but it will diffuse rapidly when it encounters turbulent flow. Reynolds [11], in his classic experiments with flow of water and transition in pipes, made use of the phenomenon. He located transition beautifully, by introducing a filament of dyed water into the pipe. When the transition was reached, the filament suddenly diffused.

1.10 Turbulence Simulation

By convention, turbulence "modeling" is the development and solution of empirical equations for the Reynolds stresses that result when the Navier-Stokes equations are averaged, with respect to time or otherwise. Various models developed for this purpose will be discussed in this book, but this is a convenient point to introduce turbulence "simulation". Turbulence simulation is the solution of the complete three-dimensional time-dependent Navier–Stokes equations, either for the complete range of eddy sizes ("full simulation") or for the large, long-wavelength eddies only, with a model for the small eddies ("large eddy simulation"), see flow chart, Fig. 2.1. Turbulence covers a very wide range of wavelengths; a full simulation must be carried out in a volume (the domain of integration) large enough to enclose the largest eddies, with a finite-difference grid spacing, or equivalent, small enough to resolve the smallest eddies.

The ratio of the length scale l ($\equiv k^{3/2}/\epsilon$) of the large, energy-containing eddies to the length scale of the smallest, viscous-dependent eddies, Eq. (1.4.1), is of the order of $(k^{1/2}/\nu)^{3/4}$, so the number of finite-difference points in the domain of integration must be somewhat larger than $(k^{1/2}l/\nu)^{9/4}$. Even for low Reynolds number flows, such as a boundary layer at a momentum-thickness Reynolds number of 1000, several million grid points are needed. Therefore, full simulation is restricted to low Reynolds numbers: enormous increases in computer memory and processing speed will be needed before full simulations at flight Reynolds numbers become possible. At present, full analysis turbulence simulations are a research technique rather than a design tool. After a period during which some experimentalists and others had doubts about the realism of the simulations on the grounds that the results of gross numerical instability look rather like turbulence, simulation results now have about the same status as experimental data. That is, errors due to poor numerical resolution (or other causes) may occur, but in principle, solutions of the exact equations describing a phenomenon are equivalent to measurements of the phenomenon. Simulations not only can give information on a much finer mesh than could be achieved in an experiment, but can also include the pressure fluctuations within the fluid. These pressure fluctuations cannot currently be reliably measured but play a vital part in the behavior of the Reynolds stresses. Therefore, simulation results are now a very useful source of information in the development of turbulence models.

Particularly at large Reynolds numbers, the statistics of the small-scale motion are almost independent of the details of the large-scale motion that produces most of the Reynolds stresses. At a high enough Reynolds number, the small-scale statistics depend only on the rate of transfer of turbulent kinetic energy from the large scales to the small scales, which is equal to the eventual rate of dissipation of turbulent energy into thermal internal energy by fluctuating viscous stresses in the smallest eddies. Therefore, acceptable predictions of the large-scale eddies in high Reynolds number flows can be obtained by modeling the small-scale eddies; the wavelength which defines the boundary between "large" and "small" eddies is just twice the finite-difference grid size. In principle it is chosen small enough to ensure that the contribution of the sub-grid-scale eddies to the Reynolds stresses (or turbulent heat-flux rates) is negligible. Calculations for turbulent wall flows are made more difficult by the fact that the Reynolds stresses near a solid surface are produced entirely by small eddies, which requires a fine grid near the surface unless the whole of the near-surface region can be modeled.

An indirect disadvantage of the large number of points needed is the high cost of performing calculations in complicated geometries, since coordinate-transformation metrics must be computed or stored at each mesh point. To date few simulations have been done with fine resolution in complex geometries, although a great deal of ingenuity has been shown in choosing simple geometries to represent complex flow behavior.

Much of the early work on simulations was done using "spectral" codes in which the computations are done in wave-number space. An advantage is that spectral codes can represent spatial derivatives exactly for all wavelengths down to twice the effective grid spacing, whereas finite-difference derivative formulas become seriously inaccurate at wavelengths smaller than 4 or 6 grid spacings. (A factor of two on grid spacing means a factor of 8 on the total number of grid points in three-dimensional space.) However, spectral codes are almost impossible to use in general geometries (simple analytically-specified coordinate stretching in one dimension is the most that is ever attempted in practice) and higher-order finite-difference codes are now coming into more general use.

Models for the sub-grid-scale motion yield apparent turbulent stresses, applied instantaneously by the sub-grid-scale motion to the resolved eddies. That is, the sub-grid-scale effects are averaged over the length and time scales of sub-grid-scale motion but are seen by the resolved motion as fluctuating stresses, in the same way that real turbulence sees fluctuating viscous stresses. Sub-grid-scale models are usually quite simple, partly because the computing cost of complicated models would be unacceptable and partly because imposing a sharp boundary between resolved and modeled wavelengths is unphysical. Specifically, turbulent eddies, however defined, are not simple Fourier modes, so a given eddy with a size near the cutoff wavelength would make contributions to both the resolved and the sub-grid-scale fields. Indeed, the most common sub-grid-scale model was developed forty years ago for use in atmospheric calculations [27]; it relates the total sub-grid-scale contribution to a given turbulent stress to the rate of strain in the resolved motion. It is closely equivalent to a "mixing length" formula (see Section 4.3) in which the mixing length is proportional to the grid spacing (i.e. proportional to the size of the larger sub-grid-scale eddies, which is plausible in principle). Several alternative suggestions have been made: a recent proposal which avoids rather than solves the problem is the scaling technique of Germano [28], in which the influence of sub-grid-scale eddies in a grid of size Δ (minimum resolvable wavelength Δ) is deduced from the resolved-scale motion in the range of wavelengths 2Δ to 4Δ (say).

It is possible for even full simulations to give poor results for the higher-order statistics of the smallest-scale eddies. In real life the smallest-scale eddies adjust themselves so that fluctuating viscous stresses dissipate the turbulent kinetic energy handed down to them by the larger eddies. In a simulation this dissipation is carried out partly by real viscosity and partly by "numerical viscosity", arising from finite-difference errors and generally proportional to the mesh size. That is, the total dissipation is correct (or the intensity of the smallest eddies would decrease or increase without limit), but the actual statistics of the smallest eddies may be incorrect to some extent. Mansour et al. [29] report an 8% discrepancy in the dissipation-equation balance in the viscous wall region: this is satisfactorily small.

Simulations of heat transfer or other scalar transfer simply involve adding transport equations for thermal energy or species concentration, at the expense of greater storage and longer computing times but without other special difficulties. However, if the Prandtl number (ratio of viscosity to thermal conductivity) is large, the smallest scales in the temperature field may be much less than those in the velocity field, so the grid size must be reduced; cost considerations currently limit simulations to Prandtl numbers (or Schmidt numbers for scalar transfer) near unity.

Turbulent combustion is obviously a very difficult phenomenon to study experimentally and is an active topic in simulation work. Even the simplest reactions have many intermediate steps, ignored in elementary chemical formulas, and concentration equations must be solved for each intermediate species together with rate equations for each step. Simulations have so far been confined to instantaneously two-dimensional flow ($w' = 0$). Since chemical reactions depend on mixing at a molecular level, full simulations covering the whole range of eddy sizes are essential.

Numerical methods for simulations are in principle the same as for any other three-dimensional time-dependent Navier-Stokes solution, but are in practice simpler because they are confined to simple geometries. Most of the spectral codes are based on the work of Rogallo [30] (see also Kim and Moin [31] and finite-difference methods discussed by Moin and Rai [32]).

Detailed analysis of simulation results can take as much computer time as the simulation itself, and a good deal more human time. Like the analysis of experimental data, it falls into two categories: (1) studies of eddy structure and behavior in which statistics are used as an adjunct to a computerized form of flow visualization (inspection of computer graphics views of the flowfield), and (2) the study of contributions to the Reynolds-stress transport equations (Chapter 6) and other equations that are the subject of Reynolds-averaged turbulence modeling.

Problems

1.1 Do your own ow visualization experiment to complement the photographs in this chapter. Fill the largest available clear-glass container nearly to the brim with water and leave it for several minutes for the water to come to rest. Then pour in a small quantity of colored liquid (a teaspoonful at most: very strong instant coffee seems to be best, but milk or orange juice also work quite well). Some experimenting will be needed to adjust quantity and ow rate, but it should be possible to see a cloud-like boundary to the descending jet, and possibly some of the internal structure as the colored liquid is gradually diluted. The liquid will mostly collect near the bottom of the container. If possible, leave it over-night and see how very slow molecular diffusion is compared to turbulent mix-ing (in liquid mixtures, the diffusivity is usually very small compared to the viscosity or thermal conductivity but even molecular diffusion of heat is small compared to turbulent heat transfer).

1.2 A 3/4″ water pipe will pass a ow of about one U.S. gallon (8 lb) of water per minute. Show that the ow is almost certainly turbulent.

1.3 A Boeing 747 is 230 ft long and cruises at 33,000 ft (10,000 m) at a speed of 880 ft/sec. (Mach number 0.9) The International Standard Atmosphere, using metric units, gives the density at this altitude as 0.413 kg/m^3 and the molec-ular viscosity as 0.0000146 N sec/m^2, so that the kinematic viscosity is 3.53×10^{-5} m^2/sec. (Note that a Newton, symbol N, is the force required to accel-erate a mass of 1 kg at 1 m/sec^2 and therefore has units of kg m^{-1}sec^2.) Calcu-late the Reynolds number based on body length. Note that it would be more logical to evaluate the viscosity at the wall temperature than at the free-stream temperature, because the direct effect of viscosity on turbulent stresses and skin friction is felt only very close to the wall. In this case the absolute temperature at the wall will be about 1.15 times that in the free stream, about 60 $\leq$ F greater.

1.4 Using the "representative" value of dissipation given in Eq. (1.3.3) and assuming that $u_e/v \approx 80{,}000$ as in Klebanoff's experiments (see the cited figures in Chapter 4) show that $\eta/\delta \approx 1{,}1 \times 10^{-5}$. The thickness of Klebanoff's boundary layer was about 3 in, so that η was about 30 μ in. The smalles signif-icance wavelength in the flow are roughly 5η and the wavelengths that contribute most to the dissipation are roughly 50η.

1.5 The "theoretical" (strictly, empirical) fit to the frequency spectral function data in Fig. 1.19 has an asymptotic form at high frequency proportional to n^{-2}. Show, using the formulas in Sect. 1.6, that the corresponding microscale is zero and the dissipation infinite. Using Eq. (1.6.5) and referring to to Fig. 1.14, explain what is wrong with the data fit. Note that a *best* fit to the data obviously has a negative slope that increases all the way up to the highest frequency resolved.

1.6 Use Eq. (1.9.4) to show that, in two-dimensional incompressible flow with u_e independent of x, the displacement thickness is indeed the distance by which the external streamlines are displaced outwards by the reduction in flow rate within the boundary layer.

References

[1] L. Prandtl, Turbulent flow. NACA Tech. Memo, 435, originally delivered to 2nd Int, Congr. Appl. Mech. Zurich (1926).

[2] H.Tennekes, J.L. Lumley, A First Course in Turbulence, MIT Press, Cambridge, MA.

[3] M. Lesieur, La Turbulence, Press Universitaires de Grenoble, 1994.

[4] S.B. Pope, Turbulent Flows, Cambridge Univ. Press, 2000.

[5] P.A. Durbin, B.A. Petterson, Statistical Theory and Modeling of Turbulent Flows, John Wiley and Sons, New York, 2001.

[6] W.C. Wilcox, Turbulence Modeling for CFD, DCW Industries, La Canada, CA, 1998.

[7] T. von Kármán, Turbulence. Twenty-fifth Wilbur Wright Memorial Lecture, J. Roy. Aeronaut. Soc. 41 (1937) 1109.

[8] J.O. Hinze, Turbulence, an Introduction to Its Mechanism and Theory, McGraw-Hill, New York, 1959.

[9] G. Hagen, On the motion of water in narrow cylindrical tubes (German), Pogg. Ann. 46 (1839) 423.

[10] L. Prandtl, O.G. Tietjens, Applied Hydro- and Aeromechanics, Dover, New York, 1934, p. 29.

[11] O. Reynolds, An experimental investigation of the circumstances which determine whether the motion of water will be direct or sinuous and the law of resistance in parallel channels, Phil. Trans. Roy. Soc. London 174 (1883) 935.

[12] J.H. Preston, The minimum Reynolds number for a turbulent boundary layer and selection of a transition device, J. Fluid Mech 3 (1957) 373.

[13] C.S. Wells, Effects of free-stream turbulence on boundary-layer transition, AIAA J 5 (1967) 172.

[14] A.M. Kolmogorov, Equations of turbulent motion of an incompressible fluid. Izvestia Academy of Sciences, USSR, Physics 6 (1 and 2) (1942) 56–58.

[15] A.M.O. Smith, D.W. Clutter, The smallest height of roughness capable of affecting boundary-layer transition in low-speed flow, Douglas Aircraft Co (1957). Rep. ES 26803, AD 149 907.

[16] G.K. Batchelor, The Theory of Homogeneous Turbulence, Cambridge Univ. Press, London New York, 1953.

[17] P.S. Klebanoff, Characteristics of turbulence in a boundary layer with zero pressure gradient, NACA Tech. Note 3178 (1954).

[18] H. Eckelmann, Experimentelle Untersuchungen in einer turbulenten Kanalströmung mit starken viskosen Wandschichten, Mitt. No. 48, Max-Planck-Inst. for Flow Res., Göttingen (1970).

[19] G.B. Schubauer, P.S. Klebanoff, Investigation of separation of the turbulent boundary layer, NACA Tech. Note No. 2133 (1950).

[20] G.K. Batchelor, A.A. Townsend, Decay of turbulence in the final period, Proc. Roy Soc. 194A (1948) 527.

[21] A. Favre, J. Gaviglio, R. Dumas, Quelques mesures de correlation dans le temps et l'espace en soufflerie, Rech. Aeronaut 32 (1953) 21.

[22] H.W. Liepmann, On the application of statistical concepts to the buffeting problem, J. Aeronaut Sci. 19 (1952) 793.

[23] G.K. Batchelor, An Introduction to Fluid Dynamics, Cambridge Univ. Press, London New York, 1967.

[24] S. Corrsin, A.L. Kistler, The free-stream boundaries of turbulent flows, NACA Tech. Note No. 3133 (1954).

[25] O.M. Phillips, The irrotational motion outside a free turbulent boundary layer, Proc. Camb. Phil. Soc. 51 (1955) 220.

[26] G.I. Taylor, Diffusion by continuous movements, Proc. London Math. Soc. 20 (1921) 196.

[27] J. Smagorinsky, General Circulation Experiments with the Primitive Equations, I. The Basic Experiment. Monthly Weather Review 91 (1963) 99.

[28] M. Germano, U. Piomelli, P. Moin, W.H. Cabot, A Dynamic Subgrid-Scale Eddy-Viscosity Model, Phys. Fluids A3 (1991) 1760.

[29] N.N. Mansour, J. Kim, P. Moin, Reynolds-Stress and Dissipation-Rate Budgets in a Turbulent Channel Flow, J. Fluid Mech. 194 (1988) 15.

[30] R.S. Rogallo, Numerical Experiments in Homogeneous Turbulence, NASA TM 81315 (1981).

[31] J. Kim, P. Moin, Application of a Fractional Step Method to Incompressible Navier-Stokes Equation, J. Comp. Phys. 59 (308) (1985). 1985 and NASA TM85898, N8422328.

[32] M.M. Rai, P. Moin, Direct Simulations of Turbulent Flow Using Finite-Difference Schemes, J. Comp. Phys. 96 (1991) 15.

Chapter 2

Conservation Equations for Compressible Turbulent Flows

Chapter Outline Head

2.1 Introduction

In this chapter we consider the Navier-Stokes equations for a compressible fluid and show how they can be put into a form more convenient for turbulent flows. We follow

Analysis of Turbulent Flows with Computer Programs. http://dx.doi.org/10.1016/B978-0-08-098335-6.00002-1

the procedure first introduced by Reynolds in incompressible flows: we regard the turbulent motion as consisting of the sum of the mean part and a fluctuating part, introduce the sum into the Navier–Stokes equations, and time[1] average the resulting expressions. The equations thus obtained give considerable insight into the character of turbulent motions and serve as a basis for attacking mean-flow problems, as well as for analyzing the turbulence to find its harmonic components. However, before these governing conservation equations for compressible turbulent flows are obtained, it is appropriate to write down the conservation equations for mass, momentum, and energy.

In the following sections we shall discuss the conservation equations and their reduced forms in terms of rectangular coordinates, and for convenience we shall use the summation notation. For a discussion of the conservation equations in terms of another coordinate system, the reader is referred to [1].

2.2 The Navier–Stokes Equations

The well-known Navier–Stokes equations of motion for a compressible, viscous, heat-conducting, perfect gas may be written in the following form [2]:

Continuity

$$\frac{\partial \varrho}{\partial t} + \frac{\partial}{\partial x_j}(\varrho u_j) = 0, \tag{2.2.1}$$

Momentum

$$\frac{\partial}{\partial t}(\varrho u_i) + \frac{\partial}{\partial x_j}(\varrho u_i u_j) = -\frac{\partial p}{\partial x_i} + \frac{\partial \tau_{i,j}}{\partial x_j}, \tag{2.2.2}$$

Energy

$$\frac{\partial}{\partial t}(\varrho H) + \frac{\partial}{\partial x_j}(\varrho u_j H) = \frac{\partial p}{\partial t} + \frac{\partial}{\partial x_j}(u_j \tau_{ij} - q_j), \tag{2.2.3}$$

where the stress tensor τ_{ij}, heat-flux vector q_j, and total enthalpy H are given by

$$\tau_{ij} = \lambda \delta_{ij} \frac{\partial u_l}{\partial x_l} + \mu\left(\frac{\partial u_i}{\partial x_j} + \frac{\partial u_j}{\partial x_i}\right), \tag{2.2.4}$$

$$q_j = -k\frac{\partial T}{\partial x_j}, \tag{2.2.5}$$

[1]See Section 2.3 for a discussion of various kinds of averaging.

$$H = h + \frac{1}{2}u_i u_i. \tag{2.2.6}$$

In these equations, λ is the bulk viscosity ($= -2/3\mu$), μ the dynamic viscosity, k the thermal conductivity, and h the static enthalpy. In Eq. (2.2.4), δ_{ij} is the Kronecker delta, having the value 1 for $i = j$ and 0 for $i \neq j$. A summation is understood for repeated indices.

Sometimes it is more convenient to express the energy equation in terms of static enthalpy h, rather than total enthalpy H. Then using Eqs. (2.2.1) and (2.2.2) Eq. (2.2.3) becomes

$$\frac{\partial}{\partial t}(\varrho h) + \frac{\partial}{\partial x_j}(\varrho h u_j) = \frac{\partial p}{\partial t} + u_j \frac{\partial p}{\partial x_j} + \tau_{ij}\frac{\partial u_i}{\partial x_j} - \frac{\partial q_j}{\partial x_j}, \tag{2.2.7}$$

where $\tau_{ij}\partial u_i/\partial x_j$ is the dissipation function.

Equations (2.2.1)–(2.2.7) apply to laminar as well as to turbulent flows. For the latter, however, the values of the dependent variables are to be replaced by their instantaneous values. A direct approach to the turbulence problem, namely the solution of the full time-dependent Navier–Stokes equations, then consists in solving the equations for a given set of boundary or initial values and computing mean values over the ensemble for solutions, as discussed in Section 1.10. Even for the most restricted problem – turbulence of an incompressible fluid that appears to be a hopeless undertaking, because of the nonlinear terms in the equations. Thus, the standard procedure is to average over the equations rather than over the solutions. The averaging can be done either by the conventional time-averaging procedure or by the mass-weighted averaging procedure. Both are discussed in the next section.

2.3 Conventional Time-Averaging and Mass-Weighted-Averaging Procedures

In order to obtain the governing conservation equations for turbulent flows, it is convenient to replace the instantaneous quantities in the equations of Section 2.2 by their mean and their fluctuating quantities. In the conventional time-averaging procedure, for example, the velocity and pressure are usually written in the following forms:[2]

$$u_i(x_i, t) = \overline{u_i(x_i)} + u_i''(x_i, t), \tag{2.3.1}$$

$$p(x_i, t) = \overline{p(x_i)} + p''(x_i, t), \tag{2.3.2}$$

[2]The usual single prime on fluctuating quantities, for example, u', p', is reserved for later use.

where $\overline{u_i(x_i)}$ and $\overline{p(x_i)}$ are the time averages of the bulk velocity and pressure, respectively, and $u_i''(x_i, t)$ and $p''(x_i, t)$ the superimposed velocity and pressure fluctuations, respectively. The time average or "mean" of any quantity $q(t)$ is defined by

$$\bar{q} = \lim_{\Delta t \to \infty} (1/\Delta t) \int_{t_0}^{t_0+\Delta t} q(t)\, dt. \tag{2.3.3}$$

In practice, ∞ is taken to mean a time that is long compared to the reciprocal of the predominant frequencies in the spectrum of q; in wind-tunnel experiments, averaging times of a few seconds to a minute are usual. Clearly, the time average is useful only if it is independent of t_0; a random process whose time averages are all independent of t_0 is called "statistically stationary." A nonstationary process (e.g., an air jet from a high-pressure reservoir of finite size) must be analyzed by means of "ensemble averages." The ensemble average q is the average of a large number of instantaneous samples of q, of which one sample is taken during each run of the process at time t_0 after the process starts. A process with periodicity imposed on turbulence, like the flow in a turbomachine, can be analyzed by phase averages – take averages at a given point in space over many events in which a blade is at a given position relative to that point. For further discussions, see [3].

For a fluctuating quantity $q''(t)$, the average, $\overline{q''(t)}$, is zero, that is,

$$\overline{q''(t)} = \lim_{\Delta t \to \infty} (1/\Delta t) \int_{t_0}^{t_0+\Delta t} q''(t)\, dt = 0. \tag{2.3.4}$$

Average values similar to Eqs. (2.3.1) and (2.3.2) can be written for the other flow quantities, such as density, temperature, and enthalpy, as follows:

$$\varrho(x_i, t) = \overline{\varrho(x_i)} + \varrho''(x_i, t), \tag{2.3.5}$$

$$h(x_i, t) = \overline{h(x_i)} + h''(x_i, t), \tag{2.3.6}$$

$$H(x_i, t) = \overline{H(x_i)} + H''(x_i, t), \tag{2.3.7}$$

$$T(x_i, t) = \overline{T(x_i)} + T''(x_i, t), \tag{2.3.8}$$

where $\overline{\varrho''} = \overline{h''} = \overline{H''} = \overline{T''} = 0$.

As an example, let us consider the continuity and the momentum equations in the forms given by Eqs. (2.2.1) and (2.2.2), respectively, and show how they can be obtained by using the conventional time-averaging procedure for compressible turbulent flows. If we substitute the expressions given by Eqs. (2.3.1), (2.3.2), and (2.3.5) into (2.2.1) and (2.2.2) and take the time average of the terms appearing in the resulting equations, we obtain the mean continuity and the mean momentum equations in the following forms:

$$\frac{\partial \bar{\varrho}}{\partial t} + \frac{\partial}{\partial x_j}\left(\bar{\varrho}\, \bar{u}_j + \overline{\varrho'' u_j''}\right) = 0, \tag{2.3.9}$$

$$\begin{aligned} &\frac{\partial}{\partial t}\left(\bar{\varrho}\, \bar{u}_i + \overline{\varrho'' u_i''}\right) + \frac{\partial}{\partial x_j}\left(\bar{\varrho}\, \bar{u}_i \bar{u}_j + \bar{u}_i \overline{\varrho'' u_j''}\right) \\ &= -\frac{\partial \bar{p}}{\partial x_i} + \frac{\partial}{\partial x_j}\left(\bar{\tau}_{ij} - \bar{u}_j \overline{\varrho'' u_i''} - \bar{\varrho}\overline{u_i'' u_j''} - \overline{\varrho'' u_i'' u_j''}\right). \end{aligned} \tag{2.3.10}$$

For incompressible flows, $d\varrho = 0$. As a result Eqs. (2.3.9) and (2.3.10) can be simplified considerably; they become

$$\frac{\partial \bar{u}_j}{\partial x_j} = 0, \tag{2.3.11}$$

$$\varrho \frac{\partial \bar{u}_i}{\partial t} + \varrho \frac{\partial}{\partial x_j}(\bar{u}_i \bar{u}_j) = -\frac{\partial \bar{p}}{\partial x_i} + \frac{\partial}{\partial x_j}\left(\bar{\tau}_{ij} - \varrho \overline{u_i'' u_j''}\right). \tag{2.3.12}$$

We see from Eqs. (2.3.9)–(2.3.12) that the continuity and the momentum equations obtained by this procedure contain mean terms that have the same form as the corresponding terms in the instantaneous equations. However, they also have terms representing the mean effects of turbulence, which are additional unknown quantities. For that reason, the resulting conservation equations are undetermined. Consequently, the governing equations in this case, continuity and momentum, do not form a closed set. They require additional relations, which have to come from statistical or similarity considerations. The additional terms enter the governing equation as turbulent-transport terms such as $\bar{\varrho}\overline{u_i'' u_j''}$ and as density-generated terms such as $\overline{\varrho'' u_j''}$ and $\overline{\varrho'' u_i'' u_j''}$. In incompressible flows, the density-generated terms disappear, as is shown in Eqs. (2.3.11) and (2.3.12). In compressible flows, the continuity equation (2.3.9) has a source term, $(\partial/\partial x_j)\overline{\varrho'' u_j''}$, which indicates that a mean mass interchange occurs across the mean streamlines defined in terms of $\bar{u}_i$. It also indicates that the splitting of u_i according to Eq. (2.3.1) is not convenient; it is not consistent with the usual concept of a streamline. For that reason, we shall replace the conventional time-averaging procedure by another procedure that is well known in the studies of gas mixtures, the *mass-weighted-averaging* procedure, which was used by Van Driest [4], Favre [5], and Laufer and Ludloff [6]. Mass-weighted averaging eliminates the mean-mass term $\overline{\varrho'' u_j''}$ and some of the momentum transport terms such as $\bar{u}_i \overline{\varrho'' u_j''}$ and $\overline{\varrho'' u_i'' u_j''}$ across mean streamlines. We define a mass-weighted mean velocity

$$\tilde{u}_i = \overline{\varrho u_i}/\bar{\varrho}, \tag{2.3.13}$$

where the bar denotes conventional time averaging and the tilde denotes mass-weighted averaging. The velocity may then be written as

$$u_i(x_i,t) = \tilde{u}_i(x_i) + u'_i(x_i,t), \tag{2.3.14}$$

where $u'_i\,(x_i, t)$ is the superimposed velocity fluctuation. Multiplying Eq. (2.3.14) by the expression for $\varrho(x_i, t)$ given by Eq. (2.3.5) gives

$$\varrho u_i = (\bar{\varrho} + \varrho'')(\tilde{u}_i + u'_i) = \bar{\varrho}\tilde{u}_i + \varrho''\tilde{u}_i + \varrho'' u'_i + \bar{\varrho} u'_i.$$

Time averaging and noting the definition of $\varrho(x_i, t)$, we get

$$\overline{\varrho u_i} = \bar{\varrho}\tilde{u}_i + \overline{\varrho u'_i}.$$

From the definition of $\tilde{u}_i$, given by Eq. (2.3.13) it follows that

$$\overline{\varrho u'_i} = 0. \tag{2.3.15}$$

Note the important differences between the two averaging procedures. In the conventional time averaging, $\overline{u''_i} = 0$ and $\overline{\varrho u''_i} \neq 0$; in the mass-weighted averaging, $\overline{u'_i} \neq 0$ and $\overline{\varrho u'_i} = 0$.

Similarly, we can define the static enthalpy, static temperature, and total enthalpy thus:

$$h(x_i,t) = \tilde{h}(x_i) + h'(x_i,t), \tag{2.3.16}$$

$$T(x_i,t) = \tilde{T}(x_i) + T'(x_i,t), \tag{2.3.17}$$

$$H(x_i,t) = \tilde{H}(x_i) + H'(x_i,t), \tag{2.3.18}$$

where

$$\begin{aligned} \tilde{T} &= \overline{\varrho T}/\bar{\varrho}, \quad \tilde{h} = \overline{\varrho h}/\bar{\varrho}, \quad \tilde{H} = \tilde{h} + \frac{1}{2}\tilde{u}_i\tilde{u}_i + \frac{1}{2}\,\overline{\varrho u'_i u'_i}/\bar{\varrho}, \\ H' &= h' + \tilde{u}_i u'_i + \frac{1}{2}u'_i u'_i - \frac{1}{2}\,\overline{\varrho u'_i u'_i}/\bar{\varrho}. \end{aligned} \tag{2.3.19}$$

Also,

$$\overline{\varrho T'} = \overline{\varrho h'} = \overline{\varrho H'} = 0. \tag{2.3.20}$$

The expressions for $\tilde{H}$ and H' in Eq. (2.3.19) follow from the definitions of H, u_i, and $\tilde{H}$. Multiplying both sides of Eq. (2.2.6) by ϱ, and introducing the definition of u_i given by Eq. (2.3.14) into the resulting expression, we can write

$$\varrho H = \varrho h + \frac{1}{2}\,\varrho\tilde{u}_i\tilde{u}_i + \tilde{u}_i\varrho u'_i + \frac{1}{2}\varrho u'_i u'_i.$$

Mass averaging the above expression gives

$$\overline{\varrho H} = \overline{\varrho h} + \frac{1}{2}\,\bar{\varrho}\tilde{u}_i\tilde{u}_i + \frac{1}{2}\,\overline{\varrho u_i' u_i'}. \tag{2.3.21}$$

Since

$$\overline{\varrho H} = \bar{\varrho}\tilde{H} \quad \text{and} \quad \overline{\varrho h} = \bar{\varrho}\tilde{h},$$

Eq. (2.3.21) can be written as

$$\tilde{H} = \tilde{h} + \frac{1}{2}\tilde{u}_i\tilde{u}_i + \frac{1}{2}\,\overline{\varrho u_i' u_i'}/\bar{\varrho}.$$

Also,

$$H = \bar{H} + H' = \bar{h} + h' + \frac{1}{2}\left(\bar{u}_i + u_i'\right)^2 = \bar{h} + h' + \frac{1}{2}\,\bar{u}_i\bar{u}_i + \bar{u}_i u_i' + \frac{1}{2} u_i' u_i'.$$

Substituting the expression for $\bar{H}$ into the above expression, we get

$$H' = h' + \bar{u}_i u_i' + \frac{1}{2} u_i' u_i' - \frac{1}{2}\,\overline{\varrho u_i' u_i'}/\bar{\varrho}.$$

The definitions given by Eqs. (2.3.14)–(2.3.21) are also convenient for turbulence measurements because in hot-wire anemometry, the quantities measured at low speeds are the fluctuations of ϱu_i, and of T, and those measured at supersonic speeds are the fluctuations of ϱu_i, and of a quantity that is very close to the total enthalpy. Mean pressure is directly measurable. For that reason, the conventional time average of pressure is convenient; we shall use the definition given by Eq. (2.3.2). Furthermore, we shall also use the conventional time-averaging procedures for the stress tensor τ_{ij} and for the heat-flux vector q_j as given by Eqs. (2.2.4) and (2.2.5), respectively.

2.4 Relation Between Conventional Time-Averaged Quantities and Mass-Weighted-Averaged Quantities

A relationship between $\bar{u}_i$ and $\tilde{u}_i$ can be established as follows. Using Eq. (2.3.5), we can write Eq. (2.3.15) as

$$\overline{\varrho u_i'} = \overline{(\bar{\varrho} + \varrho'')u_i'} = 0.$$

That expression can also be written as

$$\overline{u_i'} = -\overline{\varrho'' u_i'}/\bar{\varrho}. \tag{2.4.1}$$

Taking the mean value of Eq. (2.3.14) and rearranging, we get

$$\tilde{u}_i - \bar{u}_i = -\overline{u'_i}. \tag{2.4.2}$$

Hence,

$$\tilde{u}_i - \bar{u}_i = \overline{\varrho'' u'_i}/\bar{\varrho}. \tag{2.4.3a}$$

It follows from the definitions of u_i (x_i, t) that for $i = 1$

$$u_1(x_i, t) = u(x, t) = \tilde{u} + u' = \bar{u} + u''.$$

Multiplying both sides of that expression by ϱ and averaging, we get

$$\bar{\varrho}\tilde{u} + \overline{\varrho u'} = \bar{\varrho}\bar{u} + \overline{\varrho u''}.$$

If we note that $\overline{\varrho u'} = 0$ by definition and that $\overline{\varrho u''}$ can be written as $\overline{(\bar{\varrho} + \varrho'')u''} = \overline{\varrho'' u''}$, we can combine the above expression and Eq. (2.4.3a) as

$$\tilde{u} - \bar{u} = \overline{\varrho'' u'}/\bar{\varrho} = \overline{\varrho'' u''}/\bar{\varrho}. \tag{2.4.3b}$$

Similar relationships between $\bar{h}$ and $\tilde{h}$, $\bar{T}$ and $\tilde{T}$, $\bar{H}$ and $\tilde{H}$, etc., can be established by a similar procedure. For example, in order to find the relation $\bar{T}$ and $\tilde{T}$, we rewrite the first term in Eq. (2.3.20) in the form

$$\overline{\varrho T'} = \overline{(\bar{\varrho} + \varrho'')T'} = 0. \tag{2.4.4}$$

Taking the mean value of Eq. (2.3.17), rearranging, and substituting the value of $\overline{\varrho T'}$ from Eq. (2.4.4) into the resulting expression, we get

$$\tilde{T} - \bar{T} = \overline{\varrho'' T'}/\bar{\varrho}. \tag{2.4.5a}$$

It follows from the definition of T (x_i, t) that

$$T(x_i, t) = \tilde{T} + T' = \bar{T} + T''. \tag{2.4.5b}$$

Multiplying both sides of Eq. (2.4.5b) by ϱ and averaging, we get

$$\bar{\varrho}\tilde{T} + \overline{\varrho T'} = \bar{\varrho}\bar{T} + \overline{\varrho T''}. \tag{2.4.5c}$$

If we note that $\overline{\varrho T'} = 0$ and that $\overline{\varrho T''} = \overline{\varrho'' T''}$ and if we make use of Eqs. (2.4.5a)–(2.4.5c), we can write

$$\tilde{T} - \bar{T} = T'' - T' = \overline{\varrho'' T''}/\bar{\varrho} = \overline{\varrho'' T'}/\bar{\varrho}. \tag{2.4.6}$$

From Eqs. (2.4.3) and (2.4.5) we see that the difference between the two average velocities depends on the density–velocity correlation term $\overline{\varrho'' u'_i}$ or $\overline{\varrho'' u''_i}$. Similarly, the difference between the two average temperatures depends on the density–temperature correlation term $\overline{\varrho'' T'}$ or $\overline{\varrho'' T''}$. A discussion of the magnitude of these quantities will be given later, in subsection 3.2.2.

2.5 Continuity and Momentum Equations

If we substitute the expressions given by Eqs. (2.3.2), (2.3.5), and (2.3.14) into Eqs. (2.2.1) and (2.2.2), we obtain

$$\frac{\partial}{\partial t}\left(\bar{\varrho}+\varrho''\right)+\frac{\partial}{\partial x_j}\left(\varrho\tilde{u}_j+\varrho u_j'\right)=0, \tag{2.5.1}$$

$$\begin{aligned}&\frac{\partial}{\partial t}\left(\varrho\tilde{u}_i+\varrho u_i'\right)+\frac{\partial}{\partial x_j}\left(\varrho\tilde{u}_i\tilde{u}_j+\varrho u_j'\tilde{u}_i+\varrho u_i'\tilde{u}_j+\varrho u_i' u_j'\right)\\&=-\frac{\partial\bar{p}}{\partial x_i}-\frac{\partial p''}{\partial x_i}+\frac{\partial\tau_{ij}}{\partial x_j}.\end{aligned} \tag{2.5.2}$$

Taking the time average of the terms appearing in these equations, we obtain the mean continuity and mean momentum equations for compressible turbulent flow:

$$\frac{\partial\bar{\varrho}}{\partial t}+\frac{\partial}{\partial x_j}\left(\bar{\varrho}\tilde{u}_j\right)=0, \tag{2.5.3}$$

$$\frac{\partial}{\partial t}\left(\bar{\varrho}\tilde{u}_i\right)+\frac{\partial}{\partial x_j}\left(\bar{\varrho}\tilde{u}_i\tilde{u}_j\right)=-\frac{\partial\bar{p}}{\partial x_i}+\frac{\partial}{\partial x_j}\left(\bar{\tau}_{ij}-\overline{\varrho u_i' u_j'}\right). \tag{2.5.4}$$

A comparison of the continuity equation (2.5.3) and the momentum equation (2.5.4) obtained by the mass-weighted averaging with those obtained by the conventional time averaging, namely, Eqs. (2.3.9) and (2.3.10), shows that with mass-weighted averaging the final equations have simpler form. In fact, with the mass-weighted averaging, they have the same form, term by term, as those for incompressible flows, with two exceptions: The viscous stresses τ_{ij} and the so-called Reynolds stresses $-\overline{\varrho u_i' u_j'}$ include fluctuations in viscosity and in density, respectively.

2.6 Energy Equations

If we substitute the expressions given by Eqs. (2.3.2), (2.3.5) and (2.3.14)–(2.3.20) into Eqs. (2.2.3) and (2.2.7), we obtain

$$\begin{aligned}&\frac{\partial}{\partial t}\left(\varrho\tilde{H}+\varrho H'\right)+\frac{\partial}{\partial x_j}\left(\varrho\tilde{H}\,\tilde{u}_j+\varrho H'\tilde{u}_j+\varrho u_j'\tilde{H}+\varrho H' u_j'\right)\\&=\frac{\partial}{\partial t}\left(\bar{p}+p''\right)+\frac{\partial}{\partial x_j}\left(u_i\tau_{ij}-q_j\right),\end{aligned} \tag{2.6.1}$$

$$\frac{\partial}{\partial t}(\varrho \tilde{h} + \varrho h') + \frac{\partial}{\partial x_j}(\varrho \tilde{h} \tilde{u}_j + \varrho h' \tilde{u}_j + \varrho u'_j \tilde{h} + \varrho u'_j h')$$
$$= \frac{\partial}{\partial t}(\bar{p} + p'') + (\tilde{u}_j + u'_j)\frac{\partial}{\partial x_j}(\bar{p} + p'') + \tau_{ij}\frac{\partial u_i}{\partial x_j} - \frac{\partial q_j}{\partial x_j}. \tag{2.6.2}$$

In those equations, it is convenient not to replace u_i, by its average and fluctuating component when it or its derivative $\partial u_i/\partial x_j$ is multiplied by τ_{ij}. By taking the time average of the terms appearing in the equations, we obtain the mean energy equations in terms of total enthalpy, Eq. (2.6.1),

$$\frac{\partial}{\partial t}(\bar{\varrho}\tilde{H}) + \frac{\partial}{\partial x_j}(\bar{\varrho}\tilde{H}\tilde{u}_j)$$
$$= \frac{\partial \bar{p}}{\partial t} + \frac{\partial}{\partial x_j}\left(-\bar{q}_j - \overline{\varrho H' u'_j} + \tilde{u}_i\overline{\tau_{ij}} + \overline{u'_i\tau_{ij}}\right), \tag{2.6.3}$$

and static enthalpy, Eq. (2.6.2),

$$\frac{\partial}{\partial t}(\bar{\varrho}\tilde{h}) + \frac{\partial}{\partial x_j}(\bar{\varrho}\tilde{h}\tilde{u}_j)$$
$$= \frac{\partial \bar{p}}{\partial t} + \tilde{u}_j\frac{\partial \bar{p}}{\partial x_j} + \overline{u'_j\frac{\partial p}{\partial x_j}} + \frac{\partial}{\partial x_j}\left(-\bar{q}_j - \overline{\varrho h' u'_j}\right) + \overline{\tau_{ij}\frac{\partial u_i}{\partial x_j}}. \tag{2.6.4}$$

2.7 Mean-Kinetic-Energy Equation

The equation for kinetic energy of the mean motion can be obtained by considering the scalar product of $\tilde{u}_j$ and the mean momentum equation for $\tilde{u}_i$ [see Eq. (2.5.4)],

$$\tilde{u}_j\left[\frac{\partial}{\partial t}(\bar{\varrho}\tilde{u}_i) + \frac{\partial}{\partial x_k}(\bar{\varrho}\tilde{u}_i\tilde{u}_k)\right] = \tilde{u}_j\left[-\frac{\partial \bar{p}}{\partial x_i} + \frac{\partial}{\partial x_k}\left(\bar{\tau}_{ik} - \overline{\varrho u'_i u'_k}\right)\right], \tag{2.7.1a}$$

and the scalar product of $\tilde{u}_i$ and the mean momentum equation for $\tilde{u}_j$,

$$\tilde{u}_i\left[\frac{\partial}{\partial t}(\bar{\varrho}\tilde{u}_j) + \frac{\partial}{\partial x_k}(\bar{\varrho}\tilde{u}_j\tilde{u}_k)\right] = \tilde{u}_i\left[-\frac{\partial \bar{p}}{\partial x_j} + \frac{\partial}{\partial x_k}\left(\bar{\tau}_{jk} - \overline{\varrho u'_j u'_k}\right)\right], \tag{2.7.1b}$$

Adding Eqs. (2.7.1a) and (2.7.1b) and rearranging, we obtain

$$\begin{aligned}
&\frac{\partial}{\partial t}(\bar{\varrho}\tilde{u}_i\tilde{u}_j) + \frac{\partial}{\partial x_k}(\bar{\varrho}\tilde{u}_i\tilde{u}_j\tilde{u}_k) \\
&\quad = -\tilde{u}_j\frac{\partial \bar{p}}{\partial x_i} - \tilde{u}_i\frac{\partial \bar{p}}{\partial x_j} + \tilde{u}_j\frac{\partial}{\partial x_k}(\bar{\tau}_{ik} - \overline{\varrho u_i' u_k'}) \\
&\qquad + \tilde{u}_i\frac{\partial}{\partial x_k}(\bar{\tau}_{jk} - \overline{\varrho u_j' u_k'})
\end{aligned} \tag{2.7.2}$$

For $i = j$, that equation becomes

$$\begin{aligned}
&\frac{\partial}{\partial t}\left(\frac{1}{2}\bar{\varrho}\tilde{u}_i\tilde{u}_i\right) + \frac{\partial}{\partial x_k}\left(\frac{1}{2}\,\bar{\varrho}\tilde{u}_i\tilde{u}_i\tilde{u}_k\right) \\
&= -\tilde{u}_i\frac{\partial \bar{p}}{\partial x_i} + \frac{\partial}{\partial x_k}\left[\tilde{u}_i\left(\bar{\tau}_{ik} - \overline{\varrho u_i' u_k'}\right)\right] + \overline{\varrho u_i' u_k'}\frac{\partial \tilde{u}_i}{\partial x_k} - \bar{\tau}_{ik}\frac{\partial \tilde{u}_i}{\partial x_k},
\end{aligned} \tag{2.7.3a}$$

which can also be written as

$$\underset{\text{I}}{\frac{D}{Dt}\left(\bar{\varrho}\frac{\tilde{u}_i\tilde{u}_i}{2}\right)} = \underset{\text{II}}{-\tilde{u}_i\frac{\partial \bar{p}}{\partial x_i}} + \underset{\text{III}}{\tilde{u}_i\frac{\partial \bar{\tau}_{ik}}{\partial x_k}} \underset{\text{IV}}{- \tilde{u}_i\frac{\partial}{\partial x_k}\left(\overline{\varrho u_j' u_k'}\right)} \tag{2.7.3b}$$

Equation (2.7.3) is the kinetic energy equation of the mean motion. The terms I to IV in the equation can be given the following meaning:

$$\text{I}: \quad \frac{D}{Dt}\left(\bar{\varrho}\frac{\tilde{u}_i\tilde{u}_i}{2}\right) = \frac{\partial}{\partial t}\left(\bar{\varrho}\frac{\tilde{u}_i\tilde{u}_i}{2}\right) + \frac{\partial}{\partial x_k}\left[\tilde{u}_k\left(\bar{\varrho}\frac{\tilde{u}_i\tilde{u}_i}{2}\right)\right] \tag{2.7.4}$$

represents the rate of change of the kinetic energy of the mean motion. Sometimes this is called the gain of kinetic energy of the mean motion by advection.

$$\text{II}: \quad -\tilde{u}_i(\partial \bar{p}/\partial x_i) \tag{2.7.5}$$

represents the flow work done by the mean pressure forces acting on the control volume to produce kinetic energy of the mean motion.

$$\text{III}: \quad \tilde{u}_i(\partial \bar{\tau}_{ik}/\partial x_k) \tag{2.7.6}$$

represents the action of viscosity, which takes the form of a dissipation and a spatial transfer.

$$\text{IV}: \quad -\tilde{u}_i\frac{\partial}{\partial x_k}\overline{\varrho u_i' u_k'} = -\frac{\partial}{\partial x_k}\left(\tilde{u}_i\overline{\varrho u_i' u_k'}\right) + \overline{\varrho u_i' u_k'}\frac{\partial \tilde{u}_i}{\partial x_k}.$$

The first term on the right-hand side of IV represents the spatial transport of mean kinetic energy by the turbulent fluctuations; it is sometimes called the "gain from energy flux" or "the divergence of the energy flux transmitted by the working of the mean flow against the Reynolds stress."

The second term represents the "loss to turbulence" or the production of turbulent energy from the mean flow energy.

2.8 Reynolds-Stress Transport Equations

In the preceding sections, we have discussed the mean continuity, momentum, energy, and kinetic-energy equations for compressible turbulent flow. In this section we shall discuss the equations for the mean products of velocity fluctuation components known as Reynolds-stress transport equations.

Let us consider the scalar product of u_j and momentum equation for u_i [see Eq. (2.2.2)],

$$u_j\left[\frac{\partial}{\partial t}(\varrho u_i)+\frac{\partial}{\partial x_k}(\varrho u_i u_k) = -\frac{\partial p}{\partial x_i}+\frac{\partial \tau_{ik}}{\partial x_k}\right], \tag{2.8.1a}$$

and the scalar product of u_i and the momentum equation for u_j,

$$u_i\left[\frac{\partial}{\partial t}(\varrho u_j)+\frac{\partial}{\partial x_k}(\varrho u_j u_k) = -\frac{\partial p}{\partial x_j}+\frac{\partial \tau_{jk}}{\partial x_k}\right]. \tag{2.8.1b}$$

The sum of the two equations is

$$\begin{aligned}&\frac{\partial}{\partial t}(\varrho u_i u_j)+\frac{\partial}{\partial x_k}(\varrho u_i u_j u_k)\\ &\quad = -u_j\frac{\partial p}{\partial x_i}-u_i\frac{\partial p}{\partial x_j}+u_j\frac{\partial \tau_{ik}}{\partial x_k}+u_i\frac{\partial \tau_{jk}}{\partial x_k}.\end{aligned} \tag{2.8.2}$$

Using Eq. (2.3.14), we can write Eq. (2.8.2) as

$$\begin{aligned}&\frac{\partial}{\partial t}\left[\varrho(\tilde{u}_i+u_i')(\tilde{u}_j+u_j')\right]+\frac{\partial}{\partial x_k}\left[\varrho(\tilde{u}_i+u_i')(\tilde{u}_j+u_j')(\tilde{u}_k+u_k')\right]\\ &= -(\tilde{u}_j+u_j')\frac{\partial p}{\partial x_i}-(\tilde{u}_i+u_i')\frac{\partial p}{\partial x_j}+(\tilde{u}_j+u_j')\frac{\partial \tau_{ik}}{\partial x_k}\\ &\quad +(\tilde{u}_i+u_i')\frac{\partial \tau_{jk}}{\partial x_k},\end{aligned}$$

where now $\tau_{ik} = \bar{\tau}_{ik}+\tau''_{ik}$ and $\tau_{jk} = \bar{\tau}_{jk}+\tau''_{jk}$. Taking the time average of the terms, we obtain

$$\frac{\partial}{\partial t}\left(\bar{\varrho}\tilde{u}_i\tilde{u}_j + \overline{\varrho u_i' u_j'}\right)$$
$$+\frac{\partial}{\partial x_k}\left(\bar{\varrho}\tilde{u}_i\tilde{u}_j\tilde{u}_k + \tilde{u}_k\overline{\varrho u_i' u_j'} + \tilde{u}_i\overline{\varrho u_j' u_k'} + \tilde{u}_j\overline{\varrho u_i' u_k'} + \overline{\varrho u_i' u_j' u_k'}\right)$$
$$= -\tilde{u}_j\frac{\partial \bar{p}}{\partial x_i} - \overline{u_j'\frac{\partial p}{\partial x_i}} - \tilde{u}_i\frac{\partial \bar{p}}{\partial x_j} - \overline{u_i'\frac{\partial p}{\partial x_j}} + \tilde{u}_j\frac{\partial \bar{\tau}_{ik}}{\partial x_k} + \overline{u_j'\frac{\partial \tau_{ik}''}{\partial x_k}}$$
$$+\tilde{u}_i\frac{\partial \bar{\tau}_{jk}}{\partial x_k} + \overline{u_i'\frac{\partial \tau_{jk}''}{\partial x_k}}. \tag{2.8.3}$$

If we subtract Eq. (2.7.2) from that equation and rearrange, we obtain the equations for the components of the Reynolds stress, $-\overline{\varrho u_i' u_j'}$,

$$\frac{D}{Dt}\left(\overline{\varrho u_i' u_j'}\right) + \frac{\partial}{\partial x_k}\left(\overline{\varrho u_i' u_j' u_k'}\right)$$
$$= -\overline{u_j'\frac{\partial p}{\partial x_i}} - \overline{u_i'\frac{\partial p}{\partial x_j}} + \overline{u_j'\frac{\partial \tau_{ik}''}{\partial x_k}} + \overline{u_i'\frac{\partial \tau_{jk}''}{\partial x_k}}$$
$$-\overline{\varrho u_i' u_k'}\frac{\partial \tilde{u}_j}{\partial x_k} - \overline{\varrho u_j' u_k'}\frac{\partial \tilde{u}_i}{\partial x_k}, \tag{2.8.4}$$

where

$$\frac{D}{Dt}\left(\overline{\varrho u_i' u_j'}\right) = \frac{\partial}{\partial t}\left(\overline{\varrho u_i' u_j'}\right) + \frac{\partial}{\partial x_k}\left(\tilde{u}_k\overline{\varrho u_i' u_j'}\right).$$

Equation (2.8.4) is the transport equation for the Reynolds stress $-\overline{\varrho u_i' u_j'}$; it expresses the rate of change of the Reynolds stress along the mean streamline as the balance of generation by interaction between the turbulence and the mean flow, the gain or loss by convective movements of the turbulence and by the action of pressure gradients, and destruction by viscous forces. It is clear that the six independent components of Eq. (2.8.4), when taken together with Eqs. (2.5.3) and (2.5.4) form a set with more unknowns than equations. By itself, Eq. (2.8.4) does not lead to a knowledge of the distribution of the Reynolds stress and so to a solution of the mean-flow problem; its value lies in the restrictions it puts on the nature of turbulent transfer processes.

The meaning of the terms in Eq. (2.8.4) will be discussed later. First, we shall discuss the meaning of these terms for $i = j$.

For $i = j$, Eq. (2.8.4) becomes the mean-kinetic-energy equation of the fluctuations, often called the "turbulent-energy" equation:

$$\underbrace{\frac{D}{Dt}\frac{1}{2}\overline{\varrho u_i' u_i'}}_{\text{I}} + \underbrace{\frac{\partial}{\partial x_k}\overline{u_k'\left(\frac{1}{2}\varrho u_i' u_i'\right)}}_{\text{II}} = \underbrace{-\overline{u_i'\frac{\partial p}{\partial x_i}}}_{\text{III}} + \underbrace{\overline{u_j'\frac{\partial \tau_{ik}''}{\partial x_k}}}_{\text{IV}} - \underbrace{\overline{\varrho u_i' u_k'}\frac{\partial \tilde{u}_i}{\partial x_k}}_{\text{V}}. \tag{2.8.5a}$$

Sometimes the equation is written in the form

$$\frac{D}{Dt}\frac{1}{2}\overline{\varrho u_i' u_i'}$$

$$= -\overline{u_i'\frac{\partial p}{\partial x_i}} - \overline{\tau_{ik}''\frac{\partial u_i'}{\partial x_k}} + \frac{\partial}{\partial x_k}\overline{\left[u_i'\left(\tau_{ik}'' - \frac{1}{2}\varrho u_i' u_k'\right)\right]} - \overline{\varrho u_i' u_k'}\,\frac{\partial \tilde{u}_i}{\partial x_k}. \tag{2.8.5b}$$

The meanings of the terms in Eq. (2.8.5a) are as follows:

$$\text{I}: \quad \frac{D}{Dt}\frac{1}{2}\overline{\varrho u_i' u_i'} = \frac{\partial}{\partial t}\frac{1}{2}\overline{\varrho u_i' u_i'} + \frac{\partial}{\partial x_k}\tilde{u}_k\left(\frac{1}{2}\overline{\varrho u_i' u_i'}\right) \tag{2.8.6}$$

represents the rate of change of the kinetic energy of the turbulence. The term

$$\frac{\partial}{\partial t}\frac{1}{2}\overline{\varrho u_i' u_i'} \tag{2.8.7}$$

represents the unsteady growth of turbulent energy, and the term

$$\frac{\partial}{\partial x_k}\tilde{u}_k\left(\frac{1}{2}\overline{\varrho u_i' u_i'}\right)$$

represents the kinetic energy of the fluctuation motion that is convected by the mean motion. That term is also often called the "advection" or gain of energy by mean stream advection (gain following the mean flow).

$$\text{II}: \quad \frac{\partial}{\partial x_k}\overline{u_k'\left(\frac{1}{2}\varrho u_i' u_i'\right)}$$

represents the kinetic energy of the fluctuations convected by the fluctuations, that is, the diffusion of the fluctuation energy.

$$\text{III}: \quad -\overline{u_i'\frac{\partial p}{\partial x_i}} \tag{2.8.8}$$

represents the work due to turbulence.

$$\text{IV}: \quad \overline{u_i'\frac{\partial \tau_{ik}''}{\partial x_k}} = \frac{\partial}{\partial x_k}\overline{u_i'\tau_{ik}''} - \overline{\tau_{ik}''\frac{\partial u_i'}{\partial x_k}} \tag{2.8.9}$$

represents the work of viscous stresses due to the fluctuation motion. The term

$$\frac{\partial}{\partial x_k}\left(\overline{u_i'\tau_{ik}''}\right) \tag{2.8.10}$$

represents the spatial transport of turbulent energy by viscous forces, and the term

$$-\overline{\tau_{ik}''\frac{\partial u_i'}{\partial x_k}} \tag{2.8.11}$$

represents the viscous "dissipation" by the turbulent motion.

$$\text{V}: \quad -\overline{\varrho u_i' u_k'} \frac{\partial \tilde{u}_i}{\partial x_k} \tag{2.8.12}$$

represents the product of a turbulent stress and mean rate of strain. That term links the mean flow to the turbulent fluctuations through the "Reynolds stresses" $-\overline{\varrho u_i' u_k'}$. It is generally called the "production of turbulent energy" term. Of course, all the terms on the right-hand side of Eq. (2.8.5a) except the transport terms act to produce or to destroy (depending upon their sign) the kinetic energy of the turbulence in the control volume. Any term expressible as the spatial gradient of a time-averaged quantity is a *transport* term (its integral over the flow volume must be zero). Terms not so expressible are source/sink terms. The distinction between term V and the other terms on the right-hand side of Eq. (2.8.5a) is that V is the only one containing the mean velocity gradient; the others contain only the fluctuation quantities. Thus, only term V can act to take energy from the mean motion.

Let us now turn our attention to the meaning of the terms appearing in the Reynolds-stress transport equation, Eq. (2.8.4). From the previous discussion of the meaning of each term for $i = j$, it is apparent that the meanings of the terms in Eq. (2.8.4) for $i \neq j$ are similar to the ones for $i = j$. For example, the first term on the left-hand side of Eq. (2.8.4),

$$(D/Dt)\left(\overline{\varrho u_i' u_j'}\right) \tag{2.8.13}$$

represents the variation of the correlation $\overline{\varrho u_i' u_j'}$ as the fluid element moves along a streamline, rather than the variation of the kinetic energy of the fluctuations, $\frac{1}{2}\overline{\varrho u_i' u_i'}$.

The first two terms on the right-hand side of Eq. (2.8.4) deserve special attention. If we write them in the form

$$\begin{aligned} &-\left[\overline{u_j' \frac{\partial p}{\partial x_i}} + \overline{u_i' \frac{\partial p}{\partial x_j}}\right] \\ &= -\left[\frac{\partial}{\partial x_i}\left(\overline{u_j' p}\right) + \frac{\partial}{\partial x_j}\left(\overline{u_i' p}\right)\right] + \overline{p\left(\frac{\partial u_j'}{\partial x_i} + \frac{\partial u_i'}{\partial x_j}\right)}, \end{aligned} \tag{2.8.14}$$

the terms

$$\overline{\frac{\partial}{\partial x_i}(u_j' p) + \frac{\partial}{\partial x_j}(u_i' p)} \tag{2.8.15}$$

represent the so-called general pressure-diffusion terms. The second term on the right-hand side of Eq. (2.8.14),

$$\overline{p\left(\frac{\partial u_j'}{\partial x_i}+\frac{\partial u_i'}{\partial x_j}\right)} \tag{2.8.16}$$

is sometimes called the "redistribution" or "return-to-isotropy" term. It describes the redistribution of energy among the terms $\overline{u_i^2}$ $(i = 1, 2, 3)$ that approaches the statistically most probable state, in which all of the components of $\overline{u_i^2}$ are equal.

2.9 Reduced Forms of the Navier–Stokes Equations

The solution of the complete three-dimensional time dependent Navier–Stokes equations for turbulent flows, Eqs. (2.2.1)–(2.2.3), is rather difficult due to the wide range of length and time scales that the turbulent flows posses (see Section 1.5). Analytical solutions to even the simplest turbulent flows do not exist. A complete description of turbulent flow, where the flow variables (e.g. velocity, temperature and pressure) are known as a function of space and time can therefore only be obtained by numerically solving the Navier–Stokes equations. These numerical solutions are termed direct numerical simulations (DNS).

The instantaneous range of scales in turbulent flows increases rapidly with the Reynolds number. As a result most engineering problems, e.g. the flow around a wing, have too wide a range of scales to be directly computed using DNS. The engineering computation of turbulent flows therefore relies on simpler descriptions: instead of solving for the instantaneous flow field, the statistical evolution of the flow is sought. Approaches based on the Reynolds averaged Navier–Stokes (RANS) equations, (2.5.3), (2.5.4) and on their reduced forms (see Fig. 2.1) are the most prevalent. Another approximation, large eddy simulation (LES), is intermediate in complexity between DNS and RANS (see Fig. 2.1). Large eddy simulation (see Section 1.10) directly computes the large energy-containing scales, while modeling the influence of the small scales.

For both laminar and turbulent flows, the Navier–Stokes equations can be reduced to simpler forms by examining the relative magnitudes of the terms in the equations. In the application of this procedure, known as "order-of-magnitude" analysis, it is common to introduce length and velocity scales in order to estimate the relative magnitudes of the mean and fluctuating components as described, for example, in [7]. Figure 2.1 shows the hierarchy of the simplification of the Navier–Stokes equations.

Since the momentum equation expresses a balance among inertia forces, pressure forces, and viscous forces (in formulating the equations, the body forces were neglected), one such simplifications arises when some of the relative magnitudes of these forces are small in comparison with others. For example, at Reynolds numbers much smaller than unity, the inertia accelerating terms in Eq. (2.5.4) become small in

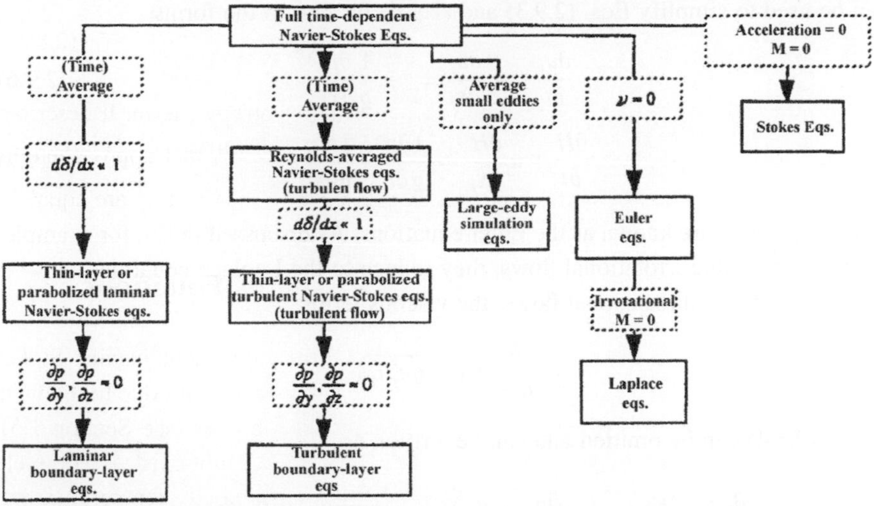

Fig. 2.1 Simplification of the Navier–Stokes equations. Dashed boxes denote simplifying approximations.

comparison with pressure and viscous terms. The resulting momentum equation together with the continuity equation are known as the Stokes flow equations; they are given by

$$\frac{\partial u_j}{\partial x_j} = 0 \tag{2.9.1}$$

$$\frac{\partial p}{\partial x_i} = \frac{\partial \tau_{ij}}{\partial x_j} \tag{2.9.2}$$

Another simplification of Eq. (2.5.4) arises when the viscous forces are negligible with respect to the inertia and the pressure forces. In such cases, the momentum equation, Eq. (2.5.4), and the energy equation, for example, Eq. (2.6.3), can be simplified considerably; they reduce to

$$\frac{\partial}{\partial t}(\varrho u_i) + \frac{\partial}{\partial x_j}(\varrho u_i u_j) = -\frac{\partial p}{\partial x_i}, \tag{2.9.3}$$

$$\frac{\partial}{\partial t}(\varrho H) + \frac{\partial}{\partial x_j}(\varrho H u_j) = \frac{\partial p}{\partial t} - \frac{\partial q_j}{\partial x_j}. \tag{2.9.4}$$

The continuity equation, Eq. (2.5.3),

$$\frac{\partial \varrho}{\partial t} + \frac{\partial}{\partial x_j}(\varrho u_j) = 0 \tag{2.9.5}$$

can be used to simplify Eqs. (2.9.3) and (2.9.4) further, to the forms

$$\frac{\partial u_i}{\partial t} + u_j \frac{\partial u_j}{\partial x_j} = -\frac{1}{\varrho}\frac{\partial p}{\partial x_i}, \tag{2.9.6}$$

$$\frac{\partial H}{\partial t} + u_j \frac{\partial H}{\partial x_j} = \frac{1}{\varrho}\frac{\partial p}{\partial t} - \frac{1}{\varrho}\frac{\partial q_j}{\partial x_j}. \tag{2.9.7}$$

These equations are known as the Euler equations. As discussed in [7], for example, for incompressible irrotational flows, they reduce to the Laplace equation.

In some three-dimensional flows, the viscous terms

$$\frac{\partial}{\partial x_1}\left(\overline{\tau}_{i1} - \overline{\varrho u_i' u_1'}\right)$$

in Eq. (2.5.4) can be omitted and can be written as

$$\frac{\partial}{\partial t}(\overline{\varrho}\tilde{u}_i) + \frac{\partial}{\partial x_j}\left(\overline{\varrho}\tilde{u}_i\tilde{u}_j\right) = -\frac{\partial \overline{p}}{\partial x_i} + \frac{\partial}{\partial x_2}\left(\overline{\tau}_{i2} - \overline{\varrho u_i' u_2'}\right) + \frac{\partial}{\partial x_3}\left(\overline{\tau}_{i3} - \overline{\varrho u_i' u_3'}\right). \tag{2.9.8}$$

Similarly, the energy equation, Eq. (2.6.4), it can be written as

$$\begin{aligned}\frac{\partial}{\partial t}\left(\overline{\varrho}\tilde{h}\right) + \frac{\partial}{\partial x_j}\left(\overline{\varrho}\tilde{h}\tilde{u}_j\right) &= \frac{\partial \overline{p}}{\partial t} + \tilde{u}_j \frac{\partial \overline{p}}{\partial x_j} + \overline{u_j' \frac{\partial p}{\partial x_j}} \\ &+ \frac{\partial}{\partial x_2}\left(-\overline{q}_2 - \overline{\varrho h' u_2'}\right) + \frac{\partial}{\partial x_3}\left(-\overline{q}_3 - \overline{\varrho h' u_3'}\right) \\ &+ \overline{\tau_{i2}\frac{\partial u_i}{\partial x_2}} + \overline{\tau_{i3}\frac{\partial u_i}{\partial x_3}}\end{aligned} \tag{2.9.9}$$

The momentum and energy equations resulting from this approximation, Eqs. (2.9.8) and (2.9.9) together with the continuity equation, Eq. (2.5.3), are known as the parabolized Navier–Stokes equations.

In other flows, the Navier–Stokes equations can be simplified further by retaining only the viscous terms with derivatives in the coordinate direction normal to the body surface x_2 or, for free shear flows, the direction normal to the thin layer. Momentum and energy equations become

$$\frac{\partial}{\partial t}\left(\overline{\varrho}\tilde{u}_i\right) + \frac{\partial}{\partial x_j}\left(\overline{\varrho}\tilde{u}_i\tilde{u}_j\right) = -\frac{\partial \overline{p}}{\partial x_i} + \frac{\partial}{\partial x_2}\left(\overline{\tau}_{i2} - \overline{\varrho u_i' u_2'}\right) \tag{2.9.10}$$

$$\begin{aligned}\frac{\partial}{\partial t}\left(\overline{\varrho}\tilde{h}\right) + \frac{\partial}{\partial x_j}\left(\overline{\varrho}\tilde{h}\tilde{u}_j\right) &= -\frac{\partial \overline{p}}{\partial t} + \tilde{u}_j \frac{\partial \overline{p}}{\partial x_j} + \overline{u_j' \frac{\partial p}{\partial x_j}} + \frac{\partial}{\partial x_2}\left(-\overline{q}_2 - \overline{\varrho h' u_2'}\right) \\ &+ \overline{\tau_{i2}\frac{\partial u_i}{\partial x_2}}\end{aligned} \tag{2.9.11}$$

The above equations together with Eq. (2.5.3) are known as the thin-layer Navier–Stokes equations.

Another simplification of the Navier–Stokes equations occurs at high Reynolds number. The resulting equations are known as the boundary-layer equations. They are discussed in the next chapter.

Problems

2.1 Show that the continuity equation (2.2.1) can be regarded as a "transport" equation for the density, in the same sense that Eq. (2.2.2) is a transport equation for the momentum per unit volume, ϱu_i.

2.2 Take the x_i derivative of Eq. (2.2.2), i.e. "take the divergence" of the set of equations for the three components of momentum, and show that if ϱ is constant the result is nominally a transport equation for the divergence $\partial u_i/\partial x_i$. Further show that if $\partial u_i/\partial x_i$ is set to zero (which Eq. (2.2.1) shows is the correct value in constant-density flow) the transport equation reduces to a Poisson equation for the pressure.

2.3 Show that the rate of viscous dissipation given in Eq. (2.8.11) is the mean product of the fluctuating viscous stress and the fluctuating rate of strain, $\partial u'_i/\partial x_k + \partial u'_k/\partial x_i)/2$. Show that this is the mean rate at which the turbulance does work against viscous stresses, and that the first law of thermodynamics confirms that this is truly the rate at which turbulent kinetic energy is converted into thermal internal energy.

References

[1] L.H. Back, Conservation equations of a viscous, heat-conducting fluid in curvilinear orthogonal coordinates. Published in Handbook of Tables for Applied Engineering Science. 415, in: R.E. Boze, G.L. Tuve (Eds.), Chem. Rubber Co., Cleveland, Ohio, 1970.

[2] H.W. Liepmann, A. Roshko, Elements of Gas Dynamics. 332, Wiley, New York, 1952.

[3] P. Bradshaw, An Introduction to Turbulence and Its Measurement, Pergamon, Oxford, 1971.

[4] E.R. Van Driest, Turbulent Boundary Layer in Compressible Fluids, J. Aeronaut. Sci. 18 (1951) 145.

[5] A. Favre, Statistical equations of turbulent gases. SIAM Problems of Hydrodynamics and Continuum Mechanics (Sedov 60th birthday volume). Also, Equations des gaz turbulents compressibles, J. Mec. 4 (1969). 361(part I), 391 (part II).

[6] J. Laufer, K.G. Ludloff, Conservation equations in a compressible turbulent fluid and a numerical scheme for their solution, McDonnell Douglas Paper WD 1355 (1970).

[7] T. Cebeci, J. Cousteix, Modeling and Computation of Boundary-Layer Flows, Horizons Publ, Long Beach and Springer-Verlag, Berlin, 1998.

Chapter 3

Boundary-Layer Equations

Chapter Outline Head

Analysis of Turbulent Flows with Computer Programs. http://dx.doi.org/10.1016/B978-0-08-098335-6.00003-3

3.1 Introduction

Another important simplification of the Navier-Stokes equations arises when the flow of a fluid past a solid at high Reynolds number is considered. In such cases, there is a very narrow region close to the surface in which the fluid velocity and possibly the temperature (or enthalpy) deviate considerably from their values far away from the surface. For example, the velocity of a fluid flowing past a stationary body changes rapidly from zero velocity at the surface to its value in the body of the fluid (except for very-low-pressure gases, when the mean free path of the gas molecules is large relative to the body). In that narrow region, the velocity gradient may be so large that, even if the fluid viscosity is small, the viscous forces may be of the same order as the inertia forces. That region is called the boundary region, and the layer of affected fluid is called the *boundary layer.* There, because gradients perpendicular to the surface are much larger than gradients parallel to the surface, some of the terms in the Navier-Stokes equations can be neglected, which simplifies the equations considerably. It is on this basis that Prandtl, in 1904, proposed his boundary-layer theory. According to that theory, the flow field may be separated into two regions: the main, inviscid flow, which is described by Eqs. (2.9.5)–(2.9.7), and the boundary region described by the simplified momentum and energy equations, called boundary-layer equations. The simplifications are discussed in the next section.

3.2 Boundary-Layer Approximations for Compressible Flows

For simplicity, we consider a two-dimensional, unsteady, compressible turbulent flow. The external flow has one velocity component, u_e, that depends on the time coordinate t (unsteady flow) and on one coordinate x in the wall surface. The flow within the boundary layer possesses two velocity components, u and v, that depend on t and two space coordinates, x and y. As is standard in boundary-layer theory, x is taken to be the distance measured along the surface (which may be curved) and y is the distance normal to the surface. The turbulence is three dimensional, with velocity components u', v', and w' in the x, y, and z directions, respectively. The total enthalpy within the boundary layer H is a function of x, y, and t. The conservation equations for mass, momentum, and energy as given by Eqs (2.5.3), (2.5.4), and (2.6.3), respectively, become

Continuity

$$\frac{\partial \bar{\varrho}}{\partial t} + \frac{\partial}{\partial x}(\bar{\varrho}\tilde{u}) + \frac{\partial}{\partial y}(\bar{\varrho}\tilde{v}) = 0, \tag{3.2.1}$$

x Momentum

$$\frac{\partial}{\partial t}(\bar{\varrho}\tilde{u}) + \frac{\partial}{\partial x}(\bar{\varrho}\tilde{u}\tilde{u}) + \frac{\partial}{\partial y}(\bar{\varrho}\tilde{u}\tilde{v}) = -\frac{\partial \bar{p}}{\partial x} + \frac{\partial}{\partial x}(\bar{\tau}_{xx} - \overline{\varrho u'u'}) + \frac{\partial}{\partial y}(\bar{\tau}_{xy} - \overline{\varrho u'v'}), \tag{3.2.2}$$

y Momentum

$$\frac{\partial}{\partial t}(\bar{\varrho}\tilde{v}) + \frac{\partial}{\partial x}(\bar{\varrho}\tilde{v}\tilde{u}) + \frac{\partial}{\partial y}(\bar{\varrho}\tilde{v}\tilde{v}) = -\frac{\partial \bar{p}}{\partial y} + \frac{\partial}{\partial x}(\bar{\tau}_{yx} - \overline{\varrho u'v'}) + \frac{\partial}{\partial y}(\bar{\tau}_{yy} - \overline{\varrho v'v'}), \tag{3.2.3}$$

Energy (Total Enthalpy)

$$\frac{\partial}{\partial t}(\bar{\varrho}\tilde{H}) + \frac{\partial}{\partial x}(\bar{\varrho}\tilde{H}\tilde{u}) + \frac{\partial}{\partial y}(\bar{\varrho}\tilde{H}\tilde{v}) = \frac{\partial \bar{p}}{\partial t} + \frac{\partial}{\partial x}\left[-\bar{q}_x - \overline{\varrho H'u'} + \overline{u\tau_{xx}} + \overline{v\tau_{yx}}\right] + \frac{\partial}{\partial y}\left[-\bar{q}_y - \overline{\varrho H'v'} + \overline{u\tau_{xy}} + \overline{v\tau_{yy}}\right] \tag{3.2.4}$$

where

$$\bar{\tau}_{xx} = 2\mu\overline{\frac{\partial u}{\partial x}}, \quad \bar{\tau}_{yy} = 2\mu\overline{\frac{\partial v}{\partial y}}, \quad \bar{\tau}_{xy} = \bar{\tau}_{yx} = \mu\overline{\left(\frac{\partial u}{\partial y} + \frac{\partial v}{\partial x}\right)}, \tag{3.2.5a}$$

and

$$\bar{q}_x = -k\overline{\frac{\partial T}{\partial x}}, \quad \bar{q}_y = -k\overline{\frac{\partial T}{\partial y}}. \tag{3.2.5b}$$

Note that in Eq. (3.2.5) we have neglected the product of the second viscosity coefficient λ and the divergence term $\partial u_l/\partial x_l$ given in Eq. (2.2.4), which is permissible within the boundary-layer approximations. As shown by the discussions that follow in the next sections, the stress term $\lambda(\partial u_l/\partial x_l)$ is of the order of δ^2 and is small compared with some of the other stress terms.

3.2.1 Laminar Flows

The conservation equations given by Eqs. (3.2.1)–(3.2.4) can be simplified considerably by using Prandtl's boundary-layer approximations, often referred to as thin-shear-layer approximations. They are applicable to both wall shear layers and free shear layers, provided that the layers are thin. By convention x is taken as the distance along the solid surface, or along the axis of a free shear layer, and the layer is "thin" in the y-direction. The approximations are made by estimating the order of magnitude of the various principal mean quantities, such as $\tilde{u}$, $\tilde{v}$, $\tilde{T}$, $\tilde{H}$, and $\overline{p}$, and the order of magnitude of various statistical averages of fluctuating quantities, such as $\overline{u'^2}$, $\overline{v'^2}$, $\overline{\varrho u'v'}$, $\overline{\varrho H'v'}$, etc. Before we discuss the Prandtl approximations for turbulent flow and apply them to Eqs. (3.2.1)–(3.2.4), we shall first discuss the boundary-layer approximations for laminar flow. Since the fluctuating quantities are zero for laminar flow, the bars are not needed. Equations (3.2.1)–(3.2.4) can be written as

Continuity

$$\frac{\partial \varrho}{\partial t} + \frac{\partial}{\partial x}(\varrho u) + \frac{\partial}{\partial y}(\varrho v) = 0, \tag{3.2.6}$$

x Momentum

$$\frac{\partial}{\partial t}(\varrho u) + \frac{\partial}{\partial x}(\varrho uu) + \frac{\partial}{\partial y}(\varrho vu) = -\frac{\partial p}{\partial x} + \frac{\partial}{\partial x}(\tau_{xx}) + \frac{\partial}{\partial y}(\tau_{xy}), \tag{3.2.7}$$

y Momentum

$$\frac{\partial}{\partial t}(\varrho v) + \frac{\partial}{\partial x}(\varrho uv) + \frac{\partial}{\partial y}(\varrho vv) = -\frac{\partial p}{\partial y} + \frac{\partial}{\partial x}(\tau_{yx}) + \frac{\partial}{\partial y}(\tau_{yy}), \tag{3.2.8}$$

Energy (Total Enthalpy)

$$\begin{aligned} &\frac{\partial}{\partial t}(\varrho H) + \frac{\partial}{\partial x}(\varrho uH) + \frac{\partial}{\partial y}(\varrho vH) \\ &= \frac{\partial p}{\partial t} + \frac{\partial}{\partial x}\left(-q_x + u\tau_{xx} + v\tau_{yx}\right) + \frac{\partial}{\partial y}\left(-q_y + u\tau_{xy} + v\tau_{yy}\right). \end{aligned} \tag{3.2.9}$$

In essence, Prandtl's boundary-layer approximations depend on the assumption that gradients of quantities such as u and H across a "principal flow direction" y, i.e. in the y-direction, are at least an order of magnitude larger than gradients along x. That assumption permits the neglect of some terms in the governing differential equations. In accordance with the boundary-layer approximations, we assume that

$$\begin{aligned} &\varrho = O(1),\ u = O(1),\ H = O(1),\ T = O(1),\ h = O(1), \\ &\partial/\partial t = O(1),\ \partial/\partial x = O(1),\ \partial/\partial y = O(\delta^{-1}). \end{aligned} \tag{3.2.10}$$

In Eq. (3.2.10), δ, the thickness of the boundary layer, is a function of x and t only. It is assumed to be small relative to a reference length L, that is, $\delta/L \ll 1$.

Let us first consider the continuity equation (3.2.6). Introducing the appropriate orders of magnitude in Eq. (3.2.10) into (3.2.6), we see that the *velocity component normal to the surface*, v, is of $O(\delta)$. Since δ is small, v is also small.

We next consider the x-momentum equation (3.2.7). Clearly, the left-hand side of Eq. (3.2.7) is of $O\,(1)$. Since $\partial/\partial_y = O\,(\delta^{-1})$ and $\partial\tau_{xy}/\partial_y$ is of $O\,(1)$ at most, τ_{xy} must be of $O\,(\delta)$. From the definition of τ_{xy}, it can then follow that μ is of $O\,(\delta^2)$ and that, since $\partial v/\partial x$ is small compared to $\partial u/\partial y$, $\tau_{xy} \approx \mu\,(\partial u/\partial y)$. From the definition of τ_{xx} and from the fact that $\mu, = O\,(\delta^2)$, we see that τ_{xx} is small compared to τ_{xy} and is of $O\,(\delta^2)$.

At the edge of the boundary layer, the viscous terms are zero. Equation (3.2.7), with the continuity equation (3.2.6), reduces to the well-known Euler equation,

$$\varrho_e \frac{\partial u_e}{\partial t} + \varrho_e u_e \frac{\partial u_e}{\partial x} = -\frac{\partial p}{\partial x}. \tag{3.2.11}$$

From Eq. (3.2.11) we see that the streamwise pressure-gradient term $\partial p/\partial x$ is of $O\,(1)$.

With these approximations, the x-momentum equation (3.2.2) becomes

$$\frac{\partial}{\partial t}(\varrho u) + \frac{\partial}{\partial x}(\varrho uu) + \frac{\partial}{\partial y}(\varrho uv) = -\frac{\partial p}{\partial x} + \frac{\partial}{\partial y}(\tau_{xy}). \tag{3.2.12a}$$

With the use of Eq. (3.2.6), we can also write Eq. (3.2.12a) as

$$\varrho \frac{\partial u}{\partial t} + \varrho u \frac{\partial u}{\partial x} + \varrho v \frac{\partial u}{\partial y} = -\frac{\partial p}{\partial x} + \frac{\partial}{\partial y}(\tau_{xy}), \tag{3.2.12b}$$

where $\tau_{xy} = \mu(\partial u/\partial y)$.

Turning our attention to the y-momentum equation (3.2.8), we see that the left-hand side is of $O\,(\delta)$. On the right-hand side, the larger stress term $(\partial/\partial y)(\tau_{yy})$ is also of $O\,(\delta)$ and $(\partial/\partial x)(\tau_{yx})$ is of $O\,(\delta^2)$. Therefore $\partial p/\partial y$ is also of $O\,(\delta)$. Thus the pressure variation across the boundary layer is of $O\,(\delta^2)$ and can be neglected within the boundary-layer approximations. Then Eq. (3.2.8) reduces to

$$p\,(x, y, t) \approx p(x, t). \tag{3.2.13}$$

According to that expression, pressure is a function of only x and t. Hence, for steady flows the pressure-gradient term in Eq. (3.2.12) becomes an ordinary derivative rather than a partial derivative.

We now consider the energy equation for total enthalpy, Eq. (3.2.9). According to Eq. (3.2.10), the left-hand side of the equation is of $O\,(1)$. Also,

$$u\tau_{xy} \gg u\tau_{xx},\ \ v\tau_{yy},\ \ v\tau_{yx}.$$

Since $\partial/\partial_y = O\,(\delta^{-1})$ and $\partial q_y/\partial y$ is at most of $O\,(1)$, q_y is of $O\,(\delta)$. Then because $\partial T/\partial y$ is of $O\,(\delta^{-1})$, the thermal-conductivity coefficient k is of $O\,(\delta^2)$. It follows from the definition of q_x that the streamwise heat transfer is of $O\,(\delta^2)$, which is small compared to the heat transfer normal to the main flow, q_y.

With those approximations, the energy equation (3.2.9) becomes

$$\frac{\partial}{\partial t}(\varrho H) + \frac{\partial}{\partial x}(\varrho u H) + \frac{\partial}{\partial y}(\varrho v H) = \frac{\partial p}{\partial t} + \frac{\partial}{\partial y}(-q_y + u\tau_{xy}). \tag{3.2.14a}$$

Again making use of Eq. (3.2.6), we can write Eq. (3.2.14a) as

$$\varrho\frac{\partial H}{\partial t} + \varrho u\frac{\partial H}{\partial x} + \varrho v\frac{\partial H}{\partial y} = \frac{\partial p}{\partial t} + \frac{\partial}{\partial y}(-q_y + u\tau_{xy}), \tag{3.2.14b}$$

where $q_y = -k\,(\partial T/\partial y)$ and $u\tau_{xy} = u\mu(\partial u/\partial y)$.

At the edge of the boundary layer, the heat transfer normal to the main flow, q_y, and the work done by the viscous forces, $u\tau_{xy}$, are zero. Equation (3.2.14b) then reduces to the unsteady inviscid energy equation

$$\varrho_e\frac{\partial H_e}{\partial t} + \varrho_e u_e\frac{\partial H_e}{\partial x} = \frac{\partial p}{\partial t}. \tag{3.2.15}$$

From the definitions of total enthalpy $H = c_pT + u^2/2$, Prandtl number $\Pr = \mu c_p/k$, and heat transfer normal to the flow q_y, we can write

$$\frac{\partial H}{\partial y} = \frac{\partial}{\partial y}(c_pT) + u\frac{\partial u}{\partial y} = -\frac{\Pr}{\mu}q_y + u\frac{\partial u}{\partial y}.$$

Solving for q_y, we get

$$-q_y = \frac{\mu}{\Pr}\frac{\partial H}{\partial y} - \frac{\mu}{\Pr}u\frac{\partial u}{\partial y} = \frac{\mu}{\Pr}\frac{\partial H}{\partial y} - \frac{1}{\Pr}u\tau_{xy}. \tag{3.2.16}$$

Substitution of that expression for q_y into Eq. (3.2.14b) gives

$$\begin{aligned}\varrho\frac{\partial H}{\partial t} + \varrho u\frac{\partial H}{\partial x} + \varrho v\frac{\partial H}{\partial y} \\ = \frac{\partial p}{\partial t} + \frac{\partial}{\partial y}\left[u\tau_{xy}\left(1 - \frac{1}{\Pr}\right) + \frac{\mu}{\Pr}\frac{\partial H}{\partial y}\right].\end{aligned} \tag{3.2.17}$$

For air, the value of the Prandtl number Pr does not vary much with temperature, and a constant value of about 0.72 has generally been assumed. But considerable simplification results if Pr is assumed to be unity and many theories have been developed on that basis.

We can easily see from Eq. (3.2.17) that for steady state when $\Pr = 1$ the total-enthalpy energy equation always has a solution.

$$H = \text{const},$$

which corresponds to the case of an adiabatic wall whose temperature is constant. For the case of steady state and of zero pressure gradient, $(dp/dx = 0)$, we see further, by comparison of Eqs. (3.2.12b) and (3.2.17), that Eq. (3.2.17) always has a solution

$$H = (\text{const})u \quad (\text{Pr} = 1),$$

provided that the wall has a uniform temperature. The general solution is then

$$H = A + Bu.$$

Since $H = H_w$ and $u = 0$ when $y = 0$, and since $H = H_e$ and $u = u_e$ as $y \to \infty$, the general solution becomes

$$H = H_w - [(H_w - H_e)/u_e]u. \tag{3.2.18}$$

Equation (3.2.18) is often referred to as the Crocco integral.

Equation (3.2.18) is very important, because when Pr = 1, the solution of the total-enthalpy energy equation (3.2.17) is given by Eq. (3.2.18) for the case of zero pressure gradient, and there remains only to solve Eq. (3.2.12) for u.

3.2.2 Turbulent Flows

We now discuss the boundary-layer approximations for turbulent flows. Although turbulent shear flows generally spread more rapidly than the corresponding laminar flows at the same Reynolds number, it is found empirically that Prandtl's boundary-layer approximations are also fairly good in turbulent cases and become better as Reynolds number increases. The approximations involve principal mean quantities and mean fluctuating quantities. For the principal mean quantities, we use the same approximations we have used for laminar flows, that is, the relations given by Eq. (3.2.10), except that now the quantities such as ϱ, u, etc. are averaged quantities, for example, $\bar{\varrho}$ and $\tilde{u}$.

Relationship between Temperature and Velocity Fluctuations. According to experimental data – for example, Kistler [1] and Morkovin [2] – the Crocco integral also holds true for turbulent flows. However, it is acceptable only for *adiabatic* walls and for flows with small heat transfer. Figure 3.1 shows the measured total-temperature (T_0) profiles in non-dimensional coordinates for adiabatic compressible turbulent flows at four Mach numbers. The measurements were made by Morkovin and Phinney, as cited in Morkovin [2], and by Kistler [1]. The experimental results show that a large fraction of the total temperature variation through the boundary layer occurs quite close to the wall and that, remarkably, the total temperature remains nearly constant in the rest of the boundary layer.

For convenience, we now use conventional time averages. In accordance with the definition of total enthalpy, $H = h + \frac{1}{2}u^2$, we can write

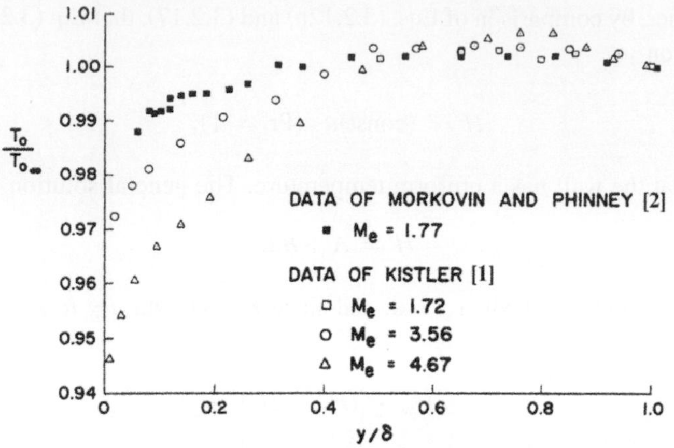

Fig. 3.1 Mean total-temperature distribution across adiabatic turbulent boundary layers.

$$\bar{H} + H'' = \bar{h} + h'' + \frac{1}{2}(\bar{u} + u'')^2 = \bar{h} + h'' + \frac{1}{2}(\bar{u})^2 + \bar{u}u'' + \frac{1}{2}(u'')^2.$$

Then the mean total enthalpy $\bar{H}$ is

$$\bar{H} = \bar{h} + \frac{1}{2}(\bar{u})^2 \tag{3.2.19a}$$

and fluctuating total enthalpy H'' is

$$H'' = h'' + \bar{u}u'' + \frac{1}{2}(u'')^2. \tag{3.2.19b}$$

Since experiments have shown that for an adiabatic turbulent flow total temperature, or total enthalpy, is constant or nearly constant, total fluctuating enthalpy must be small and can be neglected, that is,

$$H'' = h'' + \bar{u}u'' + \frac{1}{2}(u'')^2 \approx 0.$$

The second-order term $(u'')^2$ in the above expression is small compared to $\bar{u}u''$ and can be neglected. With c_p = constant, the resulting expression can be written as

$$T''/\bar{T} = -(\gamma - 1)\bar{M}^2(u''/\bar{u}), \tag{3.2.20}$$

where $\bar{M} = \bar{u}/(\mu R\bar{T})^{1/2}$ is the local Mach number within the boundary layer.

Experiments carried out in supersonic boundary layers and wakes by Kistler [1] and by Demetriades [3] support the assumption that $H'' \approx 0$. Figure 3.2 shows the distribution of total-temperature fluctuations at three Mach numbers for an adiabatic

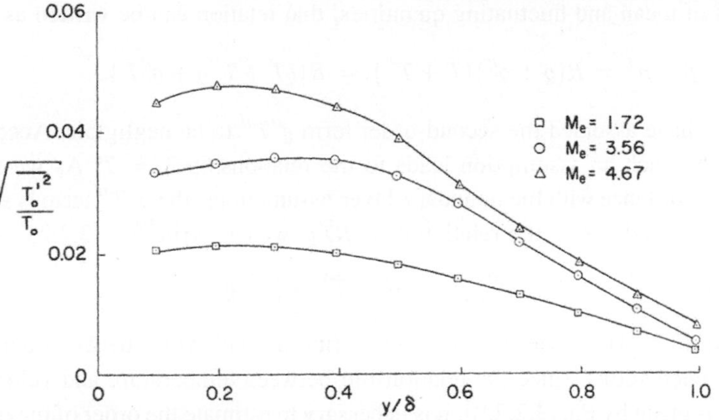

Fig. 3.2 Distribution of total-temperature fluctuations according to Kistler's measurements [1] for adiabatic walls.

turbulent boundary layer. We see that maximum total-temperature fluctuation is less than 5% at $M_e = 4.67$, which is negligible.

Equation (3.2.20) is for an adiabatic flow. By using the Crocco integral, Eq. (3.2.18), it can also be generalized to include the case of heat transfer at the surface of a boundary layer. Since

$$\bar{H} + H'' = A + B(\bar{u} + u'') = A + B\bar{u} + Bu''$$

and

$$H'' = h'' + \bar{u}u'',$$

we can write

$$h'' + \bar{u}u'' = Bu''.$$

Using the definition of B and $\bar{M}$ in the above equation, we can obtain the following relationship between the temperature and velocity fluctuations:

$$T''/\bar{T} = -\alpha(u''/\bar{u}), \tag{3.2.21}$$

where

$$\alpha \equiv (\gamma - 1)\bar{M}^2 + \left[(T_w - T_0)/\bar{T}\right](\bar{u}/u_e). \tag{3.2.22}$$

Relationship between Density and Velocity Fluctuations. The equation of state for a perfect gas is

$$p = R\varrho T.$$

In terms of mean and fluctuating quantities, that relation can be written as

$$\bar{p}+p'' = R(\bar{\varrho}+\varrho'')(\bar{T}+T'') = R(\bar{\varrho}\bar{T}+T''\bar{\varrho}+\varrho''\bar{T}), \tag{3.2.23}$$

where we have assumed the second-order term $\varrho''T''$ to be negligible. According to Eq. (2.4.6), such an assumption leads to the relationship $\tilde{T} = \bar{T}$. As we shall see later, in accordance with the boundary-layer assumptions, the $\varrho''T''$ term is small and can be neglected. Using the relation $\bar{p} = R\overline{T\varrho}$, we can write Eq. (3.2.23) as

$$p''/\bar{p} = (T''/\bar{T}) + (\varrho''/\bar{\varrho}). \tag{3.2.24}$$

From Eq. (3.2.24) we see that in order to find a relationship between density and velocity fluctuations [since the relationship between temperature and velocity fluctuations is given by Eq. (3.2.22)], it is necessary to estimate the order of magnitude of pressure fluctuations. The pressure field is indicated to the unaided observer by both the sound field associated with the turbulence and the fluctuating force on a solid surface in contact with the turbulence. It is important to know the fluctuating pressure field on material surfaces, since, for example, when flight vehicles are operated in regimes of large dynamic pressures, the pressures can have significant effects. The random forces can even cause fatigue failure in a structure, as well as undesirable levels of structural vibration. In addition, these forces can produce sound within a structure through the intermediate step of forcing the solid surface into motion.

According to the experimental results of Kistler [1], the temperature fluctuations are essentially isobaric for adiabatic flows with Mach numbers less than 5. Consequently, Eq. (3.2.24) can be written as

$$T''/\bar{T} \approx -\varrho''/\bar{\varrho}. \tag{3.2.25}$$

Unfortunately, in the flow range above Mach 5, no detailed measurements of turbulent fluctuations have as yet been reported. It is therefore not possible to provide quantitative information on the subject. There is convincing experimental evidence, however, that in flows in the vicinity of $M_e = 5$, appreciable pressure fluctuations exist in the boundary layer. Kistler and Chen [4] reported rms pressure fluctuations of 8-10% of the mean static pressure at the wall for $M_e = 5$. Under the same conditions, Laufer's measurement [5] of the value just outside the boundary layer was $(\overline{p''^2})^{1/2}/\bar{p} \approx 1\%$).

In our discussion, we shall assume the pressure fluctuations to be negligible and make the order-of-magnitude estimates of the fluctuating quantities on that basis. Substitution from Eq. (3.2.25) into Eq. (3.2.21) gives the desired relationship between density and velocity fluctuations,

$$\varrho''/\bar{\varrho} = \alpha(u''/\bar{u}). \tag{3.2.26}$$

According to experiment, that relation is justified for compressible turbulent boundary layers at Mach numbers up to approximately 5 [2].

Equations (3.2.20) and (3.2.25) and Eqs. (P3.1) and (P3.2) in Problems 3.3 and 3.4 are also valid if we substitute root-mean-square values for instantaneous fluctuations (for example, $\sqrt{u''^2}$ or u''). The relations in Eqs. (P3.1) and (P3.2) can also be used for rough estimates of low-speed boundary layers, respectively, to assess the negligibility of the terms in the mass, momentum, and energy equations in which density fluctuations appear. The relations are likely to be less satisfactory in free shear layers, where the velocity and temperature fluctuations are less closely related, but they should still be useful for rough estimates.

The largest value of $\sqrt{u''^2}/u_e$ reached in a low-speed boundary layer in zero pressure gradient is about 0.1. Values in high-speed flows may be even lower, but if we use the low-speed figure in the root-mean-square version of the above formulas, we can make generous estimates of $\sqrt{T''^2}/T$ or $\sqrt{\varrho''^2}/\bar{\varrho}$, for high-speed boundary layers or for low-speed boundary layers on strongly-heated walls. Typical figures are given in Table 3.1; to express the temperature fluctuations in high-speed flow as fractions of T_e, we have assumed that the maximum temperature fluctuation occurs when $u/u_e = 0.5$. The figures for infinite temperature ratio or Mach number are not realistic, but they serve to show that the ratio of temperature fluctuation to local (or wall) temperature does not rise indefinitely. This is easy enough to see in the case of Eq. (P3.2), which reduces, for $T_w/T_e \to \infty$, to $\sqrt{T''^2}/Tw = \sqrt{u''^2}/u_e$, in the case of Eq. (P3.1), the explanation for the approach to an asymptotic value is that if the heat transfer from a high-speed flow to a surface is not too large, the surface temperature

TABLE 3.1 **Approximate estimations of temperature fluctuations assuming $\sqrt{u''^2}/u_e = 0.1$.**

(a) Low-speed flow over heated wall						
$(T_w - T_e)/T_e$	0.25	0.5	1	2	4	∞
$T_w - T_e$ for $T_e = 300$ K	75	150	300	600	1200	∞
$\sqrt{T''^2}/T_e$ from Eq.(P3.2)	0.025	0.05	0.1	0.2	0.4	∞
$\sqrt{T''^2}/T_w$	0.02	0.033	0.05	0.067	0.08	0.10
(b) High-speed flow over adiabatic wall (zero heat transfer to surface)						
M_e	1	2	3	4	5	∞
$(T_{aw} - T_e)/T_e$	0.178	0.712	1.6	2.85	4.45	∞
$\sqrt{T''^2}/T_e$ from Eq.(P3.1)	0.04	0.16	0.36	0.64	1.0	∞
$\sqrt{T''^2}/T_{aw}$	0.017	0.047	0.069	0.083	0.092	0.112

Note: The last line of each section of the table is the more meaningful because maximum temperature fluctuations occur near the wall.

is not much less than the total temperature of the external stream and therefore rises rapidly with Mach number. Values given in Table 3.1(b) are for the "recovery" temperature, the temperature reached by an insulated (adiabatic) surface, T_{aw} (see Section 4.7, Eq. (4.7.3)).

Approximations Involving the Molecular-Transport Terms. For a perfect gas, the fluid properties μ and k are functions only of temperature for a wide range of pressures. In order to express their variation with temperature fluctuations, we write them as

$$\mu = \bar{\mu} + \mu'' = \bar{\mu}\left[1 + \left(\mu''/\bar{\mu}\right)\right], \quad k = \bar{k} + k'' = \bar{k}\left[1 + \left(k''/\bar{k}\right)\right]. \tag{3.2.27}$$

If it is assumed that μ and k are proportional to temperature, Eq. (3.2.27) becomes

$$\mu = \bar{\mu}\left[1 + \left(T''/\bar{T}\right)\right], \quad k = \bar{k}\left[1 + \left(T''/\bar{T}\right)\right]. \tag{3.2.28}$$

3.3 Continuity, Momentum, and Energy Equations

3.3.1 Two-Dimensional Flows

Let us first consider the continuity and the x-momentum equations given by Eqs. (3.2.1) and (3.2.2). Since $\varrho = \bar{\varrho} + \varrho''$ and $\mu = \bar{\mu} + \mu''$, the three terms $\overline{\varrho u'u'}$, $\overline{\varrho u'v'}$, and τ_{xy} in Eq. (3.2.2) can be written as

$$\overline{\varrho u'u'} = \bar{\varrho}\overline{u'u'} + \overline{\varrho''u'u'}\ , \qquad \overline{\varrho u'v'} = \bar{\varrho}\overline{u'v'} + \overline{\varrho''\varrho u'v'}. \tag{3.3.1a}$$

$$\bar{\tau}_{xy} = \overline{\mu\frac{\partial u}{\partial y}} = \overline{\bar{\mu}\left(1 + \frac{\mu''}{\bar{\mu}}\right)\frac{\partial}{\partial y}\tilde{u}\left(1 + \frac{u'}{\tilde{u}}\right)}, \tag{3.3.1b}$$

Since $\partial/\partial y = O(\delta^{-1})$, the term $\bar{\varrho}\overline{u'v'}$ is of $O(\delta)$ at most. Furthermore,

$$\frac{\overline{\mu''(\partial u'/\partial y)}}{\bar{\mu}(\partial\tilde{u}/\partial y)} \ll 1 \quad \text{and} \quad \frac{\overline{\varrho''u'v'}}{\bar{\varrho}\overline{u'v'}} \ll 1. \tag{3.3.2}$$

In each case, those ratios can be assumed to be less than 5% for Mach numbers less than 5, if the coefficients of correlation between viscosity and velocity gradient and between density and $u'v'$ fluctuation are at most 0.3.

Let us now estimate the order of magnitude of $\overline{\varrho''v'}$. Multiplying both sides of Eq. (3.2.26) by v, rearranging, and assuming that $u'' = u'$, we get

$$\overline{\varrho''v'} \approx \alpha\left(\bar{\varrho}/\bar{u}\right)\overline{u'v'}. \tag{3.3.3}$$

For moderate Mach numbers and heat-transfer rates, α is of $O(1)$. Since $\overline{u'v'} = O(\delta)$ and $\bar{\varrho}/\bar{u} = O(1)$, we see that $\overline{\varrho''v'} = O(\delta)$. If we assume that v' is proportional to u', then $\overline{\varrho''u'} = O(\delta)$. Therefore, from Eq. (2.4.3b) we have

$$\tilde{u} - \bar{u} = \overline{\varrho''u'}/\bar{\varrho} = \overline{\varrho''u''}/\bar{\varrho} \quad \text{or} \quad \tilde{u} = \bar{u} \tag{3.3.4a}$$

and

$$\tilde{v} - \bar{v} = \overline{\varrho''v'}/\bar{\varrho} = \overline{\varrho''v''}/\bar{\varrho} \quad \text{or} \quad \tilde{v} = \bar{v} + \overline{\varrho''v''}/\bar{\varrho}, \tag{3.3.4b}$$

since $\tilde{u} = O(1)$ and $\tilde{v} = O(\delta)$. That means that we can interchange tildes and bars on u with the boundary-layer approximations but cannot do so on v. Also, from the definition of u', that is,

$$\overline{u'} = -\overline{\varrho''u''}/\bar{\varrho}\ ,$$

we see that $\overline{u'}$ is of $O(\delta)$. As a result, $\tilde{u}(1 + \overline{u'}/\tilde{u})$ in Eq. (3.3.1b) is approximately equal to $\tilde{u}$. Furthermore, $\bar{\tau}_{xy} \gg \bar{\tau}_{xx}$ and $\bar{\varrho}\overline{u'v'} \gg \overline{\varrho''u'u'}$.

With those approximations, the continuity and x-momentum equations given by Eqs. (3.2.1) and (3.2.2), respectively, become:

Continuity

$$\frac{\partial \bar{\varrho}}{\partial t} + \frac{\partial}{\partial x}(\bar{\varrho}\bar{u}) + \frac{\partial}{\partial y}(\overline{\varrho v}) = 0, \tag{3.3.5}$$

x Momentum

$$\frac{\partial}{\partial t}(\bar{\varrho}\bar{u}) + \frac{\partial}{\partial x}(\bar{\varrho}\bar{u}\bar{u}) + \frac{\partial}{\partial y}(\overline{\varrho v}\,\bar{u}) = -\frac{\partial \bar{p}}{\partial x} + \frac{\partial}{\partial y}\left(\bar{\mu}\frac{\partial \bar{u}}{\partial y} - \bar{\varrho}\overline{u'v'}\right). \tag{3.3.6a}$$

With the use of Eq. (3.3.5), we can write Eq. (3.3.6) as

$$\bar{\varrho}\frac{\partial \bar{u}}{\partial t} + \bar{\varrho}\bar{u}\frac{\partial \bar{u}}{\partial x} + \overline{\varrho v}\frac{\partial \bar{u}}{\partial y} = -\frac{\partial \bar{p}}{\partial x} + \frac{\partial}{\partial y}\left(\bar{\mu}\frac{\partial \bar{u}}{\partial y} - \bar{\varrho}\overline{u'v'}\right), \tag{3.3.6b}$$

where

$$\overline{\varrho v} = \bar{\varrho}\,\bar{v} + \overline{\varrho''v''} = \bar{\varrho}\tilde{v}. \tag{3.3.7}$$

Following the same line of order-of-estimate study, the y-momentum equation (3.2.3) becomes

$$-\frac{\partial \bar{p}}{\partial y} - \frac{\partial}{\partial y}\left(\bar{\varrho}\overline{v'v'}\right) = 0. \tag{3.3.8}$$

We see from Eq. (3.3.8) that for laminar flows $\partial \bar{p}/\partial y$ is of $O(\delta)$ but that for turbulent flows it is of $O(1)$. Consequently, the pressure variation across the boundary layer is of $O(\delta)$, so that in comparison with the streamwise pressure variation $p(x, t)$ it is still small and can be neglected within the boundary-layer approximations.

For incompressible flows, Eqs. (3.3.5), (3.3.6), and (3.3.8) can be simplified further, as follows:

Continuity

$$\frac{\partial \bar{u}}{\partial x} + \frac{\partial \bar{v}}{\partial y} = 0, \tag{3.3.9}$$

Momentum

$$\frac{\partial \bar{u}}{\partial t} + \bar{u}\frac{\partial \bar{u}}{\partial x} + \bar{v}\frac{\partial \bar{u}}{\partial y} = -\frac{1}{\varrho}\frac{\partial \bar{p}}{\partial x} + \nu\frac{\partial^2 \bar{u}}{\partial y^2} + \frac{\partial}{\partial y}\left(\overline{-u'v'}\right). \tag{3.3.10}$$

We next consider the energy equation. According to the approximations discussed above, we have shown that the double-correlation terms involving u', v', ϱ', such as $\overline{u'v'}$, $\overline{\varrho'v'}$, etc., are of $O(\delta)$ at most, and that the triple correlation terms such as $\overline{\varrho''u'v'}$ are small[3] compared to $\bar{\varrho}\overline{u'v'}$. We have also shown that within the boundary-layer approximations, $\tilde{u} = \bar{u}$ and $\overline{\varrho v} = \bar{\varrho}\tilde{v} = \bar{\varrho}\bar{v} + \overline{\varrho''v''}$. Before we discuss the boundary-layer simplifications for the total-enthalpy energy equation, let us first show that within the boundary-layer approximations, $\tilde{H} = \bar{H}$.

From the definitions of $H(x_i, t)$, $\tilde{H}$, and $\bar{H}$, we can write (see Section 2.4)

$$\tilde{H} - \bar{H} = H'' - H' = \overline{\varrho''H'}/\bar{\varrho} = \overline{\varrho''H''}/\bar{\varrho}. \tag{3.3.11}$$

Since

$$\overline{\varrho H''} = \overline{\varrho h''} + \bar{u}\overline{\varrho u''} = \overline{\varrho'' h''} + \bar{u}\overline{\varrho'' u''}, \tag{3.3.12}$$

and the two terms on the right-hand side of Eq. (3.3.12) are all of $O(\delta)$, we see that $\overline{\varrho H''}$, which is also equal to $\overline{\varrho''H''}$, is of $O(\delta)$. Consequently, $\tilde{H} = \bar{H}$ from Eq. (3.3.11).

By extending the boundary-layer approximations discussed in the previous sections, we can write the total-enthalpy energy equation (3.2.4) as

$$\begin{aligned} &\bar{\varrho}\frac{\partial \bar{H}}{\partial t} + \bar{\varrho}\bar{u}\frac{\partial \bar{H}}{\partial x} + \overline{\varrho v}\frac{\partial \bar{H}}{\partial y} \\ &= \frac{\partial \bar{p}}{\partial t} + \frac{\partial}{\partial y}\left(\bar{u}\bar{\mu}\frac{\partial \bar{u}}{\partial y} - \bar{q}_y - \bar{\varrho}\overline{v'H'}\right), \end{aligned} \tag{3.3.13}$$

where $\bar{q}_y = -\bar{k}(\partial \bar{T}/\partial y)$.

---

[3]It is generally assumed that the triple correlation terms are of $O(\delta^2)$. Since $\overline{\varrho''v'} = O(\delta)$ and $\overline{u'} = O(\delta)$, their product must be of $O(\delta^2)$.

From the definitions of mean total enthalpy $\bar{H}$, $\bar{q}_y$, and Prandtl number, the second term in parentheses on the right-hand side of Eq. (3.3.13) can be rewritten in a slightly different form. The resulting expression becomes

$$\begin{aligned} &\bar{\varrho}\frac{\partial \bar{H}}{\partial t} + \bar{\varrho}\bar{u}\frac{\partial \bar{H}}{\partial x} + \overline{\varrho v}\frac{\partial \bar{H}}{\partial y} \\ &= \frac{\partial \bar{p}}{\partial t} + \frac{\partial}{\partial y}\left[\bar{\mu}\bar{u}\frac{\partial \bar{u}}{\partial y}\left(1 - \frac{1}{\mathrm{Pr}}\right) + \frac{\bar{\mu}}{\mathrm{Pr}}\frac{\partial \bar{H}}{\partial y} - \bar{\varrho}\overline{H'v'}\right]. \end{aligned} \tag{3.3.14}$$

Sometimes it is more convenient to express $\overline{\varrho v' H'}$ in terms of static enthalpy fluctuation h', which can easily be done by recalling the definition of H'. Neglecting the second- and higher-order terms in Eq. (2.3.19), we can write the fluctuating total enthalpy H' as

$$H' = h' + \bar{u}u',$$

where we have replaced $\tilde{u}$ by $\bar{u}$, which is permissible within the boundary-layer approximations. Multiplying both sides of that expression by $\bar{\varrho}v'$ and averaging, we get

$$\bar{\varrho}\overline{H'v'} = \bar{\varrho}\overline{h'v'} + \bar{\varrho}\bar{u}\overline{u'v'}.$$

Substituting that expression into Eq. (3.3.14) and replacing the second term on the right-hand side of Eq. (3.3.13) by the resulting expression, we get

$$\begin{aligned} &\bar{\varrho}\frac{\partial \bar{H}}{\partial t} + \bar{\varrho}\bar{u}\frac{\partial \bar{H}}{\partial x} + \overline{\varrho v}\frac{\partial \bar{H}}{\partial y} \\ &= \frac{\partial \bar{p}}{\partial t} + \frac{\partial}{\partial y}\left\{\frac{\bar{\mu}}{\mathrm{Pr}}\frac{\partial \bar{H}}{\partial y} - \bar{\varrho}\overline{h'v'} + \bar{u}\left[\left(1 - \frac{1}{\mathrm{Pr}}\right)\bar{\mu}\frac{\partial \bar{u}}{\partial y} - \bar{\varrho}\overline{u'v'}\right]\right\}. \end{aligned} \tag{3.3.15}$$

For an incompressible flow, the total-enthalpy equation simplifies considerably. Noting that

$$\frac{\bar{\mu}}{\mathrm{Pr}}\frac{\partial \bar{H}}{\partial y} = \frac{\bar{\mu}}{\mathrm{Pr}}\left[\frac{\partial \bar{h}}{\partial y} + \frac{\partial}{\partial y}\left(\frac{\bar{u}^2}{2}\right)\right],$$

we can rewrite the total energy equation (3.3.15) as

$$\begin{aligned} &\bar{\varrho}\frac{\partial \bar{H}}{\partial t} + \bar{\varrho}\bar{u}\frac{\partial \bar{H}}{\partial x} + \overline{\varrho v}\frac{\partial \bar{H}}{\partial y} \\ &= \frac{\partial \bar{p}}{\partial t} + \frac{\partial}{\partial y}\left[\frac{\bar{\mu}}{\mathrm{Pr}}\frac{\partial \bar{h}}{\partial y} - \bar{\varrho}\overline{h'v'} + \bar{u}\left(\bar{\mu}\frac{\partial \bar{u}}{\partial y} - \bar{\varrho}\overline{u'v'}\right)\right]. \end{aligned} \tag{3.3.16}$$

Substituting $\bar{h} + \bar{u}^2/2$ for $\bar{H}$ in Eq. (3.3.15) and using Eqs. (3.3.6b) and (3.3.16), we get the *energy equation in terms of static enthalpy,*

$$\bar{\varrho}\frac{\partial \bar{h}}{\partial t} + \bar{\varrho}\bar{u}\frac{\partial \bar{h}}{\partial x} + \overline{\varrho v}\frac{\partial \bar{h}}{\partial y} = \frac{\partial \bar{p}}{\partial t} + \bar{u}\frac{\partial \bar{p}}{\partial x} + \frac{\partial}{\partial y}\left(\frac{\bar{\mu}}{\mathrm{Pr}}\frac{\partial \bar{h}}{\partial y} - \bar{\varrho}\overline{h'v'}\right) + \left(\bar{\mu}\frac{\partial \bar{u}}{\partial y} - \bar{\varrho}\overline{u'v'}\right)\frac{\partial \bar{u}}{\partial y}. \tag{3.3.17}$$

That equation, like Eq. (3.3.13), is still for a compressible flow. For an incompressible flow, the fluid properties, $\bar{\varrho}$, $\bar{k}$, Pr, and μ are constant. In addition, the pressure-work term $\bar{u}(\partial\bar{p}/\partial x)$ and the dissipation term $[\bar{\mu}(\partial\bar{u}/\partial y) - \bar{\varrho}\overline{u'v'}]\ (\partial\bar{u}/\partial y)$ are small and can be neglected. Their negligibility can be easily shown by expressing Eq. (3.3.17) in terms of dimensionless quantities defined by

$$\bar{t} = tu_\infty/L, \quad \bar{x} = x/L, \quad \bar{y} = y/L, \quad u^* = \bar{u}/u_\infty, \quad (\overline{\varrho v})^* = \overline{\varrho v}/\varrho_\infty u_\infty,$$
$$(\overline{u'v'})^* = \overline{u'v'}/u_\infty^2, \ (\overline{v'h'})^* = \overline{v'h'}/u_\infty(h_w - \bar{h}_\infty), \ p^* = \bar{p}/\varrho_\infty u_\infty^2,$$
$$\varrho^* = \bar{\varrho}/\varrho_\infty, \ h^* = (\bar{h} - \bar{h}_\infty)/(h_w - \bar{h}_\infty), \ \mu^* = \bar{\mu}/\mu_\infty, \tag{3.3.18}$$

where the bars on independent variables, *t*, *x*, and *y* denote dimensionless quantities. Using the definitions of Eq. (3.3.18) and then simplifying, we can write the static-enthalpy energy equation (3.3.17) in terms of dimensionless quantities in the following form:

$$\underbrace{\varrho^*\frac{\partial h^*}{\partial \bar{t}} + \varrho^* u^*\frac{\partial h^*}{\partial \bar{x}} + (\overline{\varrho v})^*\frac{\partial h^*}{\partial \bar{y}}}_{\mathrm{I}} = \underbrace{\mathrm{E}\left(\frac{\partial p^*}{\partial \bar{t}} + u^*\frac{\partial p^*}{\partial \bar{x}}\right)}_{\mathrm{II}} + \underbrace{\frac{1}{\mathrm{R}_L}\frac{\partial}{\partial \bar{y}}\left\{\frac{\mu^*}{\mathrm{Pr}}\frac{\partial h^*}{\partial \bar{y}}\right.}_{\mathrm{III}}\underbrace{\left. - \mathrm{R}_L\varrho^*(\overline{h'v'})^*\right\}}_{\mathrm{IV}} + \underbrace{\frac{\mathrm{E}}{\mathrm{R}_L}\left(\mu^*\frac{\partial u^*}{\partial \bar{y}}\right.}_{\mathrm{V}}\underbrace{\left. - \mathrm{R}_L\varrho^*(\overline{u'v'})^*\right)\frac{\partial u^*}{\partial \bar{y}}}_{\mathrm{VI}}. \tag{3.3.19}$$

Like the Prandtl number ($\mathrm{Pr} \equiv \mu c_p/k$), the quantities R_L and E are dimensionless quantities; they are known as Reynolds number and Eckert number, respectively, and are defined as

$$\mathrm{R}_L \equiv u_\infty \varrho_\infty L/\mu_\infty, \quad \mathrm{E} \equiv u_\infty^2/h_w - h_\infty. \tag{3.3.20}$$

Since for a perfect gas $c_p = \gamma R/(\gamma - 1)$, it follows from the definition of Eckert number that

$$\begin{aligned} E &= \frac{u_\infty^2}{c_p(T_w - T_\infty)} \\ &= \frac{u_\infty^2}{c_p T_\infty}\left(\frac{T_\infty}{T_w - T_\infty}\right) = (\gamma - 1)M_\infty^2\left(\frac{T_\infty}{T_w - T_\infty}\right). \end{aligned} \tag{3.3.21}$$

From Eq. (3.3.21) we see that if the temperature difference is of the same order of magnitude as the free-stream temperature, the Eckert number becomes equivalent to the free-stream Mach number. Thus Eckert number becomes important only for small temperature differences at high Mach numbers. The Eckert number is quite small in incompressible flows ($M \approx 0$), and since R_L, is large within the boundary-layer approximations, the ratio E/R_L, is also small. Consequently, pressure-work and dissipation terms are negligible. For incompressible flows, the energy equation (3.3.17) then becomes

$$\frac{\partial \bar{h}}{\partial t} + \bar{u}\frac{\partial \bar{h}}{\partial x} + \bar{v}\frac{\partial \bar{h}}{\partial y} = \alpha\frac{\partial^2 \bar{h}}{\partial y^2} - \frac{\partial}{\partial y}\left(\overline{h'v'}\right), \tag{3.3.22}$$

where α is the thermal diffusivity, $\alpha \equiv \nu/\mathrm{Pr}$.

That equation can be written in terms of static temperature T as

$$\frac{\partial \bar{T}}{\partial t} + \bar{u}\frac{\partial \bar{T}}{\partial x} + \bar{v}\frac{\partial \bar{T}}{\partial y} = \alpha\frac{\partial^2 \bar{T}}{\partial y^2} - \frac{\partial}{\partial y}\left(\overline{T'v'}\right). \tag{3.3.23}$$

3.3.2 Axisymmetric Flows

In principle, the governing boundary-layer equations for axisymmetric flows do not differ much from those of two-dimensional flows. Again, the external potential velocity is a function of only one space coordinate, and the velocity within the boundary region has two components. Typical examples of such flows are a flow over a body of revolution, a wake behind an axially symmetrical body, and a jet issuing from an axisymmetric body. The extent of the region in the radial direction is of the order of the thickness of the boundary layer δ and is usually much smaller than both the extent of the region in the axial direction L and the radius of the body r_0.

The boundary-layer equations of a steady, compressible fluid for both laminar and turbulent axisymmetric flows for the coordinate system shown in Fig. 3.3 can be written in a form similar to those of Eqs. (3.3.5), (3.3.6b), (3.3.13), and (3.3.15). The steady continuity, momentum, and energy (total enthalpy) equations are

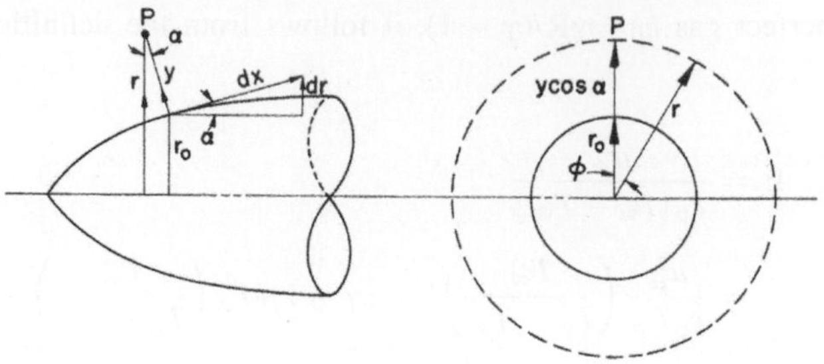

Fig. 3.3 Coordinate system for an axisymmetric flow.

Continuity

$$\frac{\partial}{\partial x}(r^k \bar{\varrho}\bar{u}) + \frac{\partial}{\partial y}(r^k \overline{\varrho v}) = 0, \tag{3.3.24}$$

Momentum

$$\bar{\varrho}\bar{u}\frac{\partial \bar{u}}{\partial x} + \overline{\varrho v}\frac{\partial \bar{u}}{\partial y} = -\frac{d\bar{p}}{dx} + \frac{1}{r^k}\frac{\partial}{\partial y}\left[r^k\left(\bar{\mu}\frac{\partial \bar{u}}{\partial y} - \bar{\varrho}\overline{u'v'}\right)\right], \tag{3.3.25}$$

Energy (Total Enthalpy)

$$\bar{\varrho}\bar{u}\frac{\partial \bar{H}}{\partial x} + \overline{\varrho v}\frac{\partial \bar{H}}{\partial y} = \frac{1}{r^k}\frac{\partial}{\partial y}\left\{r^k\left[\bar{\mu}\left(1 - \frac{1}{\mathrm{Pr}}\right)\bar{u}\frac{\partial \bar{u}}{\partial y} + \frac{\bar{\mu}}{\mathrm{Pr}}\frac{\partial \bar{H}}{\partial y} - \bar{\varrho}\overline{H'v'}\right]\right\}. \tag{3.3.26}$$

The right-hand side of Eq. (3.3.26) can also be written as [see Eq. (3.3.15)]

$$= \frac{1}{r^k}\frac{\partial}{\partial y}\left[r^k\left\{\frac{\bar{\mu}}{\mathrm{Pr}}\frac{\partial \bar{H}}{\partial y} - \bar{\varrho}\overline{h'v'} + \bar{u}\left[\left(1 - \frac{1}{\mathrm{Pr}}\right)\bar{\mu}\frac{\partial \bar{u}}{\partial y} - \bar{\varrho}\overline{u'v'}\right]\right\}\right]. \tag{3.3.27}$$

where $k = 1$ for axisymmetric flows, $k = 0$ for two-dimensional flows and, from Fig. 3.3,

$$r(x, y) = r_0(x) + y\cos\alpha\ . \tag{3.3.28}$$

Although for most axisymmetric flows the boundary-layer thickness δ is generally small compared with, say, body radius r_0, there are some flows – for example, flows over very slender cylinders or flow over the tail of a streamlined body of revolution – in which the boundary-layer thickness can be of the same order of magnitude as the radius of the body. In such cases, the so called transverse-curvature (TVC) effect must be accounted for, since such an effect strongly influences the skin friction and heat transfer. In equations of the form of Eqs. (3.3.24)–(3.3.27), the TVC effect is included.

3.3.3 Three-Dimensional Flows

In three-dimensional steady flows, the external potential flow depends on two coordinates in the wall surface, and the flow within the boundary layer possesses all three velocity components, which depend on all three space coordinates. The continuity, momentum, and energy equations for a steady compressible flow can be written as

Continuity

$$\frac{\partial}{\partial x}(\bar{\varrho}\bar{u}) + \frac{\partial}{\partial y}(\overline{\varrho v}) + \frac{\partial}{\partial z}(\bar{\varrho}\bar{w}) = 0, \tag{3.3.29}$$

x Momentum

$$\bar{\varrho}\bar{u}\frac{\partial \bar{u}}{\partial x} + \overline{\varrho v}\frac{\partial \bar{u}}{\partial y} + \bar{\varrho}\bar{w}\frac{\partial \bar{u}}{\partial z} = -\frac{\partial \bar{p}}{\partial x} + \frac{\partial}{\partial y}\left(\bar{\mu}\frac{\partial \bar{u}}{\partial y} - \bar{\varrho}\overline{u'v'}\right), \tag{3.3.30}$$

z Momentum

$$\bar{\varrho}\bar{u}\frac{\partial \bar{w}}{\partial x} + \overline{\varrho v}\frac{\partial \bar{w}}{\partial y} + \bar{\varrho}\bar{w}\frac{\partial \bar{w}}{\partial z} = -\frac{\partial \bar{p}}{\partial z} + \frac{\partial}{\partial y}\left(\bar{\mu}\frac{\partial \bar{w}}{\partial y} - \bar{\varrho}\overline{w'v'}\right), \tag{3.3.31}$$

Energy (Total Enthalpy)

$$\begin{aligned}\bar{\varrho}\bar{u}\frac{\partial \bar{H}}{\partial x} &+ \overline{\varrho v}\frac{\partial \bar{H}}{\partial y} + \bar{\varrho}\bar{w}\frac{\partial \bar{H}}{\partial z} \\ &= \frac{\partial}{\partial y}\left[\bar{\mu}\left(1-\frac{1}{\Pr}\right)\frac{\partial}{\partial y}\left(\frac{\bar{u}^2+\bar{w}^2}{2}\right) + \frac{\bar{\mu}}{\Pr}\frac{\partial \bar{H}}{\partial y} - \bar{\varrho}\overline{v'H'}\right].\end{aligned} \tag{3.3.32}$$

Another form of Eq. (3.3.32) can be obtained by using the static enthalpy in the transport terms, as follows:

$$\begin{aligned}\bar{\varrho}\bar{u}\frac{\partial \bar{H}}{\partial x} &+ \overline{\varrho v}\frac{\partial \bar{H}}{\partial y} + \bar{\varrho}\bar{w}\frac{\partial \bar{H}}{\partial z} \\ &= \frac{\partial}{\partial y}\left[\frac{\bar{\mu}}{\Pr}\frac{\partial \bar{h}}{\partial y} - \bar{\varrho}\overline{v'h'} - \bar{\varrho}\bar{u}\,\overline{v'u'} - \bar{\varrho}\bar{w}\overline{v'w'} + \bar{\mu}\frac{\partial}{\partial y}\left(\frac{\bar{u}^2+\bar{w}^2}{2}\right)\right].\end{aligned} \tag{3.3.33}$$

Here

$$\bar{H} = \bar{h} + \frac{1}{2}(\bar{u}^2 + \bar{w}^2). \tag{3.3.34}$$

Again, the pressure gradients $\partial\bar{p}/\partial x$ and $\partial\bar{p}/\partial z$ in Eqs. (3.3.30) and (3.3.31) are known from the potential flow. At the edge of the boundary layer, the two momentum equations reduce to

$$u_e\frac{\partial u_e}{\partial x}+w_e\frac{\partial u_e}{\partial z}=-\frac{1}{\bar{\varrho}_e}\frac{\partial\bar{p}}{\partial x}, \tag{3.3.35}$$

$$u_e\frac{\partial w_e}{\partial x}+w_e\frac{\partial w_e}{\partial z}=-\frac{1}{\bar{\varrho}_e}\frac{\partial\bar{p}}{\partial z}. \tag{3.3.36}$$

A special case of a three-dimensional flow arises when the external potential flow depends only on x and not on z, that is,

$$u_e=u_e(x),\quad w_e=\text{const.}$$

For a steady compressible flow, the governing equations follow from Eqs. (3.3.29)–(3.3.33).

$$\frac{\partial}{\partial x}(\bar{\varrho}\bar{u})+\frac{\partial}{\partial y}(\overline{\varrho v})=0, \tag{3.3.37}$$

$$\bar{\varrho}\bar{u}\frac{\partial u}{\partial x}+\overline{\varrho v}\frac{\partial\bar{u}}{\partial y}=-\frac{\partial\bar{p}}{\partial x}+\frac{\partial}{\partial y}\left(\bar{\mu}\frac{\partial\bar{u}}{\partial y}-\bar{\varrho}\overline{u'v'}\right), \tag{3.3.38}$$

$$\bar{\varrho}\bar{u}\frac{\partial\bar{w}}{\partial x}+\overline{\varrho v}\frac{\partial\bar{w}}{\partial y}=\frac{\partial}{\partial y}\left(\bar{\mu}\frac{\partial\bar{w}}{\partial y}-\bar{\varrho}\overline{w'v'}\right), \tag{3.3.39}$$

$$\bar{\varrho}\bar{u}\frac{\partial\bar{H}}{\partial x}+\overline{\varrho v}\frac{\partial\bar{H}}{\partial y}=\frac{\partial}{\partial y}\left[\bar{\mu}\left(1-\frac{1}{\mathrm{Pr}}\right)\frac{\partial}{\partial y}\left(\frac{\bar{u}^2+\bar{w}^2}{2}\right)+\frac{\bar{u}}{\mathrm{Pr}}\frac{\partial\bar{H}}{\partial y}-\bar{\varrho}\overline{v'H'}\right]. \tag{3.3.40}$$

$$\bar{\varrho}\bar{u}\frac{\partial\bar{H}}{\partial x}+\overline{\varrho v}\frac{\partial\bar{H}}{\partial y}=\frac{\partial}{\partial y}\left[\frac{\bar{\mu}}{\mathrm{Pr}}\frac{\partial\bar{h}}{\partial y}-\bar{\varrho}\overline{v'h'}-\bar{\varrho}\bar{u}\overline{v'u'}-\bar{\varrho}\bar{w}\overline{v'w'}+\bar{\mu}\frac{\partial}{\partial y}\left(\frac{\bar{u}^2+\bar{w}^2}{2}\right)\right]. \tag{3.3.41}$$

The above equations are known as the boundary-layer equations for laminar and turbulent flows over infinite swept wings.

3.4 Mean-Kinetic-Energy Flows

For a two-dimensional compressible, unsteady turbulent flow, the kinetic-energy equation of the mean motion, Eq. (2.7.3a), can be written as

$$\frac{\partial}{\partial t}\left[\frac{1}{2}\bar{\varrho}\left(\tilde{u}^2+\tilde{v}^2\right)\right]+\frac{\partial}{\partial x}\left[\frac{1}{2}\overline{\varrho\tilde{u}}\left(\tilde{u}^2+\tilde{v}^2\right)\right]+\frac{\partial}{\partial y}\left[\frac{1}{2}\overline{\varrho\tilde{v}}\left(\tilde{u}^2+\tilde{v}^2\right)\right]$$

$$=-\tilde{u}\frac{\partial\bar{p}}{\partial x}=\tilde{v}\frac{\partial\bar{p}}{\partial y}+\underbrace{\frac{\partial}{\partial x}\left[\tilde{u}\left(\tau_{xx}-\overline{\varrho u'u'}\right)+\tilde{v}\left(\bar{\tau}_{yx}-\overline{\varrho v'u'}\right)\right]}_{\text{I}}$$

$$+\underbrace{\frac{\partial}{\partial y}\left[\tilde{u}\left(\bar{\tau}_{xy}-\overline{\varrho u'v'}\right)+\tilde{v}\left(\bar{\tau}_{yy}-\overline{\varrho v'v'}\right)\right]}_{\text{II}} \tag{3.4.1}$$

$$+\overline{\varrho u'u'}\frac{\partial\tilde{u}}{\partial x}+\overline{\varrho v'u'}\left(\frac{\partial\tilde{v}}{\partial x}+\frac{\partial\tilde{u}}{\partial y}\right)+\overline{\varrho v'v'}\frac{\partial\tilde{v}}{\partial y}$$

$$-\bar{\tau}_{xx}\frac{\partial\tilde{u}}{\partial x}-\bar{\tau}_{yx}\frac{\partial\tilde{v}}{\partial x}-\bar{\tau}_{xy}\frac{\partial\tilde{u}}{\partial y}-\bar{\tau}_{yy}\frac{\partial\tilde{v}}{\partial y}.$$

Of the two numbered expressions in brackets, II is much larger than I. Furthermore, since

$$\tilde{u}\left(\bar{\tau}_{xy}-\overline{\varrho u'v'}\right)\gg\tilde{v}\left(\bar{\tau}_{yy}-\overline{\varrho v'v'}\right),$$

expression II becomes $(\partial/\partial y)[\tilde{u}(\bar{\tau}_{xy}-\bar{\varrho}\overline{u'v'})]$. Also,

$$\tilde{u}\frac{\partial\bar{p}}{\partial x}\gg\tilde{v}\frac{\partial\bar{p}}{\partial y},$$

$$\overline{\varrho v'u'}\frac{\partial\tilde{u}}{\partial y}\gg\overline{\partial v'u'}\frac{\partial\tilde{v}}{\partial x},\quad\overline{\varrho u'u'}\frac{\partial\tilde{u}}{\partial x},\quad\overline{\varrho v'v'}\frac{\partial\tilde{v}}{\partial y},$$

$$\bar{\tau}_{xy}\frac{\partial\tilde{u}}{\partial y}\gg\bar{\tau}_{yy}\frac{\partial\tilde{v}}{\partial y},\quad\bar{\tau}_{yx}\frac{\partial\tilde{v}}{\partial x},\quad\bar{\tau}_{xx}\frac{\partial\tilde{v}}{\partial x}.$$

With those approximations, together with the approximations discussed in the previous section, we can write Eq. (3.4.1) as

$$\frac{\partial}{\partial t}\left(\frac{1}{2}\bar{\varrho}\bar{u}^2\right)+\frac{\partial}{\partial x}\left(\frac{1}{2}\overline{\varrho u}\bar{u}^2\right)+\frac{\partial}{\partial y}\left(\frac{1}{2}\overline{\varrho v}\bar{u}^2\right)$$

$$=-\bar{u}\frac{\partial\bar{p}}{\partial x}+\bar{u}\frac{\partial}{\partial y}\left(\bar{\mu}\frac{\partial\bar{u}}{\partial y}-\bar{\varrho}\overline{u'v'}\right). \tag{3.4.2}$$

Equation (3.4.2) is the mean kinetic energy equation for unsteady, two-dimensional, compressible, turbulent boundary layers. With the use of the mean continuity equation (3.3.5), it can also be written as

$$\bar{\varrho}\frac{\partial}{\partial t}\left(\frac{1}{2}\bar{u}^2\right) + \overline{\varrho u}\frac{\partial}{\partial x}\left(\frac{1}{2}\bar{u}^2\right) + \overline{\varrho v}\frac{\partial}{\partial y}\left(\frac{1}{2}\bar{u}^2\right) \\ = -\bar{u}\frac{\partial \bar{p}}{\partial x} + \bar{u}\frac{\partial}{\partial y}\left(\frac{\partial \bar{u}}{\partial y} - \overline{\varrho u'v'}\right). \tag{3.4.3}$$

3.5 Reynolds-Stress Transport Equations

For convenience, let us write the two terms $-\overline{u'_j(\partial p/\partial x_i)}$ and $\overline{u'_j(\partial \tau''_{ik}/\partial x_k)}$ in the Reynolds-stress transport equations given by Eq. (2.8.4) as follows:

$$-\overline{u'_j\frac{\partial p}{\partial x_i}} = -\frac{\partial}{\partial x_i}\left(\overline{pu'_j}\right) + \overline{p\frac{\partial u'_j}{\partial x_i}}, \tag{3.5.1}$$

$$\overline{u'_j\frac{\partial \tau''_{ik}}{\partial x_k}} = \frac{\partial}{\partial x_k}\left(\overline{u'_j\tau''_{ik}}\right) - \overline{\tau''_{ik}\frac{\partial u'_j}{\partial x_k}}. \tag{3.5.2}$$

The first term on the right-hand side of Eq. (3.5.2) represents *turbulent viscous diffusion*; the second term represents *turbulent energy dissipation*.

Substituting these expressions into Eq. (2.8.4) and rearranging, we obtain

$$\frac{\partial}{\partial t}\left(\overline{\varrho u'_i u'_j}\right) + \frac{\partial}{\partial x_k}\left(\tilde{u}_k\overline{\varrho u'_i u'_j}\right) \\ = \overline{p\left(\frac{\partial u'_j}{\partial x_i} + \frac{\partial u'_i}{\partial x_j}\right)} - \left[\frac{\partial}{\partial x_i}\left(\overline{pu'_j}\right) + \frac{\partial}{\partial x_j}\left(\overline{pu'_i}\right)\right] - \frac{\partial}{\partial x_k}\left(\overline{\varrho u'_i u'_j u'_k}\right) \\ -\overline{\varrho u'_i u'_k}\frac{\partial \tilde{u}_j}{\partial x_k} - \overline{\varrho u'_j u'_k}\frac{\partial \tilde{u}_i}{\partial x_k} + \frac{\partial}{\partial x_k}\left(\overline{u'_j\tau''_{ik}} + \overline{u'_i\tau''_{jk}}\right) \\ -\left(\overline{\tau''_{ik}\frac{\partial u'_j}{\partial x_k} + \tau''_{jk}\frac{\partial u'_i}{\partial x_k}}\right). \tag{3.5.3}$$

It was previously shown that in a two-dimensional boundary-layer flow the mean velocity within the boundary layer has two components ($k = 1,2$) and that the fluctuation velocity components have three components ($i, j = 1,\ldots,3$). Consequently, for a two-dimensional flow, Eq. (3.5.3) yields six equations. The dependent variables for the velocity fluctuations are $\overline{u'^2}$, $\overline{v'^2}$, $\overline{w'^2}$, $\overline{u'v'}$, $\overline{u'w'}$, and $\overline{v'w'}$. But since the mean flow is two-dimensional, the last two quantities are zero by symmetry.

If we add three equations in which the dependent variables for the velocity fluctuations are $\overline{u'^2}$, $\overline{v'^2}$, and $\overline{w'^2}$, we get the turbulent kinetic energy equation (2.8.5). Of course, the same equation can be also obtained by simply writing Eq. (3.5.3) for $i=j$. Noting that within the boundary-layer approximations, $u_i'' = u_i'$, we have

$$\begin{aligned}
&\frac{\partial}{\partial t}\left(\frac{1}{2}\overline{\varrho q^2}\right) + \frac{\partial}{\partial x_k}\left[\tilde{u}_k\left(\frac{1}{2}\overline{\varrho q^2}\right)\right] \\
&= \overline{p\frac{\partial u_i'}{\partial x_i}} - \frac{\partial}{\partial x_i}\left(\overline{pu_i'}\right) - \frac{\partial}{\partial x_k}\left[\overline{u_k'\left(\frac{1}{2}\varrho q^2\right)}\right] \\
&\quad -\overline{\varrho u_i' u_k'}\frac{\partial \tilde{u}_i}{\partial x_k} + \frac{\partial}{\partial x_k}\left(\overline{u_i'\tau_{jk}''}\right) - \varepsilon\bar{\varrho},
\end{aligned} \tag{3.5.4}$$

where $q^2 = u_i'u_i'$.

In Eq. (3.5.4), ε denotes the so-called mean turbulent energy dissipation function, which is given by

$$\varepsilon = \frac{1}{\bar{\varrho}}\overline{\tau_{ik}''\frac{\partial u_i'}{\partial x_k}}. \tag{3.5.5}$$

For an unsteady, compressible, two-dimensional flow, Eq. (3.5.3) can be simplified considerably by using the boundary-layer approximations discussed in the previous section. From the resulting simplified equation we can get four equations, for $\overline{u'v'}$, $\overline{u'^2}$, $\overline{v'^2}$, and $\overline{w'^2}$.

Let us now discuss the terms in Eq. (3.5.4). Of the terms of $(\partial/\partial x_i)(\overline{pu_i'})$, we observe that

$$\frac{\partial}{\partial y}\left(\overline{pv'}\right) \gg \frac{\partial}{\partial x}\left(\overline{pu'}\right), \frac{\partial}{\partial z}\left(\overline{pw'}\right);$$

and, of the terms of $(\partial/\partial x_k)\left[\overline{u_k'\left(\frac{1}{2}\varrho q^2\right)}\right]$, we observe that

$$\frac{\partial}{\partial y}\left(\overline{v'\frac{1}{2}\varrho q^2}\right) \gg \frac{\partial}{\partial x}\left(\overline{u'\frac{1}{2}\varrho q^2}\right), \frac{\partial}{\partial z}\left[\overline{w'\left(\frac{1}{2}\varrho q^2\right)}\right];$$

and, finally, of the terms of $\overline{\varrho u_i'u_k'}(\partial\tilde{u}_i/\partial x_k)$, we again observe that

$$\overline{\varrho u'v'}\frac{\partial \tilde{u}}{\partial y} \gg \overline{\varrho u'u'}\frac{\partial \tilde{u}}{\partial x}, \overline{\varrho v'u'}\frac{\partial \tilde{v}}{\partial x}, \overline{\varrho v'v'}\frac{\partial \tilde{v}}{\partial y},$$

With the definition of τ''_{ik}, we can write $u'_i\tau''_{ik}$ as

$$\overline{u'_i\tau''_{ik}} = \bar{\mu}\left[\frac{\partial}{\partial x_k}\left(\frac{1}{2}\overline{q^2}\right) + \frac{\partial}{\partial x_i}(\overline{u'_i u'_k}) - \overline{u'_k\frac{\partial u'_i}{\partial x_i}}\right].$$

Of the terms of $(\partial/\partial x_k)\left[\overline{\mu(\partial/\partial x_k)(\frac{1}{2}q^2)}\right]$, we observe that

$$\frac{\partial}{\partial y}\left[\bar{\mu}\frac{\partial}{\partial y}\left(\frac{1}{2}\overline{q^2}\right)\right] \gg \frac{\partial}{\partial x}\left[\bar{\mu}\frac{\partial}{\partial x}\left(\frac{1}{2}\overline{q^2}\right)\right];$$

and, of the nine terms of $(\partial/\partial x_k)\left[\overline{\mu(\partial/\partial x_i)(u'_i u'_k)}\right]$, we again observe

$$\frac{\partial}{\partial y}\left[\bar{\mu}\frac{\partial}{\partial y}(\overline{v'v'})\right] \gg \frac{\partial}{\partial x}\left[\bar{\mu}\frac{\partial}{\partial x}(\overline{u'u'})\right], \frac{\partial}{\partial y}\left[\bar{\mu}\frac{\partial}{\partial x}(\overline{u'v'})\right], \text{ etc.}$$

Making use of the above relations, assuming that the divergence of the velocity fluctuations $\partial u'_i/\partial x_i$ is negligible, and noting the relationship given in Eq. (3.3.4), we can now write the turbulent kinetic energy equation (3.5.4) for a two-dimensional, unsteady, compressible boundary-layer flow as

$$\underbrace{\frac{\partial}{\partial t}\left(\bar{\varrho}\frac{\overline{q^2}}{2}\right)}_{\text{local rate of change of turbulent energy}} + \underbrace{\frac{\partial}{\partial x}\left[\bar{\varrho}\bar{u}\left(\frac{\overline{q^2}}{2}\right)\right] + \frac{\partial}{\partial y}\left[\overline{\varrho v}\left(\frac{\overline{q^2}}{2}\right)\right]}_{\text{turbulent energy convection}}$$

$$= \underbrace{-\bar{\varrho}\overline{u'v'}\frac{\partial \bar{u}}{\partial y}}_{\text{turbulent energy production}} + \underbrace{\frac{\partial}{\partial y}\left[\bar{\mu}\frac{\partial}{\partial y}\left(\frac{1}{2}\overline{q^2} + \overline{v'^2}\right)\right]}_{\text{turbulent viscous diffusion}} \tag{3.5.6a}$$

$$\underbrace{-\frac{\partial}{\partial y}\left(\overline{v'\frac{1}{2}\varrho q^2}\right)}_{\text{turbulent kinetic diffusion}} \underbrace{- \frac{\partial}{\partial y}(\overline{pv'})}_{\text{turbulent pressure diffusion}} \underbrace{- \varepsilon\bar{\varrho}.}_{\text{turbulent energy dissipation}}$$

In Eq. (3.5.6a) we have assumed that $\varrho q^2 \gg \overline{\varrho'' q^2}$.

Multiplying the continuity equation by $-\frac{1}{2}q^2$ and adding the resulting expression to Eq. (3.5.6a), we obtain

$$\bar{\varrho}\frac{\partial}{\partial t}\frac{\overline{q^2}}{2} + \bar{\varrho}\bar{u}\frac{\partial}{\partial x}\left(\frac{\overline{q^2}}{2}\right) + \overline{\varrho v}\frac{\partial}{\partial y}\left(\frac{\overline{q^2}}{2}\right)$$

$$= -\bar{\varrho}\overline{u'v'}\frac{\partial \bar{u}}{\partial y} + \frac{\partial}{\partial y}\left[\bar{\mu}\frac{\partial}{\partial y}\left(\frac{1}{2}\overline{q^2} + \overline{v'^2}\right)\right] - \frac{\partial}{\partial y}\overline{\left[v'\left(\frac{1}{2}\bar{\varrho}q^2 + p\right)\right]} - \varepsilon\bar{\varrho}, \tag{3.5.6b}$$

where ε, which is given by Eq. (3.5.5), can also be written as

$$\varepsilon \equiv \frac{1}{\bar{\varrho}}\bar{\mu}\overline{\left(\frac{\partial u_i'}{\partial x_k}+\frac{\partial u_k'}{\partial x_i}\right)\frac{\partial u_i'}{\partial x_k}}$$

$$= \bar{\nu}\left[2\overline{\left(\frac{\partial u'}{\partial x}\right)^2}+2\overline{\left(\frac{\partial v'}{\partial y}\right)^2}+2\overline{\left(\frac{\partial w'}{\partial z}\right)^2}+\overline{\left(\frac{\partial u'}{\partial y}+\frac{\partial v'}{\partial x}\right)^2}\right.$$

$$\left.+\overline{\left(\frac{\partial u'}{\partial z}+\frac{\partial w'}{\partial x}\right)^2}+\overline{\left(\frac{\partial v'}{\partial z}+\frac{\partial w'}{\partial y}\right)^2}\right], \tag{3.5.7}$$

if the $\lambda_{ik}(\partial u_l'/\partial x_l)$ term in the stress tensor is neglected.

The boundary-layer approximations for Eq. (3.5.3) are quite similar to those for Eq. (3.5.4). For $i=j=1$, Eq. (3.5.3) becomes

$$\frac{\partial}{\partial t}\left(\overline{\varrho u'^2}\right)+\frac{\partial}{\partial x_k}\left(\tilde{u}_k\overline{\varrho u'^2}\right)$$

$$= 2\overline{p\frac{\partial u'}{\partial x}}-2\frac{\partial}{\partial x}\left(\overline{pu'}\right)-\frac{\partial}{\partial x_k}\left(\overline{u_k'\varrho u'^2}\right) \tag{3.5.8}$$

$$-2\overline{\varrho u'u_k'}\frac{\partial \tilde{u}}{\partial x_k}+2\frac{\partial}{\partial x_k}\left[\overline{\mu'\left(\frac{\partial u'}{\partial x_k}+\frac{\partial u_k'}{\partial x}\right)}\right].$$

Let us now define the operator D/Dt on any function g by

$$\frac{Dg}{Dt}=\frac{\partial g}{\partial t}+\frac{\partial}{\partial x}(g\bar{u})+\frac{\partial}{\partial y}(g\tilde{v}).$$

In accordance with the boundary-layer approximations discussed for the turbulent kinetic-energy equation, Eq. (3.5.8) simplifies to

$$\frac{D}{Dt}\overline{\varrho u'^2}=2\overline{p\frac{\partial u'}{\partial x}}-\frac{\partial}{\partial y}\overline{v'\varrho u'^2}-2\overline{\varrho u'v'}\frac{\partial \bar{u}}{\partial y}$$

$$+\frac{\partial}{\partial y}\left[\bar{\mu}\frac{\partial}{\partial y}\left(\overline{u'^2}\right)\right]-2\bar{\mu}\overline{\frac{\partial u'}{\partial x_k}\frac{\partial u'}{\partial x_k}}. \tag{3.5.9}$$

Similarly, we can write an equation for each of the turbulent energies v'^2 and w'^2 by letting $i=j=2$ and $i=j=3$ in Eq. (3.5.3), and we can write a single equation for shear stress by letting $i=1, j=2$ in Eq. (3.5.3). After the application of boundary-layer approximations to these equations, we get

$$\frac{D}{Dt}\overline{\varrho v'^2} = 2\overline{p\frac{\partial v'}{\partial y}} - 2\frac{\partial}{\partial y}\left(\overline{pv'}\right) - \frac{\partial}{\partial y}\left(\overline{v'\varrho v'^2}\right)$$
$$+\frac{\partial}{\partial y}\left[\bar{\mu}\frac{\partial}{\partial y}\left(\overline{v'^2}\right)\right] - 2\bar{\mu}\overline{\frac{\partial v'}{\partial x_k}\frac{\partial v'}{\partial x_k}}, \tag{3.5.10}$$

$$\frac{D}{Dt}\overline{\varrho w'^2} = 2\overline{p\frac{\partial w'}{\partial z}} - \frac{\partial}{\partial y}\left(\overline{v'\varrho w'^2}\right) + \frac{\partial}{\partial y}\left[\bar{\mu}\frac{\partial}{\partial y}\left(\overline{w'^2}\right)\right] - 2\bar{\mu}\left(\overline{\frac{\partial w'}{\partial x_k}\frac{\partial w'}{\partial x_k}}\right), \tag{3.5.11}$$

$$\frac{D}{Dt}\overline{\varrho u'v'} = \overline{p\left(\frac{\partial u'}{\partial y} + \frac{\partial v'}{\partial x}\right)} - \frac{\partial}{\partial y}\left(\overline{\varrho u'v'^2}\right) - \overline{\varrho v'v'}\frac{\partial \bar{u}}{\partial y}$$
$$+\frac{\partial}{\partial y}\left[\bar{\mu}\frac{\partial}{\partial y}\left(\overline{u'v'}\right)\right] - \frac{\partial}{\partial y}\overline{(pu')} - 2\bar{\mu}\left(\overline{\frac{\partial u'}{\partial x_k}\frac{\partial v'}{\partial x_k}}\right). \tag{3.5.12}$$

For incompressible flows, since the flow properties are constant, Eqs. (3.5.6) and (3.5.10)–(3.5.12) can be simplified considerably.

For example, the *turbulent kinetic-energy equation* becomes

$$\frac{\partial}{\partial t}\left(\frac{\overline{q^2}}{2}\right) + \bar{u}\frac{\partial}{\partial x}\left(\frac{\overline{q^2}}{2}\right) + \bar{v}\frac{\partial}{\partial y}\left(\frac{\overline{q^2}}{2}\right)$$
$$= -\overline{u'v'}\frac{\partial \bar{u}}{\partial y} + \nu\frac{\partial^2}{\partial y^2}\left(\frac{\overline{q^2}}{2} + \overline{v'2}\right) - \frac{\partial}{\partial y}\left(\frac{\overline{v'q^2}}{2} + \frac{\overline{pv'}}{\varrho}\right) - \varepsilon. \tag{3.5.13}$$

3.6 Integral Equations of the Boundary Layer

Although the differential equations of the boundary layer discussed in the previous sections have been greatly simplified from the general differential equations of fluid flow, they are still difficult to solve, since they are nonlinear partial differential equations. Considerable simplification arises when these equations are integrated across the boundary layer. Then they are no longer partial differential equations, but just ordinary differential equations. However, they are still exact equations, at least within the boundary-layer approximations. These equations, known as the *integral equations* of the boundary layer, provide a basis for many approximate methods of boundary-layer prediction. They will now be discussed for steady two-dimensional flows.

3.6.1 Momentum Integral Equation

The momentum integral equation for a two-dimensional steady compressible flow can be obtained by integration from the boundary-layer equations (3.3.5) and (3.3.6b).[4] If we multiply Eq. (3.3.5) by $(u_e - u)$, multiply Eq. (3.3.6b) by -1, add and subtract $\varrho u(du_e/dx)$ from Eq. (3.3.6b), and add the resulting continuity and momentum equations, we can arrange the resulting expression in the form

$$\frac{\partial}{\partial x}\left[\varrho u(u_e - u)\right] + \frac{\partial}{\partial y}\left[\varrho v(u_e - u)\right] + \frac{du_e}{dx}(\varrho_e u_e - \varrho u) = -\frac{\partial}{\partial y}\left(\mu\frac{\partial u}{\partial y} - \varrho\overline{u'v'}\right).$$

Nondimensionalizing and integrating with respect to y from zero to infinity, we get

$$\frac{d}{dx}\left[\varrho_e u_e^2 \int_0^\infty \frac{\varrho u}{\varrho_e u_e}\left(1 - \frac{u}{u_e}\right)dy\right] + \frac{du_e}{dx}\varrho_e u_e\left[\int_0^\infty\left(1 - \frac{\varrho u}{\varrho_e u_e}\right)dy\right] - \varrho_w v_w u_e$$
$$= \left(\mu\frac{\partial u}{\partial y}\right)_w \equiv \tau_w, \tag{3.6.1}$$

since $(\partial u/\partial y)$ and $\overline{u'v'} \to 0$ as $y \to \infty$ and since $\overline{u'v'} \to 0$ as $y \to 0$. It is more convenient to express Eq. (3.6.1) in terms of boundary-layer thicknesses δ^* and θ. Equation (3.6.1) then becomes

$$\frac{d}{dx}\left(\varrho_e u_e^2\theta\right) + \varrho_e u_e \delta^* \frac{du_e}{dx} - \varrho_w v_w u_e = \tau_w,$$

or, in nondimensional form,

$$\frac{d\theta}{dx} + \frac{\theta}{u_e}(H+2)\frac{du_e}{dx} + \frac{\theta}{\varrho_e}\frac{d\varrho_e}{dx} - \frac{\varrho_w}{\varrho_e}\frac{v_w}{u_e} = \frac{\tau_w}{\varrho_e u_e^2}, \tag{3.6.2}$$

where H denotes the ratio δ^*/θ, which is known as the shape factor. For an ideal gas undergoing an isentropic process, we can write

$$\frac{1}{\varrho_e}\frac{d\varrho_e}{dx} = -\frac{M_e^2}{u_e}\frac{du_e}{dx}. \tag{3.6.3}$$

[4]For simplicity, we shall drop the bars from the principal mean quantities.

Substituting from that equation into Eq. (3.6.2) and rearranging, we obtain the momentum integral equation of the boundary layer for a two-dimensional compressible flow:

$$\frac{d\theta}{dx}+\frac{\theta}{u_e}(H+2-M_e^2)\frac{du_e}{dx}-\frac{\varrho_w v_w}{\varrho_e u_e}=\frac{\tau_w}{\varrho_e u_e^2}\equiv\frac{c_f}{2}, \tag{3.6.4}$$

where c_f is the local skin-friction coefficient. Note that in the case of zero mass transfer the normal velocity component at the wall v_w is zero. Then Eq. (3.6.4) becomes

$$\frac{d\theta}{dx}+\frac{\theta}{u_e}\left(H+2-M_e^2\right)\frac{du_e}{dx}=\frac{c_f}{2}. \tag{3.6.5}$$

For an incompressible flow with no mass transfer, that equation reduces to

$$\frac{d\theta}{dx}+\frac{\theta}{u_e}(H+2)\frac{du_e}{dx}=\frac{c_f}{2}. \tag{3.6.6}$$

Equations (3.6.4)–(3.6.6) are also known as the *first* momentum integral equations. They are applicable to both laminar and turbulent boundary layers.

3.6.2 Mean Energy Integral Equation

The derivation of the mean energy integral equation is similar to that of the momentum integral equation. We multiply Eq. (3.3.5) by ($u_e^2- u^2$), and Eq. (3.3.6b) by – $2u$. After the resulting expressions are added and rearranged, we obtain

$$\begin{aligned}\frac{1}{2}\frac{\partial}{\partial x}\left\{\varrho_e u_e^3\left[\frac{\varrho u}{\varrho_e u_e}\left(1-\frac{u^2}{u_e^2}\right)\right]\right\}+\frac{1}{2}\frac{\partial}{\partial y}\left[\varrho v\left(u_e^2-u^2\right)\right]\\ +\varrho_e u_e^2\left[\frac{u}{u_e}\left(1-\frac{\varrho}{\varrho_e}\right)\frac{du_e}{dx}\right]=-u\frac{\partial}{\partial y}\left(\mu\frac{\partial u}{\partial y}-\varrho\overline{u'v'}\right).\end{aligned} \tag{3.6.7}$$

The right-hand side of Eq. (3.6.7) can also be written as

$$-u\frac{\partial\tau}{\partial y}=\tau\frac{\partial u}{\partial y}-\frac{\partial}{\partial y}(u\tau), \tag{3.6.8}$$

where

$$\tau=\mu(\partial u/\partial y)-\varrho\overline{u'v'}. \tag{3.6.9}$$

Substituting from Eq. (3.6.8) into Eq. (3.6.7), integrating the resulting expression with respect to y from zero to infinity, and using the definition of energy thickness δ^{**} given by

$$\delta^{**} = \int_0^\infty \frac{\varrho u}{\varrho_e u_e}\left(1 - \frac{u^2}{u_e^2}\right) dy \tag{3.6.10}$$

we get

$$\begin{aligned} &\frac{1}{2}\frac{d}{dx}\left(\varrho_e u_e^3 \delta^{**}\right) - \frac{1}{2}\varrho_w v_w u_e^2 \\ &+ \varrho_e u_e^2 \frac{du_e}{dx}\left[\int_0^\infty \frac{u}{u_e}\left(1 - \frac{\varrho}{\varrho_e}\right) dy\right] = D, \end{aligned} \tag{3.6.11}$$

since both $\varrho v(u_e^2 - u^2)$ and $\tau \to 0$ as $y \to \infty$. In Eq. (3.6.11), D is defined by

$$D = \int_0^\infty \left(\mu \frac{\partial u}{\partial y} - \varrho \overline{u'v'}\right) \frac{\partial u}{\partial y}\, dy$$

and is called the *dissipation integral* or the *shear work integral.* It denotes the viscous work done in the boundary layer by the two shearing stresses $\mu(\partial u/\partial y)$ and $-\varrho\overline{u'v'}$.

For an incompressible flow with no mass transfer, Eq. (3.6.11) simplifies considerably, becoming

$$\frac{d}{dx}(u_e^3 \delta^{**}) = \frac{2D}{\varrho}. \tag{3.6.12a}$$

In dimensionless form, that equation becomes

$$\frac{1}{u_e^3}\frac{d}{dx}(u_e^3 \delta^{**}) = \frac{2D}{\varrho u_e^3} = 2C_D, \tag{3.6.12b}$$

where C_D is called the dissipation integral coefficient.

3.6.3 Turbulent Energy Integral Equation

Integrating Eq. (3.5.6a) with respect to y from zero to infinity, for steady state we get

$$\frac{d}{dx}\left[\int_0^\infty \varrho u \frac{\overline{q^2}}{2}\, dy\right] = \int_0^\infty -\varrho \overline{u'v'}\left(\frac{\partial u}{\partial y}\right) dy - \int_0^\infty \varrho \varepsilon\, dy, \tag{3.6.13a}$$

since, in accordance with the boundary conditions, the diffusion terms drop out. Equation (3.6.13a) is often written in the form

$$\frac{d}{dx}\left(\frac{1}{2} I Q^2\right) = P - d, \tag{3.6.13b}$$

where IQ^2 is an integral scale for the turbulence, defined by

$$IQ^2 = \int_0^\infty \varrho u \overline{q^2} dy, \tag{3.6.14}$$

and P and d represent the production and dissipation, respectively, of turbulent energy within the boundary layer, that is,

$$P = \int_0^\infty -\varrho \overline{u'v'} \left(\frac{\partial u}{\partial y}\right) dy \tag{3.6.15}$$

and

$$d = \int_0^\infty \varrho \varepsilon \, dy. \tag{3.6.16}$$

3.6.4 Energy Integral Equation

To derive the energy integral equation for two-dimensional compressible flows without body force, we start with the total enthalpy equation, Eq. (3.3.27). Using the continuity equation, Eq. (3.3.24), we can write Eq. (3.3.27) as

$$\frac{\partial}{\partial x}(\varrho u H) + \frac{\partial}{\partial y}(\overline{\varrho v} H) = \frac{\partial}{\partial y}(-\dot{q} + u\tau). \tag{3.6.17}$$

where

$$\tau \equiv \mu \frac{\partial u}{\partial y} - \varrho \overline{u'v'} - \overline{\varrho' u'v'}, \tag{3.6.18a}$$

$$\dot{q} \equiv -k \frac{\partial T}{\partial y} + c_p \varrho \overline{T'v'} + c_p \overline{\varrho' T'v'} \tag{3.6.18b}$$

We now integrate the above equation with respect to y from $y=0$ to $y=h>\delta$ to get

$$\int_0^h \frac{\partial}{\partial x}(\varrho u H) dy + \varrho_h v_h H_e = \dot{q}_w. \tag{3.6.19}$$

As in the derivation of the momentum integral equation, we substitute for $\varrho_h v_h$ from the continuity equation and write Eq. (3.6.19) as

$$\int_0^h \frac{\partial}{\partial x}[\varrho u (H_e - H)] dy = -\dot{q}_w \tag{3.6.20}$$

and then as

$$\frac{d}{dx}\left[\varrho_e u_e \int_0^\infty \frac{\varrho u}{\varrho_e u_e}\left(1-\frac{H}{H_e}\right)dy\right] = -\frac{\dot{q}_w}{H_e}, \tag{3.6.21}$$

using reasoning similar to which led to the writing of Eq. (3.6.1). Since $H_e - H = 0$ (to sufficient accuracy) for $y \geq h$, the integrand of Eq. (3.6.21) contributes only for $y < h$, and so the result is independent of h.

Equation (3.6.21) shows that the rate of increase of total-enthalpy deficit per unit span (in the z direction) is equal to the rate of heat transfer *from* the fluid to a unit area of the surface. Comparing this equation with Eq. (3.6.1), we see that the rate of increase of deficit in each case is affected by transfer into the surface (enthalpy transfer $-\dot{q}_w$, momentum transfer τ_w); the momentum integral equation contains an additional term depending on pressure gradient, but the *total* enthalpy is unaffected by pressure gradients as such.

If we introduce θ_H by

$$\theta_H = \int_0^h \frac{\varrho u}{\varrho_e u_e}\left(\frac{H-H_e}{H_w-H_e}\right)dy, \tag{3.6.22}$$

then we can write the total-enthalpy integral equation (3.6.20) for a two-dimensional compressible laminar or turbulent flow as

$$\frac{d}{dx}\left[\varrho_e u_e (H_w - H_e)\theta_H\right] = \dot{q}_w. \tag{3.6.23}$$

In Eq. (3.6.22), θ_H is a measure of total-enthalpy-flux *surplus* caused by the presence of the thermal boundary layer. For an incompressible flow, where H is equal to the static enthalpy h, the total-enthalpy thickness θ_H is equal to the static-enthalpy thickness.

$$\theta_h = \int_0^\infty \frac{\varrho u}{\varrho_e u_e}\left(\frac{h-h_e}{h_w-h_e}\right)dy, \tag{3.6.24}$$

where h can be replaced by c_pT if c_p is constant. Noting this and taking ϱ constant, we can write Eq. (3.6.23) for an incompressible flow as

$$\frac{d}{dx}\left[u_e(T_w - T_e)\theta_T\right] = \frac{\dot{q}_w}{\varrho c_p}. \tag{3.6.25}$$

When the wall temperature is *uniform*, Eq. (3.6.25) can be written as

$$\frac{1}{u_e}\frac{d}{dx}(u_e\theta_T) = \mathrm{St} \tag{3.6.26a}$$

or as

$$\frac{d\theta_T}{dx} + \frac{\theta_T}{u_e}\frac{du_e}{dx} = \mathrm{St}. \tag{3.6.26b}$$

Here St denotes the Stanton number befined by

$$\mathrm{St} = \frac{\dot{q}_w}{\varrho c_p (T_w - T_e) u_e}, \tag{3.6.27}$$

and

$$\theta_T = \int_0^h \frac{u}{u_e}\left(\frac{T - T_e}{T_w - T_e}\right) dy. \tag{3.6.28}$$

Problems

3.1 Show that in a laminar flow with heat transfer the "compression work" term $u\,\frac{\partial p}{\partial x}$ in Eq. (3.3.17) is small compared to the heat-flux term $\frac{\partial}{\partial y}\left(\frac{\mu}{\Pr}\frac{\partial h}{\partial y}\right)$ if the temperature difference across the shear layer, ΔT, is *large* compared to $\frac{\Pr u_e^2}{c_p}$.

3.2 Show that in a compressible turbulent flow the condition for the neglect of the "compression work" term $u\,\frac{\partial p}{\partial x}$ in Eq. (3.3.17) – compared to the "heat transfer" term $\frac{\partial}{\partial y}\left(\bar{\varrho}\overline{h'v'}\right)$ or $c_p\frac{\partial}{\partial y}\left(\bar{\varrho}\overline{T'v'}\right)$, say – is

$$M_e^2 \ll \frac{1}{10}\frac{1}{\gamma - 1}\frac{\Delta T}{T_e}.$$

[Hint: The factor 10 is an approximation to δ/θ in a turbulent boundary layer.]

3.3 Show that for high-speed flows

$$\frac{\varrho''}{\varrho} \cong (\gamma - 1)\bar{M}^2 \frac{u''}{u} \tag{P3.3.1}$$

3.4 In low-speed boundary layers on heated walls, the velocity and temperature fluctuations again tend to have opposite signs. However, since the driving temperature difference $T_w - T_e$ is imposed separately from the velocity difference u_e instead of being related to it as in high-speed flow, Eq. (P3.3.1) does not hold. Instead, if the analogy between heat transfer and momentum transfer were exact and the effect of departure of Pr from unity were small, show that we would have

$$\frac{T''}{T_w - T_e} = -\frac{u''}{u_e} \tag{P3.4.1}$$

or

$$\frac{\varrho''}{\varrho} = \frac{-T''}{T} = \left(\frac{T_w - T_e}{T}\right)\frac{u''}{u_e}. \tag{P3.4.2}$$

3.5 Consider the momentum equation, Eq. (3.3.5) and show that in a highspeed boundary layer,

$$\overline{\varrho''v''} = (\gamma - 1)\, M^2 \varrho \frac{\overline{\varrho''v''}}{u} = \varrho v \left[(\gamma - 1) M^2 \frac{\overline{\varrho''v''}}{uv} \right]. \tag{P3.5.1}$$

3.6 By using the arguments used to derive a typical value of $\overline{\varrho''v''}$ in Eq. (P3.5.1) show that a typical value for $\overline{\varrho''u''}$ in a nearly-adiabatic highspeed shear layer is

$$\varrho(\gamma - 1)\, M^2 \frac{\overline{u''^2}}{u} \tag{P3.6.1}$$

where M is the Mach number.

3.7 Show that in a boundary layer of thickness δ on a surface of longitudinal curvature radius R below a stream of Mach number M_e, the ratio of the pressure difference across the layer to the absolute pressure at the edge is of order $\gamma M_e^2 \delta / R$.

3.8 Show that in a turbulent boundary layer the ratio of the pressure change induced by the Reynolds normal stress to the absolute pressure at the edge is of order $\gamma M_e^2 c_f / 2$, where c_f is the skin-friction coefficient. [Hint: Assume that $\overline{\varrho v''^2}$ is of the same order as the shear stress $- \bar{\varrho}\overline{u''v''}$.]

3.9 Show that if in the decelerating turbulent boundary layer in an expanding passage (a diffuser) the skin-friction term in the momentum-integral equation, Eq. (3.6.6), is negligible and H can be taken as constant. Equation (3.6.6) gives

$$\frac{\theta}{\theta_0} = \left(\frac{u_e}{u_{e,0}}\right)^{-(H+2)}$$

where subscript 0 denotes initial conditions.

3.10 Show that for an incompressible zero-pressure gradient flow over a wall at uniform temperature, Eq. (3.6.26b) can be written as

$$\frac{d\theta_T}{dx} = \text{St}. \tag{P3.10.1}$$

3.11 Since the governing equations for two-dimensional and axisymmetric flows differ from each other only by the radial distance $r(x,y)$, the axisymmetric flow equations, Eqs. (3.3.24) to (3.3.26) can be placed in a nearly two-dimensional form by using a transformation known as the *Mangler transformation.* In the case of flow over a body of revolution of radius r_0 (a function of x), we find that if the boundary-layer thickness is small compared with r_0, so that $r(x, y) = r_0(x)$, this transformation puts them exactly in two-dimensional form. We define the Mangler transformation by

$$d\bar{x} = \left(\frac{r_0}{L}\right)^{2K} dx, \quad d\bar{y} = \left(\frac{r}{L}\right)^{K} dy \tag{P3.11.1}$$

to transform an axisymmetric flow with coordinates (x, y), into a two-dimensional flow with coordinates $(\bar{x}, \bar{y})$. In Eq. (P3.11.1) L is an arbitrary reference length. If a stream function in Mangler variables $(\bar{x}, \bar{y})$ is related to a stream function ψ in (x, y) variables by

$$\bar{\psi}(\bar{x}, \bar{y}) = \left(\frac{1}{L}\right)^{K} \psi(x, y),$$

then

(a) show that the relation between the Mangler transformed velocity components $\bar{u}$ and $\bar{v}$ in $(\bar{x}, \bar{y})$ variables and the velocity components u and v in (x,y) variables is:

$$u = \bar{u}$$
$$v = \left(\frac{L}{r}\right)^{K} \left[\left(\frac{r_0}{L}\right)^{2K} \bar{v} - \frac{\partial \bar{y}}{\partial x}\bar{u}\right]. \tag{P3.11.2}$$

(b) By substituting from Eqs (P3.11.2) into Eqs. (3.3.24)–(3.3.26), show that for laminar flows the Mangler-transformed continuity momentum and energy equations are:

$$\frac{\partial \bar{u}}{\partial \bar{x}} + \frac{\partial \bar{v}}{\partial \bar{y}} = 0$$

$$\bar{u}\,\frac{\partial \bar{u}}{\partial \bar{x}} + \bar{v}\,\frac{\partial \bar{u}}{\partial \bar{y}} = -\frac{1}{\varrho}\frac{dp}{d\bar{x}} + \nu\frac{\partial}{\partial \bar{y}}\left[(1+t)^{2K}\frac{\partial \bar{u}}{\partial \bar{y}}\right] \tag{P3.11.3}$$

$$\bar{u}\,\frac{\partial T}{\partial \bar{x}} + \bar{v}\,\frac{\partial T}{\partial \bar{y}} = \frac{k}{\varrho c_p}\frac{\partial}{\partial \bar{y}}\left[(1+t)^{2K}\frac{\partial T}{\partial \bar{y}}\right] \tag{P3.11.4}$$

where

$$t = -1 + \left(1 + \frac{2l\cos\phi}{r_0^2}\bar{y}\right)^{1/2}. \tag{P3.11.5}$$

Note that for $t = 0$, Eqs. (P3.11.3) and (P3.11.4) in the $(\bar{x}, \bar{y})$ plane are exactly in the same form as those for two-dimensional flows in the (x, y) plane.

3.12 As discussed in Section 8.2, transformed coordinates employing similarity variables are often used in the solution of the boundary-layer equations. A convenient transformation is the Falkner-Skan transformation given by Eq. (8.2.5) for two-dimensional flows. With minor changes, this transformation can also be used for axisymmetric flows.

With the transformation defined by

$$\eta = \left(\frac{u_e}{\nu \bar{x}}\right)^{\frac{1}{2}} \bar{y} \tag{P3.12.1}$$

$$\bar{\psi}(\bar{x}, \bar{y}) = (u_e \nu \bar{x})^{\frac{1}{2}} f(\bar{x}, \eta) \tag{P3.12.2}$$

show that Mangler-transformed continuity and momentum equations and their boundary conditions can be written as

$$\left(bf''\right)' + \frac{m+1}{2} f f'' + m\left[1 - (f')^2\right] = \bar{x}\left(f'\frac{\partial f'}{\partial \bar{x}} - f''\frac{\partial f}{\partial \bar{x}}\right) \tag{P3.12.3}$$

$$\eta = 0, \quad f' = 0, \quad f(\bar{x}, 0) \equiv f_w = -\frac{1}{(u_e \nu \bar{x})} \int_0^{\bar{x}} \bar{v}_w d\bar{x} \tag{P3.12.4a}$$

$$\eta = \eta_e, \quad f' = 1 \tag{P3.12.4b}$$

where

$$b = (1+t)^{2k}, \quad m = \frac{\bar{x}}{u_e} \frac{du_e}{d\bar{x}}, \tag{P3.12.5a}$$

$$t = -1 + \left[1 + \left(\frac{L}{r_0}\right)^2 \frac{2\cos\phi}{L} \left(\frac{\nu \bar{x}}{u_e}\right)^{\frac{1}{2}} \eta\right]^{\frac{1}{2}} \tag{P3.12.5b}$$

References

[1] A.L. Kistler, Fluctuation measurements in a supersonic turbulent boundary layer, Phys. Fluids 2 (1959) 290.

[2] M.V. Morkovin, W.S. Chen, Effects of compressibility on turbulent flows. The Mechanics of Turbulence, Gordon and Breach, New York, 1961, p. 367.

[3] A. Demetriades, Turbulence measurements in an axisymmetric compressible wake, Phys. Fluids 11 (1841) 1968.

[4] A.L. Kistler, W.S. Chen, The fluctuating pressure field in a supersonic turbulent boundary layer, J. Fluid Mech. 16 (1963) 41.

[5] J. Laufer, Some statistical properties of the pressure field radiated by a turbulent boundary layer, Phys. Fluids 7 (1964) 1191.

Chapter 4

General Behavior of Turbulent Boundary Layers

Chapter Outline Head

Analysis of Turbulent Flows with Computer Programs. http://dx.doi.org/10.1016/B978-0-08-098335-6.00004-5

4.1 Introduction

The development and the presentation of the governing equations of turbulent flows has been discussed in Chapter 2. In Chapter 3 we have shown that the governing conservation equations can be simplified considerably for thin shear layers. In this chapter and in the following chapters we shall discuss the solution of the thin-shear-layer equations for two-dimensional and axisymmetric turbulent boundary layers for both incompressible and compressible flows.

In this chapter we shall discuss the general behavior of turbulent boundary layers. We shall consider certain special classes of flows and discuss various empirical laws based on dimensional analysis such as "the law of the wall," "the defect law" for predicting mean velocity distribution, and similar empirical laws for predicting the mean temperature distribution in such flows.

4.2 Composite Nature of a Turbulent Boundary Layer

According to experimental data, a turbulent boundary layer can be regarded approximately as a composite layer made up of inner and outer regions. The existence of the two regions is due to the different response to shear and pressure gradient by the fluid near the wall. The reason for identifying two regions in a turbulent boundary layer can best be explained by examples.

Consider an incompressible flow past a flat plate. For a laminar boundary-layer flow, the velocity profiles are geometrically similar and reduce to a single curve if u/u_e is plotted against a dimensionless y coordinate, $\eta = (u_e/\nu x)^{1/2}y$. This is the well-known Blasius profile. The geometrical similarity is maintained, regardless of the Reynolds number of the flow or of the local skin friction. In a turbulent boundary layer there is no choice of dimensionless y coordinate that leads to the collapse of the complete velocity profiles into a single curve, because the viscous-dependent part of the profile and the Reynolds-stress-dependent part of the profile require different length-scaling parameters.

When an obstacle is placed in a laminar flat-plate boundary-layer flow, the velocity profiles downstream from the obstacle do not at first resemble the Blasius profile. However, at low Reynolds numbers, if the layer is allowed to develop far enough downstream, the velocity profiles slowly return to the Blasius profile. In turbulent boundary layers, the effect of such disturbances disappears quite soon, because of the greater diffusivity, and the velocity profiles quickly return to "normal" boundary-layer profiles. The phenomenon was experimentally investigated by Klebanoff and Diehl [1]. Analysis of the data of Fig. 4.1 shows that the inner part of the turbulent layer returns more quickly to "normal" than the outer part of the layer, which suggests that the flow close to the wall is relatively insensitive to the flow conditions away from the wall and to the upstream conditions. Figures 4.2a and 4.2b further illustrate that effect. Here, turbulent flow in a rectangular channel passes from a rough surface to a smooth one and vice versa. The figures show that in both cases the shearing stress near the wall very rapidly assumes the new value corresponding to the local surface conditions, while in layers away from the wall the shearing stress, which equals the Reynolds stress $\tau = -\varrho\overline{u'v'}$ here, changes very slowly. In fact, a new state of equilibrium is established only at rather long distances x, measured from the start of the rough surface. Although the experiment is for channel flow, the basic phenomenon applies also to boundary layers.

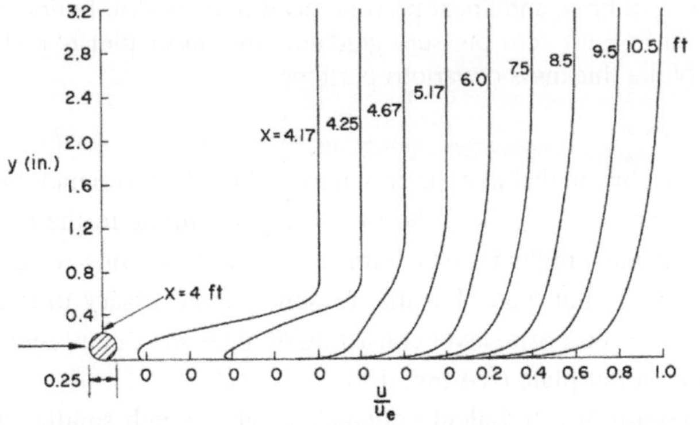

Fig. 4.1 Response of a turbulent boundary layer to wall disturbances. Mean velocity distribution of a turbulent boundary layer on a flat plate behind a cylindrical rod in contact with the surface at $x = 4$ft from the leading edge [1].

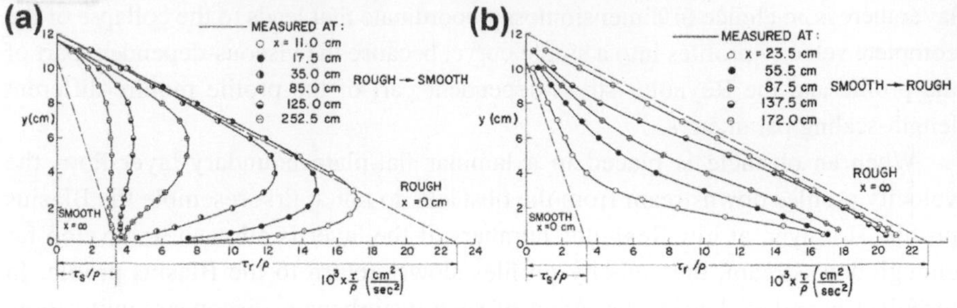

Fig. 4.2 Variation of the shearing-stress distribution in turbulent flow through a rectangular channel. Flow passes from (a) a rough surface to a smooth one and (b) a smooth to a rough one [2].

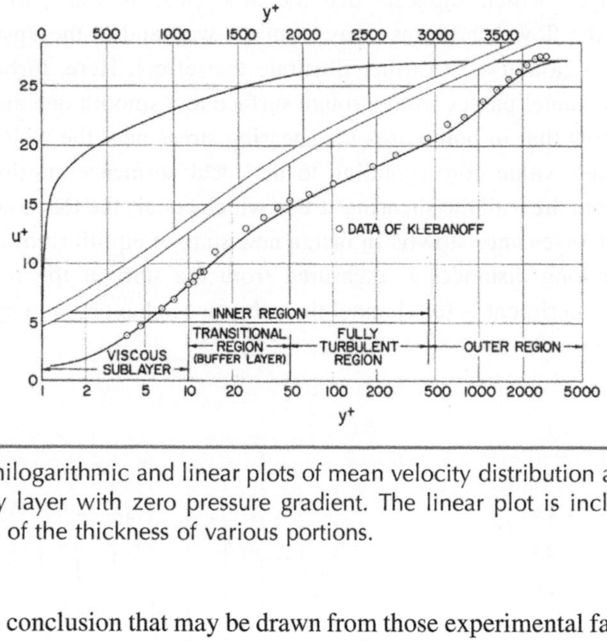

Fig. 4.3 Semilogarithmic and linear plots of mean velocity distribution across a turbulent boundary layer with zero pressure gradient. The linear plot is included to show a true picture of the thickness of various portions.

A general conclusion that may be drawn from those experimental facts is that it is fundamentally impossible to describe the flow phenomena in the entire boundary layer in terms of one single set of parameters, as can be done for certain laminar layers, especially the flat plate. For that reason, it is necessary to treat a turbulent boundary layer as a composite layer consisting of inner and outer regions, even when the flow is along a flat plate (see Fig. 4.3).

The inner region of a turbulent boundary layer is much smaller than the outer region, with thickness about 10 to 20% of the entire boundary-layer thickness.

It is generally assumed that the mean-velocity distribution in this region is completely determined by the wall shear τ_w, density ϱ, viscosity μ, and the distance y from the wall. It is given by the following expression known as the *law of the wall*:

$$u^+ \equiv u/u_\tau = \phi_1(y^+). \tag{4.2.1}$$

Here, $u_\tau = (\tau_w/\varrho)^{1/2}$ is a factor having the dimensions of velocity and hence called the friction velocity. The parameter y^+, which is a Reynolds number based on typical velocity and length scales for the turbulence, is defined as $y^+ = yu_\tau/\nu$.

In the case of the mean-temperature distribution in the inner region of a turbulent boundary layer, the version of the law of the wall for the temperature in incompressible turbulent flows is

$$\frac{T_w - T}{T_\tau} \equiv T^+ = \phi_2(y^+, \mathrm{Pr}). \tag{4.2.2}$$

Here T_τ, with $\dot{q}_w$ denoting the wall heat flux,

$$T_\tau = \frac{\dot{q}_w}{\varrho c_p u_\tau} \tag{4.2.3}$$

is called the friction temperature by analogy with the friction velocity u_τ.

The mean-velocity distribution $\phi_1(y^+)$ and mean-temperature distribution $\phi_2(y^+, \mathrm{Pr})$ depend on the condition of the wall. In Sections 4.4–4.6 we shall derive $\phi_1(y^+)$ for smooth walls, for rough walls, and for porous walls and derive $\phi_2(y^+, \mathrm{Pr})$ for smooth and rough walls. The inner region can be divided into three layers as indicated in Fig. 4.3: (1) the viscous sublayer, (2) the transitional region (sometimes called the buffer region), and (3) the fully turbulent region. In the viscous sublayer, the stresses are mainly viscous, since turbulent fluctuations, like mean velocities, become zero at the wall. The predominantly viscous region is uniform neither according to time nor according to distance along the wall. The great nonuniformity was clearly shown in Fig. 1.3. But, at any section, a time-mean value of the thickness of the region may be distinguished. We shall denote the thickness by y_s. Thus, for $y < y_s$, the flow may be assumed to be viscous.

In the region $y > y_s$ in Fig. 4.3, the effect of the viscosity on the flow decreases gradually with increasing distance from the wall. Ultimately, a region is reached where the flow is completely turbulent and the effect of viscosity is negligibly small. The intermediate region, where the total stress is partly viscous and partly turbulent, is called the transitional region (not to be confused with the standard laminar-turbulent boundary-layer transition). If we denote the average distance from the wall beyond which the flow is fully turbulent by y_t the range of the transition region is specified by $y_s < y < y_t$. In general, the thickness of either the viscous sublayer or the transitional region is quite small in comparison with that of the fully turbulent region (see Fig. 4.3).

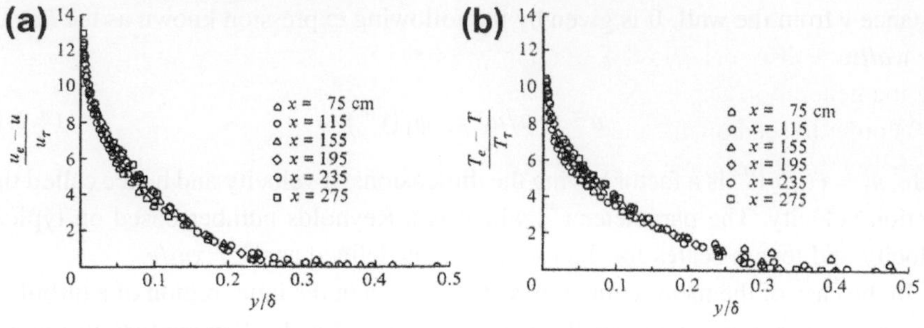

Fig. 4.4 Universal plot of turbulent (a) velocity and (b) temperature profiles in zero pressure gradient [3].

The outer region of a turbulent boundary layer contains 80–90% of the boundary-layer thickness (see Fig. 4.3). According to experiments, the mean velocity distribution in the outer region can be described by the following expression, known as the *velocity-defect law* (see Fig. 4.4a):

$$(u_e - u)/u_\tau = f(y/\delta). \tag{4.2.4}$$

This form, originally written by Darcy was soon forgotten. Much later, von Kármán [4] rediscovered it and gave it permanent importance. Equation (4.2.4) is not valid close to the wall, since there the viscosity becomes important and therefore the flow must depend on a Reynolds number ($\delta u_\tau/\nu$) as well as the ratio y/δ. Obviously, at the top of the boundary layer, when y approaches δ, the function $f(y/\delta)$ goes to zero. For flat plates and pipes, the function f has been found empirically to be independent of Reynolds number and, most significantly, of the roughness of the wall. For boundary layers on flat plates, the function f is numerically different from that for pipe flow, owing mainly to the presence of the free outer boundary. It is also markedly affected by streamwise pressure gradient, and except for specially tailored pressure gradients of which zero is one, f depends on x (see Section 4.4.5).

Similar to the representation of the mean velocity distribution in the outer region of a turbulent boundary layer, the mean-temperature distribution can be represented by

$$\frac{T_e - T}{T_\tau} = F_2(y/\delta) \tag{4.2.5}$$

which is independent of Prandtl number (see Fig. 4.4b).

According to experimental observations, as the free stream is approached, the flow at a given point becomes intermittently turbulent. Such an on-and-off character of turbulence is also observed in wake and jet flows. Figures 1.4, 1.5, 1.20, and 4.5 show the sharp boundary between a turbulent and a nonturbulent flow.

Although the behavior of turbulent flow in the inner and the outer regions of the layer is quite different, those regions are strongly coupled by the shear-stress profile and the general diffusivity of the turbulence.

In order to see how the interference takes place, it is useful to study the transport of energy in a turbulent boundary layer. That can be done by considering the mean-kinetic-energy equation (3.4.3) and the turbulent-kinetic-energy equation (3.5.6b). For a steady, two-dimensional, incompressible flow with zero pressure gradient, they can be written as follows:

Mean-Kinetic-Energy Equation

$$u\frac{\partial}{\partial x}\left(\frac{u^2}{2}\right)+v\frac{\partial}{\partial y}\left(\frac{u^2}{2}\right) = \frac{1}{\varrho}\frac{\partial}{\partial y}\left[u\left(\mu\frac{\partial u}{\partial y}-\varrho\overline{u'v'}\right)\right]+\overline{u'v'}\frac{\partial u}{\partial y}, \tag{4.2.6}$$

Turbulent-Kinetic-Energy Equation

$$u\frac{\partial}{\partial x}\left(\frac{\overline{q^2}}{2}\right)+v\frac{\partial}{\partial y}\left(\frac{\overline{q^2}}{2}\right) = -\overline{u'v'}\frac{\partial u}{\partial y}+\nu\frac{\partial^2}{\partial y^2}\left(\frac{\overline{q^2}}{2}+\overline{v'2}\right)-\frac{\partial}{\partial y}\left(\frac{\overline{v'q^2}}{2}+\frac{\overline{pv'}}{\varrho}\right)-\varepsilon. \tag{4.2.7}$$

Here, for simplicity we have again omitted the bars from the mean quantities. We note that the mean-dissipation term $\mu(\partial u/\partial y)^2$ does not appear in Eq. (4.2.6), since in incompressible flows the term is small compared to the rest of the terms. Furthermore, except for the turbulence-dissipation term ε, the two viscosity terms $(\partial/\partial y)$ $[u\mu(\partial u/\partial y)]$ and $\nu(\partial^2/\partial y^2)$ $[(\overline{q^2}/2)+\overline{v'2}]$ in Eqs. (4.2.6) and (4.2.7), respectively, are small outside the sublayer and can therefore be neglected. Then for $y > \delta_s$ where δ_s is the sublayer thickness, Eq. (4.2.6) reduces to

$$u\frac{\partial}{\partial x}\left(\frac{u^2}{2}\right)+v\frac{\partial}{\partial y}\left(\frac{u^2}{2}\right)-\overline{u'v'}\frac{\partial u}{\partial y}+\frac{\partial}{\partial y}(u\overline{u'v'}) = 0, \tag{4.2.8}$$

and Eq. (4.2.7) reduces to

$$u\frac{\partial}{\partial x}\left(\frac{\overline{q^2}}{2}\right)+v\frac{\partial}{\partial y}\left(\frac{\overline{q^2}}{2}\right)+\overline{u'v'}\frac{\partial u}{\partial y}+\frac{\partial}{\partial y}\left[\overline{v'\left(\frac{q^2}{2}+\frac{p}{\varrho}\right)}\right]+\varepsilon = 0. \tag{4.2.9}$$

Figure 4.6a shows the distribution of the three terms of Eq. (4.2.8). The experimental data are due to Klebanoff [5]. The figure shows that the loss of mean kinetic

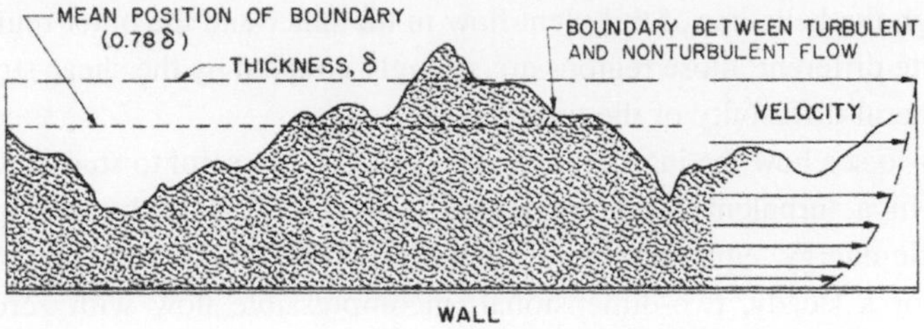

Fig. 4.5 Sketch of the turbulent boundary layer. At times the uncontaminated potential flow may extend far into the boundary layer, as shown [5].

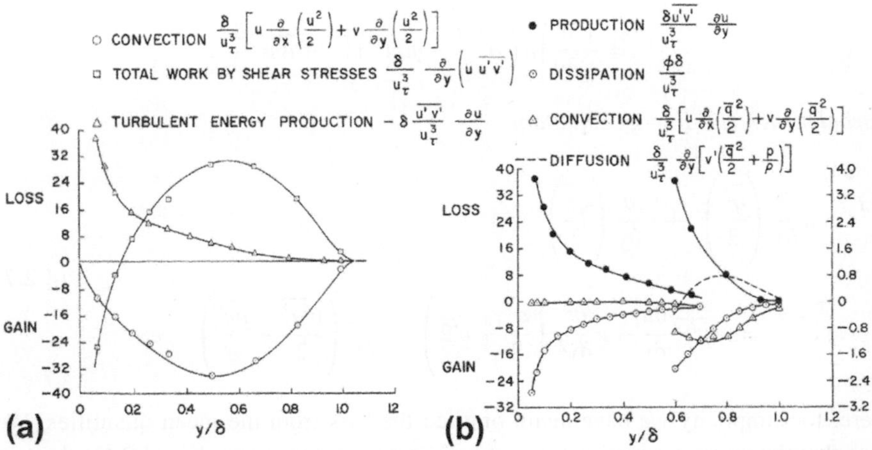

Fig. 4.6 (a) Balance of the kinetic energy of the mean flow in the boundary layer in zero pressure gradient according to Eq. (4.2.8), from the experimental data of Klebanoff [5]. (b) Balance of turbulent energy in the boundary layer according to Eq. (4.2.9), from the experimental data of Klebanoff [5]. The left-hand ordinate denotes the scale for the inner region of the boundary layer, the right-hand ordinate denotes that for the outer region.

energy is considerable, except in the region close to the wall, where turbulent energy production is most intense. The kinetic energy lost by the mean flow in the outer region is transferred by the working of the mean flow against the Reynolds stress to the inner region, where it appears as energy of the turbulent motion.

Figure 4.6b shows the variation of the four terms of Eq. (4.2.9), in dimensionless quantities, as calculated from the measurements of Klebanoff [5]. It is seen from the experimental plots that, in the inner region close to the wall, the dominant terms in Eq. (4.2.9) are the one that corresponds to the production of kinetic energy due to turbulence through action of the Reynolds stresses $(\delta\overline{u'v'}/u_\tau^3)$ $(\partial u/\partial y)$ and the one

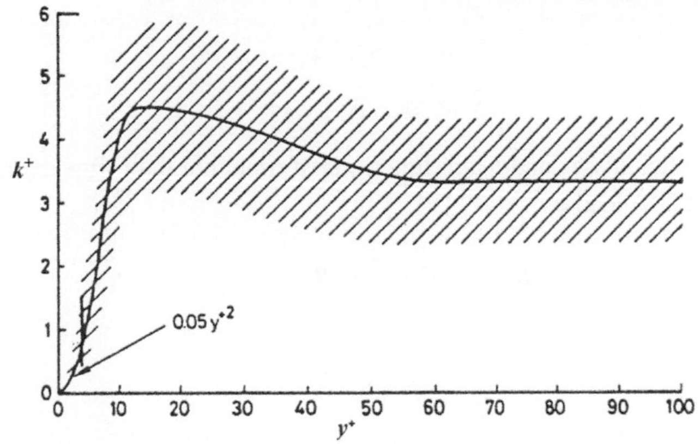

Fig. 4.7 Near-wall variation of k^+ with y^+. The equation is Eq. (4.2.11) with $A^+ = 0.05$.

that corresponds to the dissipation of energy due to viscosity, $\varepsilon\delta/u_\tau^3$. The contributions for the total energy balance are nearly equal and opposite. In addition, it may be observed that, in general, throughout the boundary layer the absolute values of these two terms are greater than the values of other contributions, except near the edge.

Consequently it is apparent that, within the inner region of the boundary layer, the energy-exchange processes are in a state of near equilibrium, with the result that the local production of energy and the local dissipation of energy are almost compensating. For that reason, the flow in the inner region is governed essentially by the local conditions.

In the outer region of the boundary layer, however, the dominant terms are those due to convection and those due to dissipation – both of which have the same sign – and the balancing contribution for the diffusion term. In that region, the turbulent-flow phenomena depend not only on the local conditions but also on the whole history of events in the flow upstream of the point in question.

Figure 4.7 shows the variation of dimensionless turbulent kinetic energy k^+ $(\equiv k/u_\tau^2)$ with y^+ according to the data compiled by Coles [6] and the data of El Telbany and Reynolds [7] as reported in [8]. In spite of the larger scatter, we see that k^+ becomes maximum around $y^+ = 15$ which corresponds to the location of the maximum production of k [9]. A representative peak value for k^+ is 4.5. In the interval $60 < y^+ < 150$, k^+ becomes nearly constant with a value of 3.3. Since in the log law region of a flat-plate boundary layer, the shear stress $-\overline{u'v'} \propto u_\tau^2$, the data suggest a value of around 0.30 for the ratio $-\overline{u'v'}/k$.

The variation of k in the immediate vicinity of the wall can be deduced from the continuity equation and the no-slip condition. Following [10], the variation of u', v' and w' with distance from the wall can be written in the form

$$u' = a_1 y + b_1 y^2 + \cdots \tag{4.2.10a}$$

$$v' = b_2 y^2 + \cdots \tag{4.2.10b}$$

$$w' = a_3 y + b_3 y^2 + \ldots \tag{4.2.10c}$$

where the coefficients a_i, and b_i are functions of time, but their time average is zero. Equations (4.2.10) lead to

$$K^+ = A^+ y^{+2} + B^+ y^{+3} + \ldots \tag{4.2.11}$$

where

$$A^+ = \frac{\frac{1}{2}\nu^2}{u_\tau^4}\left(\overline{a_1^2} + \overline{a_3^2}\right) \tag{4.2.12a}$$

$$B^+ = \frac{\nu^3}{u_\tau^5}\left(\overline{a_1 b_1 + a_3 b_3}\right) \tag{4.2.12b}$$

The data of Kreplin and Eckelmann [11] support Eqs. (4.2.10) and suggest a value of 0.035 for A^+. Sirkar and Hanratty [12] give 0.05 at a higher Reynolds number, while the data compilation of Derksen and Azad [13] suggests $0.025 < A^+ < 0.05$, with the higher value for higher Reynolds numbers.

Figure 4.8 shows the variation of dimensionless rate of dissipation $\varepsilon^+ \left(\equiv \frac{\nu\varepsilon}{u_\tau^4}\right)$ with y^+. The available data for ε defined by Eq. (6.1.7),

$$\varepsilon = \nu\left(\overline{\frac{\partial u_i'}{\partial x_k}\frac{\partial u_i'}{\partial x_k}}\right) \tag{6.1.7}$$

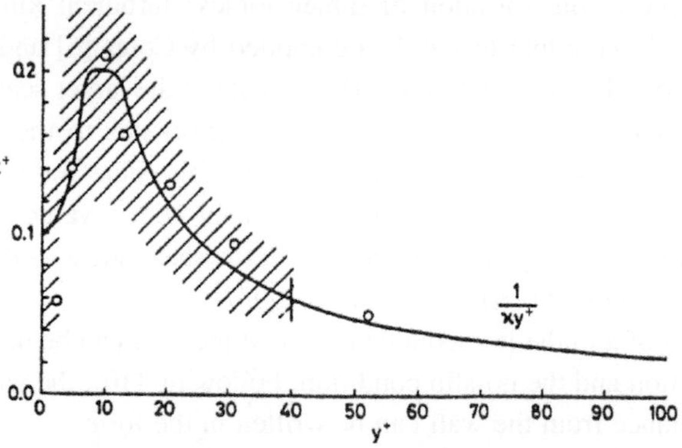

Fig. 4.8 Near-wall variation of ε^+ with y^+. The data is due to Laufer [14].

are quite limited and are subject to a large measurement uncertainty, especially in the region $0 < y^+ < 40$. Very close to the wall, the variation of ε^+ can be approximated by

$$\varepsilon^+ = 2\left(A^+ + 2B^+ y^+ + \cdots\right) \tag{4.2.13}$$

indicating a finite ε^+ at the wall value equal to $2A^+$. Experimental values of A^+ quoted earlier then indicate $0.05 < \varepsilon_w^+ < 0.10$, with a preference for the higher value at larger Reynolds numbers. If $B^+ = 0$ is assumed, Eq. (4.2.13) indicates

$$y^+ = 0, \quad \frac{\partial \varepsilon^+}{\partial y} = 0 \tag{4.2.14}$$

which can be used as a boundary condition for ε as will be discussed in Chapter 6.

4.3 Eddy-Viscosity, Mixing-Length, Eddy-Conductivity and Turbulent Prandtl Number Concepts

In order to predict the mean-velocity distribution or the mean-temperature distribution across a turbulent boundary layer, it is necessary to make an assumption for or find a model for the Reynolds stresses. Over the years, several empirical hypotheses have been used. Eddy-viscosity and mixing-length concepts are among the more popular and extensively used concepts. All these concepts relate the Reynolds stress to the local mean-velocity gradient, as will be shown in this section and later more extensively in Chapters 5 and 6. The main objection to the eddy-viscosity and mixing-length concepts is that they lack generality – they are based on local equilibrium ideas that assume the transport terms in the governing equations to be small. A more general approach, which will be discussed in Chapter 6, is to use ideas that consider the rate of change of the Reynolds stress in the governing equations. The prediction methods that use these ideas are referred to as transport-equation methods. They reduce to the methods that use eddy-viscosity or mixing-length ideas when the transport terms are small. For a detailed discussion, see Chapters 6 and 9.

Boussinesq [15] was the first to attack the problem of finding a model for the Reynolds shear stress by introducing the concept of eddy viscosity. He assumed that the turbulent stresses act like the viscosity stresses, which implies that the turbulent stresses are proportional to the velocity gradient. The coefficient of proportionality was called the "eddy viscosity" and was defined by

$$-\varrho\overline{u'v'} = \varrho\varepsilon_{\mathrm{m}}\left(\partial u/\partial y\right). \tag{4.3.1}$$

Here, ε_m like the kinematic viscosity ν, is assumed to be the product of a velocity and a length, that is,

$$\varepsilon_{\mathrm{m}} \sim \text{length} \times \text{velocity}. \tag{4.3.2}$$

The mixing-length concept was first proposed by Prandtl [16]. According to this concept the Reynolds shear stress is to be calculated from

$$-\varrho \overline{u'v'} = \varrho l^2 \left|\frac{\partial u}{\partial y}\right| \frac{\partial u}{\partial y}. \tag{4.3.3}$$

The basis of Prandtl's mixing length hypothesis is an analogy with the kinetic theory of gases, based on the assumption that turbulent eddies, like gas molecules, are discrete entities that collide and exchange momentum at discrete intervals.

By Eq. (4.3.1) we can write a relationship between eddy viscosity and mixing length:

$$\varepsilon_{\mathrm{m}} = l^2 \left|\frac{\partial u}{\partial y}\right|. \tag{4.3.4}$$

The length l *defined* by Eq. (4.3.3) is, of course, a quantity whose value is yet to be found. According to von Kármán's hypothesis [4], l is given by

$$l = \kappa \left|\frac{\partial u/\partial y}{\partial^2 u/\partial y^2}\right| \tag{4.3.5}$$

where κ is an empirical constant known as von Kármán's constant.

Equations (4.3.1) and (4.3.3) merely represent the definitions of ε_{m} and l: the assumption is that they will vary more slowly or simply than the shear stress and therefore be easier to correlate empirically. The use of eddy viscosity to predict the complete Reynolds-stress tensor is discussed later. When the characteristic length scale is readily identified and defined, for example in a jet or wall boundary layer with a moderate pressure gradient, the simplicity of the mixing-length or eddy-viscosity approach is commendable and, since it can readily be specified in algebraic form, the equations for continuity and momentum can be written with the same number of unknowns as equations. In more complex, or rapidly-changing flows, the algebraic correlations are inadequate. One alternative is to use the transport equation for l or ε_{m}. In the so-called "one-equation" models, ε_{m} is expressed as

$$\varepsilon_{\mathrm{m}} = C_\mu k^{1/2} l \tag{4.3.6}$$

where l is a length scale, still related to the shear layer thickness, and k is the turbulent kinetic energy for which a modeled partial-differential "transport" equation is solved. A more general and more popular approach requires a transport equation for l, often in the form of the rate of turbulence dissipation

$$\varepsilon = \frac{k^{3/2}}{l}. \tag{4.3.7}$$

The solution of an equation for ε, together with a transport equation for k and the eddy-viscosity expression, Eq. (4.3.6), forms the basis of the ubiquitous two-equation models suggested by, for example, Jones and Launder [17] as we shall discuss in Chapter 6. Note that this procedure simply serves to define

$$c_\mu \equiv \frac{-\overline{u'v'}}{\frac{\partial u}{\partial y}} \frac{\varepsilon}{k^2}. \tag{4.3.8}$$

However, the k-ε model is sufficiently realistic that $c_\mu = \text{constant} \simeq 0.09$ gives good predictions in many flows as we shall see in Chapter 9.

Analogous quantities can be defined for turbulent heat-transfer rates. Again using Boussinesq's eddy conductivity concept, we can write the transport of heat due to the product of time mean of fluctuating enthalpy h' and fluctuating velocity v' in the form

$$-\varrho\overline{v'h'} = \varrho\varepsilon_h\frac{\partial h}{\partial y} \tag{4.3.9a}$$

or for a perfect gas

$$-\varrho c_p\overline{T'v'} = \varrho c_p\varepsilon_h\frac{\partial T}{\partial y}. \tag{4.3.9b}$$

Note that a minus sign still appears; in heat transfer, as in momentum transfer, we expect transport *down* the gradient of the quantity in question. The eddy conductivity has the same dimensions as the eddy viscosity, namely, velocity × length.

Sometimes it has been found to be convenient to introduce a "turbulent" Prandtl number $\Pr_t$ defined by

$$\Pr_t \equiv \frac{\varepsilon_\text{m}}{\varepsilon_h} = \frac{\overline{u'v'}/\frac{\partial u}{\partial y}}{\overline{T'v'}/\frac{\partial T}{\partial y}}. \tag{4.3.10}$$

Each of the relations given by Eqs (4.3.1), (4.3.3) and (4.3.9) requires some empirical values if it is to be used for quantitative calculations. In other words, it is necessary to make assumptions for the distribution of eddy viscosity and conductivity. We shall postpone the discussion on the distribution of eddy conductivity to Chapter 5 and here concentrate on the distributions of eddy-viscosity and mixing-length.

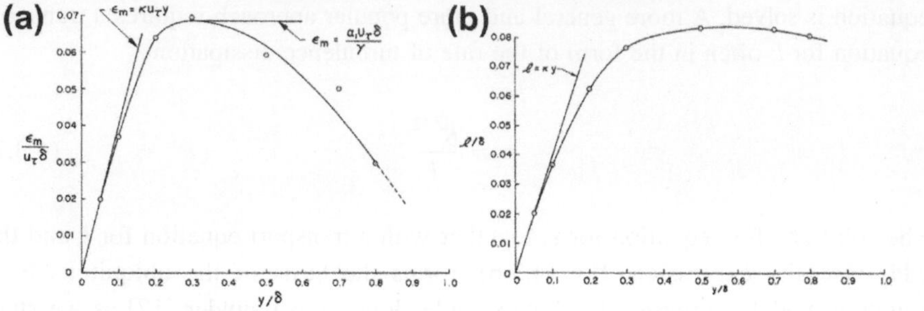

Fig. 4.9 Dimensionless (a) eddy-viscosity and (b) mixing-length distributions across a turbulent boundary layer at zero pressure gradient, according to the data of Klebanoff [5].

Figures 4.9a and 4.9b show such distributions on a flat plate according to the measurements of Klebanoff [5]. The results shown in these figures indicate that in the region $0 < y/\delta < 0.15$–0.20, the eddy viscosity and mixing length vary linearly with distance y from the wall. Both variables appear to have a maximum value anywhere from $y/\delta = 0.20$ to 0.30. Consequently, in this inner region the eddy viscosity and mixing length can be approximated by

$$\varepsilon_{\mathrm{m}} = \kappa u_\tau y, \tag{4.3.11}$$

$$l = \kappa y, \tag{4.3.12}$$

where κ is a universal constant, experimentally found to be in the region of 0.40–0.41. For y/δ greater than approximately 0.20, the eddy viscosity begins to decrease slowly, but the mixing length remains approximately constant, so it can be approximated by

$$l/\delta = \text{const}, \tag{4.3.13}$$

where the constant varies from 0.075 to 0.09, depending on the definition of boundary-layer thickness δ.

As the free stream is approached, the turbulence becomes intermittent; that is, for only a fraction γ of the time is the flow turbulent. The same phenomenon has also been observed in other shear flows that have a free boundary. The on-and-off character of the turbulence is the reason for the irregular outline of the turbulent boundary layer shown in Figs. 4.5 and 1.20. The intermittency is easily observed in oscilloscope records of the u' fluctuation in the outer region of the boundary layer, and the records can be used both to give a quantitative estimate of the factor γ and to discern some qualitative aspects of the flow. Representative sections of oscilloscope records taken at various positions across the boundary layer obtained on a flat plate by Klebanoff [5] are shown in Fig. 4.10a. It can be seen that, in the outer region of the

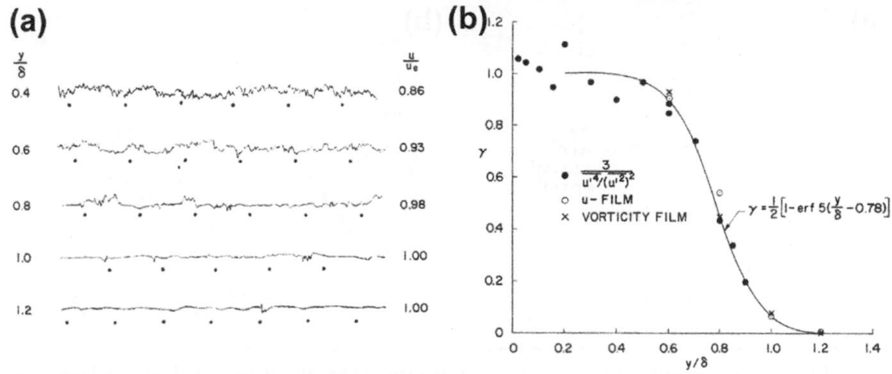

Fig. 4.10 (a) Instantaneous velocity u in a boundary layer, $u_e = 50$ ft/sec, timing dots 60/sec and (b) intermittency distribution across a turbulent boundary layer. Data represent three different techniques of measurement.

layer, $y/\delta > 0.4$, there are intervals of time when the flow is not turbulent and that these intervals become longer with increasing distance from the wall. Thus, the outer region is divided into a turbulent part and a relatively nonturbulent free-stream part, and the hot wire at a given position responds to alternate turbulent and nonturbulent flow as the pattern is swept downstream.

Intermittency factors have been obtained by Klebanoff [5] and by Corrsin and Kistler [18]. Figure 4.10b shows the distribution of intermittency factor γ according to Klebanoff's measurements for a flat-plate flow. It can be fitted approximately by the expression (see Section 1.7)

$$\gamma = \frac{1}{2}\{1 - \text{erf } 5[(y/\delta) - 0.78]\}. \quad (4.3.14)$$

If the distribution of eddy viscosity is corrected for the effect of intermittency, the dimensionless eddy viscosity $\varepsilon_m/u_\tau\delta$ becomes nearly constant across the main outer part, as is shown in Fig. 4.9a. It can be approximated by

$$\varepsilon_m = \alpha_1 u_\tau \delta, \quad (4.3.15a)$$

where α_1 is an experimental constant between 0.06 and 0.075.

It should be pointed out that the length and the velocity scales used to normalize the eddy viscosity in Fig. 4.9a are not the only possible characteristic scales. Other length and velocity scales such as δ^* and u_e, respectively, can also be used. Equation (4.3.15a) then can also be written in the form

$$\varepsilon_m = \alpha u_e \delta^*, \quad (4.3.15b)$$

where α is a constant between 0.016 and 0.02. For equilibrium boundary layers (see Section 4.4.5), the two expressions for ε_m can be shown to be the same.

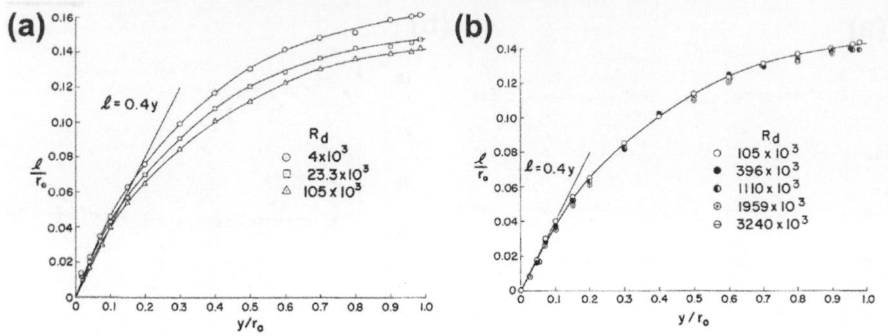

Fig. 4.11 Mixing-length distribution in a tube for (a) moderate Reynolds numbers and (b) high Reynolds numbers, $R_d > 105 \times 10^3$ according to the measurements of Nikuradse [19].

In Fig. 4.11 we show the mixing-length distribution in pipe flow according to the measurements of Nikuradse [19]. From those distributions we see that although at low Reynolds number the mixing-length distribution across the pipes varies, at high Reynolds numbers it does not. For high Reynolds numbers, the mixing-length distribution can be expressed with good approximation by the following equation:

$$l/r_0 = 0.14 - 0.08[1 - (y/r_0)]^2 - 0.06[1 - (y/r_0)]^4, \tag{4.3.16a}$$

where y denotes the distance from the wall and r_0 the radius of the pipe. Developing l as a series gives

$$l = 0.4y - 0.44(y^2/r_0) + \dots. \tag{4.3.16b}$$

Figure 4.12 shows the eddy-viscosity distribution in pipe flow according to the measurements of Laufer [14] and Nunner [20]. The experimental data indicate that for the core region of the pipe flow the distribution of the eddy viscosity resembles, both qualitatively and quantitatively, the corresponding distribution for the fully turbulent portions of the outer region of the boundary-layer flow. The eddy viscosity first increases linearly with y/r_0, then reaches a maximum at about $y/r_0 = 0.3$, and finally decreases slightly, becoming nearly constant at $y/r_0 = 0.5$.

4.4 Mean-Velocity and Temperature Distributions in Incompressible Flows on Smooth Surfaces

The momentum equation for a two-dimensional, incompressible, turbulent boundary layer with zero pressure gradient can be written as [see Eq. (3.3.10)]

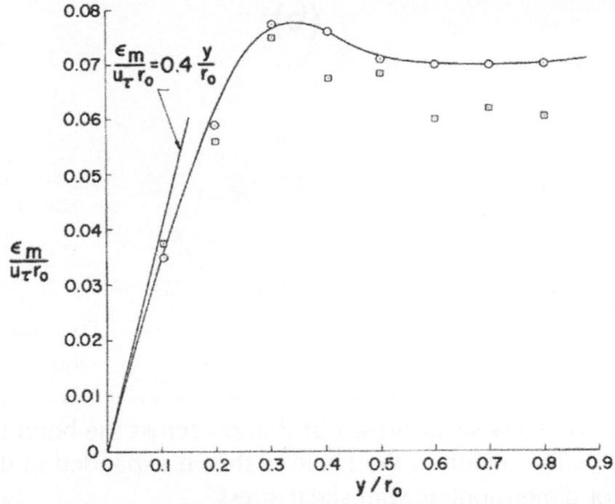

Fig. 4.12 Eddy-viscosity distribution in pipe flow according to the measurements of Laufer [14] (○) and Nunner [20] (□).

$$u\frac{\partial u}{\partial x} + v\frac{\partial u}{\partial y} = \frac{1}{\varrho}\frac{\partial \tau}{\partial y}, \tag{4.4.1}$$

where τ is the total shear stress,

$$\tau = \tau_l + \tau_t = \mu\frac{\partial u}{\partial y} - \varrho\overline{u'v'}. \tag{4.4.2}$$

For a nonporous surface, $(\partial\tau/\partial y)_w = 0$, since u and v are zero at the wall. Furthermore, by using the equation of continuity and the no-slip condition $u_w = 0$, and by differentiating Eq. (4.4.1) with respect to y, it can be shown that $(\partial^2\tau/\partial y^2)_w = 0$. Hence for some small distance from the wall we can write

$$\partial\tau/\partial y \approx 0. \tag{4.4.3}$$

That equation shows that the total shear stress τ is constant. Experiments support that relationship. Figure 4.13a shows the distribution of dimensionless Reynolds shear-stress term, $2\tau_t/\varrho u_e^2 = -2\overline{u'v'}/u_e^2$, across the boundary layer for a flat-plate flow, as measured by Klebanoff [5]. From a point very close to the wall to $y/\delta = 0.1$–0.2, the turbulent shear stress is approximately constant. As the wall is approached, the turbulent shear stress goes to zero, as shown by the experimental data of Schubauer [21] in Fig. 4.13b. In the region where turbulent shear stress begins to decrease, however, the laminar shear stress begins to increase in such a way that the total shear-stress distribution is still constant in the region $0 \le y < 0.1$–0.2, which is as it should

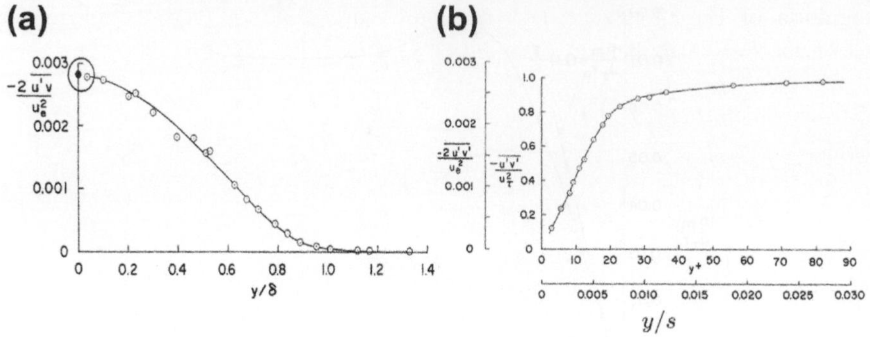

Fig. 4.13 (a) Dimensionless shear-stress distribution across the boundary layer at zero pressure gradient [5]. The region in the circle is shown expanded in (b). The solid dot denotes the value of dimensionless wall shear stress.

be, according to Eq. (4.4.3). Figure 4.13a also shows the value of the dimensionless laminar shear stress at the wall where $\tau_t = 0$:

$$\frac{2\tau_w}{\varrho u_e^2} = \frac{2\tau_l}{\varrho u_e^2} = \frac{2\nu}{u_e^2}\left(\frac{\partial u}{\partial y}\right)_w = c_f. \tag{4.4.4}$$

Integration of Eq. (4.4.3) in the region of nearly constant stress leads to $\tau = \text{const} = \tau_l + \tau_t$; hence from Eq. (4.4.2),

$$\nu\frac{du}{dy} - \overline{u'v'} = \frac{\tau}{\varrho} \approx \frac{\tau_w}{\varrho} = u_\tau^2. \tag{4.4.5}$$

From Eq. (4.4.5) it can be seen that if the variation of $-\overline{u'v'}$ with y is known or if the relationship of $-\overline{u'v'}$ to the mean flow is known, then Eq. (4.4.5) may be integrated to obtain the velocity distribution in the constant-shear-stress region.

Similarly for an incompressible flow, the energy equation can be written as

$$u\frac{\partial T}{\partial x} + v\frac{\partial T}{\partial y} = \frac{1}{\varrho c_p}\left(k\frac{\partial^2 T}{\partial y^2} - \varrho c_p \frac{\partial}{\partial y}\overline{T'v'}\right) = -\frac{1}{\varrho c_p}\frac{\partial \dot{q}}{\partial y} \tag{4.4.6}$$

where $\dot{q}$ is the total heat flux,

$$\dot{q} = \dot{q}_l + \dot{q}_t = -k\frac{\partial T}{\partial y} + \varrho c_p \overline{T'v'}. \tag{4.4.7}$$

On a smooth surface, as with the momentum equation, we can write

$$\frac{\partial \dot{q}}{\partial y} \approx 0. \tag{4.4.8}$$

Integration of Eq. (4.4.8) in the region of nearly constant heat flux leads to $q = \text{constant} = \dot{q}_l + \dot{q}_t$; hence from Eq. (4.4.7),

$$-\left(k\frac{dT}{dy} - \varrho c_p \overline{T'v'}\right) = \dot{q} \approx \dot{q}_w = T_\tau \varrho c_p u_\tau$$

or

$$\frac{\nu}{\text{Pr}\, u_\tau}\frac{dT}{dy} - \frac{\overline{T'v'}}{u_\tau} = T_\tau. \tag{4.4.9}$$

From Eq. (4.4.9) it can be seen that if the variation of $-\overline{T'v'}$ with y is known or if the relationship $-\overline{T'v'}$ to the mean temperature is known, then Eq. (4.4.9) may be integrated to obtain the temperature distribution in the constant-heat-flux region.

4.4.1 Viscous and Conductive Sublayers

The thickness of the viscous sublayer is approximately equal to 0.1–1% of the total thickness of the boundary layer. The mean-velocity distribution can be obtained from Eq. (4.4.5), which for the viscous sublayer reduces to

$$\nu(du/dy) = u_\tau^2, \tag{4.4.10}$$

since $\overline{u'v'} = 0$ at $y = 0$. Integrating that equation and expressing the result in dimensionless parameters, we obtain

$$u^+ = y^+, \tag{4.4.11}$$

which is, as it should be, a special case of Eq. (4.2.1). This is valid for y^+ less than approximately 5 (see Figs. 4.3 and 4.13b).

In the case of the temperature profile, the heat-conduction law gives, for y^+ Pr < 3 and $y^+ < 3$ (say),

$$\dot{q}_w = -k\frac{dT}{dy}$$

or

$$T_w - T = \frac{\dot{q}_w y}{k}$$

or

$$T^+ = y^+ \,\text{Pr} \tag{4.4.12}$$

which is a special case of Eq. (4.2.2).

4.4.2 Fully Turbulent Part of the Inner Region

In the part of the inner region where the flow is fully turbulent, the laminar shear stress τ_l is small compared to τ_t and can be neglected. According to Fig. 4.13b, that assumption applies approximately when $y^+ > 50$. Consequently, in the fully turbulent part of the inner region, Eq. (4.4.5) reduces to

$$\tau_t/\varrho = -\overline{u'v'} = u_\tau^2. \tag{4.4.13}$$

Substituting the expression (4.3.1) into Eq. (4.4.13) and using the eddy-viscosity relation (4.3.11), we get

$$\kappa u_\tau y(du/dy) = u_\tau^2. \tag{4.4.14}$$

Integration yields

$$u^+ = (1/\kappa)\ \ln\ y^+ + c, \tag{4.4.15}$$

where c is a constant whose value is between 4.9 and 5.5. Equation (4.4.15), called the logarithmic law for velocity, can also be derived by substituting the mixing-length formula (4.3.10), with l given by Eq. (4.3.12), in Eq. (4.4.13) and integrating the resulting expression.

Equation (4.4.15) can also be derived from Eq. (4.2.1) without using the mixing-length concept. From Eq. (4.2.1) we can write

$$\frac{du}{dy} = \frac{u_\tau^2}{\nu}\frac{d\phi_1}{dy^+}.$$

Since the right-hand side must be independent of the viscosity, it follows that

$$\frac{d\phi_1}{dy^+} = \frac{1}{\kappa y^+} \quad \text{or} \quad \frac{du}{dy} = \frac{u_\tau}{\kappa y}.$$

Integration of this expression gives Eq. (4.4.15).

An expression similar to Eq. (4.4.15) can also be derived for the temperature profile by using the eddy conductivity relations discussed in Section 4.3. It can also be obtained from Eq. (4.2.2) by writing it as

$$\frac{d}{dy}(T_w - T) = \frac{T_\tau u_\tau}{\nu}\frac{d\phi_2}{dy^+}$$

or as

$$\frac{d}{dy}(T_w - T) = \frac{T_\tau}{\kappa_h y}. \tag{4.4.16}$$

Here κ_h is an absolute constant around 0.44. Integration of Eq. (4.4.16) gives

$$\frac{T_w - T}{T_\tau} = \frac{1}{\kappa_h} \ln \frac{u_\tau y}{\nu} + c_h$$

or

$$T^+ = \frac{1}{\kappa_h} \ln y^+ + c_h \tag{4.4.17}$$

where c_h is a function of Pr.

4.4.3 Inner Region

In the transition region (buffer layer) both components of the total shear stress, namely, τ_l and τ_t, are important. Prandtl's mixing-length theory and Boussinesq's eddy-viscosity concept in their original form apply to fully turbulent flows. The flow in the buffer layer is in a state of transition. As the laminar sublayer is approached from above, the magnitude of the velocity fluctuations u', v' and, consequently, that of the turbulent shear stress, $-\varrho\overline{u'v'}$, approaches zero. As the region of fully turbulent flow is approached from below, the magnitude of the velocity fluctuations approaches the levels of the velocity fluctuations in the fully turbulent flow. So far, various assumptions have been made for the turbulent shear-stress term in Eq. (4.4.5) in order to describe the mean-velocity distribution there. Of the many proposed, one has enjoyed a remarkable success. It is the expression proposed by Van Driest [22], who assumed the following modified expression for Prandtl's mixing-length theory:

$$l = ky[1 - \exp(-y/A)], \tag{4.4.18}$$

where A is a damping-length constant defined as $26\nu(\tau_w/\varrho)^{-1/2}$. In the form given by Eq. (4.4.18), A is limited to incompressible turbulent boundary layers with negligible pressure gradient and zero mass transfer. In Chapter 5 we shall discuss its extension to turbulent boundary layers with pressure gradient and with heat and mass transfer.

If we now use Prandtl's mixing-length formula (4.3.5), together with the mixing-length expression given by Eq. (4.4.18), we can write Eq. (4.4.5) as

$$\nu(du/dy) + (\kappa y)^2[1 - \exp(-y/A)]^2 (du/dy)^2 = u_\tau^2.$$

In terms of dimensionless quantities, that equation can be written as

$$a(y^+)\left(du^+/dy^+\right)^2 + b\left(du^+/dy^+\right) - 1 = 0$$

or

$$\frac{du^+}{dy^+} = \frac{-b + (b^2 + 4a)^{1/2}}{2a},$$

where $a(y^+) = (\kappa y^+)^2 [1 - \exp(-y^+/A^+)]^2$, $A^+ = 26$, and $b = 1$. Multiplying both numerator and denominator of the du^+/dy^+ expression by $[b + (b^2 + 4a)^{1/2}]$ and formally integrating the resulting expression, we obtain

$$u^+ = \int_0^{y^+} \frac{2}{b + [b^2 + 4a(y^+)]^{1/2}} \, dy^+, \tag{4.4.19}$$

since $u^+ = 0$ at $y^+ = 0$.

Equation (4.4.19) defines a continuous velocity distribution in the inner region of the turbulent boundary layer and applies to the viscous sublayer, to the transition region, and to the region of fully turbulent flow. For example, in the viscous sublayer, $a = 0$. Then Eq. (4.4.19) reduces to the viscous-sublayer expression given by Eq. (4.4.11). In the fully turbulent region, $b = 0$ and $a = (\kappa y^+)^2$, and Eq. (4.4.19) reduces to Eq. (4.4.15), with $c = 5.24$.

Figure 4.14 shows that the mean velocity distribution calculated by Eq. (4.4.19) agrees quite well with the experimental data of Laufer [14] (□) and with the flat-plate data of Klebanoff [5] (○) and of Wieghardt [23] (△).

Equation (4.4.17), like Eq. (4.4.15) can also be extended to include the (conductive) sublayer by using a procedure similar to that used for the velocity

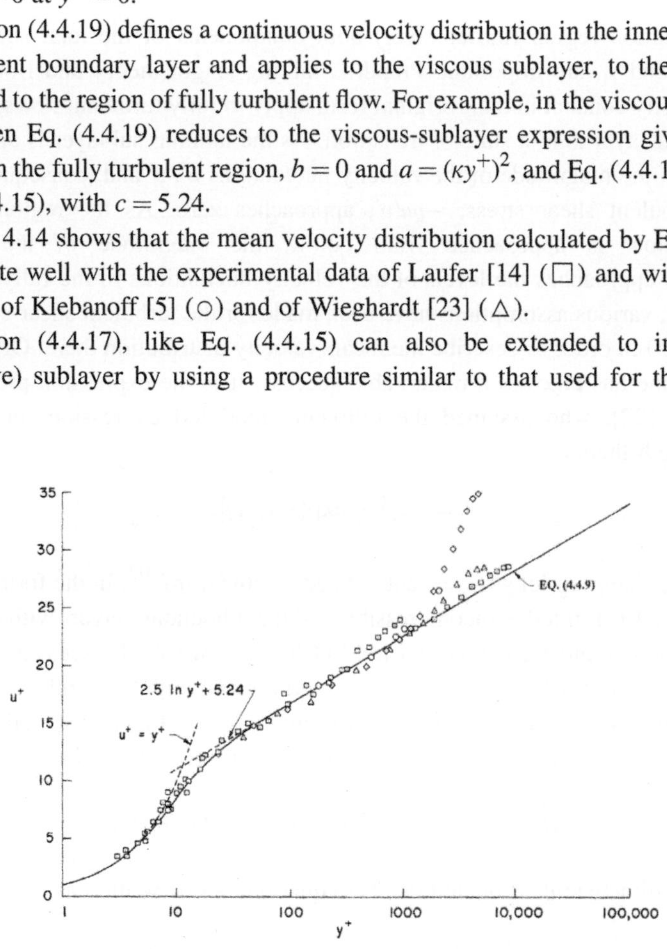

Fig. 4.14 Mean-velocity distribution in the inner region as calculated by Eq. (4.4.19).

profile. Using the definition of eddy conductivity given by Eq. (4.3.9b), we can write Eq. (4.4.9) as

$$\frac{dT^+}{dy^+} = \frac{1}{\frac{1}{\mathrm{Pr}} + \varepsilon_h^+} \tag{4.4.20}$$

where $\varepsilon_h^+ = \varepsilon_h/\nu$. To integrate this equation we need an expression for ε_h^+. Several models for ε_h^+ can be used for this purpose. According to a model developed by Cebeci [24], with turbulent number Prandtl number by

$$\mathrm{Pr_t} = \frac{\varepsilon_\mathrm{m}}{\varepsilon_h} = \frac{\kappa}{\kappa_h}\frac{1-\exp(-y/A)}{1-\exp(-y/B)}, \tag{4.4.21}$$

ε_h^+ is represented by

$$\varepsilon_h^+ = \frac{\varepsilon_\mathrm{m}^+}{\mathrm{Pr_t}} = \frac{1}{\mathrm{Pr_t}}\kappa^2 (y^+)^2 \left[1 - \exp\left(-\frac{y^+}{A^+}\right)\right]\frac{du^+}{dy^+}. \tag{4.4.22}$$

The parameter B is given by a power series in $\log_{10}$ Pr,

$$B = \frac{B^+\nu}{u_\tau}, \qquad B^+ = \frac{1}{\mathrm{Pr}^{1/2}}\sum_{i=1}^{5} C_i(\log_{10}\mathrm{Pr})^{i-1} \tag{4.4.23}$$

where $C_1 = 34.96$, $C_2 = 28.79$, $C_3 = 33.95$, $C_4 = 6.3$ and $C_5 = -1.186$. We can now integrate Eq. (4.4.20) numerically to obtain T^+ as a function of y^+ if we substitute for ε_h^+ from Eq. (4.4.22), obtaining du^+/dy^+ from Eq. (4.4.19). The variation of $\mathrm{Pr_t}$ with y^+ for a range of values of Pr is shown in Fig. 4.15, and the resulting T^+ profiles are given in Fig. 4.16.

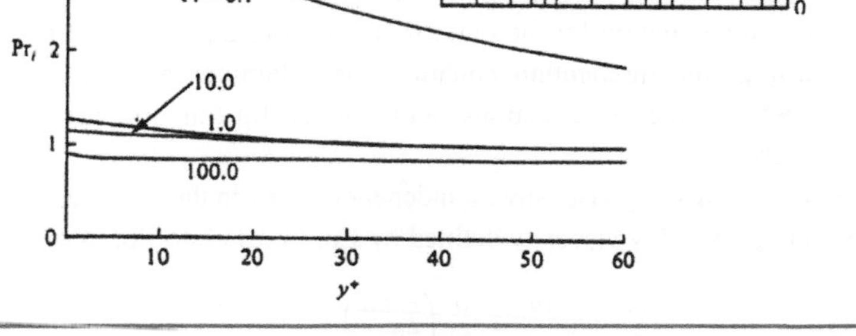

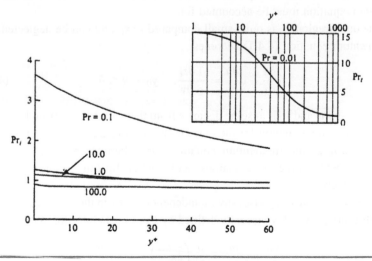

Fig. 4.15 Variation of $\mathrm{Pr_t}$ with y^+ at different values of Pr.

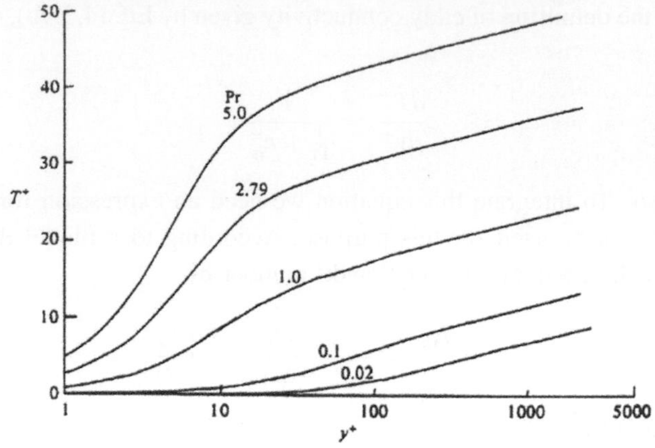

Fig. 4.16 Mean-temperature distribution across the layer as a function of Pr.

4.4.4 Outer Region

In obtaining the relation leading to Eq. (4.4.19), we have approximated the momentum equation (4.4.1) by Eq. (4.4.3), by making the assumption that u and v are small close to the wall. In other words, we have neglected the convective term in Eq. (4.4.1). According to experiments, the assumption is good only close to the wall, within approximately 20% of the boundary-layer thickness for zero-pressure-gradient flows. Since the convective term is of the same order of magnitude as the shear-stress gradient term at a greater distance from the wall, the left-hand side of the momentum equation must be accounted for.

In the outer region, τ_l is quite small compared to τ_t, and can be neglected. Then the momentum equation (4.4.1) becomes

$$u\frac{\partial u}{\partial x} + v\frac{\partial u}{\partial y} = \frac{1}{\varrho}\frac{\partial \tau_t}{\partial y}, \quad y_0 \le y \le \delta. \tag{4.4.24}$$

where $\tau_t = -\varrho\overline{u'v'}$ and y_0 is at some distance from the wall. We see from Eq. (4.4.24) that the momentum equation for the outer region of a turbulent boundary layer has the same form as the momentum equation for a laminar boundary layer. The resemblance between the two equations can be better illustrated by using the eddy-viscosity concept.

We first assume an eddy viscosity ε_m independent of y in the outer region and call it ε_0. Substituting the eddy viscosity defined by Eq. (4.3.1) into Eq. (4.4.24), we get

$$u\frac{\partial u}{\partial x} + v\frac{\partial u}{\partial y} = \frac{\partial}{\partial y}\left(\varepsilon_0\frac{\partial u}{\partial y}\right) = \varepsilon_0\frac{\partial^2 u}{\partial y^2}. \tag{4.4.25}$$

Equation (4.4.25) is identical to the laminar boundary-layer equation if we replace the kinematic eddy viscosity ε_0 by the kinematic molecular viscosity ν. The fact that the momentum equation for the outer region of a turbulent layer is similar to that for a laminar layer enables one to extract valuable information about the behavior of the mean velocity distribution of the turbulent layer from the known behavior of the laminar layer. If proper scaling variables are chosen, a close similarity between laminar and turbulent velocity profiles in the outer region can be shown to exist, as demonstrated by Clauser [25].

To illustrate, let us introduce the transformation

$$\zeta = y(u_e/\varepsilon_0 x)^{1/2}, \tag{4.4.26}$$

together with the definition of stream function $\psi(x, y)$ and the dimensionless stream function $F(\zeta \equiv \psi(x, y)\ (\varepsilon_0 u_e x)^{1/2}$ into Eq. (4.4.25). That gives the well-known Blasius equation for laminar flows, namely,

$$F''' + \frac{1}{2}FF'' = 0, \tag{4.4.27}$$

where the primes on F denote differentiation with respect to ζ. Equation (4.4.27) is subject to the following boundary conditions:

$$F(0) = 0, \qquad F'(0) = \text{const}, \qquad \lim_{\zeta \to 0} F'(\zeta) = 1. \tag{4.4.28}$$

We note that in Eq. (4.4.26) the usual kinematic molecular viscosity ν is replaced by the kinematic eddy viscosity ε_0, which is assumed to be constant.

Figure 4.17a shows the solutions of Eq. (4.4.27) for various values of $F'(0)$ as obtained by Clauser [25]. In that form of presentation, the velocity profiles do not

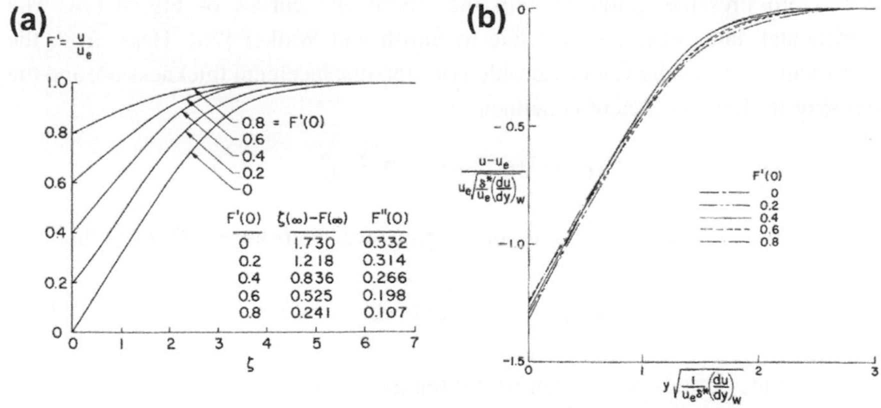

Fig. 4.17 (a) Solutions of the Blasius equation for various slip velocities and (b) replot of the solutions of the Blasius equation given in (a) in terms of new coordinates.

collapse into a single "universal" curve. However, if the solutions are to be used in describing the mean-velocity distribution in the outer region of a turbulent boundary layer, they must collapse into a universal (or nearly universal) curve, as was shown by the discussion in Section 4.2, where the velocity profiles for turbulent layers were seen to collapse into a universal curve when they were plotted as $(u_e - u)/u_\tau$ versus y/δ.

To put the "laminar" solutions in the universal form, we use $(u_e - u)/u_e$ and y/δ as variables and divide them by factors that will bring the curves of Fig. 4.17a into coincidence. The factor for y/δ is elected to make the areas above the curves equal. Two choices are available for the factor for $(u_e - u)/u_e$. Here we discuss only one. See Clauser [25] for a further discussion. By choosing the variables

$$\frac{u_e - u}{u_e\left[\left(\delta^*/u_e\right)\left(du/dy\right)_w\right]^{1/2}} \quad \text{and} \quad y\left[\frac{1}{u_e\delta^*}\left(\frac{du}{dy}\right)_w\right]^{1/2}, \tag{4.4.29}$$

the "laminar" curves of Fig. 4.17a can almost be reduced to one universal curve. Recalling the definition of δ^* and using the transformation in Eq. (4.4.26) and the fact that $F'(\zeta) = u/u_e$, we can write Eq. (4.4.29) as

$$\frac{1 - F'(\zeta)}{\left\{F''(0)[\zeta(\infty) - F(\infty)]\right\}^{1/2}} \quad \text{and} \quad \zeta\left[\frac{F''(0)}{\zeta(\infty) - F(\infty)}\right]^{1/2}. \tag{4.4.30}$$

The "laminar" curves replotted in the new coordinate system are shown in Fig. 4.17b.

We shall now compare the experimental velocity profiles for turbulent boundary layers (zero pressure gradient) with the "laminar" curves of Fig. 4.17a. The experimental data considered are due to Smith and Walker [26]. Here, from the experiment we know the velocity profile u/u_e, the displacement thickness δ^*, and the local skin-friction coefficient c_f defined as

$$c_f = 2\tau_w/\varrho u_e^2 = 2u_\tau^2/u_e^2.$$

Since we are interested only in the outer region, we see from Eq. (4.4.25) that

$$(du/dy)_w = (\tau_t/\varrho)_w(1/\varepsilon_0) = u_\tau^2/\varepsilon_0, \tag{4.4.31}$$

which with the definition of c_f can be written as

$$(du/dy)_w = (c_f/2)\left(u_e^2/\varepsilon_0\right).$$

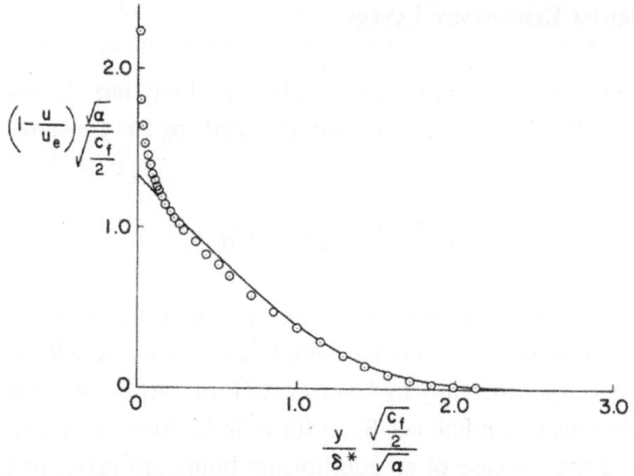

Fig. 4.18 Solutions of the Blasius equation plotted in velocity-defect form. Circles represent the data of Smith and Walker [26].

Introducing u_τ^2/ε_0 from Eq. (4.4.31) into Eq. (4.4.29) and replacing ε_0 by the value of ε_m in Eq. (4.3.15b), we obtain

$$\left(1-\frac{u}{u_e}\right)\frac{\alpha^{1/2}}{(c_f/2)^{1/2}} \quad \text{and} \quad \frac{y}{\delta^*}\frac{(c_f/2)^{1/2}}{\alpha^{1/2}}. \tag{4.4.32}$$

If a value for α is assumed, the experimental turbulent velocity profiles can be easily plotted in the coordinate system given by Eq. (4.4.32).

According to Eq. (4.4.32), it is necessary to know u/u_e, y/δ^*, and the ratio of $\alpha^{1/2}/(c_f/2)^{1/2}$, in order to use the "laminar" solutions of Eq. (4.4.27). The first two are known for a given $F'(0)$, although which value of $F'(0)$ is chosen is immaterial. The only unknown is the ratio of $\alpha^{1/2}$ to $(c_f/2)^{1/2}$ and it must be chosen to fit the experimental data.

Figure 4.18 shows a comparison of experimental velocity profiles and the solutions of the Blasius equation plotted in the coordinates given by Eq. (4.4.32). The experimental profiles of Smith and Walker [26] were plotted for $\alpha = 0.022$. The local skin-friction coefficient c_f, which was measured, was 2.1×10^{-3} at $R_{\delta^*} = 61{,}500$. The solutions of the Blasius equation correspond to the case in which $F'(0) = 0$. They[5] were plotted for a ratio of $(c_f/2)^{1/2}$ to $\alpha^{1/2}$ assumed to be 0.75. The figure shows that the calculated curves fit the experimental data quite well for the outer 80–90% of the layer. Therefore constant ε_m scaled on u_e and δ^* reproduces the observed velocity profiles in a turbulent boundary layer in zero pressure gradient.

[5]We note that for Blasius' equation $y/\delta^* = \zeta/[\zeta_\infty - F(\zeta_\infty)]$.

4.4.5 Equilibrium Boundary Layers

Equilibrium boundary layers constitute a class of boundary layers in which the mainstream velocity distribution is characterized by a constant value of the parameter

$$\beta = \left(\delta^*/\tau_w\right)\,\left(dp/dx\right), \tag{4.4.33}$$

which represents the ratio of pressure forces to shear forces in a section of the boundary layer. These flows, which are called *equilibrium* or *self-preserving* flows, were first obtained experimentally by Clauser [27] for two adverse-pressure-gradient flows, both analogous to the Falkner-Skan flows in laminar layers. A zero-pressure-gradient flow is a special case of an equilibrium boundary layer. In Clauser's equilibrium-flow experiments, long sections of two-dimensional turbulent boundary layers were subjected to various adverse pressure gradients, and by trial and error the pressure distributions were adjusted to give similar boundary-layer profiles when plotted on the basis of the velocity-defect laws. For example, at first a trial pressure distribution was adopted, and the velocity profiles at a number of x stations were measured and from them the skin-friction coefficients determined. When the results were plotted in terms of velocity-defect coordinates, that is, $(u_e - u)/u_\tau$ versus y/δ, the profiles were not similar. Thus it was necessary to alter the pressure distribution a number of times before similar profiles were obtained. When a pressure distribution was obtained for which the profiles were similar at all stations, it was found that the function f in the defect law, Eq. (4.2.4), was different from that for zero-pressure-gradient flow and was also different for each separate pressure distribution. However, the function f was the same for an arbitrary number of stations for one pressure distribution. Furthermore, it was observed that for each pressure distribution or for each equilibrium flow the parameter β remained approximately constant. From these experiments, Clauser determined that the outer part of an equilibrium boundary layer can be analyzed by assuming an eddy viscosity given by Eq. (4.3.15b).

Figure 4.19 shows the velocity profiles for the two different pressure distributions considered by Clauser [27], as well as the velocity profile for zero pressure gradient. There is a marked difference between the velocity profiles with pressure gradient and those with no pressure gradient. Furthermore, the difference increases with increasing pressure-gradient parameter β.

Equilibrium boundary layers with pressure gradient have also been measured by Bradshaw [28] and Herring and Norbury [29]. Bradshaw [28] measured equilibrium boundary layers in mild positive pressure gradient. In his experiment, the external free stream velocity varied with x as

$$u_e \sim x^{-0.15} \text{ and } u_e \sim x^{-0.255}.$$

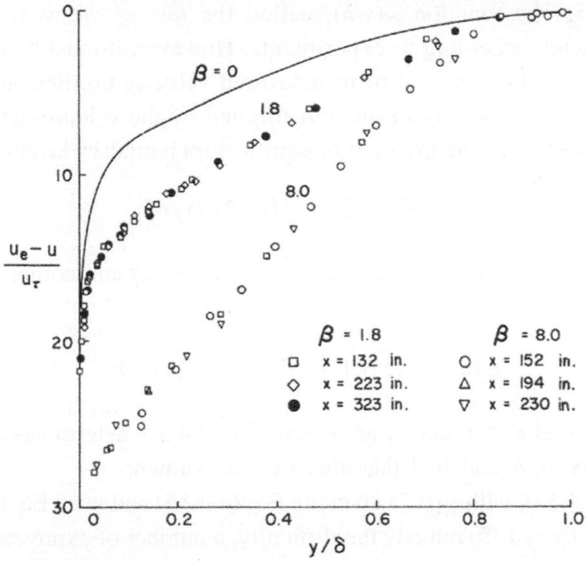

Fig. 4.19 Velocity-defect profiles for three incompressible equilibrium turbulent boundary layers. The data for $\beta = 1.8$ and 8.0 are due to Clauser [27].

Herring and Norbury [29] conducted two separate experiments, both for flows having mild negative pressure gradients. In the first experiment, the flow had a mild negative pressure gradient for which β was – 0.35. In the second, the flow had a relatively strong negative pressure gradient for which β was – 0.53.

4.4.6 Velocity and Temperature Distributions for the Whole Layer Velocity Profile

We shall now discuss a useful velocity-profile expression that can be used to predict the mean-velocity distributions in both the inner and outer regions. The expression, which was proposed by Coles [30], is

$$u^+ = \phi_1(y^+) + [\Pi(x)/\kappa] w(y/\delta). \tag{4.4.34}$$

It is applicable to flows with and without pressure gradient. If we exclude the viscous sublayer and the buffer layer, the law-of-the-wall function $\phi_1(y^+)$ is given by

$$\phi_1(y^+) = (1/\kappa) \ \ln \ y^+ + c, \ \ y^+ \geq 50. \tag{4.4.35}$$

The constants κ and c, which are independent of pressure gradient, are taken to be 0.41 and 5.0, respectively. The quantity Π is a profile parameter that is in general a function of x. The function $w(y/\delta)$, called the *law of the wake*, is of nearly universal character, according to experiments. However, it must be clearly understood that it is just an empirical fit to measured velocity profiles, and it does not imply any universal similarity of the sort implied by the velocity-defect function, Eq. (4.2.4), for zero pressure gradient or equilibrium boundary layers. It is given by

$$w(y/\delta) = 2\sin^2[(\pi/2)\,(y/\delta)]. \tag{4.4.36}$$

Evaluating Eq. (4.4.34) at the edge of the boundary layer and noting that $w(1) = 2$, we get

$$u_e/u_\tau = (1/\kappa)\ \ln\ \delta^+ + c + (2\Pi/\kappa), \tag{4.4.37}$$

where $\delta^+ = \delta u_\tau/\nu$. If κ, c, ν, and u_e are given, Eq. (4.4.37) determines any one of the three parameters u_τ, δ, and Π if the other two are known.

Equation (4.4.34), with $\phi_1(y^+)$ given by Eq. (4.4.35) and w by Eq. (4.4.36), gives $\partial u/\partial y$ nonzero at $y = \delta$. To remedy the difficulty, a number of expressions have been proposed for w. A convenient one proposed by Granville [31] uses a modification of Eq. (4.4.34) written as

$$\frac{u}{u_\tau} = \frac{1}{\kappa}\ \ln\ y^+ + c + \frac{1}{\kappa}[\Pi(1 - \cos\ \pi\eta) + (\eta^2 - \eta^3)] \tag{4.4.38}$$

From Eq. (4.4.38) and from the definitions of δ^* and θ it can be shown, provided that the logarithmic law is assumed valid to the wall, that

$$\frac{\delta^*}{\delta} = \int_0^1 \frac{u_e - u}{u_\tau}\frac{u_\tau}{u_e}\ d\eta = \frac{u_\tau}{\kappa u_e}\left(\frac{11}{12} + \Pi\right) \tag{4.4.39a}$$

$$\begin{aligned}\frac{\theta}{\delta} &= \int_0^1 \frac{u}{u_e}\left(1 - \frac{u}{u_e}\right)\ d\eta \\ &= \frac{u_\tau}{\kappa u_e}\left(\frac{11}{12} + \Pi\right) - \left(\frac{u_\tau}{\kappa u_e}\right)^2 \\ &\quad \left\{2 + 2\Pi\left[1 + \frac{1}{\pi}Si(\pi)\right] + 1.5\Pi^2 + \frac{1}{105} - \frac{7}{72} - 0.12925\ \Pi\right\}\end{aligned} \tag{4.4.39b}$$

where

$$Si(\pi) = \int_0^\pi \left[\frac{\sin\ u}{u}\right]\ du = 1.8519.$$

Equation (4.4.39b) can also be written as

$$\frac{R_\theta}{R_\delta} = \frac{u_\tau}{\kappa u_e}\left(\frac{11}{12} + \Pi\right) - \left(\frac{u_\tau}{\kappa u_e}\right)^2\left(1.9123016 + 3.0560\Pi + 1.5\Pi^2\right) \quad (4.4.39c)$$

Evaluating Eq. (4.4.38) at $\eta = 1$, we get

$$\sqrt{\frac{2}{c_f}} \equiv \frac{u_e}{u_\tau} = \frac{1}{\kappa}\left[\ln\left(\frac{\delta u_e}{\nu}\frac{u_\tau}{u_e}\right) + 2\Pi\right] + c \quad (4.4.40)$$

For given values of c_f and R_θ, Eqs. (4.4.39b) and (4.4.40) can be solved for δ and Π so that the streamwise profile u can be obtained from Eqs. (4.4.38) in the region $y^+ > 30$.

The expression (4.4.34) with $\phi_1(y^+)$ given by Eq. (4.4.35) is applicable for $y^+ \geq 30$. It can, however, be extended to include the region $0 \leq y^+ \leq 30$ by the following formula due to Thompson [32],

$$u^+ = \begin{cases} y^+, & y^+ \leq 4 \\ c_1 + c_2 \ln y^+ + c_3(\ln y^+)^2 + c_4(\ln y^+)^3, & 4 < y^+ < 30 \end{cases} \quad (4.4.41)$$

where $c_1 = 1.0828$, $c_2 = -0.414$, $c_3 = 2.2661$, $c_4 = -0.324$.

The expression (4.4.34) can also be extended to include the region $y^+ \leq 50$ by Eq. (4.4.19).

For flows with zero pressure gradient, the profile parameter Π is a constant equal to 0.55, provided that the momentum-thickness Reynolds number R_θ is greater than 5000. For $R_\theta < 5000$, the variation of Π with R_θ is as shown in Fig. 4.20. In equilibrium boundary layers, by definition, Π is constant, with its value depending

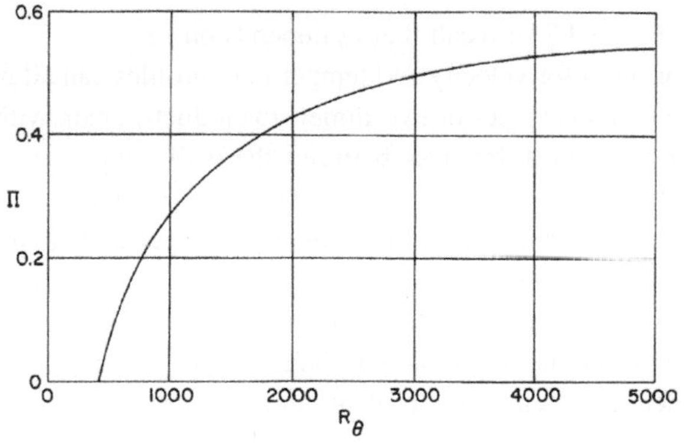

Fig. 4.20 Variation of Coles' profile parameter Π with momentum thickness Reynolds number $R\theta$ for zero-pressure-gradient flow.

on the strength of the pressure gradient. In nonequilibrium boundary layers, Π depends on x.

Temperature Profile

A form analogous to Eq. (4.4.34) can be derived for the temperature profile in a boundary layer in zero pressure gradient with uniform wall temperature or wall heat-flux rate. Again Eq. (4.4.36) provides an adequate fit to the wake function; so combining Eqs. (4.2.2) and (4.4.34), we can write

$$T^+ \equiv \frac{T_w - T}{T_\tau} = \phi_2\left(y^+, \Pr\right) + \frac{\Pi_h}{\kappa_h} w\left(\frac{y}{\delta}\right) \tag{4.4.42}$$

where Π_h is a constant differing slightly between the cases of uniform wall temperature and uniform wall heat-flux rate but in either case different from Π because $\Pr_t \neq \kappa/\kappa_h$ in the outer layer. Π_h is independent of $\Pr_t$ and Reynolds number if the Reynolds number is high. Since Π depends on Reynolds number for $u_e\theta/\nu < 5000$, we must expect Π_h to do so as well, because if viscous effects on the turbulence change the fluctuating velocity field, they will affect heat transfer as well as momentum transfer. In principle, Π_h may depend on the thermal conductivity if the Peclet number $(u_e\theta/\nu)\Pr$ is less than roughly 5000, according to the usual argument about the analogy between heat transfer and momentum transfer, but current experimental data are not sufficient to define the low-Reynolds-number and low-Peclet-number behaviors, and indeed the high-Reynolds-number value of Π_h is not known very accurately; it is about 0.3. Outside the viscous and conductive sublayers, Eq. (4.4.42) becomes

$$T^+ \equiv \frac{T_w - T}{T_\tau} = \frac{1}{\kappa_h} \ln \frac{u_\tau y}{\nu} + c_h + \frac{\Pi_h}{k_h} w\left(\frac{y}{\delta}\right) \tag{4.4.43}$$

in analogy with Eq. (4.4.34); recall that c_h depends on Pr.

The above formulas for velocity and temperature profiles can all be used in fully developed flow in circular pipes or two-dimensional ducts, again with uniform wall heat-flux rate or uniform differences between the wall temperature and the bulk-average temperature of the stream. The boundary-layer thickness is replaced by the radius of the pipe or the half-height of the duct. Values Π and Π_h are smaller than in the boundary-layer – indeed Π for a pipe flow is so small that it is often neglected. Low-Reynolds-number effects on Π and Π_h in pipe or duct flow seem to be negligible, implying that the effects in boundary layers are associated with the irregular interface between the turbulent and nonturbulent flow, the "viscous superlayer".

In arbitrary pressure gradients, Eqs. (4.4.34) and (4.4.42) are usually still good fits to experimental data but Π and Π_h depend on the pressure distribution for all positions upstream. Equation (4.4.42) is not necessarily a good fit to temperature profiles with arbitrary wall temperature or heat-flux distributions. Consider the case where a region of uniform wall heat-flux rate, in which Eq. (4.4.42) holds, is followed by a region of lower wall heat-flux rate, so that T_τ changes discontinuously while the temperature profile does *not*. Even the inner-layer formula for $T_w - T$, Eq. (4.2.2), breaks down at a step change in $\dot{q}_w$; it recovers fairly quickly, but the outer layer takes much longer.

Shear Stress Distribution

Once the velocity distribution is known, the shear-stress distribution across the boundary layer can be calculated as follows. For generality, consider a zero-pressure-gradient flow with mass transfer. First, multiply the continuity equation by u and add the resulting expression to Eq. (4.4.1) to get

$$\frac{\partial}{\partial x}(u^2) + \frac{\partial}{\partial y}(uv) = \frac{1}{\varrho}\frac{\partial \tau}{\partial y}. \tag{4.4.44}$$

Integration of Eq. (4.4.44) with respect to y yields

$$\int_0^y \frac{\partial}{\partial x}(u^2)dy + uv = \frac{1}{\varrho}(\tau - \tau_w), \tag{4.4.45}$$

and integration of the continuity equation also with respect to y yields

$$v = v_w - \int_0^y \frac{\partial u}{\partial x}\, dy. \tag{4.4.46}$$

With Eq. (4.4.46) we can write Eq. (4.4.28) in nondimensional form as

$$\int_0^y \frac{\partial g^2}{\partial x} dy + g\bar{v}_w - g\int_0^y \frac{\partial g}{\partial x}\, dy = \frac{1}{\varrho u_e^2}(\tau - \tau_w), \tag{4.4.47}$$

where $g = u/u_e$ and $\bar{v}_w = v_w/u_e$. Next, let $\eta = y/\delta$. Then, since $\partial/\partial y = (\partial/\partial\eta)(\partial\eta/\partial x) = -(\eta/\delta)\,(d\delta/dx)\,(\partial/\partial\eta)$, we can write two of the terms in Eq. (4.4.47) as follows:

$$\int_0^y \frac{\partial g^2}{\partial x} dy = -\frac{d\delta}{dx}\int_0^\eta gg'\eta\, d\eta,$$
$$\int_0^y \frac{\partial g}{\partial x} dy = -\frac{d\delta}{dx}\int_0^\eta g'\eta\, d\eta. \tag{4.4.48}$$

Integration by parts yields

$$\int_0^y \frac{\partial g^2}{\partial x} dy = -\frac{d\delta}{dx}\left(g^2\eta - \int_0^\eta g^2 d\eta\right),$$
$$\int_0^y \frac{\partial g}{\partial x} dy = -\frac{d\delta}{dx}\left(g\eta - \int_0^\eta g\, d\eta\right). \tag{4.4.49}$$

Substituting the relations given by Eq. (4.4.49) into Eq. (4.4.47) and rearranging, we get the dimensionless shear-stress distribution

$$\frac{\tau}{\tau_w} = 1 + \left(\frac{2}{c_f}\right)\left[g v_w + \frac{d\delta}{dx}\left(\int_0^\eta g^2 d\eta - g\int_0^\eta g\, d\eta\right)\right]. \tag{4.4.50}$$

Equation (4.4.50) also applies to incompressible boundary layers with pressure gradient, provided that the term $-(\delta\eta/u_e)(du_e/dx)$ is included in the bracketed expression.

To obtain an approximate expression for $d\delta/dx$, we use a power-law assumption for the velocity profiles and write the following relation, which is exact for an asymptotic layer with suction:

$$\delta/\theta = \text{const.} \tag{4.4.51}$$

Taking the derivative of Eq. (4.4.51) with respect to x gives

$$\frac{d\delta}{dx} = \frac{d\theta}{dx}\left(\frac{\theta}{\delta}\right)^{-1}. \tag{4.4.52}$$

But, by definition,

$$\frac{\theta}{\delta} = \frac{1}{\delta}\int_0^\delta \frac{u}{u_e}\left(1 - \frac{u}{u_e}\right)dy = \int_0^1 g(1-g)d\eta. \tag{4.4.53}$$

Evaluating Eq. (4.4.47) at $y = \delta$ yields the momentum integral equation

$$\frac{d\theta}{dx} = \bar{v}_w + \frac{\tau_w}{\varrho u_e^2} = \bar{v}_w + \frac{c_f}{2} = \frac{c_f}{2}(1+B), \tag{4.4.54}$$

where $B \equiv 2\bar{v}_w/c_f$. With Eqs. (4.4.53) and (4.4.54), we can write Eq. (4.4.52) as

$$\frac{d\delta}{dx} = \frac{c_f}{2}(1+B)\left[\int_0^1 g(1-g)d\eta\right]^{-1}. \tag{4.4.55}$$

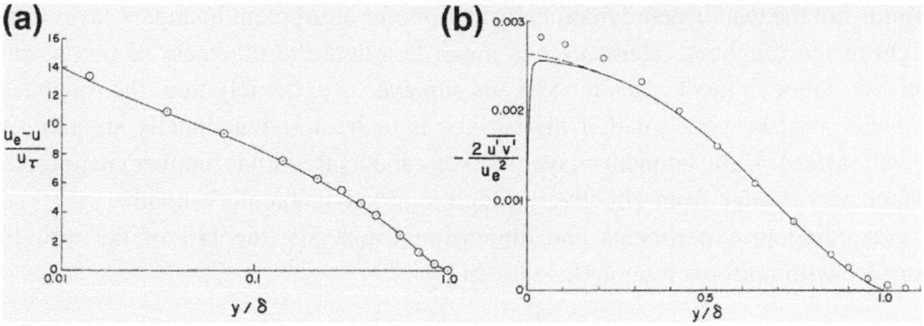

Fig. 4.21 (a) The velocity-defect law according to Coles' expression. The solid line is the calculation from Eq. (4.4.57); (b) the shear-stress distribution obtained by using Coles' velocity-profile expression. The solid line denotes the dimensionless Reynolds shear stress; the dashed line denotes the dimensionless total shear stress; circled data points are from Klebanoff [5].

Substitution of the expression (4.4.55) into Eq. (4.4.50) gives

$$\frac{\tau}{\tau_w} = 1 + Bg + (1 + B)\left(\int_0^1 g(1-g)d\eta\right)^{-1}\left(\int_0^\eta g^2 d\eta - g\int_0^\eta g\, d\eta\right). \quad (4.4.56)$$

Let us now calculate the velocity-defect and shear-stress distributions by using Coles' expression (4.4.34) and then compare them with experiment for no mass-transfer flow. Subtracting Eq. (4.4.37) from Eq. (4.4.34), we obtain the velocity defect distribution

$$(u_e - u)/u_\tau = -(1/\kappa)\ \ln\ (y/\delta) + (\Pi/\kappa)[2 - w(y/\delta)]. \quad (4.4.57)$$

Figure 4.21 shows the calculated distributions and the experimental values of Klebanoff [5]. In both cases, the agreement is excellent.

4.5 Mean-Velocity Distributions in Incompressible Turbulent Flows on Rough Surfaces with Zero Pressure Gradient

In the previous section, we have described the mean-velocity and shear-stress distributions on smooth surfaces with zero pressure gradient. In this section, we shall discuss the effect of wall roughness on mean-velocity distribution. As was discussed in Section 4.2, the velocity-defect law, Eq. (4.2.4), is valid for both smooth and rough surfaces. Roughness affects only the inner region, and hence we shall direct the discussion to that region. It is of course impossible to make a surface absolutely smooth, but the wall is aerodynamically smooth for a turbulent boundary layer if the height of the roughness elements k is much less than the thickness of the viscous sublayer. Since in most cases the viscous sublayer is extremely thin, the roughness elements must be very small if the surface is to be aerodynamically smooth. On a given surface, as the boundary-layer thickens and its Reynolds number changes, the surface may change from effectively rough to aerodynamically smooth.

According to experiments and dimensional analysis, the law of the wall for a surface with uniform roughness is given by

$$u^+ = \phi_2(y^+, k^+). \tag{4.5.1}$$

Here k^+ is a roughness Reynolds number defined by

$$k^+ = k u_\tau / \nu. \tag{4.5.2}$$

On a given surface, the surface may change from being effectively rough to aerodynamically smooth as u_τ decreases downstream.

In the fully turbulent part of the inner region, the law of the wall for a uniform rough surface is similar to that for a smooth surface except that the additive constant c in Eq. (4.4.15) is a function of the roughness Reynolds number k^+. In that region, the law of the wall can be written as

$$u^+ = (1/\kappa) \ln y^+ + B_1(k^+), \tag{4.5.3}$$

where we expect κ to be the same as for smooth surfaces. Therefore, we can write Eq. (4.5.3) as

$$u^+ = (1/\kappa) \ln y^+ + c - \left[(1/k) \ln k^+ + c - B_2\right],$$

where c is constant for a smooth surface and $B_2 = (1/\kappa) \ln k^+ + B_1 (k^+)$. Then we can write

$$u^+ = (1/\kappa) \ln y^+ + c - \Delta u^+, \tag{4.5.4}$$

where

$$\Delta u^+ \equiv \Delta u / u_\tau = (1/\kappa) \ln k^+ + B_3. \tag{4.5.5}$$

Here B_3, which is equal to $c - B_2$, is a function of roughness geometry and density.

The relation between Δu^+ and k^+ has been determined empirically for various types of roughness. The results are shown in Fig. 4.22.

We see from Eq. (4.5.4) that, since for a given roughness Δu^+ is known, the sol-e effect of the roughness is to shift the intercept $c - \Delta u^+$ as a function of k^+. For values of k^+ below approximately 5, the vertical shift Δu^+ approaches zero, except for those roughnesses having such a wide distribution of particle sizes that there are some

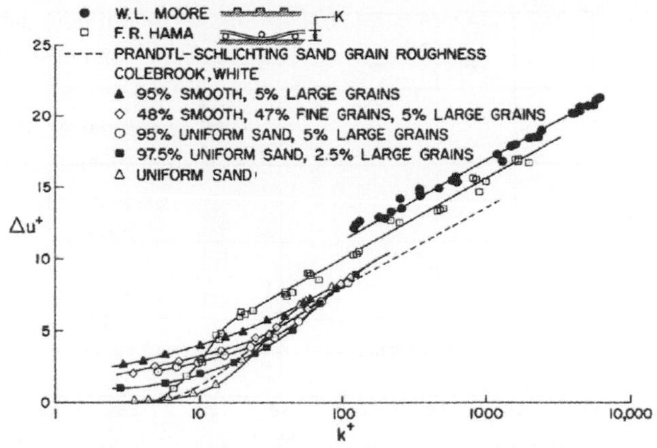

Fig. 4.22 Effect of wall roughness on universal velocity profiles [25].

particles large enough to protrude from the sublayer even though the average size is considerably less than the thickness of the sublayer. For large values of k^+, the vertical shift is proportional to $\ln k^+$, with the constant of proportionality equal to $1/\kappa$.

For surfaces covered with uniform roughness, three distinct flow regions in Fig. 4.22 can be identified. For sand-grain roughness, the boundaries of these regions are as follows:

Hydraulically smooth: $k^+ \leq 5$
Transitional: $5 \leq k^+ \leq 70$
Fully rough: $k^+ \leq 70$

The hydraulically smooth condition exists when roughness heights are so small that the roughness is buried in the viscous sublayer. The fully rough flow condition exists when the roughness elements are so large that the sublayer is completely eliminated, and the flow can be considered as independent of molecular viscosity; that is, the velocity shift is proportional to $\ln k^+$. The transitional region is characterized by reduced sublayer thickness, which is caused by diminishing effectiveness of wall damping. Because molecular viscosity still has some role in the transitional region, the geometry of roughness elements has a relatively large effect on the velocity shift, as can be seen in Fig. 4.22.

Figure 4.23 shows the variation of B_2 with k^+ according to the data of Nikuradse in sand-roughened pipes [19]. Ioselevic and Pilipenko [33] give an analytical fit to this data,

$$B_1 = 5.2; \quad k_s^+ < 2.25, \tag{4.5.6a}$$

$$B_1 = 5.2 + \left[8.5 - 5.2 - (1/\kappa)\ln k_s^+\right]\sin\left[0.4258\left(\ln k_s^+ - 0.811\right)\right];$$

$$2.25 \leq k_s^+ \leq 90, \tag{4.5.6b}$$

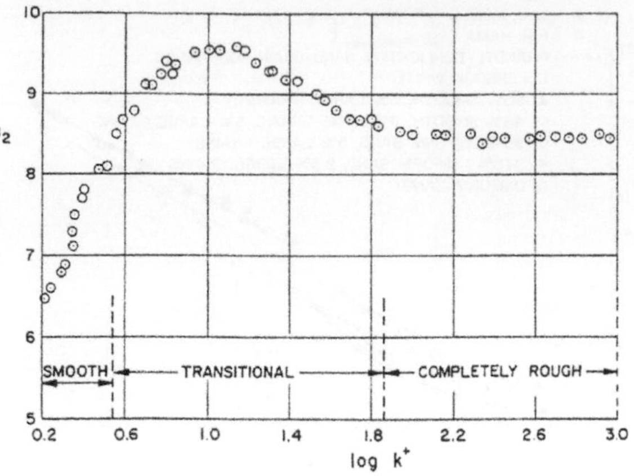

Fig. 4.23 Variation of B_2 with K^+.

$$B_1 = 8.5 - (1/\kappa)\ln k_s^+; \quad k_s^+ > 90. \tag{4.5.6c}$$

The fact that the shifts in velocity for fully rough flow are linear on the semi-logarithmic plot can be used to express different roughness geometries in terms of a reference roughness. It follows from Eq. (4.5.5) that for the same velocity shift

$$k/k_s = \exp[\kappa(B_3 - B_{3s})], \tag{4.5.7}$$

where the subscript s refers to a reference roughness, commonly taken as uniform sand-grain roughness.

Betterman [34], using two-dimensional roughness elements (rods) with varying spacing, was able to correlate his measurements in terms of Eq. (4.5.5), with the constant B_3 as a function of spacing (see Fig. 4.24). Betterman observed that for

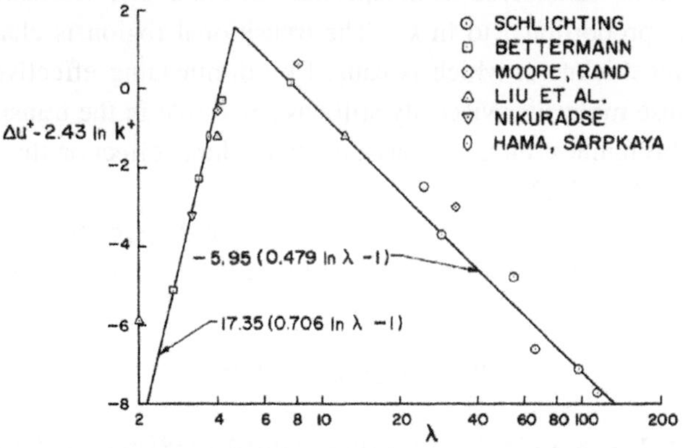

Fig. 4.24 The effect of roughness density on the law-of-the-wall intercept. The quantity λ is the ratio of the total surface area to the area covered by roughness.

a certain spacing of the rods the measured value of Δu^+ was a maximum and that as the spacing was increased or decreased Δu^+ decreased. In another series of experiments, Liu et al. [35] made the same observations.

Betterman found that in the density range $1 \leq \lambda \leq 5$, where λ is the ratio of the total surface area to the area covered by roughness, the variation of Δu^+ with roughness could be specified by

$$\Delta u^+ = 2.43 \ln k^+ + 17.35(0.706 \ln \lambda - 1), \tag{4.5.8}$$

which is plotted in Fig. 4.24.

The extension of the function Δu^+ to roughness densities greater than 5 was accomplished by Dvorak [36] from Fig. 4.24, which is based on the single set of data by Bettermann in this region and on the data obtained by Schlichting [37]. The correlation has been biased toward the two-dimensional roughness data of Bettermann, with the slope of the curve determined in conjunction with Schlichting's measurements. Dvorak's expression for Δu^+ in this region is given by the following formula:

$$\Delta u^+ = 2.43 \ln k^+ - 5.95(0.479 \ln \lambda - 1). \tag{4.5.9}$$

Numerically, Eqs. (4.5.8) and (4.5.9) are equal when the value of the density parameter λ is 4.68. It should be noted that Eq. (4.5.9), which applies only for fully rough flows, requires further verification before it can be used with confidence.

The relation given by Eq. (4.4.19) can also be extended to the prediction of the mean velocity distribution in the inner region of the boundary layer on a rough surface by using the following model discussed by Rotta [9].

In this model, the effect of roughness is considered to be equivalent to a change in the velocity jump across the viscous sublayer. Hence, it can be represented by a shift of the smooth-flow velocity profile. For rough flow, the reference plane (wall) is shifted downward by an amount Δy, and the reference plane moves with the velocity ΔU in a direction opposite to that of the main flow. Figure 4.25a shows the coordinate systems for the two flows, the smooth (y,u) and the shifted (Y,U). With the assumption that the universal law of the wall is valid for the shift, the mean-velocity distribution for the rough flow in the Y, U coordinate system is

$$U_r^+ = \phi_1(Y^+), \tag{4.5.10}$$

where the function ϕ_1 is given by the right-hand side of Eq. (4.4.19).

With the relationships $u = U - \Delta U$ and $Y = y + \Delta y$, we can write the velocity distribution for rough flow in the physical plane as

$$u_r^+ = U_r^+ - \Delta U^+ = f_1(Y^+) - f_1(\Delta y^+) = f_1(y^+ + \Delta y^+) - f_1(\Delta y^+). \tag{4.5.11}$$

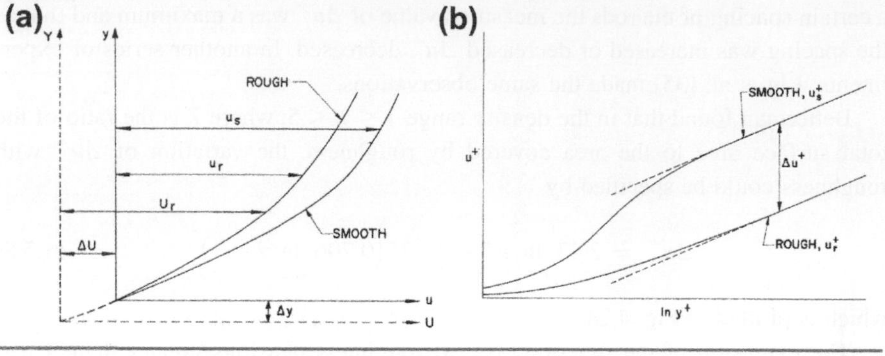

Fig. 4.25 Mean velocity distribution on smooth and rough surfaces (a) illustrating Δu and Δy shift, and (b) u_+, y^+ plot.

Obviously, the term $f_1(\Delta y^+)$ in Eq. (4.5.11) is a constant for a given Δy^+. TO determine Δy^+, we observe that for $y^+ \gg \Delta y^+$, $f(y^+ + \Delta y^+) \approx f(y^+)$. Hence, from Fig. 4.25b,

$$u_s^+ - u_r^+ = \Delta u^+ \approx f(\Delta y^+). \tag{4.5.12}$$

For a given roughness, the quantity Δu^+ is determined from experimental data as a function of k^+ (e.g., see Fig. 4.22). Our problem then is to establish a relationship between Δy^+ and k^+. For that purpose, Eq. (4.4.19) is integrated from zero to an unknown limit for a given Δu^+, thus

$$\int_0^{\Delta y^+} \frac{2}{b + [b^2 + 4a(y^+)]^{1/2}} dy^+ - \Delta u^+ = 0. \tag{4.5.13}$$

The calculated results for the case of sand-grain roughness, can be approximated by the following formula:

$$\Delta y^+ = 0.9\left[(k^+)^{1/2} - k^+ \exp\left(-k^+/6\right)\right]. \tag{4.5.14}$$

Once the shift parameter Δy^+ is known, the mean-velocity distribution for rough-wall flows can be calculated by using the relationship given by Eq. (4.5.11).

If we designate the integrand in Eq. (4.4.19) by $\phi_1(y^+)$, we can write Eq. (4.5.11) as

$$\begin{aligned} u_r^+ &= \int_0^{y^+ + \Delta y^+} \phi_1(y^+) dy^+ - \int_0^{\Delta y^+} \phi_1(y^+) dy^+ \\ &= \int_{\Delta y^+}^{y^+ + \Delta y^+} \phi_1(y^+) dy^+. \end{aligned} \tag{4.5.15}$$

If we introduce the transformation

$$y^+ = \bar{y}^+ + \Delta y^+ \tag{4.5.16}$$

into Eq. (4.5.15), we can write

$$u_r^+ = \int_0^{\bar{y}^+} \phi_1(\bar{y}^+ + \Delta y^+) d\bar{y}^+. \tag{4.5.17}$$

Comparison of Eq. (4.5.17) with the right-hand side of Eq. (4.4.19) shows that the velocity profile for rough flow can also be obtained from the smooth-flow condition by shifting the independent variable in Van Driest's formulation by Δy^+, or

$$u_r^+ = \int_0^{y^+} \frac{2\,dy^+}{1 + \left(1 + [2\kappa(y^+ + \Delta y^+)]^2\{1 - \exp[-(y^+ + \Delta y^+)/26]\}^2\right)^{1/2}}. \tag{4.5.18}$$

The approach outlined above is limited to small values of Δy^+, say less than 10, because the integrated velocity profile is slow in gaining the expected logarithmic variation for large values of Δy^+.

4.6 Mean-Velocity Distribution on Smooth Porous Surfaces with Zero Pressure Gradient

Consider an incompressible turbulent flow on a smooth porous flat surface with zero pressure gradient. Close to the wall, $\partial u/\partial x$ is small and can be neglected. The momentum equation then becomes

$$v_w \frac{du}{dy} = \frac{1}{\varrho}\frac{d\tau}{dy}, \tag{4.6.1}$$

where $\tau = \tau_l + \tau_l$. Integrating Eq. (4.6.1) and using the wall boundary condition, $\tau(0) = \tau_w,\ u(0) = 0$, we obtain

$$\tau - \tau_w + \varrho v_w u. \tag{4.6.2}$$

In the fully turbulent part of the inner region, $\tau_l \approx 0$. If τ_t is replaced by Prandtl's mixing-length expression given by Eqs. (4.3.3) and (4.3.12), integration of Eq. (4.6.2) gives

$$(2/v_w^+)\left(1 + v_w^+ u^+\right)^{1/2} = (1/\kappa)\ln y^+ + c_1, \tag{4.6.3}$$

where c_1 is an integration constant and v_w^+ is the ratio of the wall value of the normal component of velocity to the friction velocity, $v_w^+ = v_w/u_\tau$. If we subtract $2/v_w^+$ from both sides of Eq. (4.6.3), we can write

$$u_p^+ \equiv \left(2/v_w^+\right)\left[\left(1+v_w^+u^+\right)^{1/2}-1\right] = \left(1/\kappa\right)\ln y^+ + c, \tag{4.6.4}$$

where $c = c_1 - (2/v_w^+)$. Equation (4.6.4) is the law of the wall for turbulent boundary layers with mass transfer obtained by Stevenson [38].

The experimental curves for flow over both a permeable wall and an impermeable wall may now be compared on one plot if $\ln y^+$ is plotted versus u_p^+. Stevenson used this method in plotting his own experimental results for a permeable wall and found that they were close to the accepted impermeable-wall curve, Fig. 4.26. The experimental results show that the parameter c varies very little with suction or injection. Stevenson concluded that k and c in Eq. (4.6.4) were 0.41 and 5.8, respectively.

If we now write Eq. (4.6.4) for the conditions at the edge of the boundary layer, that is, $u = u_e$, $y = \delta$, we get

$$\left(2/v_w^+\right)\left[\left(1+v_w^+u_e^+\right)^{1/2}-1\right] = \left(1/\kappa\right)\ln \delta^+ + c, \tag{4.6.5}$$

where $u_e^+ \equiv u_e/u_\tau$ and $\delta^+ \equiv \delta u_\tau/\nu$. Subtracting Eq. (4.6.4) from Eq. (4.6.5) and rearranging, we obtain the modified *velocity-defect law* for incompressible turbulent boundary layers *with mass transfer* [39]:

$$\left(2/v_w^+\right)\left[\left(1+v_w^+u_e^+\right)^{1/2}-\left(1+v_w^+u^+\right)^{1/2}\right] = -\left(1/\kappa\right)\ln\left(y/\delta\right) \equiv f\left(y/\delta\right). \tag{4.6.6}$$

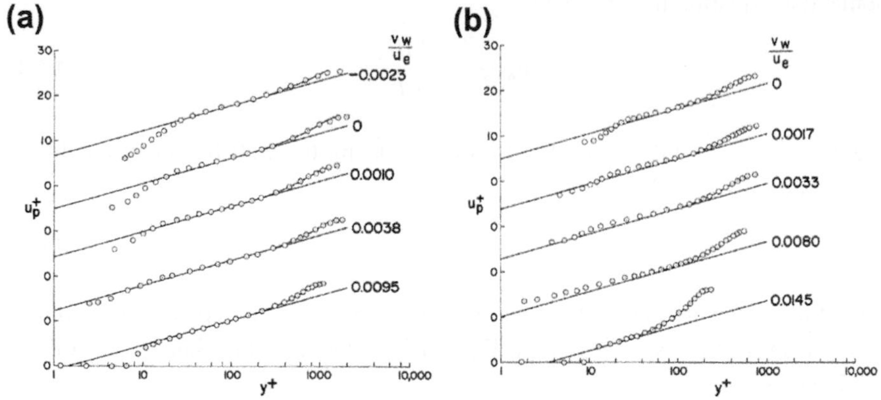

Fig. 4.26 Prediction of velocity profiles by Coles' expression, Eq. (4.6.10), for mass transfer. (a) Data of Simpson et al. [40]. (b) Data of McQuaid [42].

The results in Fig. 4.26 show that the function $f(y/\delta)$ in the equation of the modified velocity defect law, Eq. (4.2.4), is independent of v_w and u_τ in the outer region of the boundary layer.

According to Stevenson's law of the wall, *the parameter c in Eq. (4.6.4), like the parameter k, is essentially unaffected by mass transfer.* Simpson et al. [39] carried out extensive new measurements on flat plates with both blowing and suction, as a result of which they proposed a different condition, namely, that *the curve of Eq. (4.6.4) always passes through the point $u^+ = y^+ = K = 11$, regardless of the value of v_w.* Here the parameter K represents the intersection of the logarithmic profile with the linear sublayer when $v_w = 0$; that is,

$$u^+ = y^+ = (1/\kappa)\ln y^+ + c_0 \tag{4.6.7}$$

or

$$K = (1/\kappa)\ln K + c_0. \tag{4.6.8}$$

Simpson's condition for c, which is purely empirical, implies that

$$c = c_0 + \left(2/v_w^+\right)\left[\left(1 + Kv_w^+\right)^{1/2} - 1\right] - K. \tag{4.6.9}$$

Using the above expression for c, Coles [41] has shown that his profile expression (4.4.34), when generalized to the form

$$\left(2/v_w^+\right)\left[\left(1 + v_w^+ u^+\right)^{1/2} - 1\right] = \left(1/\kappa\right)\ln y^+ + c + \left(\Pi/\kappa\right)w\left(y/\delta\right), \tag{4.6.10}$$

with $c_0 = 5$ and $K = 10.805$, describes the experimental data for zero-pressure-gradient flows with mass transfer very well. As in flows with zero mass transfer, the profile parameter Π has the variation with Reynolds number given by Fig. 4.20. Figure 4.26 shows, for two sets of experimental data, the excellent prediction of pseudo-velocity profiles, u_p^+, by Eq. (4.6.10).

4.7 The Crocco Integral for Turbulent Boundary Layers

As was discussed in Chapter 3, the Crocco integral, Eq. (3.2.18),

$$(H - H_w)/(H_e - H_w) = u/u_e, \tag{4.7.1}$$

provides a good approximation for adiabatic, zero-pressure-gradient flows, both incompressible and compressible. For the compressible flow of a perfect gas, Eq. (3.2.18) can be written as

$$\frac{T}{T_e} = \frac{T_w}{T_e} - \left(\frac{T_w}{T_e} - 1\right)\frac{u}{u_e} + \frac{\gamma - 1}{2}M_e^2\frac{u}{u_e}\left(1 - \frac{u}{u_e}\right). \tag{4.7.2a}$$

Obviously, for incompressible flows it reduces to

$$\frac{T}{T_e} = \frac{T_w}{T_e} - \left(\frac{T_w}{T_e} - 1\right)\frac{u}{u_e}. \tag{4.7.2b}$$

According to experiments, the Crocco relationship does not apply in either laminar or turbulent flows with heat transfer and pressure gradient, even if the wall is isothermal and the molecular or the eddy diffusivities for momentum and heat transfer are equal or nearly equal. Instead, the relationship between total temperature and velocity is generally expected to depend upon the degree of flow acceleration (or deceleration) and the amount of wall cooling (or heating), and may be influenced to some extent by the flow speed, that is, the compressibility effect when the molecular or the eddy diffusivities for momentum are not the same (e.g., see Bertram et al. [43]; Back and Cuffel [44,45]).

Figure 4.27 shows the relationship between measured temperature and velocity profiles for an accelerating, turbulent-boundary-layer flow of air through a cooled, convergent-divergent nozzle. The data are due to Back and Cuffel [43,44]. Boundary-layer measurements were made upstream, along the convergent section, and near the end of the divergent section where the flow is supersonic. These measurements span a relatively large flow-speed range, with inlet and exit Mach

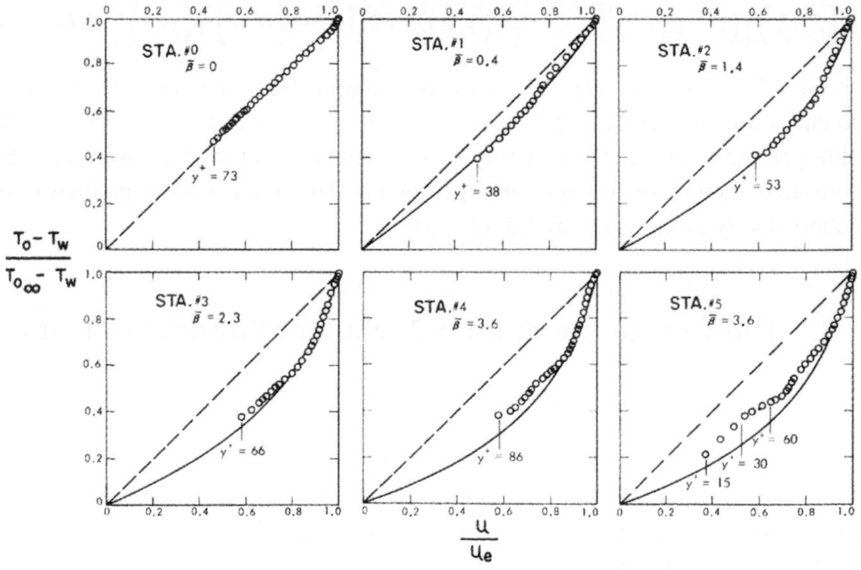

Fig. 4.27 Relationship between measured total temperature and velocity profiles for an accelerating turbulent boundary-layer flow of air through a cooled convergent-divergent nozzle [44].

numbers of 0.06 and 3.7, respectively. The operating conditions were such that the boundary layer remained essentially turbulent, that is, laminarization did not occur in the accelerating flow. The wall was cooled externally, with a ratio of wall to stagnation temperature of 0.43–0.56.

The results in Fig. 4.27 indicate that upstream of the flow acceleration region (Station 0), the temperature–velocity relationship is essentially linear in the region where the molecular transport is negligible. Such variation implies that the eddy diffusivities for momentum and heat transfer are nearly equal, and the Crocco relation applies there. The value of y^+ at the closest location to the wall is noted. At locations within the viscous sublayer, where molecular transport becomes important, the temperature profile would lie below the velocity profile, because molecular diffusivity for heat is larger than that for momentum transfer, that is, $\alpha = (1/0.7)\nu$. That is not evident in Fig. 4.27, because of the height of the probes relative to the sublayer thickness, but is apparent at lower pressures, where the measurements extend into the viscous sublayer.

At subsequent stations in the acceleration region, the temperature profiles lie below the velocity profiles at a given distance from the wall. In the representation of Fig. 4.27, the temperature-velocity relationship consequently bows progressively downward as one proceeds along the convergent section (stations 1, 2, 3, and 4). The departure of the measured total-temperature and velocity profiles from the Crocco relation near the nozzle exit (station 5), where the Mach number is 3.6, is not much different from the low-speed profile in the convergent section (station 4), where the Mach number is 0.19.

Recovery Factor

The temperature distribution according to the Crocco integral is based on the assumption that the molecular Prandtl number is unity and that the transport-of-momentum term, $-\bar{\varrho}\overline{h'v'}$, is equal to the transport-of-heat term, $-\bar{\varrho}\overline{u'v'}$. For that reason, the total enthalpy H must be constant across the boundary layer for an adiabatic flow. However, if the molecular Prandtl number is not unity, the total enthalpy is not constant. Total temperature variation for a laminar adiabatic flow with $\mathrm{Pr} = 0.75$ is shown in Fig. 4.28a, in which the energy has migrated from regions near the wall to regions near the free stream. It was obtained by Van Driest [46] by solving the governing equations for a compressible laminar flow. A manifestation of the migration is the usual experimentally observed wall temperature of adiabatic plates, which is lower than the total free-stream temperature. That experimental fact is specified by the so-called recovery factor r defined by

$$r = (T_{\mathrm{aw}} - T_{\mathrm{e}})/(T_0 - T_{\mathrm{e}}), \tag{4.7.3}$$

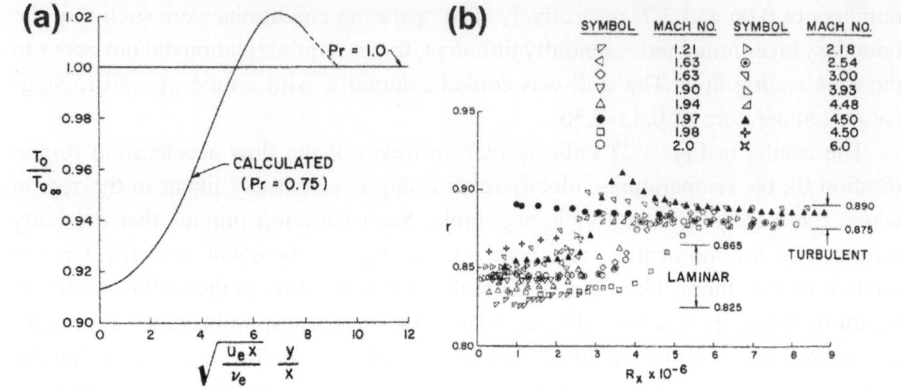

Fig. 4.28 (a) Variation of stagnation-temperature ratio across the laminar boundary layer at $M_e = 3$ [46] and (b) variation of recovery factor with Reynolds number [47].

where T_{aw} and T_0 are the adiabatic wall and the reservoir (isentropic-stagnation) temperatures, respectively. From the steady-state energy equation for an inviscid perfect-gas flow, we can write

$$T_0 - T_e = u_e^2/2c_p.$$

Substituting that expression into Eq. (4.7.3) and rearranging, we get

$$T_{aw} = T_e\{l + r[(\gamma - 1)/2]M_e^2\}. \tag{4.7.4}$$

With laminar boundary layers, the recovery factor r_{lam} is approximately equal to $(Pr)^{1/2}$ for incompressible flows.

In the case of the adiabatic fully turbulent boundary layer, the recovery factor r_t is somewhat larger than it is in laminar boundary layers. According to experiments (see Fig. 4.28b), its value is between 0.875 and 0.90 for air. It is given approximately by $(Pr)^{1/3}$.

From the Crocco integral, Eq. (4.7.1a), after rearranging we can write

$$\frac{T}{T_w} = 1 + \left[\left(1 + \frac{\gamma - 1}{2}M_e^2\right)\frac{T_e}{T_w} - 1\right]\frac{u}{u_e} - \frac{\gamma - 1}{2}M_e^2\frac{T_e}{T_w}\left(\frac{u}{u_e}\right)^2. \tag{4.7.5}$$

Replacing the relation $1 + [(\gamma - 1)/2]M_e^2$ by T_0/T_e in Eq. (4.7.5) and rearranging, we obtain

$$T = T_w + (T_0 - T_w)(u/u_e) + (T_e - T_0)(u/u_e)^2. \tag{4.7.6}$$

That expression is another form of Crocco integral. It can be extended to Prandtl numbers differing from unity by replacing the stagnation temperature T_0 by the adiabatic wall temperature T_{aw}. Then Eq. (4.7.6) can be written as

$$T = T_w + (T_{aw} - T_w)\,(u/u_e) + (T_e - T_{aw})\,(u/u_e)^2. \qquad (4.7.7)$$

In that equation, T_{aw} should be calculated from Eq. (4.7.3). A commonly used value for r is 0.89. For zero-pressure-gradient flows with small heat transfer, Eq. (4.7.7) can be used quite satisfactorily to calculate the static-temperature distribution for flows both with and without mass transfer [48,49].

4.8 Mean-Velocity and Temperature Distributions in Compressible Flows with Zero Pressure Gradient

For two-dimensional unsteady compressible boundary layers, both laminar and turbulent, the continuity, momentum, and energy equations are given by Eqs. (3.3.5), (3.3.6), and (3.3.14), respectively. For laminar layers, the equations are coupled through the variation of density and the transport properties of the gas, such as μ, k, and c_p. Although reliable experiments in variable-density flows are still few in number, there is reasonable evidence that the structure of the turbulent velocity field is not altered significantly in the presence of moderate density or temperature fluctuations. This suggests that interaction between the velocity and the temperature fluctuations is probably not strong, even in flows of moderate Mach numbers [50]. Indeed, Chu and Kovasznay [51] have shown theoretically that in a homogeneous field the interactions are second order. Thus for compressible turbulent shear flows, the main coupling between the equations occurs through the density variation only.

In this section we shall discuss the mean-velocity and temperature distributions in compressible turbulent boundary layers with zero pressure gradient, and we shall show how the various expressions developed for incompressible flows can be modified to account for the variation of density in such flows.

4.8.1 The Law-of-the-Wall for Compressible Flows

In compressible turbulent flows, the velocity profile and temperature profile in the inner part of the boundary layer depend on all the quantities that affect the velocity *or* temperature profile in compressible flows (Section 4.4), and in addition the absolute temperature (at the wall, say) must be included since, by definition, the temperature differences in compressible flows are a significant fraction of the absolute temperature. Also, if the Mach number of the flow is not small compared with unity, the speed of sound, a, and the ratio of specific heats, γ, will appear; in a perfect gas

$a = \sqrt{\gamma RT} = \sqrt{(\gamma - 1)c_p T}$, so that either a or T or both may be used as a variable. With these additions, from dimensional analysis with $u_\tau = \sqrt{\tau_w/\varrho_w}$, we can write

$$\frac{u}{u_\tau} = f_3\left(\frac{u_\tau y}{\nu_w}, \frac{\dot{q}_w}{\varrho c_p u_\tau T_w}, \frac{u_\tau}{a_w}, \gamma, \mathrm{Pr_w}\right), \tag{4.8.1}$$

$$\frac{T}{T_w} = f_4\left(\frac{u_\tau y}{\nu_w}, \frac{\dot{q}_w}{\varrho c_p u_\tau T_w}, \frac{u_\tau}{a_w}, \gamma, \mathrm{Pr_w}\right). \tag{4.8.2}$$

In Eq. (4.8.2) we have used T/T_w instead of $(T_w - T)/T_w$ for convenience, and the speed of sound at the wall, a_w, is used instead of the dimensionally correct but less meaningful quantity $\sqrt{c_p T_w}$. The quantity u_τ/a_w is called the *friction Mach number,* M_τ. The evaluation of fluid properties at the wall is adequate if, for example, ν/ν_w can be expressed as a function of T/T_w only (the pressure being independent of y in any case); this is the case if $\nu \propto T^\omega$ for some ω, which is a good approximation for common gases over a range of, say, 2:1 in temperature.

The arguments that led to the law-of-the-wall formulas for velocity and temperature in incompressible flows (Section 4.2) can be applied again to the compressible case if we are satisfied that the effects of viscosity are again small for $u_\tau y/\nu_w \gg 1$ and that the effects of thermal conductivity are again small for $(u_\tau y/\nu_w)$ Pr $\gg 1$. Provided that the effects of *fluctuations* in viscosity and thermal conductivity are small and that ν and ϱ do not differ by orders of magnitude from their values at the wall so that $u_{\tau y}/\nu_w$ is still a representative Reynolds number, molecular diffusion should indeed be small compared with turbulent diffusion if $u_\tau y/\nu_w$ is large. Since density varies in compressible flows, the local value of $(\tau/\varrho)^{1/2}$ would provide a better velocity scale than the wall value u_τ; it therefore seems logical to use the local value in compressible flows also, and in the simplest case when only ϱ and not τ, varies with y, the appropriate velocity scale is $(\tau_w/\varrho)^{1/2}$. Analogously we use $\dot{q}/\varrho$, rather than $\dot{q}_w/\varrho_w$, in the mixing-length formula for temperature; we shall see below that $\dot{q}$ always varies with y in high-speed flows. The elimination of ν_w and Pr from the lists of variables and the use of dimensional analysis on $\partial u/\partial y$ and $\partial T/\partial y$ instead of u and T give, instead of Eqs. (4.8.1) and (4.8.2)

$$\frac{\partial u}{\partial y} = \frac{(\tau/\varrho)^{1/2}}{\kappa y} f_u\left[\frac{\dot{q}}{\varrho c_p T(\tau/\varrho)^{1/2}}, \frac{(\tau/\varrho)^{1/2}}{a}, \gamma\right], \tag{4.8.3}$$

$$\frac{\partial T}{\partial y} = -\frac{\dot{q}/\varrho c_p}{(\tau/\varrho)^{1/2}\kappa_h y} f_T\left[\frac{\dot{q}}{\varrho c_p T(\tau/\varrho)^{1/2}}, \frac{(\tau/\varrho)^{1/2}}{a}, \gamma\right] \tag{4.8.4}$$

where we have consistently used *local* variables in the arguments of the *f* functions, even for τ. The analysis below will be restricted to the case of a constant-stress layer,

$\tau = \tau_w$, which, as was discussed in Section 4.2, is a good approximation to the inner 20 percent of a boundary layer in a small pressure gradient.

In order to equate the f functions to unity and recover effectively incompressible versions of the mixing-length formulas, we need to neglect the effect of density fluctuations discussed in Section 3.2. With this assumption, the heat-transfer parameter $\dot{q}/\varrho c_p(\tau/\varrho)^{1/2}T$ and the friction Mach number $(\tau/\varrho)^{1/2}/a$ – representing the two sources of density (temperature) fluctuations – do not appear in formulas (4.8.3) and (4.8.4) for the gradients of u and T; they will, however, remain in the full formulas for u and T, Eqs. (4.8.1) and (4.8.2), because they affect the temperature gradient in the viscous sublayer. Formulas for $\partial u/\partial y$ and $\partial T/\partial y$ in the viscous sublayer would nominally contain all the variables on the right-hand sides of Eq. (4.8.1) or (4.8.2). The assumption that turbulence processes are little affected by density fluctuations implies, that γ, which is a measure of the difference between adiabatic and isothermal processes, would have a negligible effect even in the viscous sublayer, but there are not enough data to check this. With the assumption that f_u and f_T are constant outside the viscous sublayer, we can now write

$$\frac{\partial u}{\partial y} = \frac{(\tau/\varrho)^{1/2}}{\kappa y}, \tag{4.8.5}$$

$$\frac{\partial T}{\partial y} = \frac{-\dot{q}/\varrho c_p}{(\tau/\varrho)^{1/2}\kappa_h y}. \tag{4.8.6}$$

Now Eq. (4.8.6) still retains the local value of $\dot{q}$, and in a high-speed flow this will differ from the wall value, even if $\tau = \tau_w$, because of viscous dissipation of mean and turbulent kinetic energy into heat. The rate at which kinetic energy is extracted from a unit volume of the mean flow by work done against viscous and turbulent stresses is $\tau \partial u/\partial y$; the part corresponding to the viscous shear stress represents direct viscous dissipation into heat, and the part corresponding to the turbulent shear stress represents production of the turbulent kinetic energy. We cannot immediately equate turbulent energy production to viscous dissipation of that turbulent energy into heat because turbulence processes include transport of turbulent kinetic energy from one place to another. However, this transport is negligible in the inner layer (outside the viscous sublayer), so that we can write a degenerate version of the energy equation, with all transport terms neglected and only y derivatives retained, as

$$\frac{\partial \dot{q}}{\partial y} = \tau_w \frac{\partial u}{\partial y}, \tag{4.8.7}$$

which simply states that the net rate of (y-component) transfer of heat leaving a control volume in the inner layer is equal to the rate at which the fluid in the control volume does work against (shear) stress. Integrating this equation, we get

$$\dot{q} = \dot{q}_w + u\tau_w. \tag{4.8.8}$$

In a low-speed flow the work done is negligible, and $\dot{q} = \dot{q}_w$, corresponding to $\tau = \tau_w$. If we divide Eq. (4.8.6) by Eq. (4.8.5), we obtain

$$\frac{\partial T}{\partial u} = \frac{-(\kappa/\kappa_h)\dot{q}}{c_p \tau_w}, \tag{4.8.9}$$

where k/k_h is the turbulent Prandtl number. Substituting for $\dot{q}$ from Eq. (4.8.8) and integrating with respect to u, we get

$$T = -\frac{(k/k_h)\dot{q}_w u}{c_p \tau_w} - \frac{(\kappa/\kappa_h)u^2}{2c_p} + \text{const.} \tag{4.8.10}$$

Here the constant of integration is *not* exactly equal to T_w because the formulas (4.8.5) and (4.8.6) are not valid on the viscous or conductive sublayers, but it is conventionally written as $c_1 T_w$, where c_1 is close to unity and is a function of $\dot{q}_w/\varrho_w c_p u_\tau T_w$, u_τ/a_w, and the molecular Prandtl number Pr. That is,

$$T = c_1 T_w - \frac{(k/k_h)\dot{q}_w u}{c_p \tau_w} - \frac{(\kappa/\kappa_h)u^2}{2c_p}. \tag{4.8.11}$$

Noting that $\varrho = \varrho_w T_w/T$, we can use Eq. (4.8.11) to eliminate ϱ from Eq. (4.8.5). The integral required to obtain u as a function of y from Eq. (4.8.5) then becomes

$$\int \frac{dy}{ky} = \int \frac{du/u_\tau}{\left[c_1 - (\kappa/\kappa_h)\dot{q}_w u/c_p T_w \tau_w - (\kappa/\kappa_h)u^2/2c_p T_w\right]^{1/2}} \tag{4.8.12}$$

Replacing $c_p T_w$ by $a_w^2/(\gamma - 1)$ and integrating Eq. (4.8.12), we obtain the law of the wall for compressible turbulent flows:

$$\frac{u}{u_\tau} = \frac{\sqrt{c_1}}{R}\sin\left(R\frac{u^*}{u_\tau}\right) - H\left[1 - \cos\left(R\frac{u^*}{u_\tau}\right)\right], \tag{4.8.13}$$

where

$$R = \frac{u_\tau}{a_w}\left[\frac{(\gamma-1)\kappa}{2\kappa_h}\right]^{1/2}, \quad H = \frac{\dot{q}_w}{\tau_w u_\tau} \equiv \frac{1}{(\gamma-1)}\frac{\dot{q}_w}{\varrho_w c_p u_\tau T_w}\left(\frac{a_w}{u_\tau}\right)^2, \tag{4.8.14a}$$

$$u^* = \frac{1}{\kappa}\ln y + \text{const.} \tag{4.8.14b}$$

Recalling from Eq. (4.8.1) that y appears in the group $u_\tau y/\nu_w$, we rewrite

$$\frac{u^*}{u_\tau} = \frac{1}{\kappa}\ln\frac{u_\tau y}{\nu_w} + c. \tag{4.8.14c}$$

If $\dot{q}_w = 0$ and $u_\tau/a_w \to 0$, then $H = 0$, $R \to 0$, and

$$\frac{u}{u_\tau} = \sqrt{c_1}\,\frac{u^*}{u_\tau}. \tag{4.8.15}$$

It can be shown that if u_τ/a_w is small, then $c_1 = 1 - O(\dot{q}_w)$, for compatibility with the logarithmic law for temperature in incompressible flows, Eq. (4.4.17), so that $u = u^*$ for small $\dot{q}_w$; thus the constant c in Eq. (4.8.14c) can be identified with the additive constant c in the logarithmic law for constant property wall layers. In general c, like c_1, is a function of the friction Mach number M_τ and of B_q defined by

$$B_q = \frac{\dot{q}_w}{\varrho_w c_p u_\tau T_w}. \tag{4.8.16}$$

The above analysis is originally due to Rotta [9] and is an extension of that of Van Driest [46].

4.8.2 Van Driest Transformation for the Law of the Wall

Simpler versions of Eq. (4.8.13) and the accompanying temperature profile Eq. (4.8.11) have been proposed by many authors. Van Driest [46] assumed $c_1 = 1$ and $\kappa/\kappa_h = 1$ (recall that κ/κ_h is a turbulent Prandtl number in the fully turbulent part of the flow, and note that $c_1 = 1$ implies that the effective Prandtl number in the viscous and conductive sublayers is unity). Van Driest presented the inverse of Eq. (4.8.3) giving u^* in terms of u. In our more general notation this is

$$\frac{u^*}{u_\tau} = \frac{1}{R}\left(\sin^{-1}\frac{R(u/u_\tau + H)}{(c_1 + R^2H^2)^{1/2}} - \sin^{-1}\frac{RH}{(c_1 + R^2H^2)^{1/2}}\right). \tag{4.8.17}$$

This formula is called the *Van Driest transformation*; it can be regarded as transforming the inner-layer part of the compressible boundary-layer profile $u(y)$ to an equivalent incompressible flow $u^*(y)$ that obeys the logarithmic formula, Eq. (4.8.14c). However, it is simpler to regard Eq. (4.8.13), with Eq. (4.8.14) as direct prediction of inner-layer similarity theory for the compressible boundary layer.

If there is no heat transfer through the surface, H is zero, the second term on the right of Eq. (4.8.13) disappears, and Eq. (4.8.17) reduces to

$$\frac{u^*}{u_\tau} = \frac{1}{R}\sin^{-1}\left(\frac{R}{\sqrt{c_1}}\frac{u}{u_\tau}\right), \tag{4.8.18}$$

which is easy to identify as the inverse of Eq. (4.8.13) without the second term on the right and of course reduces to $u^* = u$ as $R \to 0$ and $c_1 \to 1$.

The basic assumptions imply that κ and κ_h are the same as in incompressible flow. As we have seen, c_1 and c, which are constants of integration in Eqs. (4.8.9) and

(4.8.12), respectively, must be expected to be functions of the friction Mach number and the heat transfer parameter. Experimental data reviewed in great detail by Fernholz and Finley [52] support the extension of inner-layer similarity to compressible flow but fail to provide definite evidence about the variation of c and c_1. The low-speed values $c = 5.0$ and $c_1 = 1$ fit the data as a whole to within the rather large scatter, but Bradshaw [53] presents formulas for variable c and c_1 based on a selection of the more reliable data.

If we replace H by the expressions given in Eq. (4.8.14a) and (4.8.11), use the definition of Mach number, $M_e = u_e/a_e$, take $c_1 = 1$ and $\kappa/\kappa_h = 1$ following Van Driest [46], and note that $\varrho_w/\varrho_e = T_e/T_w$, then Eq. (4.8.17) can be written as

$$\frac{u^*}{u_\tau} = \frac{1}{A\sqrt{(c_f/2)(T_w/T_e)}}\left(\sin^{-1}\frac{A\sqrt{(c_f/2)(T_w/T_e)}(u/u_\tau) - B/2A}{1+(B/2A)^2}\right.$$
$$\left. + \sin^{-1}\frac{B/2A}{\sqrt{1+(B/2A)^2}}\right) \tag{4.8.19}$$

or as

$$u^* = \frac{u_e}{A}\left(\sin^{-1}\frac{2A^2(u/u_e) - B}{(B^2+4A^2)^{1/2}} + \sin^{-1}\frac{B}{(B^2+4A^2)^{1/2}}\right) \tag{4.8.20}$$

where

$$A^2 = \frac{\gamma-1}{2}\frac{M_e^2}{T_w/T_e}, \quad B = \frac{1+(\gamma-1)/2M_e^2}{T_w/T_e} - 1. \tag{4.8.21}$$

Note that the above relations assume that the recovery factor $r \equiv (T_{aw}/T_e - 1)$ $[(\gamma-1)/2M_e^2]$, where T_{aw} is the temperature of an adiabatic wall, is 1. To account for the fact that r is less than unity (about 0.89) we rewrite Eqs. (4.8.21) as

$$A^2 = \frac{[(\gamma-1)/2]M_e^2\gamma}{T_w/T_e}, \quad B^2 = \frac{1+[(\gamma-1)/2]M_e^2\gamma}{T_w/T_e} - 1. \tag{4.8.22}$$

4.8.3 Transformations for Compressible Turbulent Flows

The Van Driest Transformation for the Whole Layer

The Van Driest transformation, Eq. (4.8.17), applied to the fully turbulent part of the inner (constant-stress) layer of a compressible boundary layer produces the logarithmic profile, Eq. (4.8.14c). Applying the transformation to the outer layer of a constant-pressure compressible boundary layer, we obtain a profile that looks qualitatively like that of a constant-pressure constant-density boundary layer. In particular, the transformed profile $u^*(y)$ can be described, more or less as accurately as an incompressible profile, by the wall-plus-wake formula given by Eqs. (4.4.34)–(4.4.36),

$$\frac{u^*}{u_\tau} = \frac{1}{\kappa} \ln \frac{u_\tau y}{\nu_w} + c + \frac{\Pi}{\kappa}\left(1 - \cos \pi \frac{y}{\delta}\right). \tag{4.8.23}$$

However, this convenient data correlation is a consequence of the strong constraint on the wake profile, which has to have zero slope and zero intercept at $y = 0$, whereas the profile as a whole has zero slope at $y = \delta$ also, the "wake parameter" Π and the boundary-layer thickness δ are constants that can be adjusted to optimize the fit of Eq. (4.8.23) to any real or transformed profile. We must, therefore, not claim that the success of Eq. (4.8.23) proves the validity of Van Driest's inner-layer analysis in the *outer* layer.

As was discussed in subsection 4.4.6, Π is constant in incompressible constant-pressure flows at high Reynolds number and equal to about 0.55. The value of Π that best fits a transformed profile is expected to be a function of the friction Mach number M_τ ($\equiv u_\tau/a_w$) and of the heat-transfer parameter B_q ($\equiv \dot{q}_w/\varrho c_p u_\tau T_w$). Evaluation of this function from experimental data is hampered by the low Reynolds number of most of the compressible-flow data and uncertainty about the definition of Reynolds number that should be used in correlating low-Reynolds-number effects on the velocity-defect profile. If it is accepted that these originate at the irregular interface between the turbulent fluid and the nonturbulent "irrotational" fluid, then the fluid properties in the Reynolds number should be evaluated at freestream conditions. Now the largest-scale interface irregularities seen in flow-visualization pictures have a length scale of order δ, being the result of the largest eddies that extend across the full thickness of the shear layer. Therefore δ is the appropriate interface scale and, since it is found that the shear-stress profile plotted as $\tau/\tau_w = f(y/\delta)$ has nearly the same shape at any Reynolds number, we can use τ_w as a shear-stress scale and $(\tau_w/\varrho_e)^{1/2}$ – evaluated using the *freesteam* density – as a velocity scale. Therefore the appropriate Reynolds number is $(\tau_w/\varrho_e)^{1/2}\delta/\nu_e \equiv (u_\tau\delta/\nu_e)\cdot(\varrho_w/\varrho_e)^{1/2}$ rather than the Reynolds number $u_\tau\delta/\nu_w$ that arises naturally in the Van Driest transformation. However, Fernholz [54] has shown that the Reynolds number $\varrho_e u_e \theta/\mu_w$ gives excellent correlations of data over a wide range of Mach numbers and Reynolds numbers.

If the physics of low-Reynolds-number effects is the same in compressible flow as in incompressible flow, then the wake parameter Π of the transformed profile should be independent of Reynolds number for $(\tau_w/\varrho_e)^{1/2}\delta/\nu_e > 2000$ approximately (corresponding to $u_e\theta/\nu > 5000$ in the case of a low-speed boundary layer). Since Π is nominally a function of M_τ and B_q, it is not possible to predict the trend with

Reynolds number explicitly, but it is probably adequate to assume that the ratio of Π to its high-Reynolds-number asymptotic value is the same function of the chosen Reynolds number as in incompressible flow. In fact, although the data are rather scattered by low-speed standards, it appears that Π decreases only very slowly with increasing Mach number [55] in adiabatic-wall boundary layers ($\dot{q}_w = 0$) and therefore that $\Pi = 0.55$ is an adequate high-Reynolds number value for all M_τ.

If c and c_1 are known as functions of M_τ and B_q, and if Π is known as a function of M_τ, B_q, and Reynolds number, then putting $y = \delta$ and $u = u_e$ in Eq. (4.8.17) and using Eq. (4.8.23) to substitute for u^*, we obtain u_e/u_τ as a function of $u_\tau\delta/\nu_w$. In practice we require the skin-friction coefficient $\tau_w/\frac{1}{2}\varrho_e u_e^2$ as a function of $u_e\theta/\nu_e$ or Fernholz's variable $\varrho_e u_e \theta/\mu_w$ given M_e (and T_e) and either $\dot{q}_w$ or T_w. This requires iterative calculation, starting with an estimate of τ_w. Also, the velocity profile of the compressible flow has to be integrated at each iteration to obtain θ/δ for conversion from the "input" Reynolds number $u_e\theta/\nu_e$ to the Reynolds number $u_\tau\delta/\nu_w$ that appears in the transformation [53]. Skin-friction formulas are discussed in subsection 7.2.3.

The Van Driest transformation could be regarded as a *solution* of the compressible-flow problem only if the coefficients c, c_1 and Π were independent of Mach number and heat-transfer parameter. However, we can use the transformation, plus *compressible-flow* data for c, c_1 and Π, to correlate the mean properties of constant-pressure compressible boundary layers. As noted above the change in the coefficients is almost within the (large) experimental scatter.

In pressure gradients the transformed boundary-layer profile still fits Eq. (4.8.23) as does its true low-speed equivalent, but, as at low speeds, there is no simple formula to relate the shape parameter Π to the Iocal pressure gradient. Moreover, the variation of u_τ, Π, and δ with x will not generally correspond to any realizing low-speed boundary layer; that is, it may not be possible to choose a pressure distribution $p(x)$ for a low-speed flow that will reproduce at each x, the same velocity profile as in the compressible flow. The spirit of Van Driest's transformation, although not its details, would be retained if compressible boundary layers were calculated using the mixing-length formula to predict the shear stress and the assumption of constant turbulent Prandtl number to predict the heat transfer. We consider such calculation methods in Chapter 8.

Other Transformations

While the transformations between compressible and incompressible laminar boundary layers are rigorous but limited in application, transformations for turbulent flow are necessarily inexact because our knowledge of the time-averaged properties of turbulent motion is inexact. As in the case of laminar flow the need for transformation has decreased as our ability to do lengthy numerical calculations has increased, and the assumption that density fluctuations have negligible effect on turbulence has permitted low-speed models to be cautiously extended to compressible flow. The Van Driest transformation relies on the application of this assumption to the inner-layer (mixing-length) formula. In the outer layer where the mixing length departs from its inner-layer value κy, the Van Driest analysis is not exact and the transformation will not eliminate compressibility effects though it certainly *reduces* them, apparently within the scatter of current experimental data.

The transformations between compressible and incompressible flow fall into two classes: (1) transformations for the complete velocity profile (and by implication the shear stress profile) and (2) transformations for integral parameters only (specifically, skin-friction formulas). It is generally recognized that transformations for compressible flows in pressure gradient do not necessarily lead to realizable low-speed flows, and we will discuss only constant-pressure flows here.

The paper by Coles [56] is a useful review of previous work and presents one of the most general transformations so far proposed. The two main assumptions are that suitably defined ratios of coordinates in the original (high-speed) and transformed (low-speed) planes are functions only of x and not of y, and that the ratio of the stream functions in the transformed and original flows is equal to the ratio of the (constant) viscosity in the transformed flow to the viscosity evaluated at an, "intermediate temperature" somewhere between T_w and T_e in the compressible flow. The justification for the latter assumption is carefully discussed by Coles, but the choice of intermediate temperature is necessarily somewhat arbitrary. Coles chooses the temperature at the outer edge of the viscous sublayer, but in order to fit the experimental data for skin friction it is necessary to locate the sublayer edge at $u_\tau y/\nu = 430$ in the transformed flow, whereas the thickness of the real sublayer is only about one-tenth of this. Coles conjectures that the relevant region is perhaps not the viscous sublayer as such but the whole turbulent boundary layer at the lowest Reynolds number at which turbulence can exist; this boundary layer indeed has $u_\tau \delta/\nu$ of the order of 430. Coles' transformation should not be confused with the simpler intermediate-temperature assumption that any low-speed skin friction formula can be applied to a high-speed flow if fluid properties are evaluated at a temperature T_i somewhere between T_w and T_e [57].

4.8.4 Law of the Wall for Compressible Flow with Mass Transfer

Stevenson's law-of-the-wall expression, Eq. (4.6.4), discussed in Section 4.6, has also been extended to compressible turbulent boundary layers with mass transfer by Squire [58]. Again assuming that derivatives in the x direction are negligible compared to derivatives in the y direction, one can integrate the continuity equation to give

$$\overline{\varrho v} = \text{const} = \varrho_w v_w, \tag{4.8.24}$$

and with that result, the momentum equation can be integrated to give

$$\varrho_w v_w u = \tau - \tau_w. \tag{4.8.25}$$

By means of Prandtl's mixing-length formula, Eq. (4.8.25) can be formally integrated to give

$$\int_{u_0}^{u} \frac{(\varrho)^{1/2} du'}{(\varrho_w v_w u' + \tau_w)^{1/2}} = \frac{1}{\kappa} \ln\left(\frac{y}{y_0}\right), \tag{4.8.26}$$

where the subscript a designates values at the inner edge ot the fully turbulent region.

Equation (4.8.26) may be rearranged to give

$$\begin{aligned}\int_0^{u} \frac{(\varrho)^{1/2} du'}{(\varrho_w v_w u' + \tau_w)^{1/2}} &= \frac{1}{\kappa} \ln \frac{y u_\tau}{\nu_w} - \frac{1}{\kappa} \ln \frac{y_0 u_\tau}{\nu_w} + \int_0^{u_0} \frac{(\varrho)^{1/2} du'}{(\varrho_w v_w u' + \tau_w)^{1/2}} \\ &= \frac{1}{\kappa} \ln \frac{y u_\tau}{\nu_w} + c.\end{aligned} \tag{4.8.27}$$

Before the left-hand side of that equation can be integrated, it is necessary, as before, to know the density or the temperature. That can be obtained from Eq. (4.7.6) since $\varrho_w/\varrho = T/T_w$.

With Eq. (4.7.6), the left-hand side of Eq. (4.8.27) may be written

$$\begin{aligned}&\left(\frac{T_e}{T_w}\right)^{1/2} \int_0^{\phi} \left(\frac{\varrho_w v_w}{\varrho_e u_e}\phi' + \frac{1}{2}c_f\right)^{-1/2} \\ &\times \left(1 + \frac{T_{aw} - T_w}{T_w}\phi' + \frac{T_e - T_{aw}}{T_w}(\phi)'^2\right)^{-1/2} d\phi' \equiv u_p^+,\end{aligned} \tag{4.8.28}$$

where $\phi = u/u_e$. Thus the law of the wall for compressible turbulent boundary layers with mass transfer is

$$u_p^+ = (1/\kappa) \ln (y u_\tau/\nu_w) + c. \tag{4.8.29}$$

For incompressible, constant-temperature flows, u_p^+ reduces to

$$\int_0^{\phi} \left(\frac{v_w}{u_e}\phi' + \frac{1}{2}c_f\right)^{-1/2} d\phi' = \frac{2u_e}{v_w z}\left[\left(1 + \frac{v_w u z^2}{u_e^2}\right)^{1/2} - 1\right].$$

Thus the law of the wall for incompressible flows with mass transfer can be written as

$$\left(2/v_w^+\right)\left[\left(1 + v_w^+ u^+\right)^{1/2} - 1\right] = (1/k) \ln y^+ + c,$$

the law of the wall obtained by Stevenson [38] [see Eq. (4.6.4)].

According to Stevenson's study [38], which was discussed earlier, the parameters κ and c were virtually independent of the mass-transfer velocity v_w/u_e. However, when Danberg [48] evaluated u_p^+ for his experimental data at $M_e = 6.5$, he found that the parameter c fell with increase in blowing rate, with the result that, for a blowing rate of $\varrho_w v_w/\varrho_e u_e = 0.0012$, c was 2.5, as compared to a value of 10 for the solid wall at the same Mach number and with the same ratio of T_w/T_e. According to Squire's study, the mixing-length constant κ in a compressible turbulent boundary layer with mass transfer is independent of Mach number and mass-transfer rate. However, the additive parameter c in the law of the wall varies with both Mach number and injection, and the value depends critically on the measured skin friction. In general, the additive parameter decreases with increasing injection rate at fixed Mach number, and the rate of fall increases with increase in Mach number. Figure 4.29 shows the variation of u_p^+ with $y^+(= yu_\tau/\nu_w)$ at $M_e = 3.55$ for several blowing rates, $F \equiv \varrho_w u_w/\varrho_e u_e$.

4.9 Effect of Pressure Gradient on Mean-Velocity and Temperature Distributions in Incompressible and Compressible Flows

In Sections 4.4–4.8, we have discussed the mean-velocity distributions in flows with zero pressure gradient. Here, we shall discuss the effect of the pressure gradient on the mean-velocity and temperature distributions. We shall not consider rough surfaces. If the roughness is reasonably small compared to the boundary-layer thickness, it will just have the effect of a velocity shift in the sublayer, according to the analysis presented in Section 4.5. Therefore, nothing unusual will happen to the mean-velocity distribution since the inner region is not much affected by the pressure gradient. For that reason, the discussion for smooth surfaces also applies for rough surfaces, except for the velocity shift.

Although it is useful and important to study special flows such as equilibrium flows, in most flows with pressure gradient, the external velocity distribution does not generally vary with x in a special way, and the parameter β [see Eq. (4.4.33)] does not remain constant. Flows with arbitrary external velocity distribution are of great interest and it is best to determine the effect of pressure gradient on mean-velocity and temperature distributions by obtaining solutions of turbulent boundary-layer equations by differential methods or by Navier-Stokes methods. Integral methods can also be used for this purpose (see Section 7.3) but they are restricted to boundary and flow conditions and their accuracy is not comparable to differential methods discussed in Chapters 8 and 9.

For a given two-dimensional or three-dimensional body, which implies that the external velocity distribution can be determined and that the surface

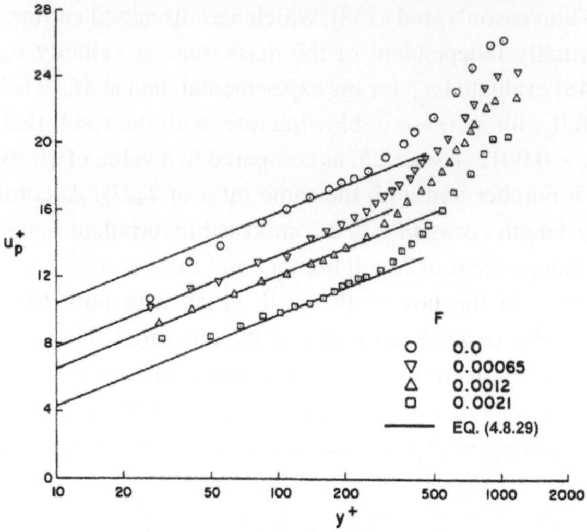

Fig. 4.29 Variation of u_p^+ with y^+ at $M_e = 3.55$ [58].

boundary condition is known, the momentum- and heat-transfer properties of the flow can be obtained by solution of the continuity, momentum and energy equations with accuracy sufficient for most engineering purposes. A general computer program for this purpose is presented and discussed in Chapter 10 for two-dimensional flows; it utilizes the eddy-viscosity formulation of Cebeci and Smith and the so-called Box scheme. This and other similar methods (Chapters 9, 10) allow the calculation of turbulent boundary layers, including free shear layers, for a wide range of boundary conditions. For more complicated flows or configurations, it may be more appropriate and necessary to use Navier-Stokes methods.

Problems

4.1 The process of dimensional analysis discussed for the law of the wall, Eq. (4.2.1), can be performed more rigorously with the "matrix elimination" method discussed by E. S. Taylor [59]; its advantage over other methods is that dimensionless groups that are already known can be inserted easily, making the analysis much shorter than in methods that start from a position of total ignorance. For an example of its use, let us apply it to the inner-layer velocity profile. We first identify all the relevant variables, namely u, y, τ_w, ϱ, and μ, and construct a matrix whose columns give the mass, length, and time dimensions of the variables in each row:

$$\begin{array}{c|ccc|} & M & L & T \\ u & 0 & 1 & -1 \\ y & 0 & 1 & 0 \\ \tau_w & 1 & -1 & -2 \\ \mu & 0 & -1 & -1 \\ \varrho & 0 & -3 & 0 \end{array}$$

The dimensions of τ_w, for example, are those of pressure or stress and can be constructed by noting that $\varrho\mu^2$ has the same dimensions as pressure, which are therefore $ML^{-3}(L^2T^{-2})$. We now eliminate the mass dimension by dividing all but one of the variables containing the mass dimension by the remaining variable containing mass; our knowledge of fluid dynamics prompts us to choose the density ϱ as the dividing variable:

$$\begin{array}{c|ccc|} & M & L & T \\ u & 0 & 1 & -1 \\ y & 0 & 1 & 0 \\ \dfrac{\tau_w}{\varrho} & 0 & 2 & -2 \\ \dfrac{\mu}{\varrho} & 0 & 2 & -1 \\ \varrho & 1 & -3 & 0 \end{array}$$

Obviously the density cannot appear in any dimensionless group in this problem except as τ_w/ϱ of μ/ϱ – there is no other way of canceling its mass dimension, and we can therefore drop it from the matrix.

Next we eliminate the time dimension, simply because in this case the T (time) column contains more zeros than the L (length) column. Again using our knowledge of fluid flow to choose physically useful combinations of variables, we get

$$\begin{array}{c|cc|} & L & T \\ \dfrac{u}{\sqrt{\tau_w/\varrho}} & 0 & 0 \\ y & 1 & 0 \\ \dfrac{u/\varrho}{\sqrt{\tau_w/\varrho}} & 1 & 0 \end{array}$$

By inspection, the length dimension can be eliminated by forming $y\sqrt{\tau_w/\varrho}/(\mu/\varrho)$. No other independent dimensionless groups can be constructed: so we have

$$f\left[\frac{u}{\sqrt{\tau_w/\varrho}}, \frac{y\sqrt{\tau_w/\varrho}}{\mu/\varrho}\right] = 0.$$

which, with our usual notation, is equivalent to Eq. (4.2.1).

Derive the "law of the wall" for temperature, Eq. (4.2.2), by using the matrix elimination procedure discussed above. Start by deriving the matrix shown below, in which the rows represent the exponents of mass, length, time and temperature (M, L, T and θ) in the dimensions of the variables shown at the left

	M	L	T	θ
$T_w - T$	0	0	0	1
y	0	1	0	0
τ_w	1	−1	−2	0
$\dot{q}_w$	1	0	−3	0
ϱ	1	−3	0	0
μ	1	−1	−1	0
k	1	1	−3	−1
c_p	0	2	−2	−1

4.2 Using the matrix elimination procedure, show that the law of the wall for velocity on a rough surface is

$$u^+ = \phi(y^+, k^+). \tag{P4.1}$$

4.3 For the inner region of a turbulent boundary layer, Reichardt [60] used the eddy-viscosity formula given by Eq. (4.3.11) and modified it to account for the viscous sublayer,

$$\varepsilon_m = \kappa u_\tau y \left[1 - \left(\frac{y_l}{y}\right) \tanh\left(\frac{y}{y_l}\right)\right] \tag{P4.2}$$

Here y_l denotes the viscous sublayer thickness. Show that ε_m is proportional to y^3 for $(y/y_l) \ll 1$.

4.4 Show that the continuity equation requires that $\overline{u'v'}$ should vary as at least the third power of y in the viscous sublayer, whereas the Van Driest formula for mixing length, Eq. (4.4.18) implies $\overline{u'v'} \sim y^4$ for small y.

4.5 Show that the "kinematic heating" parameter $\varrho u_\tau^3/\dot{q}_w$ is equal to

$$\frac{(\gamma - 1)(u_\tau/a_w)^2}{\dot{q}_w/(\varrho c_p u_\tau T_w}$$

where $a_w = \sqrt{\gamma R T_w}$.

4.6 Show that Eq. (4.6.3) reduces to Eq. (4.4.15) as $v_w \to 0$.

4.7 Show that the viscous shear stress at $y^+ = 50$ is about 5% of the wall shear stress.

4.8 If the expression for the whole velocity profile, Eq. (4.4.34) with $\phi_1(y^+)$ given by Eq. (4.4.35), is evaluated at $y = \delta$, the profile parameter Π can be related to the local skin-friction coefficient $c_f = 2\tau_w/\varrho u_e^2$ and to boundary-layer thickness δ by

$$\sqrt{\frac{2}{c_f}} \equiv \frac{u_e}{u_\tau} = \frac{1}{\kappa}\ \ln\frac{\delta u_\tau}{\nu} + c + \frac{2\Pi}{\kappa}. \qquad \text{(P4.3)}$$

Show that it can also be related to the displacement thickness δ^* and to the momentum thickness θ by

$$\kappa\frac{\delta^* u_e}{\delta u_\tau} - 1 + \Pi \qquad \text{(P4.4)}$$

and

$$\kappa^2\frac{(\delta^* - \theta)u_e^2}{\delta u_\tau^2} = 2 + 2\left[1 + \frac{1}{\pi}Si(\pi)\right]\ \Pi + \frac{3}{2}\Pi^2 \qquad \text{(P4.5)}$$

Also show that

$$\frac{\Pi}{\Pi - 1}\frac{u_\tau}{\kappa u_e} \equiv \frac{1}{\kappa G} = F(\Pi) \qquad \text{(P4.6a)}$$

$$F(\Pi) = \frac{1 + \Pi}{2 + 2[1 + 1/\pi\ Si(\pi)]\Pi + 3/2\Pi^2} \qquad \text{(P4.6b)}$$

where $Si(\pi) = \int_0^\pi\ [\sin u/u]\ du = 1.8519$ and G is the Clauser shape parameter.

$$G = \int_0^\infty \left(\frac{u - u_e}{u_\tau}\right)^2 d\left(\frac{y}{\Delta}\right),$$

$$\Delta = -\int_0^\infty \left(\frac{u - u_e}{u_\tau}\right) dy.$$

4.9 Using Eq. (P4.3), find the skin-friction coefficient in a constant pressure boundary layer at $u_e\delta^*/\nu = 15{,}000$ and then use Eq. (P4.6) to calculate $u_e\theta/\nu$. Take $k = 0.41$ and $c = 5.0$.

4.10 Find the velocity profile in the inner layer but outside the viscous layer if $\tau = \tau_w + \alpha y$, where α is a constant. On what dimensionless parameter does the final constant of integration depend?

4.11 Determine the equivalent sand-grain height of the square-bar roughness distribution tested by Moore and shown in Fig. 4.22. Assume fully rough conditions.

4.12 Consider the flat-plate problem in Problem 4.11, that is, a flat plate covered with square-bar roughness distribution. Compute the local skin-friction coefficient at $x = 1$ m for $u_e/\nu = 10^7$ m^{-1} and k (the roughness height) = 0.1 cm.

4.13 A thin flat plate is immersed in a stream of air at atmospheric pressure and at 25 °C moving at a velocity of 50 m s^{-1}. Calculate the momentum thickness, boundary-layer thickness, local skin-friction coefficient, and average skin-friction coefficient at $x = 3$ m. Assume that $\nu = 1.5 \times 10^{-5}$ m^2 s^{-1} and $R_{x_{cr}} = 3 \times 10^6$.

4.14 Consider the flat-plate problem in Problem 4.13, but assume that (a) the plate surface is covered with camouflage paint (see Table P4.1) applied in mass production conditions and (b) the plate surface is a dip-galvanized metal surface. Calculate the momentum thickness, boundary-layer thickness, local skin-friction coefficient, and average skin-friction coefficient at $x = 3$ m. As a simplification assume that roughness causes the transition to be at the leading edge so that we can neglect the contribution of laminar flow.

TABLE P4.1 Equivalent sand roughness for several types of surfaces.

Type of surface	k_s, cm
Aerodynamically smooth surface	0
Polished metal or wood	$0.05–0.2 \times 10^{-3}$
Natural sheet metal	0.4×10^{-3}
Smooth matte paint, carefully applied	0.6×10^{-3}
Standard camouflage paint, average application	1×10^{-3}
Camouflage paint, mass-production spray	3×10^{-3}
Dip-galvanized metal surface	15×10^{-3}
Natural surface of cast iron	25×10^{-3}

References

[1] P.S. Klebanoff, Z.W. Diehl, Some features of artificially thickened fully developed turbulent boundary layers with zero pressure gradient, NACA Rep. 1110 (1952).
[2] W. Jacobs, Umformung eines turbulenten Geschwindigkeitsprofils, Z. Angew. Math. Mech. 19 (1939) 87.
[3] P.H. Hoffmann, A.E. Perry, The development of turbulent thermal layers on flat plates, Int. J. Heat Mass Transfer. 22 (1979) 39.

[4] T. van Kármán, Mechanical similitude and turbulence, NACA Tech. Memo No. 611 (1931).
[5] P.S. Klebanoff, Characteristics of turbulence in a boundary layer with zero pressure gradient. NACA Tech. Note 3178 (1954).
[6] D. Coles, A Model for Flow in the Viscous Sublayer. Proceedings of the Workshop on Coherent Structure of Turbulent Boundary Layers, Lehigh University, Bethlehem, Pa, 1978.
[7] M.M.M. El Telbany, A.J. Reynolds, Turbulence in Plane Channel flows, Journal of Fiuid Mechanics 111 (1981) 283–318.
[8] V.C. Patel, W. Rodi, G. Scheuerer, Turbulence models for near-wall and low Reynolds number flows: A Review, AIAA J. 23 (No. 9) (1985) 1308–1319.
[9] J.C. Rotta, Turbulent Boundary Layers in Incompressible Flow, Progress in the Aeronautical Sciences 2 (1962) 1–219.
[10] B.E. Launder, Second Moment Closure: Methodology and Practice. Proceedings of the Ecole d'Eté d'Analyse Numerique – Modelisation Numerique de la Turbulence, Clamart, France (1982).
[11] H.-P. Kreplin, H. Eckelmann, Behaviour of the Three Fluctuating Velocity Components in the Wall Region of a Turbulent Channel Flow, Physics of Fluids 22 (1979) 1233–1239.
[12] K.K. Sirkar, T.J. Hanratty, The Limiting Behaviour of the Turbulent Transverse Velocity Component Close to a Wall, J. Fluid Mech. 44 (1970) 605–614.
[13] R.W. Derksen, R.S. Azad, Behaviour of the Turbulent Energy Equation at a Fixed Boundary, AIAA J. 19 (1981) 238–239.
[14] L. Laufer, The Structure of Turbulence in Fully Developed Pipe Flow, NACA Rept. 1174 (1954).
[15] J. Boussinesq, Theorie de l'écoulement tourbillant, 46, Mem. Pres. Acad. Sci. XXIII, Paris, 1877.
[16] L. Prandtl, Turbulent flow. NACA Tech. Memo, 435. Originally delivered to 2nd, Internat. Congr. Appl. Mech. Zürich (1926).
[17] W.P. Jones, B.E. Launder, The prediction of laminarization with a two-equation model of turbulence. Int. J. Heat. Mass. Transfer 15 (1972) 301.
[18] S. Corrsin, A.L. Kistler, The free-stream boundaries of turbulent flows, NACA Tech Note No. 3133 (1954).
[19] J. Nikuradse, Law of flow in rough pipes, Tech. Memo. NACA (1955). No. 1292.
[20] W. Nunner, Wärmeübertragung und Druckabfall in rauhen Rohren, VDI-Forschungsheft No. 455 (1956).
[21] G.B. Schubauer, Turbulent processes as observed in boundary layer and pipe, J. Appl. Phys. 25 (1954) 188.
[22] E.R. Van Driest, On turbulent flow near a wall, J. Aeronaut. Sci. 23 (1956).
[23] K. Wieghardt, Zum Reibungswiderstand rauher Platten. Kaiser-Wilhelm-Institut für Strömungsforschung, Göttngen, UM 6612 (1944).
[24] T. Cebeci, A model for eddy conductivity and turbulent Prandtl number, J. Heat Transfer 95 (1973) 227.
[25] F.H. Clauser, The turbulent boundary layer, Advan. Appl. Mech. 4 (1956) 1.
[26] D.W. Smith, J.H. Walker, Skin friction measurements in incompressible flow, NASA Tech. Rep. R-26 (1959).
[27] F.H. Clauser, Turbulent boundary layers in adverse pressure gradient, J. Aero. Sci. 21 (1954) 91.
[28] P. Bradshaw, The turbulence structure of equilibrium boundary layers. NPL Aero. Rep. 1184 (1967).
[29] T.F. Herring, J. Norbury, Some experiments on equilibrium turbulent boundary layers in favorable pressure gradients, J. Fluid Mech. 27 (1967) 541.
[30] D. Coles, The law of the wake in the turbulent boundary layer, J. Fluid Mech. 1 (1956) 191.
[31] P.S. Granville, A modified law of the wake for turbulent shear layers, J. Fluids Engineering (Sept. 1976) 578–579.

[32] B.G.J. Thompson, A Near Two-parameter Family of Mean Velocity Profiles for Incompressible Boundary layers on Smooth Walls, ARC R&M 3463 (1965).
[33] V.A. Ioselevich, V.I. Pilipenko, Logarithmic Velocity Profile for Flow at a Weak Polymer Solution Near a Rough Surface, Sov. Phys. Dokl. 18 (1974) 790.
[34] D. Betterman, Contribution a l'átude de la couche limite turbulente le long de plaques regueuses. Rapport, Centre Nat. de la Rech. Sci. Paris. France, 1965. 65–6.
[35] C.K. Liu, S.J. Kline, J.P. Johnston, An experimental study of turbulent boundary layers on rough walls. Rep. MD-15, Stanford Univ., Dept. of Mech. Eng. Stanford, California, 1966.
[36] F.A. Dvorak, Calculation of turbulent boundary layers on rough surfaces in pressure gradient, AIAA J. 7 (No. 9) (1969) 1752.
[37] H. Schlichting, Boundary-Layer Theory, McGraw-Hill, New York, 1968.
[38] T.N. Stevenson, A law of the wall for turbulent boundary layers with suction or injection, The College of Aeronautics, Cranfield Rep. Aero. No. 166. (1963). Also N64–19323.
[39] T.N. Stevenson, A modified velocity defect law for turbulent boundary layers with injection, The College of Aeronautics, Cranfield Rep. Aero. No. 170. (1963). Also N64–19324.
[40] R.L. Simpson, W.M. Kays, R.J. Moffat, The turbulent boundary layer on a porous plate: An experimental study of the fluid dynamics with injection and suction. Rep. No. HMT-2, Stanford Univ., Mech. Eng. Dept, Stanford, California, 1967.
[41] D. Coles, A survey of data for turbulent boundary layers with mass transfer, AGARD Conf. Proc. No. 93 on Turbulent Shear Flows, London, 1975. Also AD738–102.
[42] J. McQuaid, Experiments on incompressible turbulent boundary layers with distributed injection, British ARC 28735, HMT 135 (1967). Also N69–21334.
[43] M. Bertram, A. Cary, A. Whitehead, Experiments with hypersonic turbulent boundary layers on flat plates and wings. Proc. AGARD Specialists Meeting Hypersonic Boundary Layers Flow Fields, London, AGARD-CP-30 1 (1) (1968). Also N69–10186.
[44] L.H. Back, R.F. Cuffel, P.F. Massier, Effect of wall cooling on the mean structure of a turbulent boundary layer in low-speed gas flow, Int. J. Heat Mass Transfer 13 (1970) 1029.
[45] L.H. Back, R.F. Cuffel, P.F. Massier, Laminarization of a turbulent boundary layer in a nozzle flow-boundary layer and heat transfer measurements with wall cooling, J. Heat Transfer Ser. G. 92 (1970) 333.
[46] E.R. Van Driest, Turbulent boundary layer in compressible fluids, J. Aeronaut. Sci. 18 (1951) 145.
[47] L.B. Mack, An experimental investigation of the temperature recovery-factor. Jet Propulsion Lab., Rep, California, California Inst. Technol. Pasadena, 1954. 20–80.
[48] J.E. Danberg, Characteristics of the turbulent boundary layer with heat and mass transfer at $M = 6.7$. Naval Ord. Rep. NOLTR (1964) 64–99. Also AD 452–471.
[49] L.O.F. Jeromin, Compressible turbulent boundary layers with fluid injection. Ph.D. thesis, Cambridge Univ. England, 1966.
[50] J. Laufer, Thoughts on compressible turbulent boundary layers. Rand Rep. RM-5946-PR, California, Santa Monica, 1969. Also N69–31803.
[51] B.T.C. Chu, L.S.G. Kovasznay, Nonlinear interactions in a viscous heat-conducting compressible gas, J. Fluid Mech. 3 (1958) 494.
[52] H.H. Fernholz, P.J. Finley, A critical compilation of compressible turbulent boundary-layer data, AGARDographs 223 (253) (1977) 263.
[53] P. Bradshaw, An improved Van Driest skin-friction formula for compressible turbulent boundary layers, AIAA J. 15 (1977) 212.
[54] H.H. Fernholz, Ein halbempirisches Gesetz für die Wandreibung in kompressiblen turbulenten Grenzschichten bei isothermer und adiabater Wand, Z. Angew. Math. u. Mech. 51 (1971) T146.

[55] D.G. Mabey, Some observations on the wake component of the velocity profiles of turbulent boundary layers at subsonic and supersonic speeds, Aero. Quart. 30 (1979) 590.
[56] D. Coles, The turbulent boundary layer in a compressible fluid, Phys. Fluids 7 (1964) 1403.
[57] E.J. Hopkins, M. Inouye, An evaluation of theories for predicting turbulent skin friction and heat transfer on flat plates at supersonic and hypersonic Mach numbers, AIAA J. 9 (1971) 993.
[58] L.C. Squire, A law of the wall for compressible turbulent boundary layers with air injection, J. Fluid Mech. 37 (1969) 449.
[59] E.S. Taylor, Dimensional Analysis for Engineers, Clarendon, Oxford, 1973.
[60] H. Reichardt, On a theory of free turbulence, J. Roy. Aero Soc. 47 (1943) 167.

Chapter 5

Algebraic Turbulence Models

Analysis of Turbulent Flows with Computer Programs. http://dx.doi.org/10.1016/B978-0-08-098335-6.00005-7

5.1 Introduction

It was shown in Chapter 4 that an accurate calculation of velocity and temperature boundary layers in turbulent flows is complicated by the fact that the governing equations contain fluctuation terms that are at present impossible to relate correctly to the dependent variables in the equations. As discussed in Chapter 1, direct numerical solution (DNS) of the instantaneous Navier-Stokes equations for turbulent flows offers exciting possibilities. The computer requirements of DNS, however, are large, and it is unlikely that this approach can be used for turbulent flow calculations on complex bodies in the near future. For this reason, in order to proceed, it is necessary to use time-averaged equations and introduce some empiricism to them. Over the years, several approaches have been taken, and various models for the Reynolds stresses have been proposed. Algebraic turbulence models based on Prandtl's mixing-length and Boussinesq's eddy-diffusivity concepts are typical examples of such models. They are mostly justified for local equilibrium flows. Although the expressions obtained from those models do not necessarily either describe the microscopic details of a turbulent flow or provide basic information about the turbulence mechanism, they are very useful engineering tools.

In this chapter, we discuss algebraic turbulence models suitable for calculating turbulent boundary layers, transport coefficients that account for various effects such as pressure gradient and heat and mass transfer. They have been used in many calculations methods based on the solutions of the boundary-layer and Navier-Stokes equations and found to give results that usually agree well with experiment. Other turbulence models based on the solution of transport equations such as Reynolds stresses, turbulent kinetic energy, dissipation of equations will be discussed in the following chapter. The calculation methods employing these models are more general than those that employ algebraic models because they can handle a larger class of flows than algebraic models.

5.2 Eddy Viscosity and Mixing Length Models

The conservation equations for a compressible turbulent flow were derived in Chapter 2. Let us now consider an incompressible flow and write the continuity and momentum equations for it. From Eqs. (2.5.3) and (2.5.4) we can write

$$\partial \bar{u}_j / \partial x_j = 0, \tag{5.2.1}$$

$$\varrho\left(\frac{\partial \bar{u}_i}{\partial t} + \bar{u}_j \frac{\partial \bar{u}_i}{\partial x_j}\right) = -\frac{\partial p}{\partial x_i} + \frac{\partial}{\partial x_j}(\bar{\tau}_{ij} + \overline{R}_{ij}), \tag{5.2.2}$$

where $\overline{R}_{ij}$ is the Reynolds stress tensor defined by

$$\bar{\mathrm{R}}_{ij} = -\varrho\overline{u_i'u_j'}. \tag{5.2.3}$$

With Boussinesq's eddy-viscosity concept discussed in the previous chapter we can write the Reynolds stress tensor as

$$\bar{\mathrm{R}}_{ij} = \varrho\overset{*}{\varepsilon}_{\mathrm{m}}\left(\frac{\partial\bar{u}_i}{\partial x_j} + \frac{\partial\bar{u}_j}{\partial x_i}\right), \tag{5.2.4}$$

which is similar to τ_{ij} defined in Eq. (2.2.4), that is,

$$\overline{\tau}_{ij} = \mu\left(\frac{\partial u_i}{\partial x_j} + \frac{\partial u_j}{\partial x_i}\right).$$

Equation (5.2.4) is a definition $\overset{*}{\varepsilon}_{\mathrm{m}}$, but see Eq. (5.2.5).

According to Boussinesq's concept, the eddy viscosity $\overset{*}{\varepsilon}_{\mathrm{m}}$ has a scalar value. Originally, Boussinesq assumed directional constancy, but in the application of the theory to turbulent flow through channels, he assumed that $\overset{*}{\varepsilon}_{\mathrm{m}}$ was spatially constant also. Such a constant value can be expected to occur only if the turbulent flow field is at least homogeneous. In a few cases of free shear flows that are not homogeneous and show a pronounced velocity gradient and a shear stress in one direction, it is possible to describe the overall flow field in a satisfactory way on the assumption of a constant eddy viscosity. However, a constant value cannot be expected as a general rule. For example, as was discussed in Section 4.3, in turbulent flows near walls, $\overset{*}{\varepsilon}_{\mathrm{m}}$ is not a constant in the boundary layer, but varies approximately linearly with y.

If the concept of a scalar eddy viscosity is to be used, a more accurate procedure would be to extract the average turbulence pressure from the turbulence stresses as a separate term and write the Reynolds stress tensor as

$$\mathrm{R}_{ij} \equiv \frac{\bar{\mathrm{R}}_{ij}}{\varrho} = \frac{2}{3}k\delta_{ij} + \varepsilon_{\mathrm{m}}\left(\frac{\partial\bar{u}_i}{\partial x_j} + \frac{\partial\bar{u}_j}{\partial x_i}\right) \tag{5.2.5}$$

where $k = q^2/2$.

Another closure approach to model the Reynolds stress tensor is the "mixing length" approach, in which Eq. (5.2.5) is written in the form

$$\mathrm{R}_{ij} = \frac{2}{3}k\delta_{ij} + l^2\left(\frac{\partial\bar{u}_i}{\partial x_j} + \frac{\partial\bar{u}_j}{\partial x_i}\right)\left(\frac{\partial\bar{u}_i}{\partial x_j} + \frac{\partial\bar{u}_j}{\partial x_i}\right). \tag{5.2.6}$$

Occasionally the approaches are mixed. Comparison of Eq. (5.2.6) with Eq. (5.2.5) shows that

$$\varepsilon_{\mathrm{m}} = l^2\left(\frac{\partial \bar{u}_i}{\partial x_j} + \frac{\partial \bar{u}_j}{\partial x_i}\right), \tag{5.2.7}$$

where l is a turbulence length scale.

For a two-dimensional, incompressible, steady flow, the boundary-layer forms of Eqs. (5.2.1) and (5.2.2) are

$$\frac{\partial u}{\partial x} + \frac{\partial v}{\partial y} = 0, \tag{5.2.8}$$

$$u\frac{\partial u}{\partial x} + v\frac{\partial u}{\partial y} = -\frac{1}{\varrho}\frac{dp}{dx} + \nu\frac{\partial^2 u}{\partial y^2} - \frac{\partial}{\partial y}\overline{u'v'}. \tag{5.2.9}$$

The relation between Reynolds shear stress and mean velocity gradient is

$$-\overline{u'v'} = \varepsilon_{\mathrm{m}}\frac{\partial u}{\partial y} = l^2\left(\frac{\partial u}{\partial y}\right)^2. \tag{5.2.9b}$$

For the sake of simplicity, we have neglected the bars over the mean quantities in these equations. For convenience, we shall call $-\overline{u'v'}$ the Reynolds shear stress (actually $-\varrho\overline{u'v'}$ is the shear stress).

Information on the distribution of l and ε_{m} in turbulent flows comes from experimental data. The distribution of l across a boundary layer can conveniently be described by two separate empirical functions. In the fully turbulent part of the inner region, l is proportional to y, and in the outer region it is proportional to δ. Therefore,

$$l_i = \kappa y \qquad y_0 \le y \le y_{\mathrm{c}}, \tag{5.2.10a}$$

$$l_{\mathrm{o}} = \alpha_1 \delta \qquad y_{\mathrm{c}} \le y \le \delta, \tag{5.2.10b}$$

where y_0 is a small distance from the wall and y_{c} is obtained from the continuity of l. The empirical parameters κ and α_1 vary slightly, according to the experimental data. Here, we shall take them to be 0.40 and 0.075, respectively. Later, we shall discuss the universality of these parameters (see subsection 5.3.1).

Similarly, according to experiments with equilibrium boundary layers, ε_{m} also varies linearly with y in the inner region and is nearly constant (except for the intermittency) in the outer region (see Section 4.3). Its variation across the boundary layer can conveniently be described by the following formulas:

$$(\varepsilon_m)_i = l^2 |\partial u/\partial y| \quad y_0 \leq y \leq y_c, \tag{5.2.11a}$$

$$(\varepsilon_m)_o = \alpha u_e \delta_k^* \gamma \quad y_c \leq y \leq \delta. \tag{5.2.11b}$$

Here δ_k^* is defined by

$$\delta_k^* = \int_0^\delta \left(1 - \frac{u}{u_e}\right) dy$$

and γ is given by Eq. (4.3.14) which can be approximated by

$$\gamma = \left[1 + 5.5\left(\frac{y}{\delta_0}\right)^6\right]^{-1}$$

where δ_0 is defined as the location where $u/u_e = 0.995$.

Although it varies somewhat with Reynolds number when $R_\theta < 5000$, the parameter α in Eq. (5.2.11b) is generally assumed to be a constant equal to 0.016–0.0168. Later, we shall also discuss its universality (see Section 5.4).

The mixing-length and eddy-viscosity expressions given by Eqs. (5.2.10) and (5.2.11) apply in the fully turbulent part of the boundary layer, excluding the sublayer and buffer layer close to the wall. They can be modified, in order to make them applicable over the entire boundary layer, by using various empirical expressions. Here, we use the expression proposed by Van Driest [1] and write the mixing length as

$$L = l[1 - \exp(-y/A)]. \tag{5.2.12}$$

where

$$A = A^+ \nu (\tau_w/\varrho)^{-1/2}. \tag{5.2.13}$$

Here A^+ is a dimensionless constant. The parameter A is often referred to as the Van Driest damping parameter.

For the inner region of the boundary layer, where the law of the wall applies, the mixing length l is proportional to the distance y from the wall: $l = \kappa y$. Taking $\kappa = 0.40$ and comparing his model with experimental data at high Reynolds number ($R_\theta > 5000$), Van Driest empirically determined the constant A^+ in Eq. (5.2.13) to be 26. With those assumptions and constants, the eddy viscosity for the inner region becomes

$$(\varepsilon_m)_i = (\kappa y)^2 [1 - \exp(-y/A)]^2 |\partial u/\partial y|. \tag{5.2.14}$$

Over the years several algebraic eddy viscosity and mixing-length models have been developed and their accuracy have been explored for a range of turbulent shear flows: the Cebeci-Smith (CS), Mellor-Herring (MR), Patankar-Spalding (PS) and Michel-Quemard-Durant (MQD) models are typical examples. The CS [2] and MH [3] models use the eddy viscosity approach and the PS [4] and MQD [5] models use the mixing-length approach. All four treat the boundary layer as a composite layer characterized by inner and outer regions and use separate expressions for ε_m or l in the two regions. The main differences lie in the numerical method used to solve the equations. Also the treatment of the empirical functions ε_m or l^2 in the inner region very close to the wall and in the outer region differ.

Of these algebraic turbulence models, the CS model has been extended and tested for a wide-range of turbulent flows, mostly for boundary-layer flows, and is discussed here in some detail for momentum and heat transfer.

5.3 CS Model

The CS model is based on a two-layer eddy-viscosity formulation given by Eqs. (5.2.11), (5.2.12) and (5.2.13) and contains several modifications to the inner and outer expressions. For example, the expression (5.2.13) was obtained for a flat-plate flow with no mass transfer and should not be used for a turbulent boundary layer with strong pressure gradient and heat and mass transfer. That there must be no strong pressure gradient is quite obvious, since for a flow with an adverse pressure gradient, τ_w may approach zero (flow separation), in which case the inner eddy viscosity predicted by the Van Driest formula will be zero.[6] For that reason the expression was extended by Cebeci [6,7] to flows with pressure gradient, mass transfer, and heat transfer. According to this extension, the damping-length constant A in Eq. [5.2.13] is given by

$$A = A^{+}\frac{\nu}{N}\left(\frac{\tau_w}{\varrho_w}\right)^{-1/2}\left(\frac{\varrho}{\varrho_w}\right)^{1/2}, \tag{5.3.1}$$

where

$$N = \left\{\frac{\mu}{\mu_e}\left(\frac{\varrho_e}{\varrho_w}\right)^2\frac{p^+}{v_w^+}\left[1-\exp\left(11.8\frac{\mu_w}{\mu}v_w^+\right)\right]+\exp\left(11.8\frac{\mu_w}{\mu}v_w^+\right)\right\}^{1/2} \tag{5.3.2a}$$

---

[6] Writing the ratio of y/A in the exponential term as y^+/A^+, we see that, if $\tau_w = 0$, $[1-\exp(-y/A)]$ will be zero. Hence $(\varepsilon_i)_m = 0$.

$$p^+ = \left(\nu_e u_e/u_\tau^3\right)\left(du_e/dx\right), \quad v_w^+ = v_w/u_\tau, \quad u_\tau = (\tau_w/\varrho_w)^{1/2}. \tag{5.3.2b}$$

For flows with no mass transfer, N can be written as

$$N = \left[1 - 11.8(\mu_w/\mu_e)(\varrho_e/\varrho_w)^2 p^+\right]^{1/2}. \tag{5.3.2c}$$

5.3.1 Effect of Low Reynolds Number

After an extensive survey of mean-velocity-profile measurements in flows with zero pressure gradient, Coles [8] showed that the mean-velocity distribution across an incompressible boundary layer outside the sublayer at low Reynolds number can accurately be described by the expression

$$u^+ = (1/\kappa)\ln y^+ + c + (\Pi/\kappa)w(y/\delta), \tag{4.4.23}$$

with $\kappa = 0.41$ and $c = 5.0$, provided that the profile parameter Π varies with R_θ according to the curve of Fig. 4.20. There, it can be seen that for $R_\theta > 5000$, Π is a constant equal to 0.55. The variation of Π with R_θ can be approximated by

$$\Pi = 0.55\left[1 - \exp\left(-0.243z_1^{1/2} - 0.298z_1\right)\right], \tag{5.3.3}$$

where $z_1 = (R_\theta/425 - 1)$.

There has been a number of studies conducted to see whether the parameters κ and c in Eq. (4.4.23) were not constant. On the basis of his experimental data, Simpson [9] reported that for values of $R_\theta < 6000$, they varied with R_θ by the following empirical formulas:

$$\kappa = 0.40(R_\theta/6000)^{-1/8}, \tag{5.3.4a}$$

$$c = R_\theta^{1/8}[7.90 - 0.737\ln|R_\theta|]. \tag{5.3.4b}$$

Furthermore, the parameter α_1 in the outer mixing-length formula (5.2.10b) and the parameter α in the outer eddy-viscosity formula (5.2.11b) were not constant; for values of $R_\theta < 6000$, they varied with R_θ.

Simpson approximated the variation of α with R_θ by the following expression:

$$\alpha = 0.016R_\theta^{-1/4}. \tag{5.3.5}$$

In their eddy-viscosity method for calculating compressible turbulent boundary layers, Herring and Mellor [3] observed that, if κ was kept constant and α varied with Reynolds number as

$$\alpha = 0.016\left[1 + (1100/R_s)^2\right], \tag{5.3.6}$$

the calculated results agreed much better with experiment than those obtained with constant α (= 0.016). In Eq. (5.3.6), R_s is defined by

$$R_s = u_e\delta_k^*/\nu_s, \tag{5.3.7}$$

where ν_s is the kinematic viscosity at the edge of the sublayer.

Cebeci and Mosinskis [10] varied κ and the damping-length constant A^+ and showed that when Eq. (5.2.11b) (with $\alpha = 0.0168$) and Eq. (5.2.14) were used as the eddy-viscosity formulation in the solution of the boundary-layer equations by the CS method, the agreement with experiment was improved. The variation of κ and A^+ were related to Reynolds number by the following interpolation formulas:

$$\kappa = 0.40 + \frac{0.19}{1 + 0.49z_2^2}, \quad A^+ = 26 + \frac{14}{1 + z_2^2}, \tag{5.3.8}$$

where $z_2 = R_\theta \times 10^{-3} > 0.3$.

Huffman and Bradshaw [11] obtained a correlation in terms of A^+ and $\partial\tau^+/\partial y^+$ that is valid for a number of flows ranging from axisymmetric wall jets to two-dimensional boundary layers. They concluded that the von Kármán constant κ in the mixing-length formula is a universal constant.

Bushnell and Morris [12] analyzed measurements in hypersonic turbulent boundary layers at low Reynolds numbers. They observed variations with Reynolds number of the parameters κ and α in the inner and the outer eddy-viscosity formulas similar to those in Eqs. (5.3.5) and (5.3.8).

The universality of the parameters κ and α was also studied by Cebeci [13]. The study showed that α is not a universal constant at low Reynolds numbers, but that it varies with Reynolds number. According to his study, α is given by

$$\alpha = \alpha_0(1 + \Pi_0)/(1 + \Pi), \tag{5.3.9}$$

where $\Pi_0 = 0.55$, $\alpha = 0.0168$ and Π is given by Eq. (5.3.3).

Figure 5.1 shows a comparison of calculated local skin-friction values for a flat plate for a range of R_θ of 425 to 10,000. The calculations were made by using the eddy-viscosity formulation given by Eqs. (5.2.11b), (5.2.14), and (5.3.9) with $\kappa = 0.40$, $A^+ = 26$. In one set of calculations, α was kept constant (= 0.0168), and in another set α was varied according to Eq. (5.3.9). The results in Fig. 5.1 show that when α is constant, the calculated skin-friction values differ considerably from those given by Coles [8]. However, when α varies according to Eq. (5.3.9), the calculated c_f values are in very close agreement with those of Coles.

Figure 5.2a shows the calculated R_θ and H values as functions of R_x obtained with the same values of α as in Fig. 5.1. As can be seen, the value of α has an important

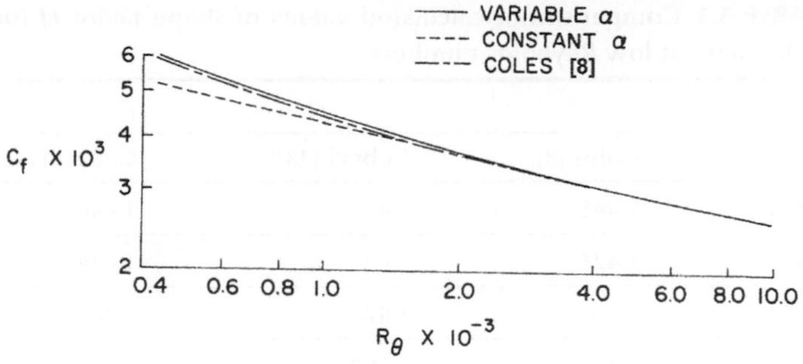

Fig. 5.1 Effect of a on local skin-friction coefficient. The calculations were made by the CS method [2].

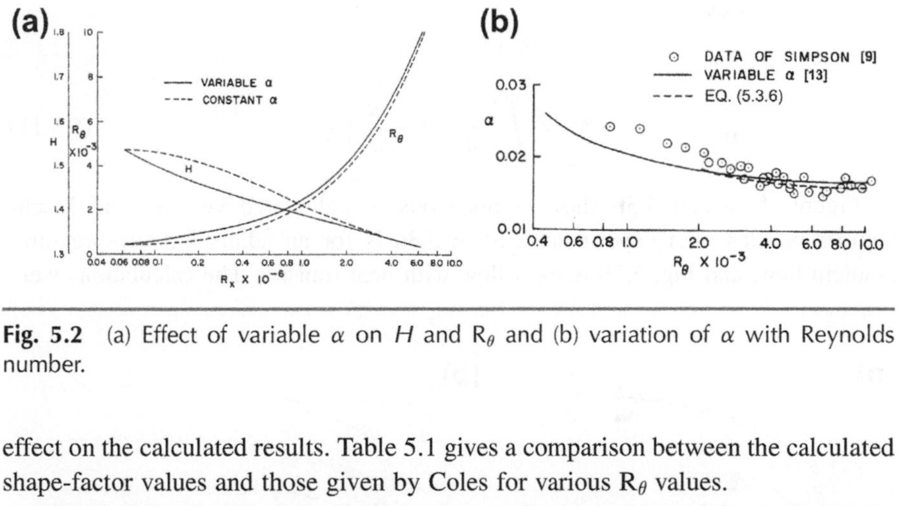

Fig. 5.2 (a) Effect of variable α on H and R_θ and (b) variation of α with Reynolds number.

effect on the calculated results. Table 5.1 gives a comparison between the calculated shape-factor values and those given by Coles for various R_θ values.

Figure 5.2b shows the variation of α with momentum-thickness Reynolds number R_θ. The experimental data of Simpson [9] and the curve calculated by the Herring and Mellor [3] formula (5.3.6) are shown in the same figure. That formula was used to improve their computed results only for values of R_θ higher than 2000. For that reason, in Fig. 5.2b the formula was not used for values of R_θ below 2000.

Cebeci [13] has also extended Eq. (5.3.9) to compressible flows by replacing the momentum-thickness Reynolds number in Eq. (5.3.3) by the kinematic quantity R_{θ_k} defined by

$$R_{\theta_k} = u_e \theta_k / \nu_w, \tag{5.3.10}$$

where ν_w is the wall kinematic viscosity and θ_k is the kinematic momentum thickness defined by

TABLE 5.1 Comparison of calculated values of shape factor H for a flat plate at low Reynolds numbers

	H		
R_θ	Coles [8]	Cebeci [13][a]	Cebeci [14][b]
1150	1.445	1.472	1.446
1450	1.425	1.447	1.424
2050	1.403	1.420	1.402
2650	1.390	1.397	1.387
4150	1.365	1.365	1.363
5650	1.350	1.346	1.350

[a] α = constant.
[b] α = variable.

$$\theta_{\mathrm{k}} = \int_0^\infty \frac{u}{u_{\mathrm{e}}}\left(1 - \frac{u}{u_{\mathrm{e}}}\right)dy. \tag{5.3.11}$$

Figures 5.3a and 5.3b show comparisons of calculated velocity and Mach-number profiles with experiment. Figure 5.3a is for an adiabatic, zero-pressure-gradient flow, and Fig. 5.3b is for a flow with heat transfer. The calculations were

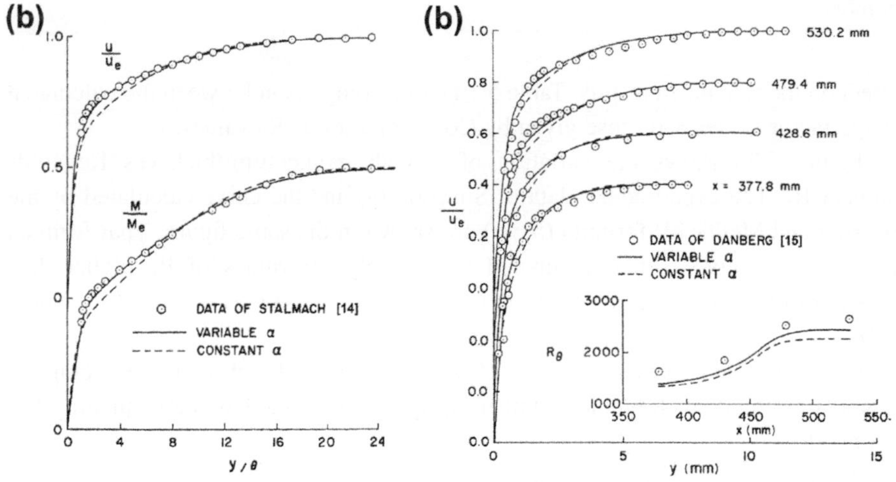

Fig. 5.3 Effect of α on the (a) velocity and Mach profiles for an adiabatic compressible flow, $M_e = 3.4$, $R_\theta = 2.76 \times 10^3$, and (b) velocity profiles for a compressible flow with heat transfer, $M_e = 6.6$, $T_w/T_e = 5.2$.

made by the CS method. In one set of calculations, the parameter α was held constant ($= 0.0168$); in the other, it varied according to Eq. (5.3.9), with R_{θ_k} given by Eq. (5.3.10). Clearly, in all cases the agreement with experiment is better if α is varied.

5.3.2 Effect of Transverse Curvature

First-order boundary-layer theory is based on the assumption that the boundary-layer thickness δ is small in comparison with a characteristic length L. There are some flows for which this assumption fails. Typical examples are long, slender bodies of revolution with slender tails. In flows past these bodies, the thickness of the boundary layer may be of the same order as, or larger than, a characteristic length, for example, the body radius r_0. As an example, consider an axial flow at zero incidence along a cylinder. If the radius of the cylinder is large in comparison with the thickness of the boundary layer δ, the flow is essentially two-dimensional and is not significantly different from flow past a flat plate. However, if the radius of the cylinder is small in comparison with δ, we may expect the flow to differ from the two-dimensional case, since in that case the flow tends to wrap itself around the body. The effect, called the transverse curvature (TVC) effect, strongly influences the skin-friction and heat-transfer characteristics and must be taken into consideration in calculating the flow.

If thick axisymmetric turbulent boundary layers are to be calculated by means of an eddy-viscosity concept, the question of the applicability of the two-dimensional eddy-viscosity distribution immediately arises. If that distribution is not applicable, how can the proper distribution for such flows be found?

The question was considered by Cebeci [16]. According to his study, the inner eddy-viscosity distribution in such layers differs from the two-dimensional eddy-viscosity distribution, but the outer eddy-viscosity distribution does not change, that is, it is given by Eqs. (5.2.11b), (5.3.9), and (5.3.10). For thick axisymmetric turbulent boundary layers, the inner eddy-viscosity formula is

$$(\varepsilon_m)_i = L^2 \frac{r_0}{r} \frac{du}{dy}, \tag{5.3.12}$$

where

$$L = 0.4 r_0 \ln (r/r_0) \left\{ 1 - \exp\left[-\frac{r_0}{A} \ln\left(\frac{r}{r_0}\right)\right]\right\}. \tag{5.3.13}$$

As before, the damping length parameter, A, in Eq. (5.3.13) is given by Eq. (5.3.1).

Figures 5.4a and 5.4b show the comparison of calculated and experimental velocity and Mach-number profiles. The experimental data are due to Richmond [17]. The calculations were made by the CS method with the two-dimensional

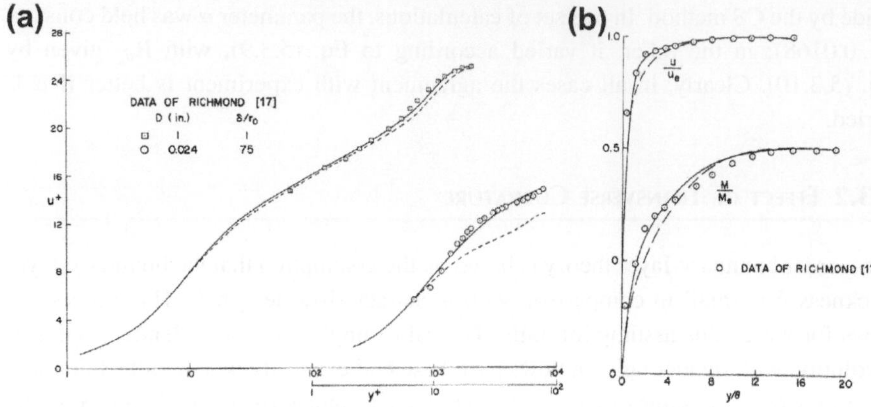

Fig. 5.4 Comparison of calculated and experimental dimensionless (a) velocity profiles for two cylinders in incompressible flows and (b) velocity and Mach-number profiles for compressible adiabatic flows. $M_e = 5.825$, $R_\theta = 4390$, cylinder diam. = 0.024 in. In (a) the experimental u^+ values are obtained by normalizing the measured u values by the calculated friction velocity u_τ. The lower y^+ scale refers to the 0.024 in. cylinder. – refers to the modified eddy viscosity formulation and - - - to the 2-d formulation.

eddy-viscosity distribution given by Eqs. (5.2.11b), (5.2.14), (5.3.1), (5.3.9), and (5.3.10) and the extension of that formulation to thick axisymmetric boundary layers, namely, Eqs. (5.2.11b), (5.3.12), (5.3.13), (5.3.19), and (5.3.55). Figure 5.4a shows comparisons of velocity profiles for two cylinders with diameters of 0.024 in. and 1 in. for an incompressible flow. Figure 5.4b shows the comparisons for a cylinder with a diameter of 0.024 in. for an adiabatic compressible turbulent boundary layer for $M_e = 5.825$. As can be seen, modifying the two-dimensional eddy-viscosity distribution for thick axisymmetric boundary layers improves the calculations.

5.3.3 Effect of Streamwise Wall Curvature

Streamwise wall curvature may increase or decrease the intensity of the turbulent mixing, depending on the degree of the wall curvature, and it can strongly affect the skin friction and the heat-transfer rates. For example, Thomann [18] showed that the rate of heat transfer in a supersonic turbulent boundary layer on a concave wall was increased by the streamwise curvature of the wall. For the configuration he investigated, the pressure was held constant along the wall, and the increase of about 200% was therefore due only to the wall curvature. Under the same conditions, he found a comparable decrease for a convex wall.

To some extent, the streamwise curvature effect can be incorporated into the eddy-viscosity expressions [19] by multiplying the right-hand side of Eq. (5.2.11)

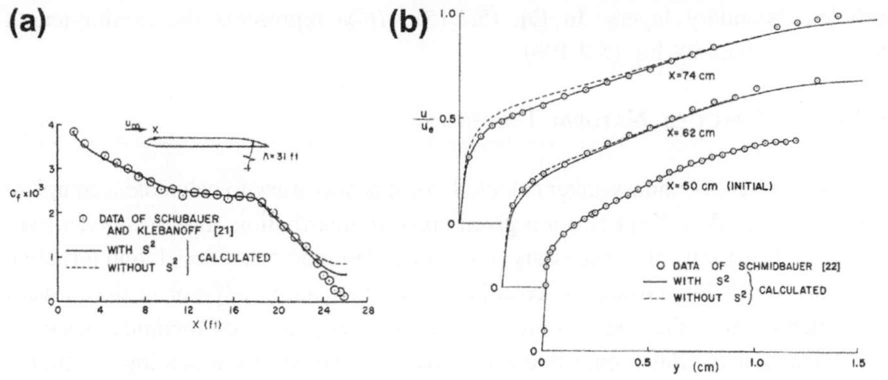

Fig. 5.5 Comparison of calculated (a) local skin-friction coefficients and (b) velocity profiles with experiment.

with the inner eddy-viscosity expression (5.2.14) by S^2, an expression given by Bradshaw [20]. Bradshaw's expression, which is based on an analogy between streamline curvature and buoyancy in turbulent shear flows, is

$$S = \frac{1}{1+\beta \mathrm{Ri}}, \qquad \mathrm{Ri} = \frac{2u}{\Lambda}\left(\frac{\partial u}{\partial y}\right)^{-1}, \tag{5.3.14}$$

where Ri is analogous to the Richardson number and Λ is the longitudinal radius of curvature. The parameter β is equal to 7 for a convex surface and 4 for a concave surface, according to meteorological data and the use of the above analogy. The radius of curvature is positive for a convex surface and negative for a concave surface. According to Bradshaw, the effects of curvature on the mixing length or eddy viscosity are appreciable if the ratio of boundary-layer thickness to radius of curvature, $\delta : \Lambda$, exceeds roughly 1 : 300.

Figure 5.5a shows the effect of wall-curvature modification on the computed skin-friction for the experimental data of Schubauer and Klebanoff [21]. Figure 5.5b shows the effect of wall-curvature modification on the computed velocity for profiles for the data of Schmidbauer [22]. In the former case, $\delta : \Lambda$ is around 1 : 100; in the latter case, it is around 1 : 75. The wall-curvature correction seems to improve the calculations.

Bradshaw's expression for curvature effect has also been used in the mixing-length expressions. For example, Bushnell and Alston [23] modified the mixing-length expression by using

$$l/\delta = (l/\delta)_0(1 - \beta \mathrm{Ri}) \tag{5.3.15}$$

and obtained better agreement with the experimental data of Hoydysh and Zakkay [24] than without any correction. Their calculations were made for hypersonic turbulent boundary layers. In Eq. (5.3.15), $(l/\delta)_0$ represents the mixing-length distribution given by Eq. (5.2.10b).

5.3.4 The Effect of Natural Transition

In most practical boundary-layer calculations, it is necessary to calculate a complete boundary-layer flow. That is, for a given pressure distribution and for a given transition point (natural), it is necessary to calculate laminar, transitional, and turbulent boundary layers by starting the calculations at the leading edge or at the forward stagnation point of the body. In most boundary-layer prediction methods, however, the calculation of transitional boundary layers is avoided by assuming the transitional region to be just a switching point between laminar and turbulent regions. In general, especially at low Reynolds numbers, that is not a good procedure, and it can lead to substantial errors. The point can best be described by an example. Consider the flow past a turbine or compressor blade and assume two blade Reynolds numbers, $R_b = 10^5$ and 10^6. The extent of the transitional region on the blade at each of those two Reynolds numbers can be estimated by using a correlation given by Chen and Thyson [25]:

$$R_{\Delta x} = R_{x_t} - R_{x_{tr}} = CR_{x_t}^{2/3}, \tag{5.3.16}$$

where $R_{\Delta x}$ is the extent of the transition region, $R_{x_{tr}}$ is the Reynolds number based on the distance to the start of the transition, and R_{x_t} is the Reynolds number based on the completion of transition. C is an empirical expression given by

$$C = 60 + 4.86M_e^{1.92} \quad 0 < M_e < 5. \tag{5.3.17}$$

The expressions (5.3.16) and (5.3.17) are based on the correlation of incompressible and compressible adiabatic data for Mach numbers less than 5. If we assume that transition starts at two points, namely, at the 10% and 50% chord points, the extents of the transitional Reynolds number $R_{\Delta x}$ for two blade Reynolds numbers R_b, according to Eq. (5.3.16), are shown in Table 5.2.

The tabulated values of $R_{\Delta x}$ clearly show that the transitional region is very important and that it must be accounted for in order to make accurate boundary-layer calculations. For example, for $R_b = 1 \times 10^5$ and transition starting at $x/c = 0.5$, the transition region is $0.81c$ in length, which means that the flow on the body from the start of transition right up to the trailing edge is in a transitional state.

Naturally developing transition does not occur as a sharp, continuous front. Instead, random spots of turbulence arise. Outside of these spots, the flow is still fully laminar. The spots grow because fluid in contact with them is contaminated. While growing, they are carried along by the flow. The net result is that they sweep out wedges of about 8 or 9° half angle. As more and more spots are formed throughout

TABLE 5.2 Extent of transitional Reynolds number $R_{\Delta x}$ at two blade chord Reynolds numbers R_b, with $M_e = 0$

R_b	Start of transition $R_{x_{tr}} \times 10^{-5}$	Length of transition $R_{\Delta x} \times 10^{-5}$	End of transition $R_{x_t} \times 10^{-5}$
1	0.1	0.28	0.38
1	0.5	0.81	1.31
10	1.0	1.30	2.30
10	5.0	3.80	8.80

the transition region and as the existing ones grow in size, the flow reaches a point where no laminar gaps are left, so that it has become fully turbulent. Emmons [26] first identified these spots and the intermittency by observation of water flow in a shallow channel. In his paper he laid down the foundations of a statistical theory for analyzing their effect and the coalescing process. He introduced an intermittency factor γ, such that $\gamma = 0$ corresponds to fully laminar and $\gamma = 1$ to fully turbulent flow.

The eddy-viscosity distribution given by Eq. (5.2.11) can be modified to account for the transition region in both incompressible and compressible flows [19]. The transition region can be accounted for by multiplying Eq. (5.2.11) by an intermittency expression given by Chen and Thyson [25]. That expression was developed from the point of view of intermittent production of turbulent spots and is a further extension to compressible flow with pressure gradient of Emmons' spot theory and Dhawan and Narasimha's intermittency expression [27] for incompressible flows. According to Chen and Thyson, the intermittency factor γ_{tr} is given by

$$\gamma_{tr} = 1 - \exp\left[G r_0 (x_{tr}) \left(\int_{x_{tr}}^{x} \frac{dx}{r_0}\right) \left(\int_{x_{tr}}^{x} \frac{dx}{u_e}\right)\right], \tag{5.3.18}$$

where G is a spot-formation-rate parameter

$$G = \left(3/C^2\right)\left(u_e^3/\nu^2\right) R_{x_{tr}}^{-1.34} \tag{5.3.19}$$

and x_{tr} is the location of the start of transition. The transition Reynolds number is defined as $R_{x_{tr}} = u_e x_{tr}/\nu$. For simple shapes, Eq. (5.3.18) can be simplified considerably. For example, for a straight tube or for a flat plate, it becomes

$$\gamma_{tr} = 1 - \exp\left[-\left(G/u_e\right)\left(x - x_{tr}\right)^2\right]. \tag{5.3.20}$$

For a cone in supersonic flow,

$$\gamma_{tr} = 1 - \exp\left[-Gx_{tr}\left(\ln \frac{x}{x_{tr}}\right)\left(\frac{x - x_{tr}}{u_e}\right)\right]. \tag{5.3.21}$$

It should be pointed out that for incompressible two-dimensional flows, the start of transition can be satisfactorily calculated by using several empirical correlations. One such useful expression is based on a combination of Michel's method [28] and Smith and Gamberoni's e^9 correlation curve [29]. It is given by Cebeci et al. [30] as

$$R_{\theta_{tr}} = 1.174\left[1 + \left(22400/R_{x_{tr}}\right)\right]R_{x_{tr}}^{0.46}. \tag{5.3.22}$$

Figures 5.6a and 5.6b show comparisons of calculated and experimental results of using the CS method in Chapter 8 for two different flows. Figure 5.6a is for an incompressible flow at relatively low Reynolds number. Figure 5.6b is for a supersonic adiabatic flow. In both cases, the calculations that use the product of the intermittency distribution given by Eq. (5.3.18) and the eddy-viscosity formulation given by Eqs. (5.2.11), (5.2.14), (5.3.1), and (5.3.9) seem to account for the transition region rather well.

Studies of methods of calculating the transition region between the laminar part and the turbulent part of a boundary layer have also been conducted by Adams [33] and by Harris [34]. Both authors used Dhawan and Narasimha's intermittency expression and obtained good agreement with experiment. Figure 5.7 shows a comparison of calculated and experimental velocity profiles for laminar, transitional, and turbulent boundary-layer flows over a hollow cylinder. The calculations were made by Harris [34]. The experimental data are due to O'Donnell [35]. For a unit Reynolds number of 2.2×10^6 per meter, the boundary layer was laminar throughout the measured area. The velocity profiles are similar, and the agreement

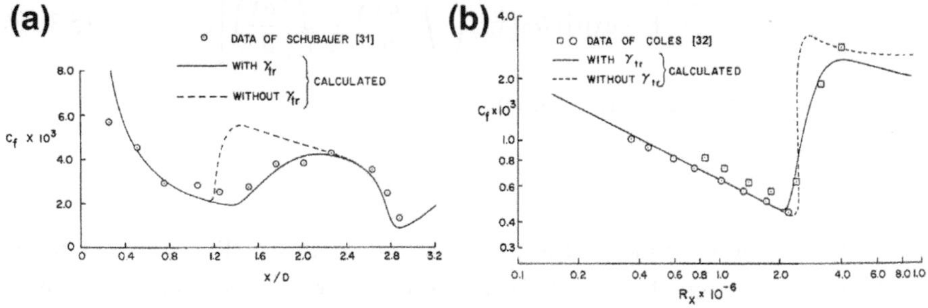

Fig. 5.6 Comparison of calculated laminar, transitional, and turbulent local skin-friction coefficients with experiment for (a) incompressible flows and (b) for adiabatic compressible flows, $M_e = 1.97$.

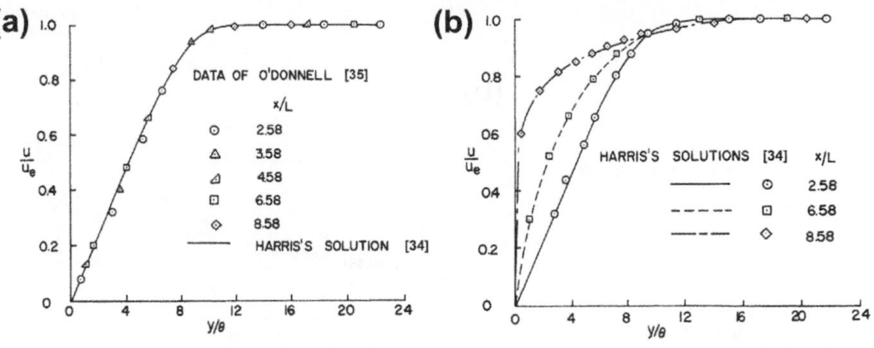

Fig. 5.7 Comparison of calculated and experimental velocity profiles for laminar, transitional, and turbulent boundary flows over a hollow cylinder. The calculations were made by Harris [34]; $L = 2.54$cm. (a) Laminar flow, $R/m = 2.2 \times 10^6$. (b) Laminar, transitional, and turbulent flow, $R/m = 9.45 \times 10^6$.

between the calculated results and experiment is very good (see Fig. 5.7a). For unit Reynolds numbers of 9.45×10^6 per meter, laminar, transitional, and turbulent flow occurred. Again the agreement between the calculated results and experiment is very good, as is shown in Fig. 5.7b.

Separation-Induced Transition

The length of the transition region is also susceptible to the degree of freestream turbulence, especially of large scale, flow separation and surface roughness, decreasing rapidly as these features of the flow become more pronounced. While these features are difficult to incorporate into expressions like Eq. (5.3.18), or to correlate with data, in the case of two-dimensional low Reynolds-number flows, Eq. (5.3.18) was extended by Cebeci [36] to model the transition region in separation bubbles. The parameter C in Eq. (5.3.19) was expressed in terms of $R_{x_{tr}}$ and its variation with $R_{x_{tr}}$, with the onset of transition obtained from the e^n-method [38], is shown in Fig. 5.8a, together with the experimental data obtained for four airfoils. The data encompass a typical low Reynolds number range from chord Reynolds $R_c = 2.4 \times 10^5$ to 2×10^6. They fall conveniently on a straight line on a semilog scale and can be represented by the equation

$$C^2 = 213\left(\log R_{x_{tr}} - 4.7323\right). \qquad (5.3.23)$$

Figure 5.8b shows the results obtained with this modification to the CS model for the ONERA-D airfoil examined by Cousteix and Pailhas [37] in a wind tunnel with a chord Reynolds number, R_c, of 3×10^5 at zero angle of attack. The calculations were made by using the interactive boundary layer method described in [38], with the onset of transition calculated with the e^n-method.

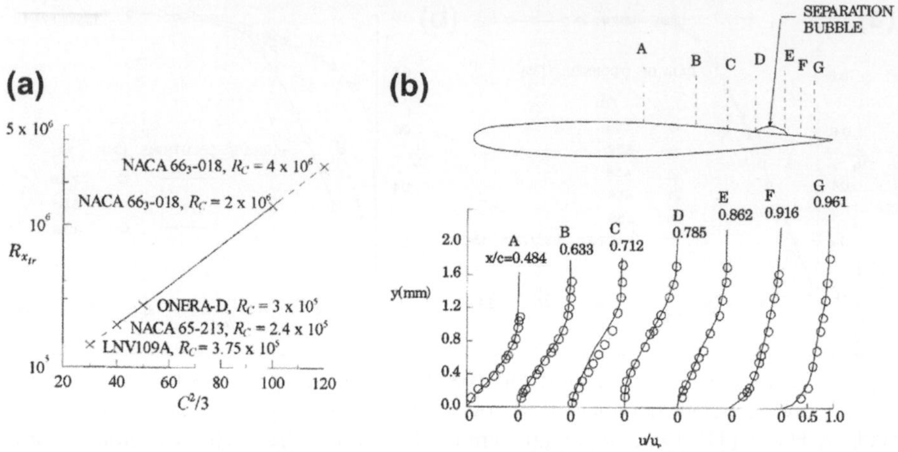

Fig. 5.8 (a) Variation of $C^2/3$ with R_{xtr} and (b) comparison of calculated (solid lines) and measured (symbols) velocity profiles for the ONERA-D airfoil for $\alpha = 0°$, $R_c = 3.0 \times 10^5$.

The airfoil and mean velocity profiles shown in Fig. 5.8b indicate excellent agreement between the measured and calculated results. For this flow, transition occurred within the separated flow region and caused reattachment shortly thereafter. The calculations revealed transition at $x/c = 0.81$ with the e^n-method in comparison with measurement which revealed transition at $x/c = 0.808$. Additional comparisons between calculations and experimental data are given in [38].

5.3.5 Effect of Roughness

The CS model discussed in the previous subsections was also extended by Cebeci and Chang [39] to represent flow over rough walls without and with pressure gradients. This was done by modifying the inner eddy viscosity formula with the help of Rotta's model [40] which recognized that the velocity profiles for smooth and rough walls can be similar, provided that the coordinates are displaced; we rewrite L in Eq. (5.2.12) with l given by Eq. (5.2.10a) as

$$L = 0.4(y + \Delta y)[1 - \exp\{-(y + \Delta y)/A\}] \tag{5.3.24}$$

and express Δy as a function of an equivalent sand-grain roughness parameter $k_s^+ (\equiv k_s u_\tau / \nu)$, i.e.

$$\Delta y = 0.9(\nu/u_\tau)\left[\sqrt{k_s^+} - k_s^+ \exp\left(-k_s^+/6\right)\right]$$

given by Eq. (4.5.14) which is valid for

$$4.535 < k_s^+ < 2000$$

with the lower limit corresponding to the upper bound for hydraulically smooth surface.

It remains to provide a link between k_s^+ and the geometry of a particular rough surface. As was discussed in Section 4.5, Schlichting determined experimentally equivalent sand roughness for a large number of roughnesses arranged in a regular fashion. Dvorak established a correlation between the velocity shift Δu and the roughness density from which the equivalent sand roughness can be determined [41]. For the roughness elements other than the ones investigated by Schlichting and Dvorak, the equivalent sand roughness must be determined experimentally or by some empirical methods.

As we shall discuss later, for turbulent flows, it is sometimes more convenient to use "wall" boundary conditions at some distance y_0 away from the wall. Usually this y_0 is taken to be the distance, given by

$$y_0 = (\nu/u_\tau)y_0^+$$

with y_0^+ given by 50 for smooth surfaces. In that case, the "wall" boundary conditions for u and v can be represented by

$$u_0 = u_\tau\left[\frac{1}{\kappa}\ln\frac{y_0 u_\tau}{\nu} + c\right] \tag{5.3.25a}$$

$$v_0 = -\frac{u_0 y_0}{u_\tau}\frac{du_\tau}{dx}. \tag{5.3.25b}$$

Here c is a constant equal to 5.2. Equation (5.3.25b) results from integrating the continuity equation with u given by Eq. (4.2.1). The shear stress at y_0, namely τ_0, is obtained from

$$\tau = \tau_w + \frac{dp}{dx}y + \nu\frac{du_\tau}{dx}\int_0^{y^+}\left(\frac{u}{u_\tau}\right)^2 dy^+ \tag{5.3.26}$$

In the viscous sublayer and in the buffer layer ($y^+ \leq 30$), u/u_τ can be obtained from Thompson's velocity profile given by Eq. (4.4.41). For $y^+ > 30$ we can use the logarithmic velocity formula, Eq. (9.3.1a). See subsection 9.3.1.

The above equations are also applicable to flows over rough walls provided we replace c in Eq. (5.3.25a) with B, given by Eq. (4.5.6). The use of these boundary conditions are especially advantageous for fully rough flow conditions because they do not directly require "low Reynolds number" modifications.

Cebeci and Chang modification to the CS model to account for wall roughness has been investigated for several flows by using the differential boundary layer method described in Chapter 8 and in [38]. Here, of the several flows considered in that study, two are presented below to demonstrate the accuracy of this model for flows over rough walls.

Figure 5.9 presents measured values of momentum thickness, displacement thickness, and skin-friction coefficients, reported by Betterman [42], together with lines corresponding to the Cebeci-Chang calculations. The measurements correspond to values of roughness density of 2.65 to 4.18 and roughness height between 2.4 and 4.0 mm. The equivalent sand roughness heights were calculated as described in Section 4.5 based on the correlated results presented by Dvorak [41]. The measured c_f and R_θ at the 0.4-m station, where initial perturbations have died down, were used to generate initial data. As can be seen, the agreement is generally very good, with the maximum discrepancy in skin-friction coefficient and integral thicknesses amounting to approximately 5% and corresponding to the results obtained with the longest roughness height.

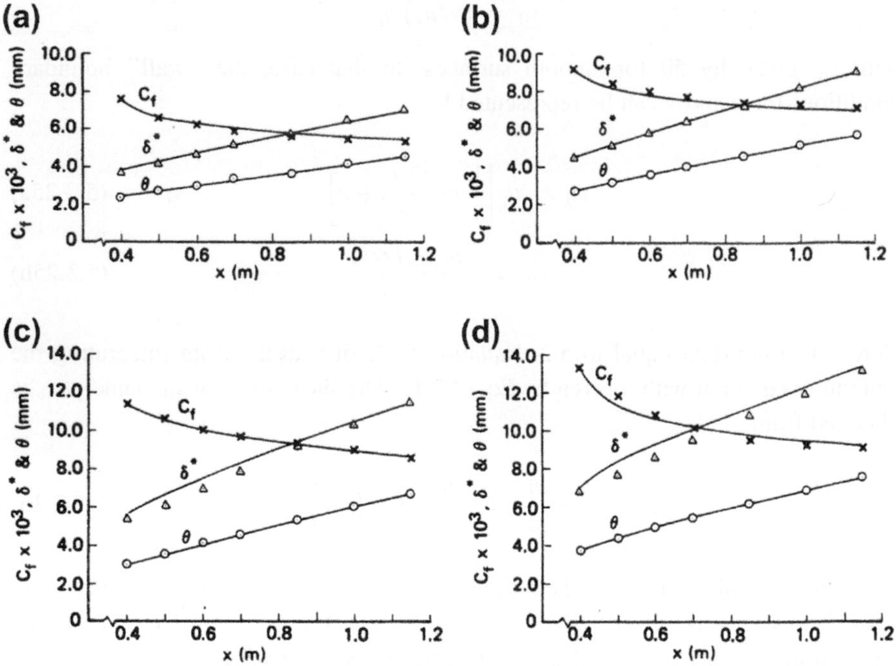

Fig. 5.9 Results for Bettermann's data (— computed, × Δ ○ data): (a) $k = 3$ mm, $\lambda = 2.65$, $k_s = 1.26$ mm; (b) $k = 3$ mm, $\lambda = 3.30$, $k_s = 3.8$ mm; (c) $k = 2.4$ mm, $\lambda = 4.13$, $k_s = 9.26$ mm; (d) $k = 4.0$ mm, $\lambda = 4.18$, $k_s = 14.0$ mm. In all cases, $u_e = 30$ m/s, $v = 1.44 \times 10^{-5}$ m²/s.

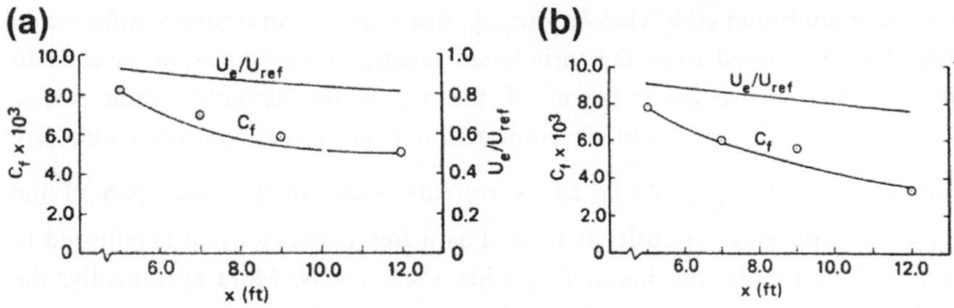

Fig. 5.10 Results for Perry and Joubert's data for two different external velocity distributions (— computed, o data). In both cases, $k = 0.0104$ ft, $\lambda = 4.0$, $k_s = 0.0346$ ft, $\nu = 1.56 \times 10^{-4}$ ft²/s, $u_{ref} = 100$ ft/s.

Perry and Joubert's data serve as another test for flows over rough walls with adverse pressure gradient [43]. They measured the boundary layers in a closed-circuit-type wind tunnel over rough surface made of 0.125-in.-square bar elements with roughness density of 4 held to the plate by strips of double-coated adhesive tape. Because of difficulty in accurately determining c_f for flows with pressure gradients from the measured velocity profiles based on Clauser's plot, they deduced the wall shear stress based on Coles' wake function. In general, the agreement between the calculated and experimental results is good, as shown in Fig. 5.10.

5.4 Extension of the CS Model to Strong Pressure-Gradient Flows

Extensive studies, mostly employing boundary-layer equations, show that while many wall boundary-layer flows can satisfactorily be calculated with the CS model discussed in the previous section, improvements are needed for flows which contain regions of strong pressure gradient and flow separation, for example, flows either approaching stall or post-stall. The main weakness in this model is the parameter α used in the outer eddy viscosity formula, Eq. (5.2.11b), taken as 0.0168. Experiments indicate that in strong pressure gradient flows, the extent of the law of the wall region becomes smaller; to predict flows under such conditions, it is necessary to have a smaller value of α in the outer eddy viscosity formula. The question is how to relate α to the flow properties so that the influence of strong pressure gradient is included in the variation of α.

5.4.1 Johnson–King Approach

One approach developed by Johnson and King [44] and Johnson and Coakley [45] is to adopt a nonequilibrium eddy-viscosity formulation ε_m in which the CS model serves as an equilibrium eddy viscosity $(\varepsilon_m)_{eq}$ distribution. An ordinary differential equation (ODE), derived from the turbulence kinetic energy equation, is used to describe the streamwise development of the maximum Reynolds shear stress, $-(\overline{\varrho u'v'})_m$, or $(-\overline{u'v'})_m$ for short, in conjunction with an assumed eddy-viscosity distribution which has $\sqrt{(-\overline{u'v'})_m}$ as its velocity scale. In the outer part of the boundary layer, the eddy viscosity is treated as a free parameter that is adjusted to satisfy the ODE for the maximum Reynolds shear stress. More specifically, the nonequilibrium eddy-viscosity distribution is defined again by separate expressions in the inner and outer regions of the boundary layer. In the inner region, $(\varepsilon_m)_i$ is given by

$$(\varepsilon_m)_i = (\varepsilon_{m_i})_1(1-\gamma_2) + (\varepsilon_{m_i})J-K\gamma_2 \tag{5.4.1}$$

where $(\varepsilon_{m_i})_1$ is given either by $(\kappa y)^2 \partial u/\partial y$ or $u_\tau y$. The expression $(\varepsilon_{m_i})\,J-K$ is

$$(\varepsilon_{m_i})_{J-K} = D^2 \kappa y u_m \tag{5.4.2}$$

where

$$u_m = \max\left(u_\tau, \sqrt{\left(-\overline{u'v'}\right)_m}\right) \tag{5.4.3a}$$

and D is a damping factor similar to that defined by Eq. (5.2.12)

$$D = 1 - \exp\left(\sqrt{\left(-\overline{u'v'}\right)_m}\,\frac{y}{\nu A^+}\right) \tag{5.4.3b}$$

with the value of A^+ equal to 17 rather than 26, as in Eq. (5.2.12). The parameter γ_2 in Eq. (5.4.1) is given by

$$\gamma_2 = \tanh\left(\frac{y}{L'_c}\right) \tag{5.4.3c}$$

where, with y_m corresponding to the y-location of maximum turbulent shear stress, $(-\overline{u'v'})_m$,

$$L'_c = \frac{u_\tau}{u_\tau + u_m} L_m \tag{5.4.4}$$

with

$$L_m = \begin{cases} 0.4y_m & y_m \le 0.225\delta \\ 0.09\delta & y_m > 0.225\delta. \end{cases} \tag{5.4.5}$$

In the outer region, $(\varepsilon_m)_o$ is given by

$$(\varepsilon_m)_o = \sigma(0.0168 u_e \delta^* \gamma) \tag{5.4.6}$$

where σ is a parameter to be determined. The term multiplying σ on the righthand side of Eq. (5.4.6) is the same as the expression given by Eq. (5.2.11b) without γ_{tr} and with $\alpha = 0.0168$.

The nonequilibrium eddy viscosity across the whole boundary-layer is computed from

$$\varepsilon_{\mathrm{m}} = (\varepsilon_{\mathrm{m}})_{\mathrm{o}} \tanh\left[\frac{(\varepsilon_{\mathrm{m}})_{\mathrm{i}}}{(\varepsilon_{\mathrm{m}})_{\mathrm{o}}}\right]. \tag{5.4.7}$$

The maximum Reynolds shear stress $(-\overline{u'v'})_{\mathrm{m}}$ is computed from the turbulence kinetic energy equation using assumptions similar to those used by Bradshaw et al. (see subsection 6.3.1). After the modeling of the diffusion, production and dissipation terms and the use of

$$\frac{\left(-\overline{u'v'}\right)_{\mathrm{m}}}{k_{\mathrm{m}}} = a_1 = 0.25$$

the transport equation for $(-\overline{u'v'})_{\mathrm{m}}$ with u_{m} now denoting the streamwise velocity at y_{m}, is written as

$$\frac{d}{dx}\left(-\overline{u'v'}\right)_{\mathrm{m}} = \frac{a_1\left(-\overline{u'v'}\right)_{\mathrm{m}}}{L_{\mathrm{m}}u_{\mathrm{m}}}\left[\left(-\overline{u'v'}\right)_{\mathrm{m,eq}}^{1/2} - \left(\overline{u'v'}\right)_{\mathrm{m}}^{1/2}\right] - \frac{a_1}{u_{\mathrm{m}}}D_{\mathrm{m}} \tag{5.4.8}$$

where, with $c_{\mathrm{dif}} = 0.5$, the turbulent diffusion term along the path of maximum $(-\overline{u'v'})$ is given by

$$D_{\mathrm{m}} = \frac{c_{\mathrm{dif}}}{a_1\delta\left[0.7 - (y/\delta)_{\mathrm{m}}\right]}\left(-\overline{u'v'}\right)_{\mathrm{m}}^{3/2}\left\{1 - \left[\frac{\left(-\overline{u'v'}\right)_{\mathrm{m}}}{\left(-\overline{u'v'}\right)_{\mathrm{m,eq}}}\right]^{1/2}\right\}. \tag{5.4.9}$$

To use this closure model, the continuity and momentum equations are first solved with an equilibrium eddy viscosity $(\varepsilon_{\mathrm{m}})_{\mathrm{eq}}$ distribution such as in the CS model, and the maximum Reynolds shear stress distribution is determined based on $(\varepsilon_{\mathrm{m}})_{\mathrm{eq}}$, which we denote by $(-\overline{u'v'})_{\mathrm{m,eq}}$. Next the location of the maximum Reynolds shear stress is determined so that y_{m} and u_{m} can be calculated. The transport equation for $(-\overline{u'v'})_{\mathrm{m}}$ is then solved to calculate the nonequilibrium eddy-viscosity distribution ε_{m} given by Eq. (5.4.7) for an assumed value of σ so that the solutions of the continuity and momentum equations can be obtained. The new maximum shear stress term is then compared with the one obtained from the solution of Eq. (5.4.8). If the new computed value does not agree with the one from Eq. (5.4.8), a new value of σ is used to compute the outer eddy viscosity and eddy-viscosity distributions across the whole boundary-layer so that a new $(-\overline{u'v'})_{\mathrm{m}}$ can be computed from the solution of the continuity and momentum equations. This iterative procedure of determining σ is repeated until $(-\overline{u'v'})_m$ is computed from the continuity and momentum equations agrees with that computed from the transport equation, Eq. (5.4.8).

5.4.2 Cebeci-Chang Approach

Cebeci and Chang used another approach in order to improve the predictions of the CS model for flows with strong adverse pressure gradient and separation [46]. They related the parameter α to a parameter F, according to the suggestion of Simpson et al. [47] by

$$\alpha = \frac{0.0168}{F^{1.5}}. \tag{5.4.10}$$

Here $(1 - F)$ denotes the ratio of the production of the turbulence energy by normal stresses to that by shear stress, evaluated at the location where shear stress is maximum, that is

$$F = 1 - \left[\frac{\left(\overline{u'^2} - \overline{v'^2}\right)\partial u/\partial x}{-\overline{u'v'}\,\partial u/\partial y}\right]_m. \tag{5.4.11}$$

Before Eq. (5.4.10) can be used in Eq. (5.2.11b), an additional relationship between $(\overline{u'^2} - \overline{v'^2})$ and $(-\overline{u'v'})$ is needed. For this purpose, the ratio in Eq. (5.4.11)

$$\beta = \left[\frac{\overline{u'^2} - \overline{v'^2}}{-\overline{u'v'}}\right]_m \tag{5.4.12}$$

is assumed to be a function of $R_t = \tau_w/(-\varrho\overline{u'v'})_m$ which, according to the data of Nakayama [48], can be represented by

$$\beta = \frac{6}{1 + 2R_t(2 - R_t)} \tag{5.4.13a}$$

for $R_t < 1.0$. For $R_t \geq 1.0$, β is taken to be

$$\beta = \frac{2R_t}{1 + R_t} \tag{5.4.13b}$$

Introducing the above relationships into the definition of F and using Eq. (5.4.7) results in the following expression for α

$$\alpha = \frac{0.0168}{[1 - \beta(\partial u/\partial x)/(\partial u/\partial y)]_m^{1.5}} \tag{5.4.14}$$

where β is given by Eq. (5.4.13).

Separation-Induced Transition

Another improvement to the CS model was made by replacing the intermittency parameter γ in Eq. (5.2.11b) by another intermittency expression recommended by Fiedler and Head [49]. According to the experiments conducted by Fiedler and Head, it was found that the pressure gradient has a marked effect on the distribution of intermittency. Their experiments indicated that in the boundary-layer proceeding to separation, the intermittent zone decreased in width and moved further from the surface as shape factor H increased. The reverse trend was observed with decreasing H in a favorable pressure gradient.

In the improved CS model the intermittency expression of Fiedler and Head is written in the form

$$\gamma = \frac{1}{2}\left[1 - \operatorname{erf}\frac{y - Y}{\sqrt{2}\sigma}\right] \tag{5.4.15}$$

where Y and σ are general intermittency parameters with Y denoting value of y for which $\gamma = 0.5$ and σ, the standard deviation. The dimensionless intermittency parameters Y/δ^*, σ/δ^* and δ/δ^* expressed as functions of H are shown in Fig. 5.11.

The predictions of the original and modified Cebeci-Smith turbulence models were investigated for several airfoils [46] by using the interactive boundary-layer

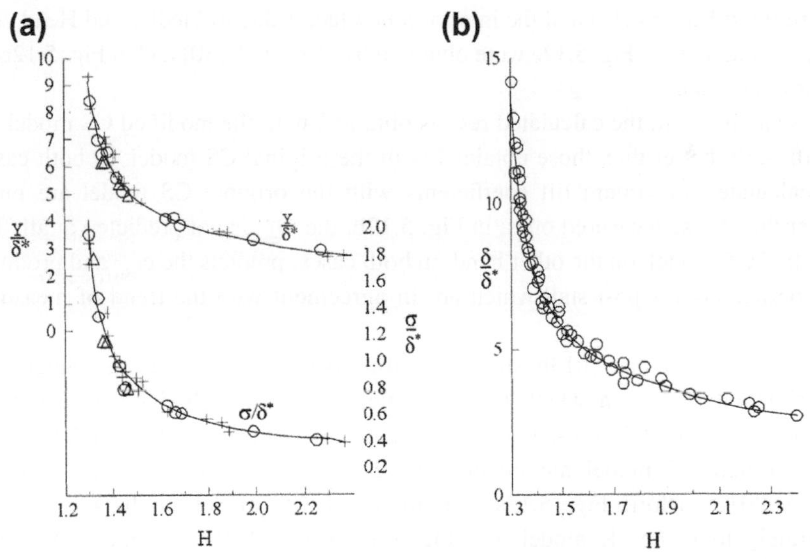

Fig. 5.11 Variation of Y/δ^*, σ/δ^* and δ/δ^* with H according to the data of Fiedler and Head [49].

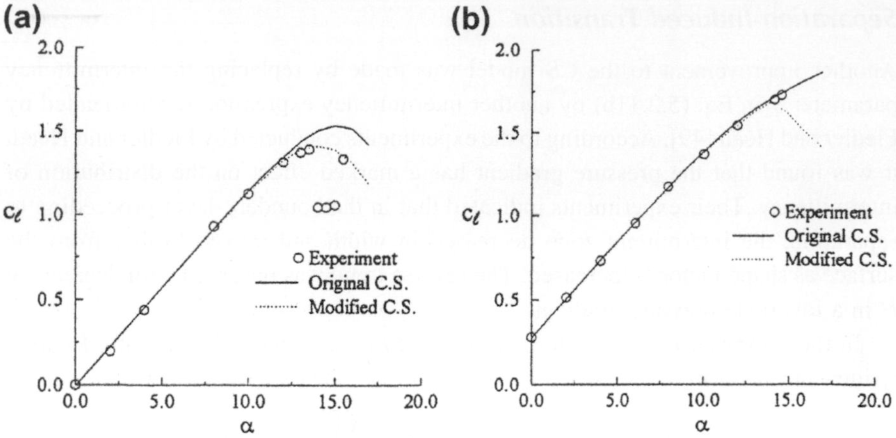

Fig. 5.12 IBL results for the (a) NACA 0012 airfoil and (b) Boeing airfoil, $M_\infty = 0.3$.

(IBL) method of Cebeci [38]. For each airfoil, the onset of the transition location was computed with Michel's correlation [28], and the calculated lift coefficients were compared with data for a range of angles of attack, including stall and post-stall.

Figures 5.12a and 5.12b show a sample of results obtained with the original and modified CS models, the latter corresponding to the one in which α is computed according to Eq. (5.4.14) and the intermittency factor due to Fiedler and Head. The experimental data in Fig. 5.12a were obtained by Carr et al. [50] and in Fig. 5.12b by Omar et al. [51].

As can be seen, the calculated results obtained with the modified CS model are significantly better than those obtained with the original CS model. In both cases, the calculated maximum lift coefficients with the original CS model are much higher than those measured ones; in Fig. 5.12b, the $c_{l_{max}}$ is not predicted at all. The modified CS model, on the other hand, in both cases, predicts the $c_{l_{max}}$ and produces lift coefficients for post stall which are in agreement with the trend of measured values.

Figures 5.13a and 5.13b show a comparison between the calculated and experimental results in which the calculated ones were obtained by using the modified CS and Johnson-King (JK) models. In both cases, the predictions of the modified CS model are better than the JK model. For example, for the NACA 0012 airfoil, Fig. 5.13a, the modified CS model predicts $c_{l_{max}}$ more accurately than the JK model. For the Boeing airfoil, the modified CS model appears to predict post stall better than the JK model. For additional comparisons, see [38].

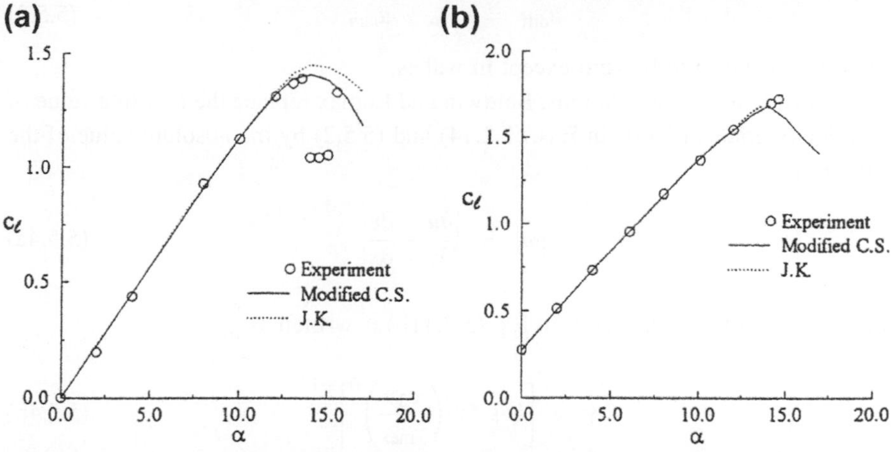

Fig. 5.13 IBL results for the (a) NACA 0012 airfoil and (b) Boeing airfoil, $M_\infty = 0.3$.

5.5 Extensions of the CS Model to Navier–Stokes Methods

Due to its simplicity and its good success in external boundary-layer flows, the CS model with modifications has also been used extensively in the solution of the Reynolds-averaged Navier–Stokes equations for turbulent flows. These modifications are described below.

Baldwin and Lomax [52] adopt the CS model, leave the inner eddy viscosity formula given by Eq. (5.2.14) essentially unaltered, but in the outer eddy viscosity, Eq. (5.2.11b), use alternative expressions for the length scale δ^* of the form

$$(\varepsilon_m)_o = \alpha c_1 \gamma y_{max} F_{max} \tag{5.5.1a}$$

or

$$(\varepsilon_m)_o = a c_1 \gamma c_2 u^2_{diff} \frac{y_{max}}{F_{max}} \tag{5.5.1b}$$

with $c_1 = 1.6$ and $c_2 = 0.25$. The quantities F_{max} and y_{max} are determined from the function

$$F = y\left(\frac{\partial u}{\partial y}\right)\left[1 - e^{-y/A}\right] \tag{5.5.2}$$

with F_{max} corresponding to the maximum value of F that occurs in a velocity profile and y_{max} denoting the y-location of F_{max}. u_{diff} is the difference between maximum and minimum velocity in the profile

$$u_{\text{diff}} = u_{\max} - u_{min} \tag{5.5.3}$$

where $u_{\min}$ is taken to be zero except in wakes.

In Navier–Stokes calculations, Baldwin and Lomax replace the absolute value of the velocity gradient $\partial u/\partial y$ in Eqs. (5.2.14) and (5.5.2) by the absolute value of the vorticity $|\omega|$,

$$|\omega| = \left|\frac{\partial u}{\partial y} - \frac{\partial v}{\partial x}\right| \tag{5.5.4a}$$

and the intermittency factory γ in Eq. (5.2.11b) is written as

$$\gamma = \left[1 + 5.5\left(\frac{c_3 y}{y_{\max}}\right)^6\right]^{-1} \tag{5.5.4b}$$

with $c_3 = 0.3$. The studies conducted by Stock and Haase [53] clearly demonstrate that the modified algebraic eddy viscosity formulation of Baldwin and Lomax is not a true representation of the CS model since their incorporation of the length scale in the outer eddy viscosity formula is not appropriate for flows with strong pressure gradients.

Stock and Haase proposed a length scale based on the properties of the mean velocity profile calculated by a Navier–Stokes method. They recommend computing the boundary-layer thickness δ from

$$\delta = 1.936 y_{\max} \tag{5.5.5}$$

where $y_{\max}$ is the distance from the wall for which $y|\partial u/\partial y|$ or F in Eq. (5.5.2) has its maximum. With δ known, u_e in the outer eddy viscosity formula, Eq. (5.2.11b) is the u at $y = \delta$, and γ is computed from

$$\gamma = \left[1 + 5.5\left(\frac{y}{\delta}\right)^6\right]^{-1} \tag{5.5.6}$$

based on Klebanoff's measurements on a flat plate flow and not from Eq. (5.5.4b). The displacement thickness δ^* for attached flows is computed from its definition,

$$\delta* = \int_0^\delta \left(1 - \frac{u}{u_e}\right) dy \tag{5.5.7a}$$

and, for separated flows from

$$\delta* = \int_{y_{u=0}}^\delta \left(1 - \frac{u}{u_e}\right) dy \tag{5.5.7b}$$

either integrating the velocity profile from $y = 0$, or $y = y_{u=0}$ to δ, or using the Coles velocity profile. The results obtained with this modification to the length scale in the outer CS eddy viscosity formula improve the predictions of the CS model in Navier–Stokes methods as discussed in Stock and Haase [53].

A proposal which led to Eq. (5.5.5) was also made by Johnson [54]. He recommended that the boundary-layer thickness δ is calculated from

$$\delta = 1.2 y_{1/2} \tag{5.5.8}$$

where

$$y_{1/2} = y \text{ at } \frac{F}{F_{\max}} = 0.5. \tag{5.5.9}$$

The predictions of the original and modified CS models were also investigated by Cebeci and Chang by using the Navier-Stokes method of Swanson and Turkel [55] as well as by the interactive boundary-layer method of Cebeci (see subsection 5.4.2). The models considered include the original CS model, BL model, modifications to the BL model and the JK model.

Figures 5.14a and 5.14b show the results obtained with the original CS and BL models. In the former case, the length scale δ^* in the outer eddy-viscosity formula was computed based on the definition of the boundary-layer thickness δ given by Stock and Haase [53] and Johnson [54].

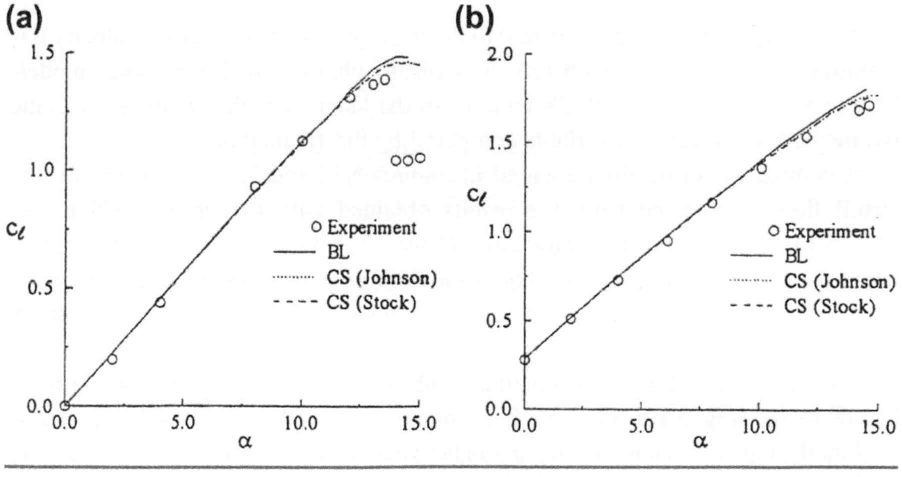

Fig. 5.14 Navier–Stokes results for the (a) NACA 0012 airfoil and (b) Boeing airfoil, $M_\infty = 0.3$.

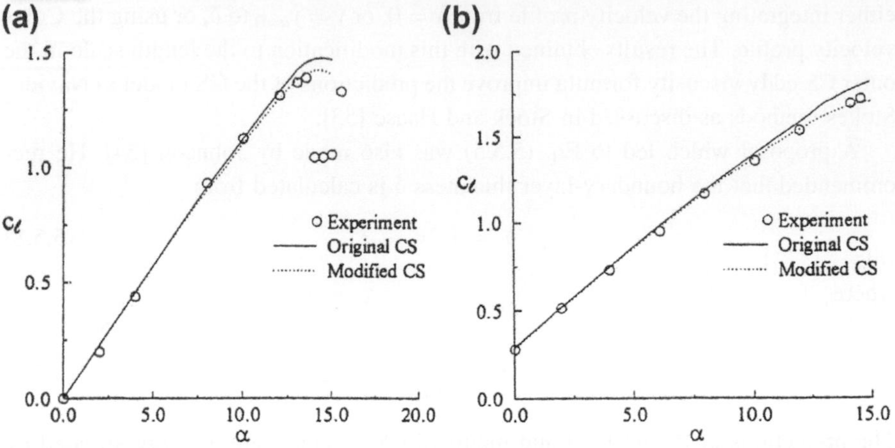

Fig. 5.15 Navier–Stokes results for the (a) NACA 0012 airfoil and (b) Boeing airfoil, $M_\infty = 0.3$.

Figures 5.15a and 5.15b show similar comparisons with turbulence models corresponding to the original CS and modified CS models. In the latter case the boundary-layer thickness was computed from

$$\delta = 1.5y_{1/2}$$

or from

$$\delta = y_m$$

if $1.5y_{1/2} > y_m$, with y_m corresponding to the location where streamwise velocity u is maximum. Figures 5.16a and 5.16b show results obtained with turbulence models based on modified CS and BL-JK models. In the latter case, the parameter α in the BL method was taken as a variable computed by the JK method.

A comparison of results presented in Figures 5.14 and 5.15 shows that for the airfoil flows considered here, the results obtained with the original CS model (Fig. 5.14) with δ defined by Stock and Haase [53] and Johnson [54] are slightly better than those given by the BL model and the results with the modified CS model (Fig. 5.15) are much better than all the other modified versions of the original CS model.

A comparison of the results obtained with the modified CS model and with the BL-JK model (Fig. 5.16) show that both models essentially produce similar results.

Finally, Fig. 5.17 shows a comparison between the predictions of the IBL and NS methods. In both methods, the turbulence model used is the modified CS model. As can be seen, the predictions of both methods are identical at low and moderate angles

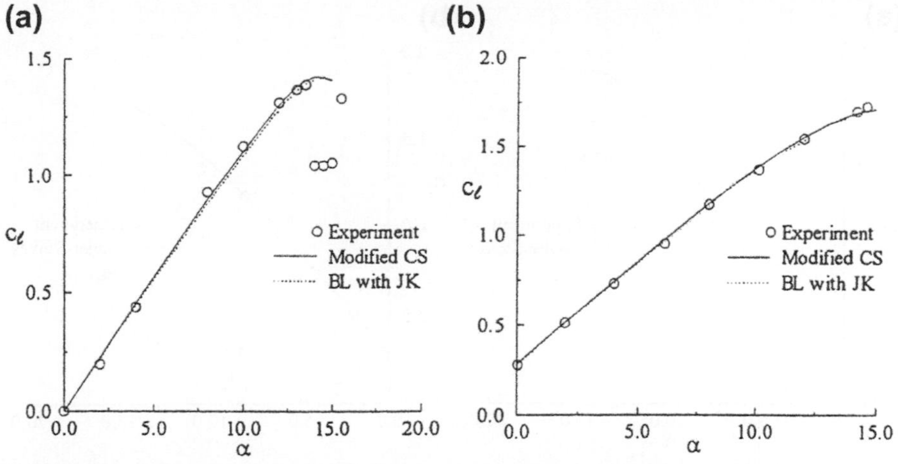

Fig. 5.16 Navier–Stokes results for the (a) NACA 0012 airfoil and (b) Boeing airfoil, $M_\infty = 0.3$.

of attack. At higher angles, especially near stall and post stall, while there are some differences, both methods predict the stall angle well. The requirements for the computer resources for the IBL method, however, are considerably less than those provided by the Navier–Stokes method.

5.6 Eddy Conductivity and Turbulent Prandtl Number Models

Using Boussinesq's eddy-conductivity concept, we can write the transport of heat due to the product of time mean of fluctuating enthalpy h' and fluctuating velocity v' in the form

$$-\varrho\overline{v'h'} = \varrho\varepsilon_{\mathrm{h}}(\partial h/\partial y). \tag{5.6.1}$$

Sometimes it has been found to be convenient to introduce a "turbulent" Prandtl number $\mathrm{Pr_t}$ defined by

$$\mathrm{Pr_t} = \varepsilon_{\mathrm{m}}/\varepsilon_{\mathrm{h}}. \tag{5.6.2}$$

It is obvious from Eq. (5.6.1) that in order to predict temperature distribution within a boundary layer, it is necessary to describe the distribution of ε_{h} in the layer. For that reason, various assumptions have been made and several models have been proposed for ε_{h}. One assumption that has been used extensively is due to Reynolds [56]. According to his assumption, heat and momentum are transferred by the same

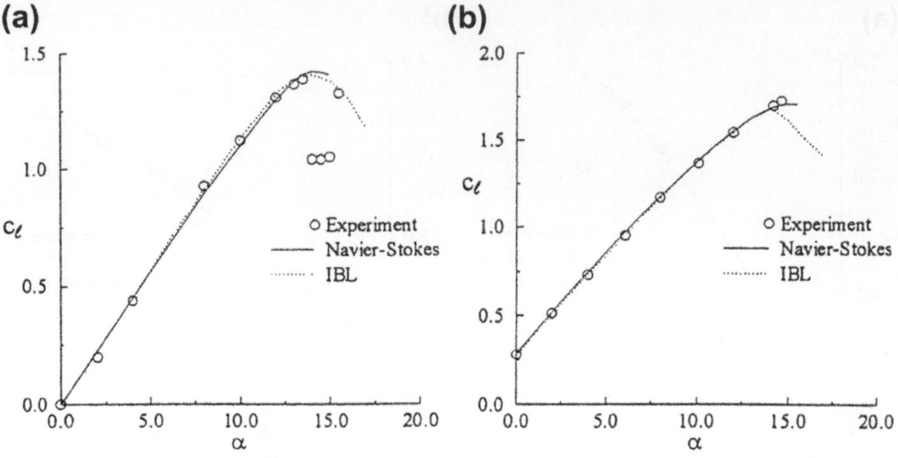

Fig. 5.17 Comparison of NS and IBL results obtained with the modified CS model for the (a) NACA 0012 airfoil and (b) Boeing airfoil, $M_\infty = 0.3$.

process, which means that the eddy coefficients for momentum and heat transport are the same. That assumption leads to a turbulent Prandtl number of unity. The literature discusses the relationship between those coefficients at great length (see, e.g., Kestin and Richardson [57]) without definite conclusions. According to mercury experiments in pipes, $\mathrm{Pr_t} > 1$; according to gas experiments in pipes, $\mathrm{Pr_t} < 1$. From the experiments, it is not clear whether or not the eddy conductivity and, consequently, the turbulent Prandtl number are completely independent of the molecular Prandtl number.

One of the first proposals for a modification of the Reynolds analogy was made by Jenkins [58], who took into consideration the heat conduction to or from an element of an eddy during its movement transverse to the main flow. He assumed that if the temperature of the eddy did not change in transit, the definition of mixing length, $l = u_\tau\,(\partial u/\partial y)^{-1}$, and the definition of eddy conductivity, Eq. (5.6.1), would give $\varepsilon_\mathrm{h} = lv'$, since $T' = l(\partial T/\partial y)$. However, if heat were lost during transit, the fluctuation temperature T' would actually be less than that, because of molecular thermal conductivity. Then the eddy conductivity would be

$$\varepsilon_\mathrm{h} = lv'\frac{T_\mathrm{f} - T_\mathrm{i}}{l(\partial T/\partial y)}, \tag{5.6.3}$$

where T_f and T_i are the final and the initial eddy mean temperatures, respectively. In order to obtain an expression for $(T_\mathrm{f} - T_\mathrm{i})/l(\partial T/\partial y)$, Jenkins assumed that the eddies were spheres of radius l, the mixing length, with the surface temperature of the particles varying linearly with time during their movement. The interval of time between an eddy's creation and its destruction was taken to be l/v'. Using Carslaw and Jaeger's formula for the average temperature of a sphere under those conditions, he obtained an expression for Eq. (5.6.3).

Treating the effects of molecular viscosity on an eddy in movement in the same way as the effects of molecular thermal conductivity, Jenkins obtained the following expression for the ratio of eddy conductivity to eddy viscosity:

$$\frac{\varepsilon_\mathrm{h}}{\varepsilon_\mathrm{m}} \equiv \frac{1}{\mathrm{Pr}_t}$$

$$= \frac{1}{\mathrm{Pr}}\left\{\frac{\frac{2}{15}-\left(12/\pi^6\right)\varepsilon_\mathrm{m}^+\left\{\sum_{n=1}^{\infty}(1/n^6)\left[1-\exp(-n^2\pi^2/\varepsilon_\mathrm{m}^+)\right]\right\}}{\frac{2}{15}-\left(12/\pi^6\right)\mathrm{Pr}\varepsilon_\mathrm{m}^+\left\{\sum_{n=1}^{\infty}(1/n^6)\left[1-\exp(-n^2\pi^2/\mathrm{Pr}\varepsilon_\mathrm{m}^+)\right]\right\}}\right\}. \tag{5.6.4}$$

The variation of $\varepsilon_\mathrm{h}/\varepsilon_\mathrm{m}$ with Pr for various values of $\varepsilon_m^+(\equiv\varepsilon_m/v)$ according to Eq. (5.6.4) is shown in Fig. 5.18. Calculations made at low Prandtl numbers with the data of Fig. 5.18 for the relationship between ε_h and ε_m are in fair agreement with experiment [59], although the more recent experimental data for liquid metals suggest that the loss of heat by an eddy in transit is not as great as that predicted by Jenkins. Furthermore, according to the experimental data of Page et al. [60] for air at Pr = 0.7, the eddy conductivity is greater than the eddy viscosity. Jenkins' result

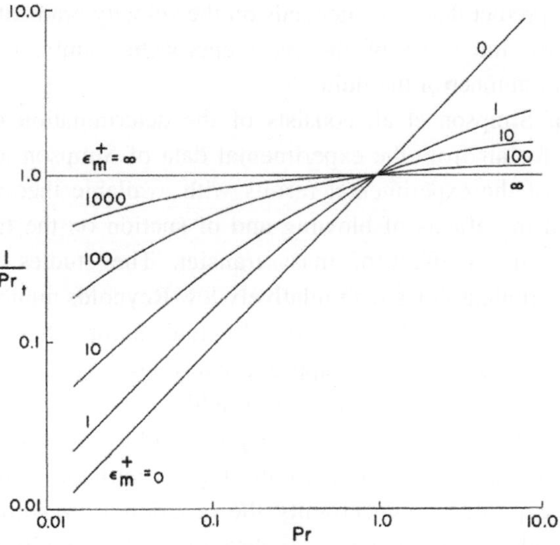

Fig. 5.18 Variation of reciprocal of turbulent Prandtl number with molecular Prandtl number for various values of ε_m^+, according to Eq. (5.6.4).

gives an opposite effect; some analysts have therefore preferred to use Jenkins' diffusivity ratio multiplied by a constant factor such as 1.10 or 1.20, to bring the results more into line with measurements at Pr = 0.7. At high Prandtl numbers, the Jenkins model predicts values for $\varepsilon_h/\varepsilon_m$ that have no upper bound as the Prandtl number increases, which also is not found by experiment.

Rohsenow and Cohen [61] also derived an expression for $\varepsilon_h/\varepsilon_m$. They expressed the ratio of the two eddy coefficients as

$$\frac{\varepsilon_h}{\varepsilon_m} = 416\mathrm{Pr}\left[\frac{1}{15} - \frac{6}{\pi^4}\sum_{n=1}^{\infty}\frac{1}{n^6}\exp\left(-\frac{0.0024\pi^2 n^2}{\mathrm{Pr}}\right)\right]. \qquad (5.6.5)$$

Their analysis assumes that when an eddy passes through the fluid, a temperature gradient is set up in it and that the surface heat-transfer coefficient is infinite. Again, that expression leads to infinite values of $\varepsilon_h/\varepsilon_m$ as the Prandtl number increases without limit.

Studies of the problem have also been made by Deissler [62], by Simpson et al. [63], and by Cebeci [64]. Deissler's first method is based on a modified mixing-length theory, and his second method is based on correlation coefficients. Neither leads to an expression for the eddy diffusivity ratios that can be compared directly with those given by other authors, but the modified mixing-length theory does seem successful in predicting heat transfer in the low-Prandtl-number range. In his second method, Deissler derived from the momentum and the energy equations the correlation between velocities and temperatures at two points in a homogeneous turbulent fluid. His results predict that $\varepsilon_h/\varepsilon_m$ depends on the velocity gradient and that, as the gradient increases, the value of the ratio approaches unity, regardless of the molecular Prandtl number of the fluid.

The study of Simpson et al. consists of the determination of the turbulent Prandtl number for air from the experimental data of Simpson and Whitten and the comparison of the experimental results with available theories. In addition, they investigated the effects of blowing and of suction on the turbulent Prandtl number and found no effect of mass transfer. The studies were made for incompressible turbulent flows with relatively low Reynolds number (R_x), ranging from 1.3×10^5 to 2×10^6. The minimum Reynolds number based on momentum thickness with no mass transfer is approximately 600. The results of the study show that (1) their experimental turbulent Prandtl number results agree, within the experimental uncertainties, with Ludwieg's pipe results [65], which were obtained for $0.1 < y/\delta < 0.9$, and that (2) in the inner region the Jenkins model is found to describe, within experimental uncertainty, the variation of Pr_t with ε_m^+. In the outer region, a new model for Pr_t and ε_m^+ was developed. The results for both models fall within the uncertainty envelope of their experimental results and indicate no dependence of Pr_t on blowing or suction.

The approach taken by Cebeci [64] is based on the mixing-length concept. It differs from others in that his eddy-conductivity expression (1) provides a continuous temperature distribution across the boundary layer, (2) accounts for the mass transfer and the pressure gradient, and (3) accounts for both low-and-high-Prandtl-number fluids. According to this model, the turbulent Prandtl number is given by

$$\mathrm{Pr_t} = \frac{\kappa[1-\exp(-y/A)]}{\kappa_h[1-\exp(-y/B)]}. \tag{5.6.6a}$$

At the wall,

$$\mathrm{Pr_t} = \frac{\kappa}{\kappa_h}\frac{B}{A} = \frac{\kappa}{\kappa_h}\frac{B^+}{A^+}. \tag{5.6.6b}$$

Note that as y becomes larger, the exponential terms in Eq. (5.6.6a) approach zero. The turbulent Prandtl number then becomes

$$\mathrm{Pr_t} = \kappa/\kappa_h. \tag{5.6.6c}$$

We also note from Eq. (5.6.6) that the molecular Prandtl number plays a strong role in $\mathrm{Pr_t}$ close to the wall, since $B^+ = B^+(\mathrm{Pr})$, and has no effect on $\mathrm{Pr_t}$ away from the wall.

The damping constant in Eq. (5.6.6) is for air, whose Prandtl number is approximately 0.7. For fluids other than air it varies since B^+ is a function of the molecular Prandtl number. If we assume that κ, κ_h, and A^+ are 0.40, 0.44, and 26, respectively, B^+ can be calculated from Eq. (5.6.6b), provided that the $\mathrm{Pr_t}$ is known at the wall. Following that procedure and using the experimental values of $\mathrm{Pr_t}$, Na and Habib [66] expressed the variation of B^+ with Pr, for a range of Pr from 0.02 to 15, by

$$B = \frac{B^+ \nu}{u_\tau}, \quad B^+ = B^{++}/(\mathrm{Pr})^{1/2}, \tag{5.6.7}$$

where B^{++} is represented by the following formula:

$$B^{++} = \sum_{i=1}^{5} C_i(\log_{10}\mathrm{Pr})^{i-1}, \tag{5.6.8}$$

with $C_1 = 34.96$, $C_2 = 28.79$, $C_3 = 33.95$, $C_4 = 6.33$, and $C_5 = -1.186$.

To account for the low Reynolds number effects ($R_\theta < 5000$) for air with $\mathrm{Pr} = 1$, κ_h and B^+ are represented by

$$\kappa_h = 0.44 + \frac{0.22}{1+0.42z_2^2}, \quad B^+ = 35 + \frac{25}{1+0.55z_2^2}, \tag{5.6.9}$$

where $z_2 \equiv R_\theta \times 10^{\ 3} \geq 0.3$.

Equation (5.6.6) with B and A given by (5.6.7) and (5.2.13) is restricted to incompressible flows without pressure gradient. It can easily be extended to compressible flows by replacing A by Eq. (5.3.1) with N given by Eq. (5.3.2c) and by replacing B by

$$B = B^{+}\frac{\nu}{N}\left(\frac{\tau_w}{\varrho_w}\right)^{-1/2}\left(\frac{\varrho}{\varrho_w}\right)^{1/2}. \tag{5.6.10}$$

The expression for the turbulent Prandtl number given by Eq. (5.6.6) has been evaluated for flows with and without mass transfer by comparing its predictions with experiment and with other predictions.

For an incompressible flow with no mass transfer, Figures 5.19a and 5.19b show the variation of the turbulent Prandtl number with y^+ and ε_m^+ respectively, according to the Cebeci model for $R_\theta = 1000$ and 4000 and according to Jenkins' model. Also shown is the uncertainty envelope and the variation of the mean turbulent Prandtl number determined by Simpson et al. [63] from their experimental data. We note that the values of Pr_t calculated by Eq. (5.6.6) show a slight Reynolds-number effect for $R_\theta < 4000$, an effect that was also observed by Simpson et al. The predicted results fall within the uncertainty envelope and agree well in both inner and outer regions, with the predictions of Jenkins in the inner region ($y^+ < 10^2$), and with the experimental data of Simpson et al.

Figure 5.20a shows the variation of turbulent Prandtl number with y/δ for R_θ values of 1000 and 4000. It also shows the uncertainty envelope of Simpson et al. and the experimental data of Johnson [67] and of Ludwieg [65]. The experimental data of Johnson are for flat-plate flow at high Reynolds numbers. Johnson studied the temperature distribution downstream of an unheated starting length where the thermal boundary layer was contained at all times in an inner fraction of the momentum boundary layer, which provided no information about the outer region. He compared the turbulent shear stress and the heat flux obtained by hot-wire

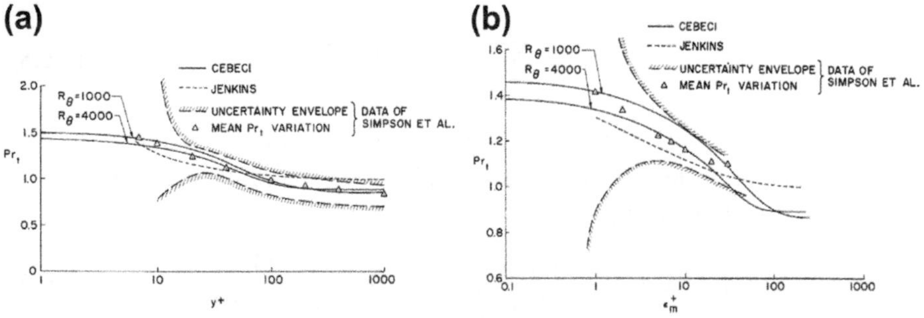

Fig. 5.19 Variation of turbulent Prandtl number (a) with y^+ and (b) with ε_m^+.

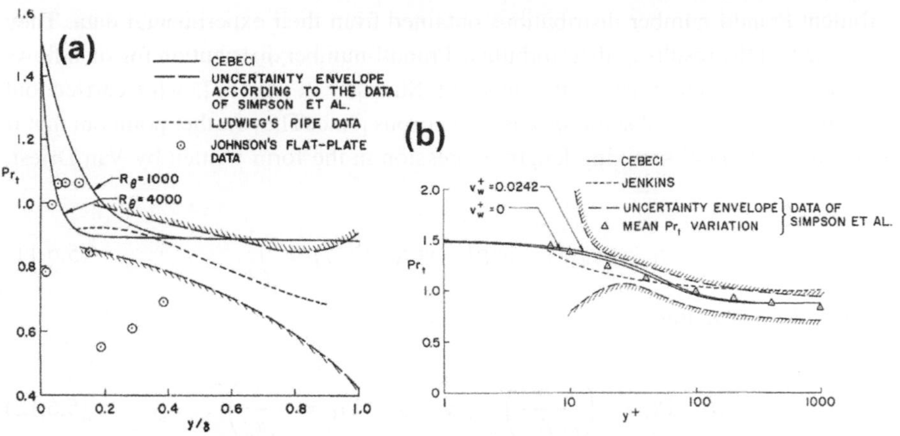

Fig. 5.20 (a) Comparison of calculated turbulent Prandtl number with experiment and (b) effect of mass transfer on turbulent Prandtl number.

measurements with those generated from mean-velocity and -temperature distributions and found a 50% discrepancy in the shearing stresses and good agreement for the heat fluxes. His values of skin-friction coefficient obtained by several independent methods did not agree. The anomalous behavior was attributed to three-dimensionality of the flow. The experimental data of Ludwieg are based on measurements in a pipe, again at high Reynolds numbers. According to Kestin and Richardson's study [57], Ludwieg's results are the most reliable for air flowing in a pipe.

The comparisons in Fig. 5.20a show that the results obtained by Eq. (5.6.6), especially one obtained for $R_\theta = 4000$, agree reasonably well with Ludwieg's results for $0.1 \leq y/\delta \leq 0.4$ and differ slightly from his results within the uncertainty envelope of Simpson et al. It is interesting to note that the predicted results for the region near the wall also agree well with Johnson's data, although the discrepancy is significant away from the wall.

Next we study the effect of mass transfer on turbulent Prandtl number. We use the experimental data of Simpson and calculate Pr_t at various values of y^+ and y/δ for given values of R_θ and v_w^+. Figure 5.20b shows the results calculated for $v_w^+ = 0$ and $v_w^+ = 0.0242$ for $R_\theta = 2000$ by using Eq. (5.6.6), together with the uncertainty envelope and the variation of the mean turbulent Prandtl number of Simpson et al. and the predictions of Jenkins' model. Considering the fact that the calculations were made for a low Reynolds number, it can be said that the results agree reasonably well with the findings of Simpson et al. and show no appreciable effect of mass transfer on the turbulent Prandtl number.

Next we compare the present model with the experimental data of Meier and Rotta [68]. Those authors present temperature distributions in supersonic flows and turbulent Prandtl number distributions obtained from their experimental data. They point out that the results of their turbulent-Prandtl-number distribution for such flows are in excellent agreement with those of Simpson et al. [63], who carried out measurements at low subsonic speeds on a porous plate. They further point out that if they express Prandtl's mixing-length expression in the form written by Van Driest, that is,

$$l_\mathrm{h} = \kappa_\mathrm{h} y \left[1 - \exp\left(-y(\tau_\mathrm{w}\varrho)^{1/2}/\mu B^+\right)\right], \tag{5.6.11}$$

and use the restrictions

$$y = 0, \quad \mathrm{Pr_t} = \left(\frac{\kappa}{\kappa_\mathrm{h}}\frac{B^+}{A^+}\right)^2 \quad y \to \delta, \quad \mathrm{Pr_t} = \left(\frac{\kappa}{\kappa_\mathrm{h}}\right)^2, \tag{5.6.12}$$

then the calculated temperature distributions are in excellent agreement with their experimental data, provided that they choose the constants in Eqs. (5.6.11) and (5.6.12) as $\kappa = 0.40$, $\kappa_\mathrm{h} = 0.43$, $A^+ = 26$, and $B^+ = 33.8$, four empirical constants that compare reasonably well with those used in Eq. (5.6.6). Figures 5.21a and 5.21b show the experimental $\mathrm{Pr_t}$ variation for Meier and Rotta's experiment, together with calculated and experimental temperature distributions taken from Meier and Rotta [68].

We now show the heat-transfer results obtained for pipe flow by using the Cebeci model. The calculations were made by Na and Habib [66] for fluids with low, medium, and high Prandtl numbers ($\mathrm{Pr} = 0.02-15$). Figure 5.22 shows comparisons of calculated and experimental values of Nusselt number Nu, defined as

$$\mathrm{Nu} = hd/k, \tag{5.6.13}$$

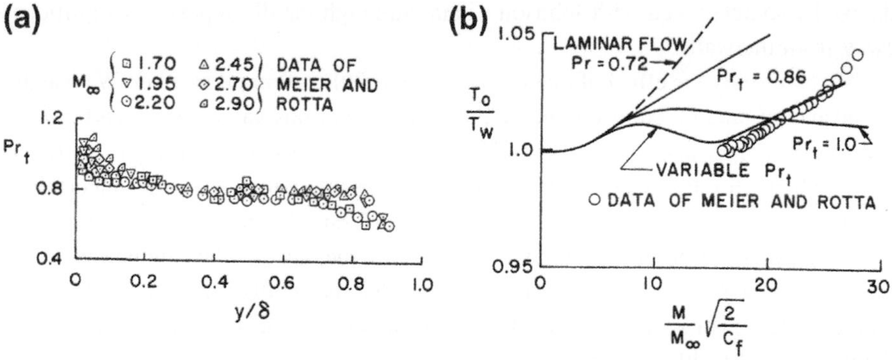

Fig. 5.21 (a) $\mathrm{Pr_t}$ variation in the boundary layer and (b) effect of variable $\mathrm{Pr_t}$ on the calculated temperature distributions [68].

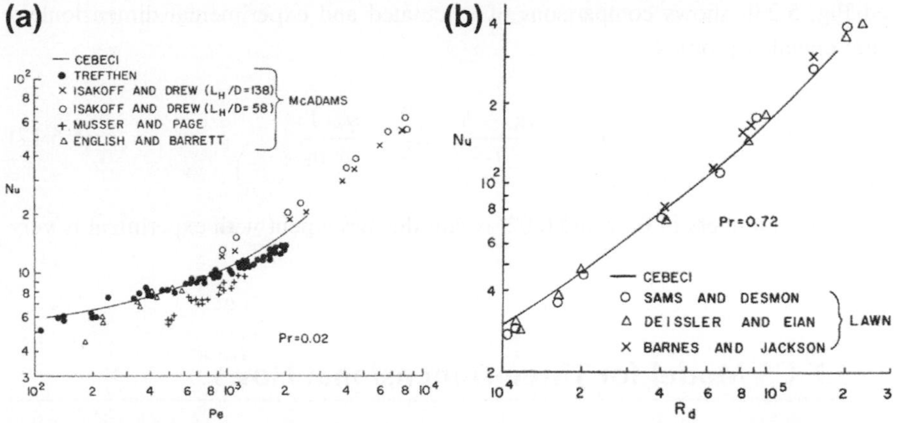

Fig. 5.22 Comparison of calculated and experimental values of Nusselt number for a turbulent pipe flow at different values of (a) Peclet number and (b) Reynolds number.

for various values of Peclet number Pe, defined as

$$\text{Pe} = \text{R}_d \, \text{Pr}, \tag{5.6.14}$$

for two different fluids. Figure 5.22a shows the results for mercury and Figure 5.22b for air. In Eq. (5.6.13) h is the heat-transfer-film coefficient, d is the pipe diameter, and k is the thermal conductivity of the fluid. In Eq. (5.6.14) R_d is the Reynolds number defined as

$$\text{R}_d = \bar{u}d/\nu, \tag{5.6.15}$$

where $\bar{u}$ is the mean velocity.

Figure 5.23a shows comparisons of calculated and experimental Stanton numbers St, defined as

$$\text{St} = \text{Nu}/(\text{R}_d \, \text{Pr}). \tag{5.6.16}$$

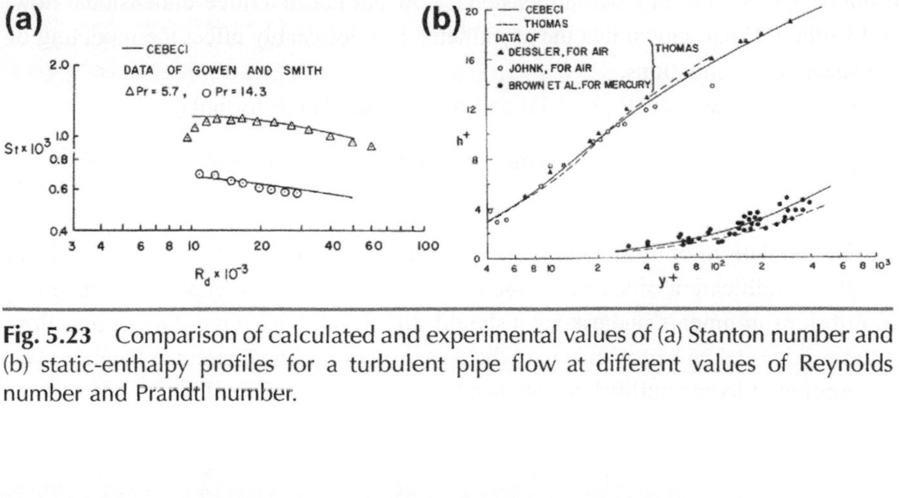

Fig. 5.23 Comparison of calculated and experimental values of (a) Stanton number and (b) static-enthalpy profiles for a turbulent pipe flow at different values of Reynolds number and Prandtl number.

and Fig. 5.23b shows comparisons of calculated and experimental dimensionless static-enthalpy profiles h^+

$$h^+ = \frac{h_w - h}{h_\tau}, \quad h_\tau = \frac{q_w}{\varrho}\frac{1}{u_\tau} \tag{5.6.17}$$

for Prandtl numbers of 0.72 and 0.02. Again the agreement with experiment is very good.

5.7 CS Model for Three-Dimensional Flows

With the eddy viscosity concept, the boundary-layer equations for three-dimensional turbulent flows can be expressed in the same form as those for laminar flows,

$$-\varrho\overline{u'v'} = \varrho\varepsilon_m\frac{\partial u}{\partial y}, \quad -\varrho\overline{w'v'} = \varrho\varepsilon_m\frac{\partial w}{\partial y} \tag{5.7.1}$$

Here $-\varrho\overline{u'v'}$ denotes the shear stress acting in the x-direction on a plane parallel to the xz-plane, and $-\varrho\overline{w'v'}$, usually written as $-\varrho\overline{v'w'}$, the shear stress acting in the z-direction on the same plane. Almost all workers have inferred, from the fact that the choice of direction of the axes in the xz-plane is arbitrary, that the assumptions made for $-\varrho\overline{v'w'}$ should be closely analogous to those made for $-\varrho\overline{u'v'}$. Mathematically, they assume that the turbulence model equation for $-\varrho\overline{v'w'}$ should be obtainable from that for $-\varrho\overline{u'v'}$ by cyclic interchange of symbols. However, it is not so obvious that the equations for $-\varrho\overline{u'v'}$ can be simply derived from models used for a two-dimensional flow. The argument commonly used is that turbulence, being instantaneously three-dimensional, should not be seriously affected by moderate three-dimensionality of the mean flow. There is, of course, a loss of symmetry, for instance, $\overline{v'w'}$ is zero in a two-dimensional flow but not in a three-dimensional flow, and Rotta [69] has shown that the asymmetry can noticeably affect the modeling of the shear-stress equations.

The law of the wall [Eq (4.2.1)] and the mixing-length formula

$$\frac{\partial u}{\partial y} = \frac{(\tau/\varrho)^{1/2}}{\kappa y} \tag{5.7.2}$$

are the foundations of most methods for two-dimensional flows. Clearly, Eq. (5.2.11) requires modification since the velocity now has an extra component. The local equilibrium arguments suggest that it should still be valid in a three-dimensional flow if the x-axis is taken to coincide with the direction of the shear stress at height y. The assumption of local equilibrium between the magnitudes of τ and $\partial u/\partial y$ that leads to

Eq. (5.7.2) implies, when it is taken at face value, that there should be local equilibrium, that is, coincidence between their directions. This leads to

$$\frac{\partial u}{\partial y} = \frac{-\overline{u'v'}}{(\tau/\varrho)^{1/2}\kappa y} \tag{5.7.3a}$$

$$\frac{\partial w}{\partial y} = \frac{-\overline{v'w'}}{(\tau/\varrho)^{1/2}\kappa y} \tag{5.7.3b}$$

where

$$\tau = \left[\left(\varrho\overline{u'v'}\right)^2 + \left(\varrho\overline{v'\omega'}\right)^2\right]^{1/2}$$

The argument is not, of course, very convincing – the local equilibrium is an approximation whose limits of validity need further investigation by experiment. Experiments in three-dimensional flows, particularly measurements of $\overline{v'w'}$, are difficult, and there is evidence both for and against Eq. (5.7.3). A safe position to take is that local equilibrium concepts are not likely to fail catastrophically as soon as the mean flow becomes slightly three-dimensional, and indeed the calculation methods that use Eq. (5.7.3) seem to agree acceptably with most of the experimental data not too near separation, as we shall see in this section.

An undeniable difficulty in treating three-dimensional wall layers is that the viscous sublayer is not a local-equilibrium region; there is a transfer of turbulent energy toward the wall by the turbulent fluctuations themselves to compensate for viscous dissipation. Therefore, conditions at one value of y depend on conditions at the other values of y, and although the directions of velocity gradient and of shear stress coincide at the surface (Reynolds stresses negligible) and, according to Eq. (5.7.3), again coincide outside the sublayer, they may differ within the sublayer. As a result, the direction of the velocity outside the sublayer may differ from that of the shear stress or velocity gradient. In practical terms, the constant of integration in any velocity profile derived from Eq. (5.7.3), or the damping length constant, A in Eq. (5.2.12), will have two components. The effect will be significant only if the direction of the shear stress changes significantly across the sublayer. Since at the surface, $\partial\tau_x/\partial y = \partial p/\partial x$ and $\partial\tau_z/\partial y = \partial p/\partial z$, this will occur only if there is a significant pressure gradient normal to the wall-stress vector, as for example in a boundary layer flowing into a lateral bend ($\partial w_e/\partial x \neq 0$). Van den Berg [70] has proposed a dimensionally correct empirical correlation taking the x-axis in the direction of the wall shear stress, the z-component velocity at the outer edge of the sublayer is $12u_\tau(\nu/\varrho u_\tau^3)\partial p/\partial z$.

The outer layer, like the sublayer, is not a local equilibrium region, and the direction of the shear stress will lag behind the direction of the velocity gradient if the latter changes with x. Several experiments have shown that angles between the shear stress and velocity gradient vectors are of the same order as that of the cross-flow (i.e., the angle between the external velocity and the surface shear stress). However, the accuracy of prediction of the boundary-layer thickness and the surface shear-stress vector does not depend critically on the shear-stress direction in the outer layer, and good agreement has been obtained between the available data and an extension of the CS eddy-viscosity formulation for two-dimensional flows in which the velocity defect used in Eq. (5.2.11b) is just taken as the magnitude of the vector $(u_{te} - u_t)$ at given y. The same eddy viscosity is used in Eq. (5.7.2) so that the directions of shear stress and velocity gradient are equated. According to [71], a generalization of the CS eddy-viscosity formulation for three-dimensional boundary layers is

$$(\varepsilon_m)_i = l^2\left[\left(\frac{\partial u}{\partial y}\right)^2 + \left(\frac{\partial w}{\partial y}\right)^2\right]^{\frac{1}{2}} \quad 0 \le y \le y_c \tag{5.7.4a}$$

$$(\varepsilon_m)_o = \alpha\left|\int_0^\delta (u_{te} - u_t)dy\right| \quad y_c \le y \le \delta \tag{5.7.4b}$$

with $\alpha = 0.0168$ for small adverse pressure gradient flows. Its variation with strong pressure gradient flows can again be expressed by a generalization of Eq. (5.4.14), but this has not been attempted yet.

In Eq. (5.7.4a), the mixing length l is given by Eq. (5.2.12) with A and N defined by Eqs. (5.3.1) and (5.3.2a) except that now

$$u_\tau = \left(\frac{\tau_t}{\varrho}\right)^{\frac{1}{2}}_{\max}, \quad \left(\frac{\tau_t}{\varrho}\right)_{\max} = \nu\left[\left(\frac{\partial u}{\partial y}\right)^2 + \left(\frac{\partial w}{\partial y}\right)^2\right]^{\frac{1}{2}}_{\max} \tag{5.7.5}$$

In Eq. (5.7.4b), u_{te} and u_t are the total edge and local velocites defined by

$$u_{te} = \left(u_e^2 + w_e^2\right)^{1/2} \tag{5.7.6a}$$

$$u_t = \left(u^2 + w^2\right)^{1/2}. \tag{5.7.6b}$$

In the following subsections we present an evaluation of this turbulence model with experimental data for three-dimensional incompressible flows.

5.7.1 Infinite Swept Wing Flows

The accuracy of the CS turbulence model of this section and other models has been investigated for several infinite swept wing flows, as discussed in [71]. Here we present a sample of results taken from those studies and discuss first the results for the data of Bradshaw and Terrell [72] and then for the data of Cumpsty and Head [73].

Data of Bradshaw and Terrell

This experiment was set up especially to test the outer-layer assumptions made in extending the boundary-layer calculation method of Bradshaw et al. [74] from two dimensions to three [75]. Measurements were made only on the flat rear of the wing in a region of nominally zero pressure gradient and decaying cross flow. See the sketch in Fig. 5.24a. Spanwise and chordwise components of mean velocity and shear stress, and all three components of turbulence intensity, were measured at a number of distances $x' = 0, 4, 10, 16$ and 20 in. from the start of the flat portion of the wing (Fig. 5.24). The surface shear stress, measured with a Preston tube, was constant along a generator to the start of the flat part of the wing, except for a few inches at each end and except for small undulations of small spanwise wavelength caused by residual nonuniformities in the tunnel flow.

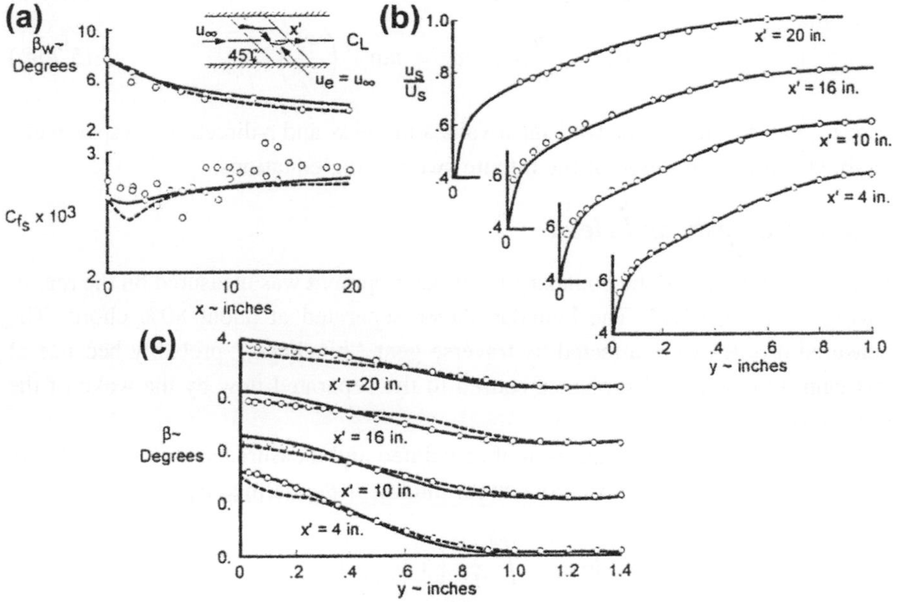

Fig. 5.24 Results for the relaxing flow of Bradshaw and Terrell: (a) wall cross-flow angle and local skin friction, (b) velocity profiles, (c) cross-flow angle distributions. The symbols denote the experimental data, the solid line the numerical solutions of Cebeci [76] and the dashed line the numerical solutions of Bradshaw et al. [74].

Figure 5.24 shows the calculated results (solid lines) with experimental results (symbols) and those obtained by Bradshaw's method [74] (dashed lines). The cross-flow angle β which represents the departure of the velocity vector within the boundary-layer from the freestream velocity vector was computed from

$$\beta \equiv \tan^{-1}\left(\frac{w}{u}\right) \tag{5.7.7a}$$

The above formula becomes indeterminate at $y = 0$; however, with the use of L'Hopital's rule, it can be written as

$$\beta_w = \tan^{-1}\left[\left(\frac{\partial w}{\partial y}\right)_w \left(\frac{\partial u}{\partial y}\right)_w^{-1}\right] \tag{5.7.7b}$$

The streamwise component of the local skin-friction coefficient c_{f_s} was calculated from

$$c_{f_s} = \frac{\tau_{w_s}}{\frac{1}{2}\varrho U_s^2} \tag{5.7.8}$$

with τ_{w_s} and U_s given by

$$\tau_{w_s} = \tau_{w_x} \cos\phi + \tau_{w_z} sin\,\phi \tag{5.7.9a}$$

$$U_s = \frac{u_e}{\cos\phi}, \quad \phi = \tan^{-1}\left(\frac{w_e}{u_e}\right) \tag{5.7.9b}$$

Here τ_{w_x} and τ_{w_z} denote the wall shear values in the x- and z-directions, respectively, obtained from the solution of the infinite swept wing equations.

Data of Cumpsty and Head

In this experiment [73] the boundary-layer development was measured on the rear of a wing swept at 61.1°. The boundary-layer separated at about 80% chord. The measured profiles were affected by traverse gear "blockage," probably because of upstream influence of disturbance caused to the separated flow by the wake of the traverse gear.

Figure 5.25 shows a comparison of calculated and measured streamwise velocity profiles u_s/U_s and streamwise momentum thickness θ_{11} defined by

$$\theta_{11} = \int_0^\delta \frac{u_s}{U_s}\left(1 - \frac{u_s}{U_s}\right) dy \tag{5.7.10}$$

where u_s/U_s is calculated from

$$\frac{u_s}{U_s} = \frac{u}{u_e}\cos^2\phi + \frac{w}{w_e}\sin^2\phi \tag{5.7.11}$$

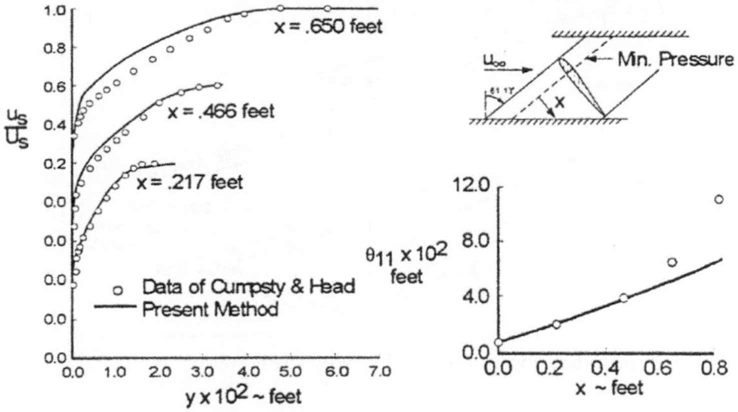

Fig. 5.25 Comparison of calculated (solid lines) and experimental (symbols) results for the data of Cumpsty and Head on the rear of a swept infinite wing.

The results in Fig. 5.25 show good agreement with experiment at two x-stations. However, with increasing distance they begin to deviate from experimental values and at $x = 0.650$ ft, the agreement becomes poor.

The above results indicate what was already observed and discussed in relation to the shortcomings of the Cebeci-Smith algebraic eddy-viscosity formulation, that is, it requires improvements for strong adverse pressure gradient flows. As discussed in subsection 5.4.2, the improvements to this formulation were made for two-dimensional flows by allowing α in the outer eddy-viscosity formula to vary. A similar improvement is needed to the formulation for three-dimensional flows.

5.7.2 Full Three-Dimensional Flows

To illustrate the evaluation of the CS model for full three-dimensional flows, we consider two flows corresponding to an external flow formed by placing an obstruction in a thick two-dimensional boundary-layer (data of East and Hoxey) and an external flow on a prolate spheroid at an incidence angle of 10° (data of Meier and Kreplin).

Data of East and Hoxey

Figure 5.26 shows a schematic drawing of East and Hoxey's test setup in which a wing is placed in a thick two-dimensional boundary layer [77]. The strong pressure gradients exposed by the obstruction caused the boundary layer to become three-dimensional and to separate. The measurements were made in the three-dimensional boundary layer upstrcam of and including the three-dimensional separation.

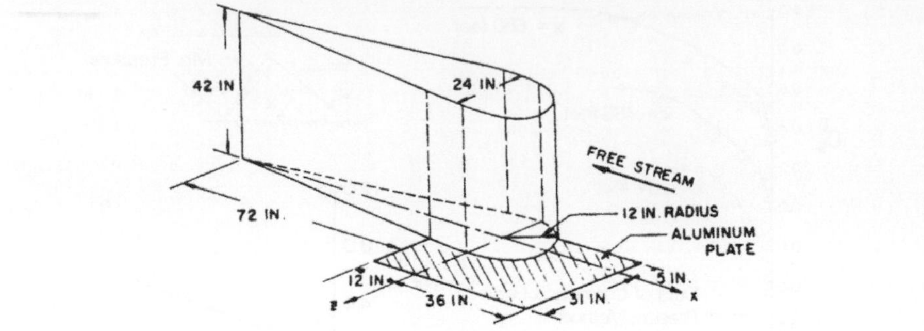

Fig. 5.26 Schematic drawing of East and Hoxey's test setup.

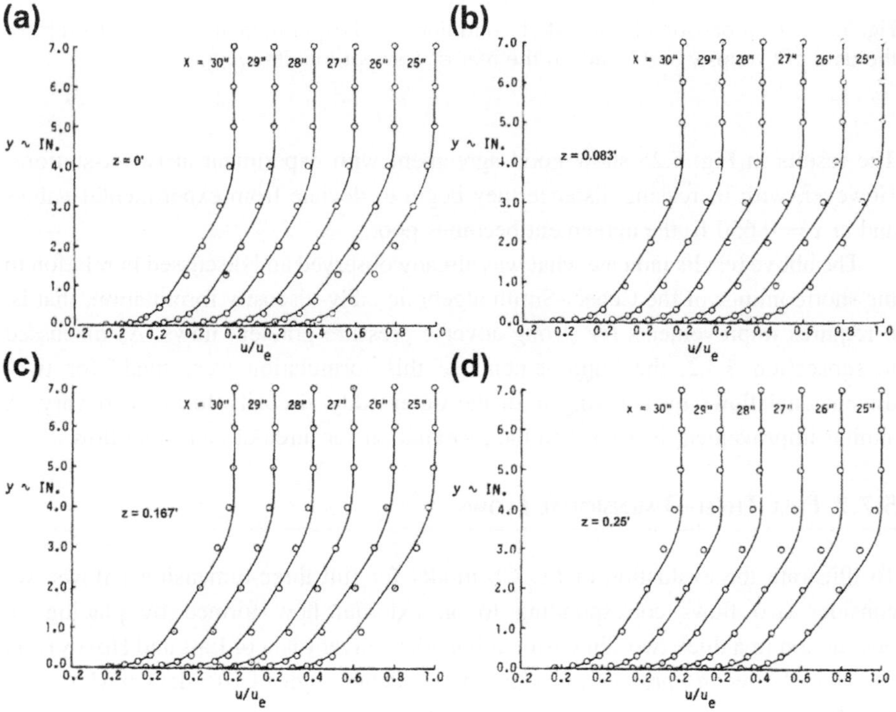

Fig. 5.27 Comparison of calculated (solid lines) and measured (symbols) velocity profiles (a) on the line of symmetry and (b, c, d) off the line of symmetry for the East and Hoxey flow.

Figure 5.27 shows a comparison between calculated and measured velocity profiles on the line of symmetry (Fig. 5.27a) and off the line of symmetry as described in [76]. In general the agreement with experiment is satisfactory.

Data of Meier and Kreplin

Meier and Kreplin's data correspond to laminar, transitional and turbulent flow on a prolate spheroid with a thickness ratio of 6 for a Reynolds number of 6.6×10^6 [78–80]. As discussed in [81], the calculations for this flow were made for freestream velocities of 45 and 55 m/s corresponding to natural and imposed transition. To account for the transitional region between a laminar and turbulent flow, the right-hand sides of Eqs. (5.7.4) are multiplied by the intermittency factor γ_{tr}, defined by Eqs. (5.3.18) and (5.3.19). Since detailed and corresponding correlation formulas for three-dimensional transitional flows are lacking, the same expression was used for three-dimensional flows by using the local similarity assumption with u_e in Eqs. (5.3.18) and (5.3.19) replaced by the total velocity.

The experimental data of Meier et al. consists of surface shear stress magnitude and direction vectors and velocity profiles over a range of angles of attack. Figure 5.28a shows a comparison of calculated surface shear stress vectors in laminar flow at $\alpha = 10°$. The magnitude of the shear stress vector is proportional to the shear intensity. The agreement between the calculation and measurements on the windward side is generally good, although there are some differences that are partly due to the use of inviscid potential flow in the calculations, whereas the measured pressure distribution shows viscous-inviscid interaction effects. It is clear that the laminar flow is separated on the leeward side of the body at some distance aft of the nose. The origin or nature of the high shear intensities leeward of the separation line cannot be determined from calculations because calculations based on external flow that is purely inviscid is not expected to account for strong interactions.

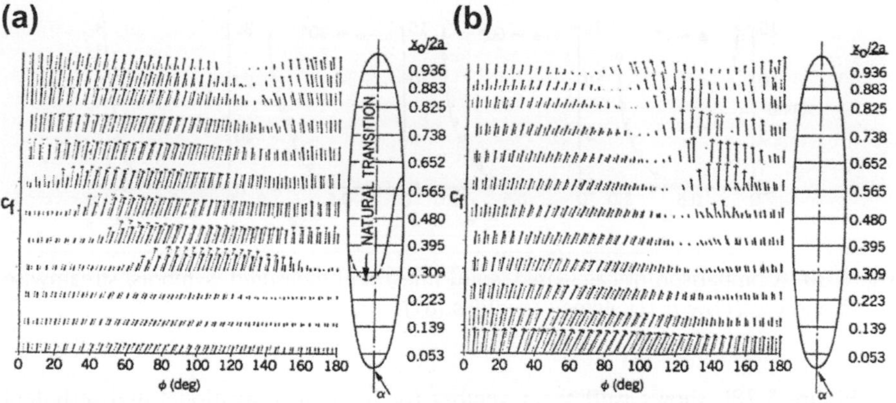

Fig. 5.28 (a) Measured (→) and calculated (→) distributions of wall shear stress vectors $\left(c_f \equiv \tau_w / \frac{1}{2}\varrho u_\alpha^2\right)$ for laminar flow and (b) for laminar, transitional and turbulent flow on a prolate spheroid at $\alpha = 10°$ [81].

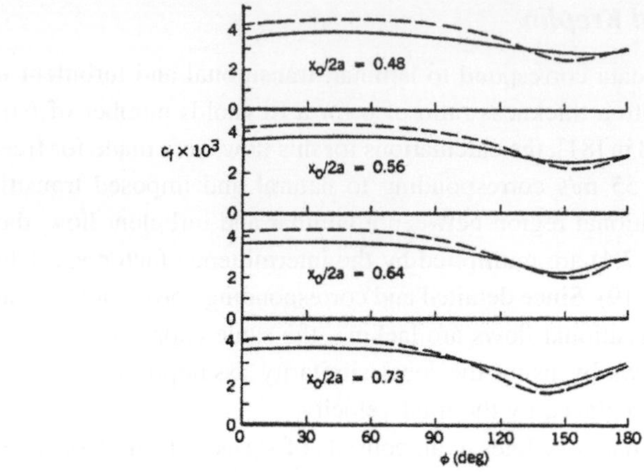

Fig. 5.29 Measured (dashed line) and calculated (solid line) resultant wall shear stress values on the prolate spheroid at $\alpha = 10°$ [81].

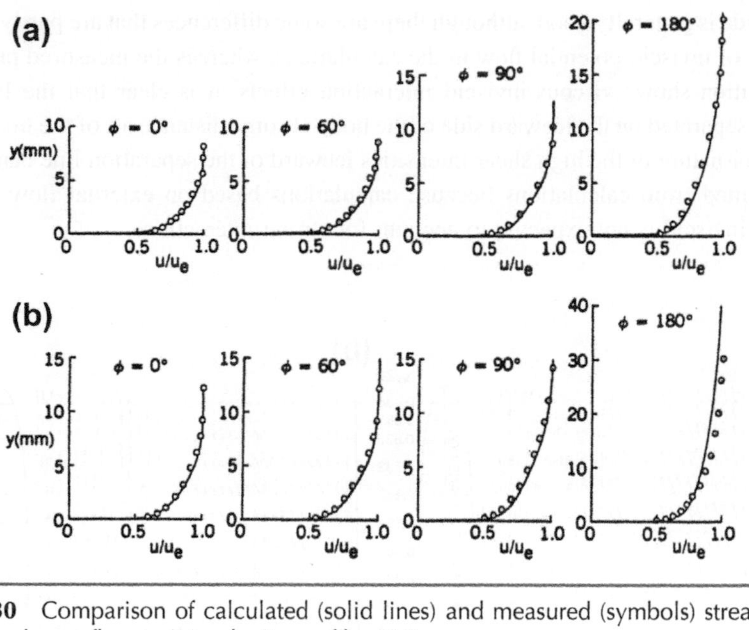

Fig. 5.30 Comparison of calculated (solid lines) and measured (symbols) streamwise u_s/U_s and crossflow u_s/U_s velocity profiles [81].

Figure 5.28b shows wall shear vectors for laminar, transitional and turbulent flow with natural transition. In general, the calculated and measured results are in agreement with discrepancies (which are small) confined to the region close to the specified transition. More quantitative comparison with the imposed transition experiment is afforded by Fig. 5.29 which displays circumferential distributions of wall shear stress at four axial locations. The calculated results display the correct trends and are within 15% of the measured values with discrepancies tending to diminish with downstream distance. A sample of the velocity profiles is shown in Fig. 5.30 and corresponds to $x/2a$ of 0.48 and 0.73, and again the agreement is within or very close to the error bounds of the measurements, except in the regions where the inviscid velocity distribution differed from the measured one. Additional comparisons of calculated and experimental data are given in [81].

5.8 Summary

In the previous sections we have discussed coefficients for transport of momentum and heat suitable for calculating two- and three-dimensional turbulent boundary-layer flows. We have shown how effects such as mass transfer, heat transfer, pressure gradient, etc., can be included in the empirical relations in order to calculate turbulent flows for a wide range of conditions. Although these relations lack rigor and do not improve any fundamental understanding of turbulence, they provide results that are very useful in engineering calculations.

On the basis of comparisons presented in this chapter and those in Chapter 8, we recommend the following eddy-viscosity formulation for calculating two-dimensional and axisymmetric and three-dimensional turbulent boundary layers: The turbulent Prandtl number distribution is given by Eqs. (5.6.6), (5.6.7) and (5.6.8).

Two-Dimensional Flows

Inner Region

$$(\varepsilon_m)_i = L^2\left|\frac{\partial u}{\partial y}\right|\gamma_{tr}, \quad 0 \le y \le y_c; \tag{5.8.12}$$

Outer Region

$$(\varepsilon_m)_o = \alpha\int_0^\infty (u_e - u)dy\,\gamma_{tr}\gamma \quad y_c \le y \le \delta. \tag{5.8.13}$$

For flows on smooth surfaces L is given by Eq. (5.2.12), A by Eq. (5.3.1), γ_{tr} by Eq. (5.3.18), α by Eq. (5.4.10) and γ by Eq. (5.4.15). For flows over surfaces with roughness, L is given by Eq. (5.3.24).

Axisymmetric Flows

For axisymmetric flows with or without transverse curvature effect, the eddy viscosity formulation given by Eqs. (5.8.12) and (5.8.13) still apply provided that L is given by Eq. (5.3.13).

Three-Dimensional Flows

The eddy viscosity formulation for three-dimensional flows is given by Eqs. (5.7.4) with $\alpha = 0.0168$ for small adverse-pressure gradient flows. In Eq. (5.7.4a), the mixing length l is given by Eq. (5.2.12) with A and N defined by Eqs. (5.3.1) and (5.3.2a) except that now u_τ and $(\tau/\varrho)_{\max}$ are defined by Eqs. (5.7.6).

Problems

5.1 By using Eq. (4.4.18), show that $(\varepsilon_m)_i$ given by Eq. (5.2.11a) is proportional to y^4 for $(y/\delta) \ll 1$.

5.2 Show that for equilibrium boundary layers at high Reynolds numbers, if α in Eq. (5.2.11b) is 0.0168, then α_1 in Eq. (4.3.15a) must be 0.0635 for $\kappa = 0.41$.

5.3 In Chapter 8 we discuss the numerical solution of the boundary-layer equations for two-dimensional incompressible flows. We use the CS model to represent the Reynolds shear stress, $-\varrho\overline{u'v'}$. In Chapter 10 we describe a computer program which utilizes this numerical method, and the accompanying CD-ROM presents the computer program. For simplicity, the turbulence model (subroutione EDDY) is limited to flows over smooth surfaces with pressure gradient.

(a) Modify subroutine EDDY to include mass transfer. Note that now for incompressible flows, the definition of N, [Eq. 5.3.2a)], becomes

$$N = \left\{\frac{p^+}{v_w^+}\left[1 - \exp\left(11.8v_w^+\right)\right] + \exp\left(11.8v_w^+\right)\right\}^{1/2}$$

(b) Compute turbulent flow on a flat plate with suction for a Reynolds number of $u_\infty c/\nu = 3 \times 10^6$ with $v_w/u_\infty = -0.15 \times 10^{-4}$. Assume transition at the leading edge. Plot the variation of δ^*/c, R_θ, c_f and f''_w with x/c and compare them with those on a nonporous surface.

(c) Repeat (b) for flow with injection for $v_w/u_\infty = 0.15 \times 10^{-4}$. Take $h_1 = 0.015$, $\kappa = 1.12$ and with $\Delta x/c = 10^{-2}$ for $0 \le x/c \le 1$, NXT = 101.

5.4 Repeat Problem 5.3, this time to acount for the roughness effects on a flat plate. Again assume transition at the leading edge and plot the variation of R_θ, c_f, δ^*/c, and f''_w for dimensionless sand-grain roughness heights of $k_s/c = 0.25 \times 10^{-3}$,

0.50×10^{-3}, 1×10^{-3}, 1.50×10^{-3} and compare your results with those given in the accompanying CD-ROM.

5.5 Repeat Problem 5.3, this time to acount for the transverse curvature effect on a circular cylinder. Remember that this flow is an axisymmetric flow; in addition to making changes in the EDDY viscosity subroutine, it is necessary to redefine b.

(a) Compute c_f with the modified eddy viscosity model, [Eqs. (5.3.12) and (5.3.13)], for $r_0/c = 0.15 \times 10^{-2}$ and 0.5×10^{-4} for $R_c = 10 \times 10^6$. Assume $(x/c)_{tr}$ at 0.075. Plot the variation of δ/r_0, c_f and R_θ with x/c and compare the results with those obtained with the 2d eddy viscosity model [no changes to the subroutine EDDY for TVC effect].

(b) Plot the variation of δ/r_0 with x/c with calculations made by using the original and modified eddy viscosity formulas.

References

[1] E.R. Van Driest, On Turbulent Flow Near a Wall, J. Aeronaut. Sci. 23 (1956).

[2] T. Cebeci, A.M.O. Smith, Computation of Turbulent Boundary Layers – 1968, in: S.J. Kline, M.V. Morkovin, G. Sovran, D.S. Cockrell (Eds.), AFOSR-IFP-Stanford Conf, 1, Stanford Univ. Press, Stanford, California, 1968, p. 346.

[3] G.L. Mellor, H.J. Herring, Computation of Turbulent Boundary Layers – 1968, in: S.J. Kline, M.V. Morkovin, G. Sovran, D.S. Cockrell (Eds.), AFOSR-IFP-Stanford Conf, 1, Stanford Univ. Press, Stanford, California, 1968, p. 247.

[4] S.V. Patankar, D.B. Spalding, Computation of Turbulent Layers – 1968, in: S.J. Kline, M.V. Morokovin, G. Sovran, D.S. Cockrell (Eds.), AFOSR-IFP-Stanford Conf, 1, Stanford Univ. Press, Stanford, California, 1968, p. 356.

[5] R. Michel, C. Quémard, R. Durant, Application d'un schéma de longueur de mélange à l'étude des couches limites d'équilibre, ONERA Note Technique No. 154 (1969).

[6] T. Cebeci, Behavior of Turbulent Flow Near a Porous Wall With Pressure Gradient, AIAA J. 8 (1970).

[7] T. Cebeci, Calculation of Compressible Turbulent Boundary Layers With Heat and Mass Transfer, AIAA J. 9 (1971) 1091.

[8] D. Coles, The Turbulent Boundary Layer in a Compressible Fluid. Rep. R-403-PR, Rand Corp, California, Santa Monica, 1962. Also AD 285–651.

[9] R.L. Simpson, Characteristics of Turbulent Boundary Layers at Low Reynolds Numbers with and without Transpiration, J. Fluid Mech. 42 (1970) 769.

[10] T. Cebeci, G.J. Mosinskis, Computation of Incompressible Turbulent Boundary Layers at Low Reynolds Numbers, AIAA J. 9 (1971).

[11] D.G. Huffman, P. Bradshaw, A Note on Van Karman's Constant in Low Reynolds Number Turbulent Flows, J. Fluid Mech. 53 (1972) 45.

[12] D.M. Busnell, D.J. Morris, Shear-Stress Eddy-Viscosity and Mixing Length Distributions in Hypersonic Turbulent Boundary Layers, NASA Tech. Memo No. X-2310 (1971). Also N71–31310.

[13] T. Cebeci, Kinematic Eddy Viscosity at Low Reynolds Numbers, AIAA J. 11 (1973) 102.

[14] T. Cebeci, Variation of the Van Driest Damping Parameter with Mass Transfer, AIAA J. 9 (1968) 1973.

[15] J.E. Danberg, Characteristics of the Turbulent Boundary Layer with Heat and Mass Transfer at $M = 6.7$, Naval Ord. Rep. NOLTR (1964) 64–99. Also AD 452–471.

[16] T. Cebeci, Eddy Viscosity Distribution in Thick Axisymmetric Turbulent Boundary Layers, J. Fluids Eng. 95 (1973) 319.

[17] R. Richmond, Experimental Investigation of Thick Axially Symmetric Boundary Layers On Cylinders at Subsonic and Hypersonic Speeds. Ph.D. Thesis, California Inst. Technology, Pasadena, California, 1957.

[18] H. Thomann, Effect of Streamwise Wall Curvature on Heat Transfer in a Turbulent Boundary Layer, J. Fluid Mech. 33 (1968) 283.

[19] T. Cebeci, Curvature and Transition Effects in Turbulent Boundary Layers, AIAA J. 9 (1968) 1971.

[20] P. Bradshaw, The Analogy Between Streamline Curvature and Buoyancy in Turbulent Shear Flow, J. Fluid Mech. 36 (1969) 177.

[21] G.B. Schubauer, P.S. Klebanoff, Investigation of Separation of the Turbulent Boundary Layer, NACA Tech. Note No. 2133 (1956).

[22] H. Shmidbauer, Turbulent Friction Along Convex Surfaces, NACA Tech. Memo 791 (1936).

[23] D.M. Bushnell, D.W. Alston, On the Calculation of Transitional Boundary-Layer Flows, AIAA J. 10 (1972) 229.

[24] W.G. Hoydysh, V. Zakkay, An Experimental Investigation of Hypersonic Turbulent Boundary Layers in Adverse Pressure Gradient. Rep. F-67-5, New York Univ., Bronx. New York, 1967.

[25] K.K. Chen, N.A. Thyson, Extension of Emmons Spot Theory to Flows on Blunt Bodies, AIAA J. 5 (1971) 821.

[26] H.W. Emmons, The Laminar-Turbulent Transition in a Boundary Layer – Part 1, J. Aeronaut. Sci. 18 (1951) 490.

[27] S. Dhawan, R. Narasimha, Some Properties of Bounday-Layer Flow During the Transition from Laminar to Turbulent Motion, J. Fluid Mach. 3 (1958) 418.

[28] R. Michel, Resultants sur la couche limite turbulent aux grandes vitesses, Memo Tech 22 (1961). Office Nat. d'Etudes et de Rech. Aeronaut.

[29] A.M.O. Smith, N. Gamberoni, Transition, In: Pressure Gradient and Stability Theory, 4, Proc. Int. Congr. Appl. Mech., 9th, Brussels, Belgium, 1956. 234. Also AD 125 559m.

[30] T. Cebeci, G.J. Mosinskis, A.M.O. Smith, Calculation of Viscous Drag and Turbulent Boundary-Layer Separation on Two-dimensional and Axisymmetric Bodies in Incompressible Flows. Rep. No. MDC J0973–01, Douglas Aircraft Co, Long Beach, California, 1970. Also N71-25868.

[31] G.B. Schubauer, Air Flow in the Boundary Layer of an Elliptic Cylinder, NACA Rep. No. 652 (1939).

[32] D. Coles, Measurements in the Boundary Layer on a Smooth Flat Plate in Supersonic Flow. III. Measurements in a Flat-Plate Boundary Layer at the Jet Propulsion Lab. Rep. No. 20–71, Jet Propulsion Lab., California Inst. Technol, Pasadena, California, 1953.

[33] J.C. Adams Jr., Implicit Finite-Difference Analysis of Compressible Laminar, Transitional and Turbulent Boundary Layers Along the Windward Streamline of a Sharp Cone at Incidence. Rep. AEDC-TR-71–235, Arnold Eng. Develop. Center, Tennessee, 1971. Also N72-19346.

[34] J.E. Harris, Numerical Solution of the Equations for Compressible Laminar, Transitional and Turbulent Boundary Layers and Comparisons with Experimental Data, NASA Tech. Rep (1971). No. R-368. Also N71–32164.

[35] R.M. O'Donnell, Experimental Investigation at a Mach Number of 2.41 of Average Skin-Friction Coefficients and Velocity Profiles for Laminar and Turbulent Boundary Layers and on Assessment of Probe Effects, NACA Tech. Note No. 3122 (1954).

[36] T. Cebeci, Essential Ingredients of a Method for Low Reynolds-Number Airfoils, AIAA J. 27 (1983) 1680–1688.

[37] J. Cousteix, G. Pailhas, Etude Exploratorie d'un Processus de Transition Laminair-Turbulent au Voisinage du Decollemente d'une Couche Limite Laminaire, T. P. No (1979-86) 213–218. Also LeRecherche Aerospatiale No. 1979-3.

[38] T. Cebeci, An Engineering Approach to the Calculation of Aerodynamic Flows, Horizons Publishing, Long Beach and Springer-Verlag, Heidelberg, 1999.

[39] T. Cebeci, K.C. Chang, Calculation of Incompressible Rough-Wall Boundary-Layers Flows, AIAA J. 16 (1978) 730.

[40] J.C. Rotta, Turbulent Boundary Layers in Incompressible Flows, Progr. Aeronaut. Sci. 2 (1962) 1.

[41] F.A. Dvorak, Calculation of Turbulent Boundary Layers on Rough Surfaces in Pressure Gradient, AIAA J. 7 (No. 9) (1969). 1752.

[42] D. Betterman, Contribution a l'étude de la couche limite turbulente le long de plaques regueuses, Rapport 65–6, Centre Nat. de la Rech. Sci, Paris, France, 1965.

[43] A.E. Perry, P.N. Joubert, Rough-Wall Boundary Layers in Adverse Pressure Gradients, J. Fluid Mech. 17 (1963) 193–211.

[44] D.A. Johnson, L.S. King, Mathematically Simple Turbulence Closure Model for Attached and Separated Turbulent Boundary Layers, AIAA J. 23 (No. 11) (1985) 1684–1692.

[45] D.A. Johnson, T.J. Coakley, Improvement to a Nonequilibrium Algebraic Turbulence Model, AIAA J. 28 (No. 11) (1990). 2000–2003.

[46] T. Cebeci, K.C. Chang, An Improved Cebeci-Smith Turbulence Model for Boundary-Layer and Navier-Stokes Methods, 20th Congress of the International Council of the Aeronautical Sciences, Sorrento, Italy, 1996, paper No. ICAS-96–1.7.3.

[47] R.L. Simpson, Y.T. Chew, B.G. Shivaprasad, The Structure of a Separating Turbulent Boundary Layer, Part 1, Mean Flow and Reynolds Stresses, J. Fluid Mech. 113 (1981) 23–51.

[48] A. Nakayama, Measurements in the Boundary Layer and Wake of Two Airfoil Models. Report No, MDC J2403, Douglas Aircraft Co. (1982).

[49] H. Fiedler, M.R. Head, Intermittency Measurements in the Turbulent Boundary Layer, J. Fluid Mech., 25, Part 4 (1966). 719–535.

[50] Carr, L. W., McCroskey, W. J., McAlister, K. W., Pucci, S. L., and Lambert, O.: An Experimental Study of Dynamic Stall on Advanced Airfoil Sections, 3, Hot-Wire and Hot-Film Measurements, NASA TM 84245.

[51] E. Omar, T. Zierten, M. Hahn, E. Szpiro, A. Mahal, Two-Dimensional Wind-Tunnel Tests of a NASA Supercritical Airfoil with Various High-Lift Systems, 2, Test Data, NASA CR-2215 (1977).

[52] B.S. Baldwin, H. Lomax, Thin Layer Approximation of Algebraic Model for Separated Turbulent Flows, AIAA paper No (1978) 78–257.

[53] H.W. Stock, W. Haase, Determination of Length Scales in Algebraic Turbulence Models for Navier-Stokes Methods, AIAA J. 27 (No. 1) (1989) 5–14.

[54] D.A. Johnson, Nonequilibrium Algebraic Turbulence Modeling Considerations for Transonic Airfoils and Wings, AIAA paper No (1992). 92–0026.

[55] R.C. Swanson, E. Turkel, A Multistage Time-Stepping Scheme for the Navier-Stokes Equations, AIAA paper No (1985). 85–0035.

[56] O. Reynolds, On the Extent and Action of the Heating Surface for Steam Boilers, Proc. Manchester Lit. Phil. Soc. 14 (1874) 7.

[57] J. Kestin, P.D. Richardson, Heat Transfer Across Turbulent Incompressible Boundary Layers, Int. J. Heat Mass Transfer 6 (1963) 147.

[58] R. Jenkins, Variation of Eddy Conductivity with Prandtl Modulus and Its Use in Prediction of Turbulent Heat Transfer Coefficients, Inst. Heat Transfer Fluid Mech. 147 (1951).
[59] W.M. Kays, Convective Heat and Mass Transfer, McGraw-Hill, New York, 1966.
[60] F. Page, W.G. Schlinger, D.K. Breaux, B.H. Sage, Point Values of Eddy Conductivity and Viscosity in Uniform Flow Between Parallel Plates, Ind. Eng. Chem. 44 (1952) 424.
[61] W.M. Rohsenow, H.Y. Choi, Heat, Mass and Momentum Transfer, Prentice-Hall, Englewood Cliffs, New Jersey, 1961.
[62] R.G. Deissler, Turbulent Heat Transfer and Temperature Fluctuations in a Field with Uniform Velocity and Temperature Gradients, Int. J. Heat Mass Transfer 6 (1963) 257.
[63] R.L. Simpson, D.C. Whitten, R.J. Moffat, An Experimental Study of the Turbulent Prandtl Number of Air with Injection and Suction, Int. J. Heat Mass Transfer 13 (1970) 125.
[64] T. Cebeci, A Model for Eddy Conductivity and Turbulent Prandtl Number, J. Heat Transfer 95 (1973) 227.
[65] H. Ludweig, Bestimmung des Verhältnisses der Austauschkoeffizienten für Wärme und Impuls bei turbulenten Grenzschichten, Z. Flugwiss. 5 (1956) 73.
[66] T.Y. Na, I.S. Habib, Heat Transfer in Turbulent Pipe Flow Based on a New Mixing Length Model, Appl. Sci. Res. 28 (4/5) (1973).
[67] D.S. Johnson, Velocity and Temperature Fluctuation Measurements in a Turbulent Boundary Layer Downstream of a Stepwise Discontinuity in Wall Temperature, J. Appl. Mech. 26 (1959) 325.
[68] Meier, H. U. and Rotta, J. C.: Temperature Distributions in Supersonic Turbulent Boundary Layers. AIAA J., 9, 2149, 1971.
[69] J. Rotta, C., A Family of Turbulence Models for Three-Dimensional Thin Shear Layers, DFVLR-AVA Göttingen (1977). IB251–76A25.
[70] B. Van den Berg, A Three-Dimensional Law of the Wall for Turbulent Shear Flows, J. Fluid Mech. 70 (1975) 149.
[71] T. Cebeci, J. Cousteix, Modeling and Computation of Boundary-Layer Flows, Horizons Publishing, Long Beach, California and Springer-Verlag, Heidelberg, 1998.
[72] P. Bradshaw, M.G. Terell, The Response of a Turbulent Boundary Layer on an "Infinite" Swept Wing to the Sudden Removal of Pressure Gradient, NFL Aero. Rept. 1305, ARC 31514 (1969).
[73] N.A. Cumpsty, M.R. Head, The Calculation of Three-Dimensional Turbulent Boundary-Layers, Part IV: Comparison of Measurements with Calculations on the Rear of a Swept Wing. C. P. 1077, Aero Research Council, London, England, 1970.
[74] P. Bradshaw, D.H. Ferriss, N.P. Atwell, Calculation of Boundary-Layer Development Using the Turbulent Energy Equation, J. Fluid Mech. 23 (1967) 593.
[75] T. Cebeci, Calculation of Three-Dimensional Boundary Layers, Part 2, Three-Dimensional Flows in Cartesian Coordinates, AIAA J. 13 (1975) 1056.
[76] P. Bradshaw, G. Mizner, K. Unsworth, Calculation of Compressible Turbulent Boundary Layers on Straight-Tapered Swept Wings, AIAA J. 14 (1976) 399.
[77] L.F. East, R.P. Hoxey, Low-Speed Three-Dimensional Turbulent Boundary-Layer Data. Pt. I, TR 69041, Royal Aircraft Establishment, Farnborough, England, 1969.
[78] H.U. Meier, H.P. Kreplin, Experimental Investigation of the Boundary Layer Transition and Separation on a Body of Revolution, Zeitschrift für Flugwissenschaften und Weltraumforschung 4 (March-April 1980) 65–71.
[79] H.U. Meier, H.P. Kreplin, A. Landhäusser, D. Baumgarten, Mean Velocity Distributions in Three-Dimensional Boundary Layers, Developing on a 1:6 Prolate Spheroid with Natural Transition ($\alpha = 10°$, $U_\infty = 45$ m/s, Cross Sections $x_0/2_a = 0.56$; 0.64 and 0.73), German Aerospace Research Establishment, Göttingen, Germany, March 1984. Rept. IB 222-84 A 10.

[80] H.U. Meier, H.P. Kreplin, A. Landhäusser, D. Baumgarten, Mean Velocity Distributions in Three-Dimensional Boundary Layers, Developing on a 1:6 Prolate Spheroid with Natural Transition ($\alpha = 10°$, $U_\infty = 55$ m/s, Cross Sections $x_0/2_a = 0.48; 0.56; 0.64$ and 0.73), German Aerospace Research Establishment, Göttingen, Germany, March 1984. Rept. IB 222-84 A 11.

[81] T. Cebeci, H.U. Meier, Turbulent Boundary Layers on a Prolate Spheroid, AIAA paper (June 1987) 87–1299.

Chapter 6

Transport-Equation Turbulence Models

Chapter Outline Head

6.1 Introduction

While the zero-equation models discussed in Chapter 5 are useful and accurate for most boundary-layer flows, these models lack generality. For turbulent shear flows other than wall boundary-layers, they require different expressions for mixing length and eddy

Analysis of Turbulent Flows with Computer Programs. http://dx.doi.org/10.1016/B978-0-08-098335-6.00006-9

viscosities. For example, for a plane jet, round jet and plane wake $l = 0.09\delta$, $l = 0.07\delta$, $l = 0.16\delta$, respectively, with δ denoting the shear layer thickness. Transport equation models have less limitations than the zero-equation models for modeling Reynolds stresses. Before we discuss several of these methods that are popular and in wide use, let us consider the Reynolds-stress transport equation (3.5.3) and the kinetic energy equation (2.8.5) which these models use. For convenience, we again call R_{ij} $(\equiv -\overline{u_i' u_j'})$ the Reynolds stress tensor, rather than the actual one, $\bar{R}_{ij}$ $(\equiv -\overline{\varrho u_i' u_j'})$, and also consider an incompressible flow. Before we rewrite Eq. (3.5.3), let us divide the last two terms in Eq. (3.5.3) by ϱ. If we denote these two terms V_{ij}, that is,

$$V_{ij} = \frac{1}{\varrho}\frac{\partial}{\partial x_k}\left(\overline{u_j' \tau_{ik}''} + \overline{u_i' \tau_{jk}''}\right) - \frac{1}{\varrho}\left(\overline{\tau_{ik}''\frac{\partial u_j'}{\partial x_k}} + \overline{\tau_{jk}''\frac{\partial u_i'}{\partial x_k}}\right), \tag{6.1.1}$$

then from the definition of the stress tensor we can write the first term as

$$\begin{aligned}
&\frac{1}{\varrho}\frac{\partial}{\partial x_k}\left(\overline{u_j' \tau_{ik}''} + \overline{u_i' \tau_{jk}''}\right) \\
&= \nu\frac{\partial}{\partial x_k}\left[\overline{u_j'\left(\frac{\partial u_i'}{\partial x_k} + \frac{\partial u_k'}{\partial x_i}\right)} + \overline{u_i'\left(\frac{\partial u_j'}{\partial x_k} + \frac{\partial u_k'}{\partial x_j}\right)}\right] \\
&= \nu\frac{\partial}{\partial x_k}\left(\frac{\partial}{\partial x_k}\left(\overline{u_i' u_j'}\right)\right) + \nu\frac{\partial}{\partial x_k}\left(\overline{u_j'\frac{\partial u_k'}{\partial x_i} + u_i'\frac{\partial u_k'}{\partial x_j}}\right).
\end{aligned} \tag{6.1.2}$$

Now we write the second term in Eq. (6.1.1) as

$$\begin{aligned}
&-\frac{1}{\varrho}\left(\overline{\tau_{ik}''\frac{\partial u_j'}{\partial x_k}} + \overline{\tau_{jk}''\frac{\partial u_i'}{\partial x_k}}\right) \\
&= -\nu\overline{\left(\frac{\partial u_i'}{\partial x_k} + \frac{\partial u_k'}{\partial x_i}\right)\frac{\partial u_j'}{\partial x_k}} - \nu\overline{\left(\frac{\partial u_j'}{\partial x_k} + \frac{\partial u_k'}{\partial x_j}\right)\frac{\partial u_i'}{\partial x_k}} \\
&= -\nu\left(2\overline{\frac{\partial u_i'}{\partial x_k}\frac{\partial u_j'}{\partial x_k}} + \overline{\frac{\partial u_k'}{\partial x_i}\frac{\partial u_j'}{\partial x_k}} + \overline{\frac{\partial u_k'}{\partial x_j}\frac{\partial u_i'}{\partial x_k}}\right).
\end{aligned} \tag{6.1.3}$$

Combining Eqs. (6.1.2) and (6.1.3), we get

$$V_{ij} = -\nu\frac{\partial^2 R_{ij}}{\partial x_k^2} - 2\nu\overline{\frac{\partial u_i'}{\partial x_k}\frac{\partial u_j'}{\partial x_k}} \tag{6.1.4}$$

since for an incompressible flow $\overline{u_j'(\partial^2 u_k'/\partial x_k \partial x_i)}$ and $\overline{u_i'(\partial^2 u_k'/\partial x_k \partial x_j)}$ are zero.

With Eq. (6.1.4) and with the continuity equation (2.5.3), Eq. (3.5.3) can be written as

$$\frac{D R_{ij}}{Dt} = \frac{\partial}{\partial x_k}\left(\nu \frac{\partial R_{ij}}{\partial x_k}\right) - R_{ik}\frac{\partial \bar{u}_j}{\partial x_k} - R_{jk}\frac{\partial \bar{u}_i}{\partial x_k} + 2\nu \overline{\frac{\partial u_i'}{\partial x_k}\frac{\partial u_j'}{\partial x_k}}$$
$$+ \frac{\partial}{\partial x_k}\left(\overline{u_i'u_j'u_k'} + \overline{\frac{p'}{\varrho}u_i'}\delta_{jk} + \overline{\frac{p'}{\varrho}u_j'}\delta_{ik}\right) \tag{6.1.5}$$
$$- \overline{\frac{p'}{\varrho}\left(\frac{\partial u_j'}{\partial x_i} + \frac{\partial u_i'}{\partial x_j}\right)}$$

where

$$\frac{D R_{ij}}{Dt} = \frac{\partial R_{ij}}{\partial t} + \bar{u}_k \frac{\partial R_{ij}}{\partial x_k}.$$

A contraction of Eq. (6.1.5), $i = j$ and with $k = -\frac{R_{ii}}{2}$ gives the following equation for the turbulence kinetic energy discussed in Section 2.8:

$$\frac{Dk}{Dt} = \frac{\partial}{\partial x_k}\left(\nu \frac{\partial k}{\partial x_k} - \overline{\frac{p'}{\varrho}u_i'}\delta_{ik} - \overline{u_k'\frac{u_i'u_i'}{2}}\right) + R_{ik}\frac{\partial \bar{u}_i}{\partial x_k} - \varepsilon \tag{6.1.6}$$

where

$$\frac{Dk}{Dt} = \frac{\partial k}{\partial t} + \bar{u}_k \frac{\partial k}{\partial x_k}$$
$$\varepsilon = \nu \overline{\left(\frac{\partial u_i'}{\partial x_k}\frac{\partial u_i'}{\partial x_k}\right)} \tag{6.1.7}$$

The left-hand side of Eq. (6.1.6) represents the rate of change of turbulence kinetic energy. The first term on the right-hand side is called molecular diffusion and represents the diffusion of turbulence energy caused by the molecular transport process of the fluid. The second term is called *pressure diffusion* and the triple velocity correlation term which represents the rate at which turbulence energy is transported through the fluid by turbulent fluctuations, is called *turbulent transport.* The fourth term is known as production and represents the rate at which turbulence kinetic energy is transferred from the mean flow to the turbulence. Finally the last term may be called "isotropic dissipation", since the actual dissipation ε is given by Eq. (3.5.5),

$$\varepsilon = \overline{\frac{1}{\varrho}\tau_{ik}''\frac{\partial u_i'}{\partial x_k}} \tag{3.5.5}$$

Another equation that is employed in most transport-equation turbulence models is the rate of dissipation of turbulent energy which is obtained from a transport equation for ε derived by Harlow and Nakayama [1],

$$\frac{D\varepsilon}{Dt} = \frac{\partial}{\partial x_k}\left(\nu\frac{\partial \varepsilon}{\partial x_k}\right) - 2\nu\frac{\partial \bar{u}_i}{\partial x_k}\left(\overline{\frac{\partial u_i'}{\partial x_l}\frac{\partial u_k'}{\partial x_l}} + \overline{\frac{\partial u_l'}{\partial x_i}\frac{\partial u_l'}{\partial x_k}}\right)$$
$$-2\nu\overline{\frac{\partial u_i'}{\partial x_k}\frac{\partial u_i'}{\partial x_l}\frac{\partial u_k'}{\partial x_l}} - 2\left[\nu\overline{\frac{\partial^2 u_i'}{\partial x_k \partial x_l}}\right]^2 \tag{6.1.8}$$
$$-\frac{\partial}{\partial x_k}\overline{u_k'\varepsilon'} - \frac{\nu}{\varrho}\frac{\partial}{\partial x_i}\left[\overline{\frac{\partial p'}{\partial x_l}\frac{\partial u_i'}{\partial x_l}}\right].$$

The following closure assumptions are made for the terms on the right-hand side of this equation.

The second term on the right-hand side, the generation term, is modeled by

$$2\nu\left(\overline{\frac{\partial u_i'}{\partial x_l}\frac{\partial u_k'}{\partial x_l}} + \overline{\frac{\partial u_l'}{\partial x_i}\frac{\partial u_l'}{\partial x_k}}\right) = \left(c_{\varepsilon_1}\frac{\overline{u_i'u_k'}}{k} + \tilde{c}_{\varepsilon_1}\delta_{ik}\right)\varepsilon \tag{6.1.9}$$

where $c_{\varepsilon 1}$ and $\tilde{c}_{\varepsilon 1}$ are constants. In fact, the term containing $\tilde{c}_{\varepsilon 1}$ vanishes when Eq. (6.1.1) is multiplied by $\partial\bar{u}_i/\partial x_k$; thus it need not be considered further.

The third and fourth terms on the right-hand side of Eq. (6.1.8) are combined into one term, modeled by

$$2\left\{\nu\overline{\frac{\partial u_i'}{\partial x_k}\frac{\partial u_i'}{\partial x_l}\frac{\partial u_k'}{\partial x_l}} + \left(\nu\overline{\frac{\partial^2 u_i'}{\partial x_k \partial x_l}}\right)^2\right\} = c_{\varepsilon_2}\frac{\varepsilon^2}{k}. \tag{6.1.10}$$

In two-equation models, the fifth term, which accounts for the diffusion of ε from velocity fluctuations, is modeled by

$$-\overline{u_k'\varepsilon'} = \frac{\varepsilon_m}{\sigma_\varepsilon}\frac{\partial \varepsilon}{\partial x_k} \tag{6.1.11}$$

which is different than the modeling used in stress-transport models, see Eq. (6.4.7).

The last term, which represents the diffusional transport of ε by pressure fluctuations, is neglected.

With these closure assumptions, the final form of the rate of dissipation of turbulent energy may be written as

$$\frac{D\varepsilon}{Dt} = \frac{\partial}{\partial x_k}\left(\nu\frac{\partial \varepsilon}{\partial x_k}\right) + c_{\varepsilon_1}\frac{\varepsilon}{k}R_{ik}\frac{\partial \bar{u}_i}{\partial x_k} - c_{\varepsilon_2}\frac{\varepsilon^2}{k} + \frac{\partial}{\partial x_k}\left(\frac{\varepsilon_m}{\sigma_\varepsilon}\frac{\partial \varepsilon}{\partial x_k}\right) \tag{6.1.12}$$

The constants $c_{\varepsilon 1}$, $c_{\varepsilon 2}$ and σ_ε, which are obtained by reference to experimental data will be discussed later.

Except for the stress-transport models discussed in Section 6.4, most of the transport equation models, with the exception of the Spalart and Allmaras model [2], use the turbulence kinetic energy equation by itself (and continuity and mean momentum equations) or with another equation, like the rate of dissipation equation (6.1.12). The latter leads to *two-equation models* discussed in Section 6.2 and the former to *one-equation models* discussed in Section 6.3. In Section 6.4 we discuss stress-transport models in which the exact transport equations for some or all of the Reynolds stresses are modeled term by term.

6.2 Two-Equation Models

Over the years a number of two-equation models have been proposed. A description of most of these models is given in detail by Wilcox [3]. Here we consider three of the more popular, accurate and widely used models. They include the k-ε model of Jones and Launder [4], the k-ω model of Wilcox [3] and the SST model of Menter which blends the k-ε model in the outer region and k-ω model in the near wall region [5]. All three models can be used for a range of flow problems with good accuracy as we shall discuss in Chapter 9.

6.2.1 k-ε Model

The k-ε model is the most popular and widely used two-equation eddy viscosity model. In this model various terms in the kinetic energy and rate of dissipation equations are modeled as follows.

Equation (6.1.6) contains four terms that require closure assumptions. The modeling of the second, third and fourth terms makes use of the eddy viscosity concept in which the Reynolds stress R_{ij} is given by Eq. (5.2.4), which can be written as

$$R_{ij} = \varepsilon_m \left(\frac{\partial \bar{u}_i}{\partial x_j} + \frac{\partial \bar{u}_j}{\partial x_i} \right) \tag{6.2.1}$$

with ε_m is written as

$$\varepsilon_m = c_2 \frac{k^2}{\varepsilon} \tag{6.2.2}$$

Here c_2 is a constant at high Reynolds number. The second and third terms in Eq. (6.1.6), namely the pressure diffusion and turbulent transport terms are related to the gradients of k,

$$-\left(\overline{u'_k\frac{p'}{\varrho}+u'_k\frac{u'_iu'_i}{2}}\right)=\frac{\varepsilon_m}{\sigma_k}\frac{\partial k}{\partial x_k} \tag{6.2.3}$$

where σ_k is a constant or a specified function. Substituting Eqs. (6.2.1) and (6.2.3) into Eq. (6.1.6) we obtain the modeled form of the *turbulence kinetic energy equation*

$$\frac{Dk}{Dt}=\frac{\partial}{\partial x_k}\left[\left(\nu+\frac{\varepsilon_m}{\sigma_k}\right)\frac{\partial k}{\partial x_k}\right]+\varepsilon_m\left(\frac{\partial \bar{u}_i}{\partial x_j}+\frac{\partial \bar{u}_j}{\partial x_i}\right)\frac{\partial \bar{u}_i}{\partial x_j}-\varepsilon \tag{6.2.4}$$

Similarly, with the relation given by Eq. (6.2.1), the *dissipation equation* (6.1.12) becomes

$$\frac{D\varepsilon}{Dt}=\frac{\partial}{\partial x_k}\left[\left(\nu+\frac{\varepsilon_m}{\sigma_\varepsilon}\right)\frac{\partial \varepsilon}{\partial x_k}\right]+c_{\varepsilon_1}\frac{\varepsilon}{k}\varepsilon_m\left(\frac{\partial \bar{u}_i}{\partial x_j}+\frac{\partial \bar{u}_j}{\partial x_i}\right)\frac{\partial \bar{u}_i}{\partial x_j}-c_{\varepsilon_2}\frac{\varepsilon^2}{k} \tag{6.2.5}$$

The first term on the right-hand side of Eq. (6.2.5) represents molecular and turbulent diffusion of dissipation, and the sum of the second and third terms represent the production and dissipation. The parameter σ_ε is a parameter to be specified.

For boundary-layer flows at high Reynolds number and with Eq. (6.2.2) now written as

$$\varepsilon_m = c_\mu\frac{k^2}{\varepsilon} \tag{6.2.6}$$

Eqs. (6.2.4) and (6.2.5) can be written as

$$u\frac{\partial k}{\partial x}+v\frac{\partial k}{\partial y}=\frac{\partial}{\partial y}\left(\frac{\varepsilon_m}{\sigma_k}\frac{\partial k}{\partial y}\right)+\varepsilon_m\left(\frac{\partial u}{\partial y}\right)^2-\varepsilon \tag{6.2.7}$$

$$u\frac{\partial \varepsilon}{\partial x}+v\frac{\partial \varepsilon}{\partial y}=\frac{\partial}{\partial y}\left(\frac{\varepsilon_m}{\sigma_\varepsilon}\frac{\partial \varepsilon}{\partial y}\right)+c_{\varepsilon_1}\frac{\varepsilon}{k}\varepsilon_m\left(\frac{\partial u}{\partial y}\right)^2-c_{\varepsilon_2}\frac{\varepsilon^2}{k} \tag{6.2.8}$$

The parameters c_μ, c_{ε_1}, c_{ε_2}, σ_k and σ_ε are given by

$$c_\mu, = 0.09,\ c_{\varepsilon_1} = 1.44,\ c_{\varepsilon_2} = 1.92,\ \sigma_k = 1.0,\ \sigma_\varepsilon = 1.3 \tag{6.2.9}$$

These equations apply only to free shear flows. For wall boundary-layer flows, they require modifications to account for the presence of the wall. Without wall functions, it is necessary to replace the true boundary conditions at $y = 0$ by new "boundary conditions" defined at some distance y_0 outside the viscous sublayer to avoid integrating the equations through the region of large y gradients near the surface. Usually this y_0 is taken to be the distance given by

$$y_0 = \left(\frac{\nu}{u_\tau}\right) y_0^+,$$

y_0^+ being a constant taken as about 50 for smooth surfaces. For the velocity field, the boundary conditions at $y = y_0$ use the law of the wall, Eq. (4.2.1), and require that

$$u_0 = u_\tau\left(\frac{1}{\kappa} \ln \frac{y_0 u_\tau}{\nu} + c\right), \tag{6.2.10a}$$

$$v_0 = -\frac{u_0 y_0}{u_\tau}\frac{du_\tau}{dx} \tag{6.2.10b}$$

Here c is a constant around 5 to 5.2. Equation (6.2.10b) results from integrating the continuity equation with u given by Eq. (4.2.1). We also use relations for the changes in shear stress between $y = 0$ and $y = y_0$ in order to calculate u_τ from

$$u_\tau^2 = \frac{\tau_0}{\varrho} - \alpha y_0 \tag{6.2.11a}$$

where τ_0 is calculated from

$$\tau_0 = \left[\left(\nu + \varepsilon_m \frac{\partial u}{\partial y}\right]_{y_0}$$

with α semiempirically given by

$$\alpha = 0.3\frac{du_0^2}{dx} - u_e\frac{du_e}{dx} \tag{6.2.11b}$$

The friction velocity u_τ is obtained from

$$\tau = \tau_w + \frac{dp}{dx}y + \nu\frac{du_\tau}{dx}\int_0^{y^+}\left(\frac{u}{u_\tau}\right)^2 dy^+ \tag{6.2.12}$$

In the viscous sublayer and in the buffer layer ($y^+ \leq 30$), u/u_τ can be obtained from Thompson's velocity profile given by Eq. (4.4.41). For $y^+ > 50$, we can use the logarithmic velocity formula, Eq. (9.3.1a). See subsection 9.3.1.

There are several ways to specify the "wall" boundary conditions for k and ε. A common one for k makes use of the relation between shear stress τ and k [see Eq. (6.3.2)],

$$y = y_0, \quad k_0 = \frac{\tau_0}{a_1} \tag{6.2.13a}$$

where $a_1 = 0.30$. With τ_0 defined by $-\overline{u'v'} = \varepsilon_m \frac{\partial u}{\partial y}$ and ε_m by Eq. (6.2.6), Eq. (6.2.13a) becomes

$$a_1 = c_\mu \frac{k_0}{\varepsilon_0} \left(\frac{\partial u}{\partial y} \right)_0 \tag{6.2.13b}$$

The boundary condition for ε can be obtained by equating the eddy viscosity given by the CS model, $(\varepsilon_m)_{\text{CS}}$, to the eddy viscosity definition used in the k-ε model, Eq. (6.2.6), which with low Reynolds number correction, can be written as

$$(\varepsilon_m)_{k-\varepsilon} = c_\mu f_\mu \frac{k^2}{\varepsilon} \tag{6.2.14}$$

Here f_μ is a specified function discussed later in this section. Thus,

$$y = y_0, \; \varepsilon_0 = \frac{c_\mu f_\mu k_0^2}{(\varepsilon_m)_{\text{CS}}} \tag{6.2.15}$$

The edge boundary conditions for the k-ε model equations, aside from the edge boundary condition for the momentum equation,

$$y \to \delta, \qquad u \to u_e(x) \tag{6.2.16}$$

are

$$y \to \delta, \qquad k = k_e, \qquad \varepsilon \to \varepsilon_e \tag{6.2.17}$$

To avoid numerical problems, k_e and ε_e should not be zero. In addition, k_e and ε_e can not be prescribed arbitrarily because their development is governed by the transport equations (6.2.7) and (6.2.8) written at the boundary-layer edge,

$$u_e \frac{dk_e}{dx} = -\varepsilon_e \tag{6.2.18a}$$

$$u_e \frac{d\varepsilon_e}{dx} = -c_{\varepsilon_2} \frac{\varepsilon_e^2}{k_e} \tag{6.2.18b}$$

The above equations can be integrated with respect to x with initial conditions corresponding to k_{e_0} and ε_{e_0} at x_0. The solution provides the evolutions of $k(x)$ and $\varepsilon(x)$ as boundary conditions for the k- and ε-equations.

Low-Reynolds-Number Effects

To account for the presence of the wall, it is necessary to include low-Reynolds-number effects into the k-ε model. Without such modifications, this model fails to predict the sharp peak in turbulence kinetic energy close to the surface for pipe and channel flow as well as fails to predict a realistic value of the additive constant c in the law of the wall.

There are several approaches that can be used to model Eqs. (6.2.7) and (6.2.8) near the wall region. For an excellent review of these models, see Wilcox [3] and Patel et al. [6].

Patel et al. [6] reviewed eight models and evaluated them against test cases, which involved a flat-plate boundary layer, an equilibrium adverse pressure gradient boundary layer, strong favorable pressure gradient (relaminarizing) boundary layers, and sink boundary layers. Their study indicated that not all of the available low Reynolds number models reproduced the most basic feature of a flat-plate boundary layer. Only the more promising versions of Launder and Sharma, LS, [7], Chien, CH, [8], Lam and Bremhorst, LB, [9] will therefore be discussed here in the context of reviewing models for low Reynolds number effects. These models, LS, CH and LB all gave comparable results and performed considerably better than the other low Reynolds number turbulence models considered in Patel et al.'s study. However, it was pointed out in their study that even these models needed further refinement if they were to be used with confidence to calculate near-wall and low Reynolds number flows.

Before we present a brief review of these models as described in [6], it is useful to write the k-ε model equations in the following general form,

$$u\frac{\partial k}{\partial x}+v\frac{\partial k}{\partial y}=\frac{\partial}{\partial y}\left[\left(\nu+\frac{\varepsilon_m}{\sigma_k}\right)\frac{\partial k}{\partial y}\right]+\varepsilon_m\left(\frac{\partial u^2}{\partial y}\right)-\left(\tilde{\varepsilon}+\varDelta\right) \tag{6.2.19}$$

$$u\frac{\partial \tilde{\varepsilon}}{\partial x}+v\frac{\partial \tilde{\varepsilon}}{\partial y}=\frac{\partial}{\partial y}\left[\left(\nu+\frac{\varepsilon_m}{\sigma_\varepsilon}\right)\frac{\partial \tilde{\varepsilon}}{\partial y}\right]+c_{\varepsilon_1}f_1\frac{\tilde{\varepsilon}}{k}\varepsilon_m\left(\frac{\partial u^2}{\partial y}\right)-c_{\varepsilon_2}f_2\frac{\tilde{\varepsilon}^2}{k}+E \tag{6.2.20}$$

where $\varDelta$ and E as well as c_{ε_1}, c_{ε_2}, f_1, f_2 are model dependent and

$$\tilde{\varepsilon}=\varepsilon-\varDelta \tag{6.2.21}$$

The parameters $\varDelta$ and E for the LS, CH and LB1 models, including those for high Reynolds, HR, numbers are summarized in Table 6.1 together with their wall boundary conditions, with ε_m defined by

$$\varepsilon_m=c_\mu f_\mu\frac{k^2}{\tilde{\varepsilon}} \tag{6.2.22}$$

Similarly the parameters f_1, f_2, f_μ, σ_k, σ_ε, c_{ε_1} and c_{ε_2} are summarized in Table 6.2.

Since the review of Patel et al. [6], another low Reynolds number correction to the k-ε model was proposed by Hwang and Lin [10] who added an F term to the right

TABLE 6.1 Parameters Δ and E and the wall boundary conditions for LS, CH, LB1 and HR models

Model	Δ	E	Boundary Conditions	
			$\tilde{\varepsilon}$	k
HR	0	0	wall functions	
LS	$2\nu\left(\frac{\partial\sqrt{k}}{\partial y}\right)^2$	$2\nu\varepsilon_m\left(\frac{\partial^2 u}{\partial y^2}\right)^2$	0	0
CH	$2\nu\frac{k}{y^2}$	$-2\nu\left(\frac{\tilde{\varepsilon}}{y^2}\right)\exp\left(-\frac{1}{2}y^+\right)$	0	0
LB1	0	0	$\frac{\partial\tilde{\varepsilon}}{\partial y}=0$	0

TABLE 6.2 Parameters f_1, f_2, f_μ, σ_k, σ_k, $c_{\varepsilon 1}$ and $c_{\varepsilon 2}$ for LS, CH, LB1 and HR models. $R_T = \frac{k^2}{\nu\tilde{\varepsilon}}$, $R_y = \sqrt{k}y/\nu$, $y^+ = yu_\tau/\nu$.

Model	f_1	f_2	f_μ	σ_k	σ_ε	c_{ε_1}	c_{ε_2}
HR	1.0	1.0	1.0	1.0	1.3	1.44	1.92
LS	1.0	$1-0.3\exp\left(-R_T^2\right)$	$\exp\left[\frac{-3.4}{(1+R_T/50)^2}\right]$	1.0	1.3	1.44	1.92
CH	1.0	$1-0.22\exp\left(-\frac{R_T^2}{36}\right)$	$1-\exp(-0.0115y^+)$	1.0	1.3	1.35	1.8
LB1	$1+\left(\frac{0.05}{f_\mu}\right)^3$	$1-\exp\left(-R_T^2\right)$	$[1-\exp(-0.0165R_y)]^2 \times \left(1+\frac{20.5}{R_T}\right)$	1.0	1.3	1.44	1.92

hand side of Eq. (6.2.19) and defined the parameters Δ, E, F and other parameters in Table 6.2 by

$$\Delta = 2\nu\left(\frac{\partial\sqrt{k}}{\partial y}\right)^2, \quad E = -\frac{\partial}{\partial y}\left(\nu\frac{\tilde{\varepsilon}}{k}\frac{\partial k}{\partial y}\right)$$

$$F = -\frac{1}{2}\frac{\partial}{\partial y}\left(\nu\frac{k}{\tilde{\varepsilon}+\Delta}\frac{\partial\Delta}{\partial y}\right), \quad f_\mu = 1-\exp\left(-0.01y_\lambda - 0.008y_\lambda^3\right) \tag{6.2.23}$$

$$f_1 = 1.0, \quad f_2 = 1.0, \quad \sigma_k = 1.4 - 1.1\exp[-(y_\lambda/10)]$$

$$\sigma_\varepsilon = 1.3 - \exp[-(y_\lambda/10)], \quad c_{\varepsilon_1} = 1.44, \quad c_{\varepsilon_2} = 1.92$$

$$y_\lambda = \frac{y}{\sqrt{\nu k/\tilde{\varepsilon}}}$$

The wall boundary conditions for this model are same as those for LS and CH models. The edge boundary conditions given by Eq. (6.2.18) remain the same.

The application of this model to fully developed channel flows, turbulent plane Couette-Poisseuille flow and turbulent flow over a backward-facing step show very good agreement with data. Calculated results with this model show a much better agreement with measurements than those calculated with the models of CH and LS.

In the methods that use the *k*-*ε* model the coefficient c_μ in Eq. (6.2.22) is still 0.09, but in a recent investigation Marvin and Huang [11] propose that to account for adverse pressure gradient effect, c_μ should be

$$c_\mu = 0.09\left\{ \max\left(1, 0.29\left|\frac{\partial u}{\partial y}\right|\frac{k}{\varepsilon}\right)\right\}^{-1} \tag{6.2.24}$$

A preliminary study shows promise but it still needs to be examined further.

Another approach to include the low-Reynolds-number effects in the *k*-*ε* model is to employ a simple model near the wall (a mixing-length model [12] or a one equation model [13] which is valid only near the wall region) and a transport equation model in the outer region of the boundary layer; the two solutions are matched at a certain point in the boundary layer as discussed by Arnal et al. [12]. This approach, sometimes referred to as the *two-layer method* or the *zonal method* will be discussed in Section 9.2.

Other Extensions of the k-ε *Model*

Another extension of the *k*-*ε* model was developed by Yakhot et al. [14]. With techniques from renormalization group theory they proposed the so-called RNG *k*-*ε* model. In this model, *k* and *ε* are still given by Eqs. (6.2.7) and (6.2.8). The only different occurs in the definitions of the parameters given by Eq. (6.2.9). In the RNG *k*-*ε* model, they are given by

$$c_{\varepsilon_1} = 1.42, \qquad c_{\varepsilon_2} = 1.68 + \frac{c_\mu \lambda^3(1-\lambda/\lambda_0)}{1+0.012\lambda^3}$$

$$\lambda = \frac{k}{\varepsilon}\sqrt{2s_{ij}s_{ji}}, \qquad \lambda_0 = 4.38, \qquad c_\mu = 0.085 \tag{6.2.25}$$

$$\sigma_k = 0.72, \qquad \sigma_\varepsilon = 0.72, \qquad s_{ij} = \frac{1}{2}\left(\frac{\partial u_i}{\partial x_j} + \frac{\partial u_j}{\partial x_i}\right)$$

6.2.2 K-ω MODEL

Like the *k*-*ε* model discussed in the previous subsection, *k*-*ω* model is also very popular and widely used. Over the years, this model has gone over many changes and improvements as described in [3]. The most recent model is due to Wilcox [3] and is given by the following defining equations.

With ε_m defined by

$$\varepsilon_m = \frac{k}{\omega} \tag{6.2.26}$$

the turbulence kinetic energy and specific dissipation rate equations are

$$\frac{Dk}{Dt} = \frac{\partial}{\partial x_k}\left[\left(\nu + \frac{\varepsilon_m}{\sigma_k}\right)\frac{\partial k}{\partial x_k}\right] + R_{ik}\frac{\partial \bar{u}_i}{\partial x_k} - \beta^* k\omega \tag{6.2.27}$$

$$\frac{D\omega}{Dt} = \frac{\partial}{\partial x_k}\left[\left(\nu + \frac{\varepsilon_m}{\sigma_\omega}\right)\frac{\partial \omega}{\partial x_k}\right] + \alpha\frac{\omega}{k}R_{ik}\frac{\partial \bar{u}_i}{\partial x_k} - \beta\omega^2 \tag{6.2.28}$$

where R_{ik} is given by Eq. (6.2.1) and

$$\alpha = \frac{13}{25},\ \beta = \beta_0 f_\beta,\ \beta^* = \beta_0^* f_{\beta^*},\ \sigma_k = 2,\ \sigma_\omega = 2 \tag{6.2.29a}$$

$$\beta_0 = \frac{9}{125},\ f_\beta = \frac{1+70\chi_\omega}{1+80\chi_\omega},\ \chi_\omega = \left|\frac{\Omega_{ij}\Omega_{jk}S_{ki}}{(\beta_0^*\omega)^3}\right| \tag{6.2.29b}$$

$$\beta_0^* = \frac{9}{100},\ f_{\beta^*} = \begin{cases} 1, & \chi_k \le 0 \\ \dfrac{1+680\chi_k^2}{1+400\chi_k^2}, & \chi_k > 0 \end{cases},\quad \chi_k = \frac{1}{\omega^3}\frac{\partial k}{\partial x_j}\frac{\partial \omega}{\partial x_j} \tag{6.2.29c}$$

The tensors Ω_{ij} and S_{ki} appearing in Eq. (6.2.29b) are the mean rotation and mean-strain-rate tensors, respectively, defined by

$$\Omega_{ij} = \frac{1}{2}\left(\frac{\partial \bar{u}_i}{\partial x_j} - \frac{\partial \bar{u}_j}{\partial x_i}\right),\quad S_{ki} = \frac{1}{2}\left(\frac{\partial \bar{u}_k}{\partial x_i} + \frac{\partial \bar{u}_i}{\partial x_k}\right) \tag{6.2.30}$$

The parameter χ_ω is zero for two-dimensional flows. The dependence of β on χ_ω has a significant effect for round and radial jets [3]. This model takes the length scale in the eddy viscosity as

$$l = \frac{\sqrt{k}}{\omega} \tag{6.2.31a}$$

and calculates dissipation ε from

$$\varepsilon = \beta^* \omega k \tag{6.2.31b}$$

Wilcox's model equations have the advantage over the k-ε model that they can be integrated through the viscous sublayer, without using damping functions. At the wall the turbulent kinetic energy k is equal to zero. The specific dissipation rate can be specified in two different ways. One possibility is to force ω to fullfill the solution of Eq. (6.2.28) as the wall is approached [5]:

$$\omega \rightarrow \frac{6\nu}{\beta y^2} \text{ as } y \rightarrow 0 \tag{6.2.32}$$

The other [5] is to specify a value for ω at the wall which is larger than

$$\omega_w > 100\Omega_w$$

where Ω_w is the vorticity at the wall.

Menter [5] applied the condition of Eq. (6.2.32) for the first five grid points away from the wall (these points were always below $y^+ = 5$). He repeated some of his computations with $\omega_w = 1000\Omega_w$ and obtained essentially the same results. He points out that the second condition is much easier to implement and does not involve the normal distance from the wall. This is especially attractive for computations on unstructured grids [5].

The choice of freestream values for boundary-layer flows are

$$\omega_\infty > \lambda \frac{u_\infty}{L}, \quad (\varepsilon_m)_\infty < 10^{-2}(\varepsilon_m)_{max}, \quad k_\infty = (\varepsilon_m)_\infty \omega_\infty \tag{6.2.33}$$

where L is the approximate length of the computational domain and u_∞ is the characteristic velocity. The factor of proportionality $\lambda = 10$ has been recommended [5].

Free shear layers are more sensitive to small freestream values of ω_∞ and larger values of ω are needed in the freestream. According to [5], a value of at least $\lambda = 40$ for mixing layers, increasing up to $\lambda = 80$ for round jets is recommended.

According to [5], in complex Navier-Stokes computations it is difficult to exercise enough control over the local freestream turbulence to avoid small freestream ω ambiguities in the predicted results.

For boundary-layer flows, Eq. (6.2.27) reduces to Eq. (6.2.19) with

$$\Delta = 0, \quad \tilde{\varepsilon} = 0.09\,\omega k \tag{6.2.34}$$

The specific dissipation rate equation, Eq. (6.2.28), becomes

$$u\frac{\partial \omega}{\partial x} + v\frac{\partial \omega}{\partial y} = \frac{\partial}{\partial y}\left[\left(\nu + \frac{\varepsilon_m}{\sigma_\omega}\right)\frac{\partial \omega}{\partial y}\right] + \alpha\left(\frac{\partial u}{\partial y}\right)^2 - \beta_0 \omega^2 \tag{6.2.35}$$

6.2.3 SST Model

The SST model of Menter [5] combines several desirable elements of existing two-equation models. The two major features of this model are a zonal weighting of model coefficients and a limitation on the growth of the eddy viscosity in rapidly strained flows. The zonal modeling uses Wilcox's k-ω model near solid walls and Launder and Sharma's k-ε model near boundary layer edges and in free shear layers. This switching is achieved with a blending function of the model coefficients. The shear stress transport (SST) modeling also modifies the eddy viscosity by forcing the turbulent shear stress to be bounded by a constant times the turbulent kinetic energy inside boundary layers. This modification, which is similar to the basic idea behind the Johnson-King model, improves the prediction of flows with strong adverse pressure gradients and separation.

In order to blend the k-ω model and the k-ε model, the latter is transformed into a k-ω formulation. The differences between this formulation and the original k-ω model are that an additional cross-diffusion term appears in the ω-equation and that the modeling constants are different. Some of the parameters appearing in k-ω model are multiplied by a function F_1 and some of the parameters in the transformed k-ε model by a function $(1 - F_1)$ and the corresponding equations of each model are added together. The function F_1 is designed to be a value of one in the near wall region (activating the k-ω model) and zero far from the wall. The blending takes place in the wake region of the boundary layer.

The SST model also modifies the turbulent eddy viscosity function to improve the prediction of separated flows. Two-equation models generally under-predict the retardation and separation of the boundary layer due to adverse pressure gradients. This is a serious deficiency, leading to an underestimation of the effects of viscous-inviscid interaction which generally results in too optimistic performance estimates for aerodynamic bodies. The reason for this deficiency is that two-equation models do not account for the important effects of transport of the turbulent stresses. The Johnson-King model (subsection 5.4.1) has demonstrated that significantly improved results can be obtained with algebraic models by modeling the transport of the shear stress as being proportional to that of the turbulent kinetic energy. A similar effect is achieved in the SST model by a modification in the formulation of the eddy viscosity using a blending function F_2 in boundary layer flows [5].

In this model, the eddy viscosity expression, Eq. (6.2.26), is modified,

$$\varepsilon_m = \frac{a_1 k}{\max\left(a_1 \omega, \Omega F_2\right)} \tag{6.2.36}$$

where $a_1 = 0.31$. In turbulent boundary layers, the maximum value of the eddy viscosity is limited by forcing the turbulent shear stress to be bounded by the turbulent kinetic energy times a_1, see Eq. (6.3.2). This effect is achieved with an auxiliary function F_2 and the absolute value of the vorticity Ω. The function F_2 is defined as a function of wall distance y as

$$F_2 = \tanh\left(\arg_2^2\right) \tag{6.2.37a}$$

where

$$\arg_2 = \max\left(2\frac{\sqrt{k}}{0.09\omega y};\quad \frac{500\nu}{y^2\omega}\right) \tag{6.2.37b}$$

The two transport equations of the model for compressible flows are defined below with a blending function F_1 for the model coefficients of the original ω and ε model equations.

$$\frac{D\varrho k}{Dt} = \frac{\partial}{\partial x_k}\left[(\mu + \varrho\varepsilon_m\sigma_k)\frac{\partial k}{\partial x_k}\right] + R_{ik}\frac{\partial \bar{u}_i}{\partial x_k} - \beta^*\varrho\omega k \tag{6.2.38}$$

$$\begin{aligned}\frac{D\varrho\omega}{Dt} = {} & \frac{\partial}{\partial x_k}\left[(\mu + \varrho\varepsilon_m\sigma_\omega)\frac{\partial \omega}{\partial x_k}\right] + \frac{\gamma}{\varepsilon_m}R_{ik}\frac{\partial \bar{u}_i}{\partial x_k} - \beta\varrho\omega^2 \\ & +2(1-F_1)\,\varrho\sigma_{\omega_2}\frac{1}{\omega}\frac{\partial k}{\partial x_k}\frac{\partial \omega}{\partial x_k}\end{aligned} \tag{6.2.39}$$

where

$$R_{ik} = \varrho\varepsilon_m\left(\frac{\partial \bar{u}_i}{\partial x_k} + \frac{\partial \bar{u}_k}{\partial x_i} - \frac{2}{3}\frac{\partial u_j}{\partial x_j}\delta_{ik}\right) - \frac{2}{3}\varrho k\delta_{ik} \tag{6.2.40}$$

The last term in Eq. (6.2.39) represents the cross-diffusion (CD) term that appears in the transformed ω-equation from the original ε-equation. The function F_1 is designed to blend the model coefficients of the original k-ω model in boundary layer zones with the transformed k-ε model in free shear layer and freestream zones. This function takes the value of one on no-slip surfaces and near one over a larger portion of the boundary layer, and goes to zero at the boundary layer edge. This auxiliary blending function F_1 is defined as

$$F_1 = \tanh\left(\arg_1^4\right) \tag{6.2.41}$$

$$\arg_1 = \min\left[\max\left(\frac{\sqrt{k}}{0.09\omega y};\ \frac{500\nu}{y^2\omega}\right);\ \frac{4\varrho\sigma_{\omega_2}k}{\mathrm{CD}_{k\omega}y^2}\right] \tag{6.2.42}$$

where $\mathrm{CD}_{k\omega}$ is the positive portion of the cross-diffusion term of Eq. (6.2.39):

$$\mathrm{CD}_{k\omega} = \max\left(2\varrho\sigma_{\omega_2}\frac{1}{\omega}\frac{\partial k}{\partial x_k}\frac{\partial \omega}{\partial x_k},\ 10^{-20}\right) \tag{6.2.43}$$

The constants of the SST model are

$$\beta^* = 0.09, \qquad \kappa = 0.41 \tag{6.2.44}$$

The model coefficients β, γ, σ_k and σ_ω denoted with the symbol ϕ are defined by blending the coefficients of the original k-ω model, denoted as ϕ_1, with those of the transformed k-ε model, denoted as ϕ_2.

$$\phi = F_1\phi_1 + (1 - F_1)\phi_2 \tag{6.2.45}$$

where

$$\phi = \{\sigma_k, \sigma_\omega, \beta, \gamma\}$$

with the coefficients of the original models defined as

inner model coefficients

$$\begin{aligned} \sigma_{k_1} &= 0.85\,, \qquad \sigma_{\omega_1} = 0.5, \qquad \beta_1 = 0.075 \\ \gamma_1 &= \frac{\beta_1}{\beta^*} - \sigma_{\omega_1}\frac{\kappa^2}{\sqrt{\beta^*}} = 0.553 \end{aligned} \tag{6.2.46}$$

outer model coefficients

$$\begin{aligned} \sigma_{k_2} &= 1.0, \qquad \sigma_{\omega_2} = 0.856, \qquad \beta_2 = 0.0828 \\ \gamma_2 &= \frac{\beta_2}{\beta^*} - \frac{\sigma_{\omega_2}\kappa^2}{\sqrt{\beta^*}} = 0.440 \end{aligned} \tag{6.2.47}$$

The boundary conditions of the SST model equations are the same as those described in the previous subsection for the k-ω model.

For the numerical implementation of the SST model equations to Navier-Stokes equations, the reader is referred to [5].

For incompressible boundary-layer flows, Eq. (6.2.38) is same as the kinetic energy equation given by Eqs. (6.2.19) and (6.2.27). Equation (6.2.39) is same as Eq. (6.2.35) except that its righthand side contains the cross diffusion term,

$$+2(1 - F_1)\sigma_{\omega_2}\frac{1}{\omega}\frac{\partial k}{\partial y}\frac{\partial \omega}{\partial y} \tag{6.2.48}$$

6.3 One-Equation Models

In this section we discuss one-equation models. Of the several methods that fall in this group, we only consider two methods due to Bradshaw et al. [15] and Spalart and

Allmaras [2]. The former method has only been used for boundary-layer flows and is not used much anymore. It has, however, some important features that have been employed in other methods. The Spalart and Allmaras method employs a single transport equation for eddy viscosity, is very popular for wall boundary-layer and free shear flows and is used in both boundary-layer and Navier-Stokes methods.

6.3.1 Bradshaw's Model

Bradshaw's model [15] is also based on the turbulent kinetic energy equation, which for two-dimensional flows without the molecular diffusion term $\nu \frac{\partial k}{\partial y}$ can be written as

$$u\frac{\partial k}{\partial x} + v\frac{\partial k}{\partial y} = -\frac{\partial}{\partial y}\overline{p'v'} - \overline{u'v'}\frac{\partial u}{\partial y} - \varepsilon \tag{6.3.1}$$

Whereas the two-equation models discussed in Section 6.2 use the turbulent kinetic energy equation to form an eddy viscosity, Bradshaw's model uses that equation to form a relation to the Reynolds shear stress,

$$a_1 = \frac{-\overline{u'v'}}{k} \tag{6.3.2}$$

where $a_1 \cong 0.30$. The pressure diffusion term is written as

$$\overline{p'v'} = G\left(-\overline{u'v'}\right)\left(-\overline{u'v'}\right)_{\max}^{1/2} \tag{6.3.3}$$

The use of $(-\overline{u'v'})_{\max}^{1/2}$ is suggested by physical arguments about the large eddies that effect most of the diffusion of turbulent energy.

The dissipation term ε is modeled by

$$\varepsilon = \frac{\left(-\overline{u'v'}\right)^{3/2}}{l} \tag{6.3.4}$$

The parameter G and length scale l are prescribed as functions of the position across the boundary layer (see Fig. 6.1). With the relations given by Eqs. (6.3.2)–(6.3.4), the turbulent energy equations becomes

$$\frac{D}{Dt}\left(-\frac{\overline{u'v'}}{a_1}\right) = -\frac{\partial}{\partial y}\left[G\left(-\overline{u'v'}\right)\left(-\overline{u'v'}\right)_{\max}^{1/2}\right] + \left(-\overline{u'v'}\right)\frac{\partial u}{\partial y} - \frac{\left(-\overline{u'v'}\right)^{3/2}}{l} \tag{6.3.5}$$

The "wall" boundary conditions for this equation are given by Eq. (6.2.13a), (6.2.10) and (6.2.12). The edge boundary conditions are

$$y \to \delta, \quad u \to u_e(x), \quad \tau\left(\equiv -\overline{u'v'}\right) \to 0 \tag{6.3.6}$$

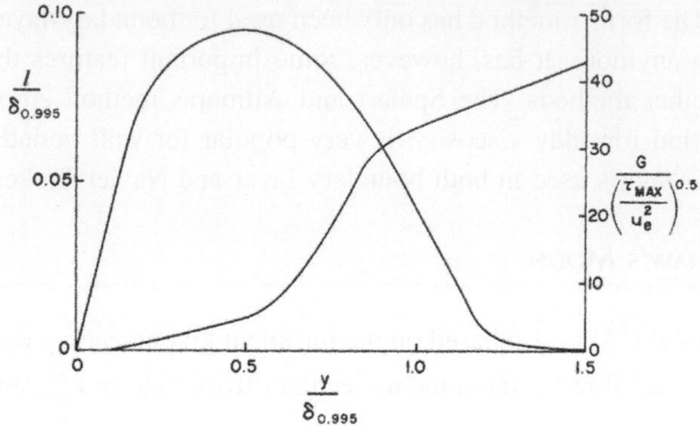

Fig. 6.1 Empirical functions used in Bradshaw's method.

It should be mentioned that the closure assumption, (6.2.3), is of considerable importance; with this assumption, Eq. (6.2.4) is *parabolic*, while with Eq. (6.3.3), Eq. (6.3.5) is *hyperbolic* with three real characteristic lines. Thus, there is another important difference between, for example, *k-ε* method, which uses a system of parabolic equations, and Bradshaw's method which uses a system of hyperbolic equations.

Note that Eq. (6.3.5) could equally well be thought of as a directly modeled version of the exact $\overline{u'v'}$ transport equation, which has terms whose effect is similar to that of the terms in Eq. (6.3.1). The advantage of this method is that a_1, l can all be measured, except for the term in G, which seems to be small. Although currently available measurements of turbulence quantities are not as accurate as the predictions of $\overline{u'v'}$ must be, they are much better than nothing. They define the error band within which a_1, l and G can be arbitrarily adjusted and, what is even more important, they give advance warning of breakdown of the correlations in difficult cases.

It should also be mentioned that the Johnson and King model discussed in Section 5.4 can be regarded as a simplified version of Bradshaw's method, using an eddy viscosity to give the shear stress profile shape but an ordinary differential equations for $(-\overline{u'v'})_{\max}$ to specify the shear-stress level. Both models use algebraic correlation for length scale and are therefore restricted to shear layers with a well behaved thickness. However, the Johnson and King model as well as the Cebeci-Chang model have been used successfully for separated flows as described in Section 5.4.

6.3.2 Spalart-Allmaras Model

Unlike the Cebeci-Smith model which uses algebraic expressions for eddy viscosity, this model uses a transport equation for eddy viscosity. Unlike most one-equation models, this model is local (i.e., the equation at one point does not depend on the solution at other points), and therefore compatible with grids of any structure and Navier-Stokes solvers in two or three-dimensions. It is numerically forgiving, in terms of near-wall resolution and stiffness, and yields fairly rapid convergence to steady state. The wall and freestream boundary conditions are trivial. The model yields relatively smooth laminar-turbulent transition at the specified transition location. Its defining equations are as follows.

$$\varepsilon_m = \tilde{\nu}_t f_{\nu_1} \tag{6.3.7}$$

$$\begin{aligned}\frac{D\tilde{\nu}_t}{Dt} &= c_{b_1}\left[1 - f_{t_2}\right]\tilde{S}\tilde{\nu}_t - \left(c_{w_1} f_w - \frac{c_{b_1}}{\kappa^2} f_{t_2}\right)\left(\frac{\tilde{\nu}_t}{d}\right)^2 \\ &+ \frac{1}{\sigma}\frac{\partial}{\partial x_k}\left[(\nu + \tilde{\nu}_t)\frac{\partial \tilde{\nu}_t}{\partial x_k}\right] + \frac{c_{b_2}}{\sigma}\frac{\partial \tilde{\nu}_t}{\partial x_k}\frac{\partial \tilde{\nu}_t}{\partial x_k}\end{aligned} \tag{6.3.8}$$

Here

$$c_{b_1} = 0.1355, \quad c_{b_2} = 0.622, \quad c_{\nu_1} = 7.1, \quad \sigma = \frac{2}{3} \tag{6.3.9a}$$

$$c_{w_1} = \frac{c_{b_1}}{\kappa^2} + \frac{(1 + c_{b_2})}{\sigma}, \quad c_{w_2} = 0.3, \quad c_{w_3} = 2, \quad \kappa = 0.41 \tag{6.3.9b}$$

$$f_{\nu_1} = \frac{\chi^3}{\chi^3 + c_{\nu_1}^3}, \quad f_{\nu_2} = 1 - \frac{\chi}{1 + \chi f_{\nu_1}}, \quad f_w = g\left[\frac{1 + c_{w_3}^6}{g^6 + c_{w_3}^6}\right]^{1/6} \tag{6.3.9c}$$

$$\chi = \frac{\tilde{\nu}_t}{\nu}, \quad g = r + c_{w_2}\left(r^6 - r\right), \quad \tau = \frac{\tilde{\nu}}{\tilde{S}\kappa^2 d^2} \tag{6.3.9d}$$

$$\tilde{S} = S + \frac{\tilde{\nu}_t}{\kappa^2 d^2} f_{\nu_2}, \quad S = \sqrt{2\Omega_{ij}\Omega_{ij}} \tag{6.3.9e}$$

$$f_{t_2} = c_{t_3} e^{-c_{t_4}\chi^2}, \quad c_{t_3} = 1.1, \quad c_{t_4} = 2 \tag{6.3.9f}$$

where d is the distance to the closest wall and S is the magnitude of the vorticity, $\Omega_{ij} = \frac{1}{2}\left(\frac{\partial u_i}{\partial x_j} - \frac{\partial u_j}{\partial x_i}\right)$.

The wall boundary condition is $\tilde{\nu}_t = 0$. In the freestream and as initial condition 0 is best, and values below $\frac{\nu}{10}$ are acceptable [2].

For boundary-layer flows, Eq. (6.3.8) can be written as

$$\begin{aligned} u\frac{\partial \tilde{\nu}_t}{\partial x} + v\frac{\partial \tilde{\nu}_t}{\partial y} &= c_{b_1}\left(1 - f_{t_2}\right)\tilde{S}\tilde{\nu}_t + \frac{1}{\sigma}\left\{\frac{\partial}{\partial y}\left[(\nu + \tilde{\nu}_t)\frac{\partial \tilde{\nu}_t}{\partial y}\right] + c_{b_2}\left(\frac{\partial \tilde{\nu}_t}{\partial y}\right)^2\right\} \\ &- \left(c_{w_1} f_w - \frac{c_{b_1}}{\kappa^2} f_{t2}\right)\left(\frac{\tilde{\nu}_t}{d}\right)^2 \end{aligned} \tag{6.3.10}$$

where

$$\tilde{S} = \left|\frac{\partial u}{\partial y}\right| + \frac{\tilde{\nu}_t}{\kappa^2 d^2} f_{v_2} \tag{6.3.11}$$

As discussed in Section 9.4, this model not only predicts wall boundary-layer flows well, but it also predicts free shear flows well.

6.4 Stress-Transport Models

As pointed out by Bradshaw "it is so obvious that stress-transport models are more realistic in principle than eddy viscosity models that the improvements they give are very disappointing and most engineers have decided that the increased numerical difficulties (complexity of programming, expense of calculation, occasional instability) do not warrant changing up from eddy-viscosity models at present. Even stress-transport models often give very poor predictions of complex flows – notoriously, the effects of streamline curvature are not naturally reproduced, and empirical fixes for this have been very reliable" [16].

Of the several versions of this approach to turbulence models, we consider the Launder-Reece-Rodi (LRR) model [17] which is the best known model based on the ε-equation. Most recent stress-transport models are based on the LRR model and differ primarily in the modeling of the pressure-strain term. For an excellent review of versions closure assumptions for the terms appearing in the Reynolds stress-transport equation (6.1.5), the reader is referred of [3].

In the LRR model, the pressure-strain term is modeled by

$$\begin{aligned}\overline{\frac{p'}{\varrho}\left(\frac{\partial u_j'}{\partial x_i} + \frac{\partial u_i'}{\partial x_j}\right)} \equiv \Pi_{ij} &= c_1 \frac{\varepsilon}{k}\left(R_{ij} + \frac{2}{3}k\delta_{ij}\right) \\ &- \hat{\alpha}\left(P_{ij} - \frac{2}{3}P\delta_{ij}\right) - \hat{\beta}\left(D_{ij} - \frac{2}{3}D\delta_{ij}\right) - \hat{\gamma}ks_{ij} \\ &- \left[0.125\frac{\varepsilon}{k}\left(R_{ij} + \frac{2}{3}k\delta_{ij}\right) - 0.015(P_{ij} - D_{ij})\right]\frac{k^{3/2}}{\varepsilon n}\end{aligned} \tag{6.4.1}$$

where n denotes distance normal to the surface and

$$\begin{gathered}P_{ij} = R_{im}\frac{\partial \bar{u}_j}{\partial x_m} + R_{jm}\frac{\partial \bar{u}_i}{\partial x_m}, \quad D_{ij} = R_{im}\frac{\partial \bar{u}_m}{\partial x_j} + R_{jm}\frac{\partial \bar{u}_m}{\partial x_i} \\ P = \tfrac{1}{2}P_{kk}, \quad D = \tfrac{1}{2}D_{kk}\end{gathered} \tag{6.4.2}$$

and the closure coefficients are given by

$$\hat{\alpha} = \frac{8+c_2}{11}, \quad \hat{\beta} = \frac{8c_2-2}{11}, \quad \hat{\gamma} = \frac{60c_2-4}{55} \tag{6.4.3}$$

In their original paper, Launder et al. recommend $c_1 = 1.5$, $c_2 = 0.4$. Gibson and Launder [18], however, recommend $c_1 = 1.8$, $c_2 = 0.60$.

The turbulent transport term rewritten as

$$\varrho\overline{u_i'u_j'u_k'} + \overline{p'u_i'}\delta_{jk} + \overline{p'u_j'}\delta_{ik}$$

is modeled by

$$-c_S\frac{\varrho k}{\varepsilon}\left(R_{im}\frac{\partial R_{jk}}{\partial x_m} + R_{jm}\frac{\partial R_{ik}}{\partial x_m} + R_{km}\frac{\partial R_{ij}}{\partial x_m}\right) \tag{6.4.4}$$

where

$$c_s = 0.11$$

Because dissipation occurs at the smallest scales, most modelers, including Launder et al. [17] use the Kolmogorov [19] hypothesis of local isotropy, which implies

$$2\nu\overline{\frac{\partial u_i'}{\partial x_k}\frac{\partial u_j'}{\partial x_k}} = \frac{2}{3}\varepsilon\delta_{ij} \tag{6.4.5}$$

With these closure assumptions, for compressible flows at high Reynolds numbers, the Reynolds stress equation (6.1.5) can be written as

$$\begin{aligned}\varrho\frac{D}{Dt}R_{ij} &= -\varrho P_{ij} + \frac{2}{3}\varrho\varepsilon\delta_{ij} \\ &\quad -c_s\frac{\varrho}{\varrho x_k}\left[\frac{\varrho k}{\varepsilon}\left(R_{im}\frac{\partial R_{jk}}{\partial x_m} + R_{jm}\frac{\partial R_{ik}}{\partial x_m} + R_{km}\frac{\partial R_{ij}}{\partial x_m}\right)\right] \\ &\quad -\varrho\Pi_{ij}\end{aligned} \tag{6.4.6}$$

The dissipation ε is again from the transport equation (6.1.12), except that the diffusion of ε from velocity fluctuations is modeled by

$$\overline{u_k'\varepsilon'} = c_\varepsilon\frac{k}{\varepsilon}R_{km}\frac{\partial\varepsilon}{\partial x_m} \tag{6.4.7}$$

rather than using an isotropic eddy viscosity model as was done in two-equation models. With this change the dissipation rate equation, (6.1.12), becomes

$$\varrho\frac{D\varepsilon}{Dt} = c_{\varepsilon_1}\frac{\varepsilon}{k}R_{ij}\frac{\partial\bar{u}_i}{\partial x_j} - c_{\varepsilon_2}\varrho\frac{\varepsilon^2}{k} - c_\varepsilon\frac{\partial}{\partial x_k}\left[\varrho\frac{k}{\varepsilon}R_{km}\frac{\partial\varepsilon}{\partial x_m}\right] \tag{6.4.8}$$

where

$$c_\varepsilon = 0.18$$

With the boundary-layer approximations, the Reynolds-stress transport equations for twodimensional incompressible flows can be written as

$$\frac{D}{Dt}\overline{u'v'} = -\overline{v'^2}\frac{\partial u}{\partial y} + c_s\frac{\partial}{\partial y}\left[\frac{k}{\varepsilon}\left(\overline{u'v'}\frac{\partial \overline{v'^2}}{\partial y} + 2\overline{v'^2}\frac{\partial(\overline{u'v'})}{\partial y}\right)\right]$$
$$c_1\frac{\varepsilon}{k}\overline{u'v'} + \left[\tilde{\alpha}\overline{v'^2} + \hat{\beta}\overline{u'^2} - \frac{\hat{\gamma}}{2}k\right]\frac{\partial u}{\partial y} + (\Phi_{12})_w \tag{6.4.9}$$

$$\frac{D}{Dt}\overline{u'^2} = -2\overline{u'v'}\frac{\partial u}{\partial y} - \frac{2}{3}\varepsilon + c_s\frac{\partial}{\partial y}\left[\frac{k}{\varepsilon}\left(2\overline{u'v'}\frac{\partial}{\partial y}\overline{u'v'} + \overline{v'^2}\frac{\partial \overline{u'^2}}{\partial y}\right)\right]$$
$$-c_1\frac{\varepsilon}{k}\left(\overline{u'^2} - \frac{2}{3}k\right) + \left[2\hat{\alpha} - \frac{2}{3}(\hat{\alpha} + \hat{\beta})\right]\overline{u'v'}\frac{\partial u}{\partial y} + (\Phi_{11})_w \tag{6.4.10}$$

$$\frac{D\overline{v'^2}}{Dt} = -\frac{2}{3}\varepsilon + c_s\frac{\partial}{\partial y}\left[\frac{k}{\varepsilon}\left(3\overline{v'^2}\frac{\partial \overline{v'^2}}{\partial y}\right)\right] - c_1\frac{\varepsilon}{k}\left(\overline{v'^2} - \frac{2}{3}k\right)$$
$$+\left[2\hat{\beta} - \frac{2}{3}\left(\hat{\alpha} + \hat{\beta}\right)\right]\overline{u'v'}\frac{\partial u}{\partial y} + (\Phi_{22})w \tag{6.4.11}$$

$$\frac{D\overline{w'^2}}{Dt} = -\frac{2}{3}\varepsilon + c_s\frac{\partial}{\partial y}\left[\frac{k}{\varepsilon}\left(\overline{v'^2}\frac{\partial \overline{w'^2}}{\partial y}\right)\right] - c_1\frac{\varepsilon}{k}\left(\overline{w'^2} - \frac{2}{3}k\right)$$
$$+\left[-\frac{2}{3}(\hat{\alpha} + \hat{\beta})\right]\overline{u'v'}\frac{\partial u}{\partial y} + (\Phi_{33})_w \tag{6.4.12}$$

Here

$$(\Phi_{11})_w = \frac{k^{3/2}}{\varepsilon y}\left[0.125\frac{\varepsilon}{k}\left(\overline{u'^2} - \frac{2}{3}k\right) - 0.015\left(2\overline{u'v'}\right)\frac{\partial u}{\partial y}\right] \tag{6.4.13a}$$

$$(\Phi_{22})_w = \frac{k^{3/2}}{\varepsilon y}\left[0.125\frac{\varepsilon}{k}\left(\overline{v'^2} - \frac{2}{3}k\right) - 0.015\left(-2\overline{u'v'}\right)\frac{\partial u}{\partial y}\right] \tag{6.4.13b}$$

$$(\Phi_{33})_w = \frac{k^{3/2}}{\varepsilon y}\left[0.125\frac{\varepsilon}{k}\left(\overline{w'^2} - \frac{2}{3}k\right)\right] \tag{6.4.13c}$$

$$(\Phi_{12})_w = \frac{k^{3/2}}{\varepsilon y}\left[0.125\frac{\varepsilon}{k}\overline{u'v'} - 0.015\left(\overline{v'^2} - \overline{u'^2}\right)\frac{\partial u}{\partial y}\right] \tag{6.4.13d}$$

The dissipation rate equation for ε is

$$\frac{D\varepsilon}{Dt} = -c_{\varepsilon_1}\frac{\varepsilon}{k}\overline{u'v'}\frac{\partial u}{\partial y} - c_{\varepsilon_2}\frac{\varepsilon^2}{k} + c_\varepsilon\frac{\partial}{\partial y}\left(\frac{k}{\varepsilon}\overline{v'^2}\frac{\partial \varepsilon}{\partial y}\right) \tag{6.4.14}$$

The wall boundary conditions for the system, Eqs. (6.4.8)–(6.4.13) are satisfied at y_0^+ with the following conditions [17]

$$\overline{u'^2} = 5.1u_\tau^2, \quad \overline{u'^2} = 1.0u_\tau^2, \quad \overline{w'^2} = 2.3u_\tau^2$$

$$k = 3.5(-\overline{u'v'}), \quad \varepsilon = -\overline{u'v'}\left(\frac{\partial u}{\partial y}\right) \tag{6.4.15}$$

$$-\overline{u'v'} = \tau_w + \frac{dp}{dx}y_0, \quad u = u_\tau\left[\frac{1}{\kappa}\ \ln\ y_0^+ + c\right]$$

At the edge of the boundary layer, the following conditions prevail:

$$u = u_e, \quad u_e\frac{d\overline{u'v'}}{dx} = -c_1\frac{\varepsilon}{k}\overline{u'v'}, \quad u_e\frac{dk_e}{dx} = -\varepsilon_e \tag{6.4.16}$$

$$u_e\frac{d\varepsilon_e}{dx} = -c_{\varepsilon_2}\frac{\varepsilon^2}{k}$$

$$u_e\frac{d\overline{u'^2}}{dx} = -\frac{2}{3}\varepsilon - c_1\frac{\varepsilon}{k}\left(\overline{u'^2} = \frac{2}{3}k\right)$$

$$u_e\frac{d\overline{v'^2}}{dx} = -\frac{2}{3}\varepsilon - c_1\frac{\varepsilon}{k}\left(\overline{v'^2} = \frac{2}{3}k\right) \tag{6.4.17}$$

$$u_e\frac{d\overline{w'^2}}{dx} = -\frac{2}{3}\varepsilon - c_1\frac{\varepsilon}{k}\left(\overline{w'^2} = \frac{2}{3}k\right)$$

Problems

6.1 Using the relation given by Eq. (6.3.2) and noting that close to the wall

$$-\overline{u'v'}\frac{\partial u}{\partial y} = \varepsilon \tag{P6.1.1}$$

calculate c_μ Eq. (4.3.8).

6.2 Equations (6.2.7) and (6.2.8) can be used to estimate the kinetic energy behind a turbulence grid in a wind tunnel. Taking the mean velocity constant, $v = 0$, $w = 0$, and assuming the turbulence to be homogeneous and isotropic, we can neglect the diffusion terms and reduce Eqs. (6.2.7) and (6.2.8) to

$$u\frac{\partial k}{\partial x} = -\varepsilon$$

$$u\frac{\partial \varepsilon}{\partial x} = -c_{\varepsilon_2}\frac{\varepsilon^2}{k}$$

(a) Show that the solutions of the above equations have the form

$$k = C(x - x_0)^{-m}$$

$$\varepsilon = muC(x - x_0)^{-m-1}$$

(b) According to experiments, $m = 1.25$. Calculate the value of $c_{\varepsilon 2}$.

6.3 For boundary-layer flows, the k-ω model equations are given by Eqs. (6.2.19), (6.2.34) and (6.2.35). From these equations form an equation for the dissipation rate ε. Show that this equation is not equivalent to the ε-equation used in the k-ε model.

6.4 In Problem 6.3, study the behavior of k and ω in the vicinity of the wall, i.e., around $y = 0$. Assume that k and ω vary as $k = by^m$ and $\omega = ay^n$ and consider a simplified form of Eqs. (6.2.19), (6.2.34) and (6.2.35) in which the convection, production and turbulent diffusion terms are neglected. Show that

$$a = 6\frac{\nu}{\beta}, \quad n = 2, \quad m(m-1) = 6\frac{\beta^*}{\beta}$$

with b left undetermined. Compute the value of m and compare it with the theoretical value of $m = 2$.

6.5

(a) Study the properties of Wilcox's model equations in the logarithmic region of the boundary layer. Assume that

$$-\overline{u'v'} = u_\tau^2, \qquad \frac{\partial u}{\partial y} = \frac{u_\tau}{\kappa y}$$

and with the assumption that "*Production = Dissipation*" in the kinetic energy equation (P6.1.1), show that

$$-\frac{\overline{u'v'}}{k} = c_\mu^{1/2}$$

(b) Compute the evolution of ω as function of y. From the ω-equation show that the value of κ can be written as

$$\kappa = \frac{\left(\beta - \gamma c_\mu\right)^{1/2}}{c_\mu^{1/4}\sigma^{1/2}} = 0.41$$

References

[1] F. Harlow, P.I. Nakayama, Turbulence transport equations, Phys. Fluids 10 (1967) 2323.

[2] P.R. Spalart, S.R. Allmaras, A One-Equation Turbulence Model for Aerodynamics Flows, AIAA Paper 92 (0439) (1992).

[3] D. C. Wilcox, Turbulence Modeling for CFD. DCW Industries, Inc. 5354 Palm Drive, La Cañada, Calif. 1998.

[4] W.P. Jones, B.E. Launder, The Prediction of Laminarization with a Two-Equation Model of Turbulence, International Journal of Heat and Mass Transfer 15 (1972) 301–314.

[5] F.R. Menter, Two-Equation Eddy Viscosity Turbulence Models for Engineering Applications, AIAA J. 32 (1994) 1299–1310.

[6] V.C. Patel, W. Rodi, G. Scheuerer, Turbulence Models for Near-Wall and Low Reynolds Number Flows: A Review, AIAA J. 23 (No. 9) (1985) 1308–1319.

[7] B.E. Launder, B.I. Sharma, Application of the Energy Dissipation Model of Turbulence to the Calculation of Flow Near a Spinning Disc, Letters in Heat and Mass Transfer 1 (No. 2) (1974) 131–138.

[8] K.Y. Chien, Predictions of Channel and Boundary-Layers Flows with a Low-Reynolds-Number Turbulence Model, AIAA J. 20 (1982) 33–38.

[9] C.K.G. Lam, K.A. Bremhorst, Modified Form of the k-ε Model for Predicting Wall Turbulence, J. Fluids Eng. 103 (1981) 456–460.

[10] C.B. Hwang, C.A. Lin, Improved-Low-Reynolds Number k-ε Model Based on Direct Numerical Simulation Data, AIAA J. 36 (No. 1) (1998) 38–43.

[11] J.D. Marvin, G.P. Huang, Turbulence Modeling – Progress and Future Outlook, 15th International Conference on Numerical Methods in Fluid Dynamics, California, Monterey, June, 1996.

[12] D. Arnal, J. Cousteix, R. Michel, Couche limite se développant avec gradient de pression positif dans un écoulement turbulent, La Rech. Aérosp. 1 (1976).

[13] L.H. Norris, W.C. Reynolds, Turbulent Channel Flow with a Moving Wavy Boundary. Report FM-20, Department of Mechanical Engineering, Standford University, Stanford, California, 1975.

[14] V. Yakhot, S.A. Orszag, Renormalization Group Analysis of Turbulence, 1. Basic Theory, J. Scientific Computing 1 (1986) 3–51.

[15] P. Bradshaw, D.H. Ferriss, N.P. Atwell, Calculation of Boundary Layer Development Using the Turbulent Energy Equation, J. Fluid Mechanics 28 (1967) 593–616, pt. 3.

[16] P. Bradshaw, The Best Turbulence Models for Engineers. Modeling of Complex Flows, in: M.D. Salas, J.N. Hefner, L. Sakell (Eds.), Kluwer, Dordrecht, 1999.

[17] B.E. Launder, G.J. Reece, W. Rodi, Progress in the Development of a Reynold-Stress Turbulence Closure, J. Fluid Mechanics 68 (1975) 537–566, pt. 3.

[18] M.M. Gibson, B.E. Launder, Ground Effects on Pressure Fluctuations in the Atmospheric Boundary Layer, J. Fluid Mechanics 86 (1978) 491–511, pt. 3.

[19] A.N. Kolmogorov, Local Structure of Turbulence in Incompressible Viscous Flows for Very Large Reynolds Number, Doklady AN SSSR 30 (1941) 299–303.

Chapter 7

Short Cut Methods

Chapter Outline Head

Analysis of Turbulent Flows with Computer Programs. http://dx.doi.org/10.1016/B978-0-08-098335-6.00007-0

7.1 Introduction

Over the years many attempts have been made to calculate turbulent flows and various approaches have been taken. At first, before high-speed computers became available, almost all attempts avoided the mathematical difficulties of solving highly nonlinear Navier-Stokes and boundary-layer equations in their partial-differential form and, instead, concentrated on the solution of the "integral" forms of the boundary-layer equations, which yield ordinary differential equations. Such methods are commonly called *integral* methods.

The interest in the solution of boundary-layer equations in their differential form began early in 1960 when computers began to offer the possibility of solving complicated systems of partial differential equations numerically. As a result, since about 1960, a number of methods called *differential* methods have been development. At present there are several very efficient and accurate differential methods such as the one discussed in Chapters 8 and 9 for laminar and turbulent flows. With increase in computer power, around late 1970, interest next concentrated in the solution of the Navier-Stokes equations. As a result, at present there are several powerful methods for solving the Navier-Stokes equations for both laminar and turbulent flows.

In this chapter we discuss simple methods and formulas for calculating two-dimensional turbulent flows. These methods, which we have called "short-cut" methods do not have the accuracy of the differential boundary-layer methods discussed in Chapters 8 and 9. Furthermore, they are restricted to simple two-dimensional flows with restricted boundary conditions. Their chief advantage and usefulness lies in their simplicity; unlike differential boundary-layer and Navier-Stokes methods they either do not require computers or only small computers. They can easily be used in many practical engineering problems.

Description of short-cut methods for flows with zero-pressure gradient begins in Section 7.2 and is continued in Section 7.3 with integral methods for flows with pressure gradient. Section 7.4 discusses the prediction of flow separation in two-dimensional incompressible flows and Section 7.5 discusses the calculation of several free shear flows based on similarity concepts.

7.2 Flows with Zero-Pressure Gradient

Short-cut methods discussed here for flows with zero-pressure gradient includes incompressible flows on a smooth flat plate (subsection 7.2.1), on a rough flat plate (subsection 7.2.2), compressible flow on a smooth flat plate (subsection 7.2.3) and on a rough flat plate (subsection 7.2.4).

7.2.1 Incompressible Flow on a Smooth Flat Plate

Let us consider an incompressible flow over a smooth flat plate. If the Reynolds number is sufficiently large, we can identify three different flow regimes on such a surface. Starting from the leading edge, there is first a region ($0 < R_x < R_{x_{tr}}$) in which the flow is laminar. After a certain distance, there is a region ($R_{x_{tr}} < R_x < R_{x_{tr}}$) in which transition from laminar to turbulent flow takes place. In the third region ($R_x \geq R_{x_{tr}}$) the flow is fully turbulent. The transition Reynolds number $R_{x_{tr}}$ depends partly upon the turbulence in the free stream; $R_{x_{tr}}$ may be as low as 5×10^4 or as high as 5×10^6.

For laminar flow over a flat plate, the boundary-layer parameters can be obtained exactly from the solution of the similarity equations and can be expressed in terms of very useful formulas. For a turbulent flow, the momentum and energy equations do not reduce to similarity equations. Furthermore, the presence of the Reynolds stress terms in the equations prevents an exact solution. For that reason, it is necessary to introduce some empiricism into the equations and check their solutions with experiment.

Skin Friction Formulas

Over the years, a large number of experiments have been conducted with smooth flat plates. Velocity profiles and local skin-friction coefficients have been measured at various Reynolds numbers. The experimental data have been the basis for several useful formulas for boundary-layer parameters, as well as for several general prediction methods such as those discussed in Chapters 8 and 9 for calculating turbulent boundary-layers with and without pressure gradient. Here we shall restrict our discussion to several approximate formulas that can be used for calculating c_f, δ^*, θ, δ, etc. For simplicity, we shall assume that the transition region is a point and that the transition from laminar to turbulent flow takes place instantaneously, that is $R_{x_{tr}} = R_{x_{tr}}$.

For zero-pressure-gradient flow, the momentum integral equation (3.6.6) can be written as

$$dR_\theta / dR_x = c_f/2 \tag{7.2.1}$$

where $R_\theta = u_e\theta/\nu$ and $R_x = u_e x/\nu$. Denoting $(2/c_f)^{1/2}$ by z and using integration by parts, we can express Eq. (7.2.1) in the form

$$R_x = z^2 R_\theta - 2\int_{z_{tr}}^{z} R_\theta z \, dz + A_1, \tag{7.2.2}$$

where A_1 is an integration constant and z_{tr} is the value of the skin-friction parameter z at transition.

The integral in Eq. (7.2.2) can be integrated, provided that R_θ is expressed as a function of z. That can be done by first recalling the definition of θ and writing it as

$$\begin{aligned}\theta &= \delta\int_0^1 \frac{u}{u_e}\left(1-\frac{u}{u_e}\right)d\eta \\ &= \delta\int_0^1\left(\frac{u_e-u}{u_e}\right)d\eta - \delta\int_0^1\left(\frac{u_e-u}{u_e}\right)^2 d\eta,\end{aligned} \tag{7.2.3}$$

where $\eta = y/\delta$. But for equilibrium boundary layers at high Reynolds numbers, c_1 and c_2, defined as

$$c_1 \equiv \int_0^1\left(\frac{u_e-u}{u_\tau}\right)d\eta, \quad c_2 \equiv \int_0^1\left(\frac{u_e-u}{u_\tau}\right)^2 d\eta, \tag{7.2.4}$$

are constant (see Fig. 4.4). Substituting from Eq. (7.2.4) into Eq. (7.2.3) and non-dimensionalizing, we obtain

$$\mathrm{R}_\delta \equiv \frac{u_e\delta}{\nu} = \frac{\mathrm{R}_\theta z}{c_1 - c_2/z} \tag{7.2.5}$$

Next we consider Coles' velocity-profile expression evaluated at the edge of the boundary layer, Eq. (4.4.37), and write it as

$$z = \frac{1}{\kappa}\ln\frac{R_\delta}{z} + c + \frac{2\Pi}{\kappa}. \tag{7.2.6}$$

With the values of c and Π taken as 5.0 and 0.55, respectively, we can now integrate Eq. (7.2.2) with the relations given by Eqs. (7.2.5) and (7.2.6). This integration allows the resulting expression to be written as

$$(R_x - A_2)c_f = 0.324\exp\left[\frac{0.58}{\sqrt{c_f}}\left(1 - 8.125\sqrt{c_f} + 22.08c_f\right)\right]. \tag{7.2.7}$$

Here A_2 is an integration constant that depends on the values of c_f and R_x at transition. Figure 7.1 presents the results for three transition Reynolds numbers, $R_{x_{tr}} = 0$, 4.1×10^5 and 3×10^6, the first being the case when transition takes place at the leading edge. The value of $R_x = 4 \times 10^5$ corresponds to the approximate minimum value of R_x for which the flow can be turbulent. The highest value of R_x is a typical natural transition Reynolds number on a smooth flat plate in low-turbulence test rigs with no heat transfer. If the plate is heated, the location of natural transition in a gas flow moves upstream, decreasing the value of the transitional Reynolds number, whereas if the plate is cooled, the location of transition moves downstream. The reason is that since μ rises with gas temperature, the velocity gradient near the wall is reduced by heating,

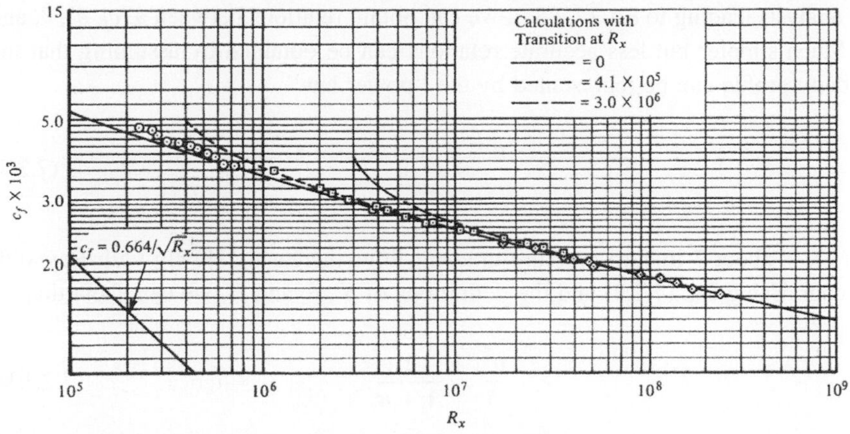

Fig. 7.1 Local skin-friction coefficient on a smooth flat plate with three transition Reynolds numbers according to Eq. (7.2.7). The variation of laminar c_f with R_x is shown by $c_f = 0.664/\sqrt{R_x}$.

distorting the profile to a more unstable shape, and for cooling, the converse holds. In liquid flows μ falls with increasing fluid temperature, and the effect is reversed.

Putting $A_2 = 0$ in Eq. (7.2.7) (i.e., assuming that the turbulent boundary layer starts at $x = 0$ with negligible thickness), taking logarithms, and making further approximations lead to formulas like

$$\frac{1}{\sqrt{c_f}} = a + b \log c_f R_x,$$

where a and b are constants chosen to get the best agreement with experiment. Such less-rigorous formulas have been derived by many previous workers. Von Karman [1] took $a = 1.7$ and $b = 4.15$; i.e.

$$\frac{1}{\sqrt{c_f}} = 1.7 + 4.15 \log c_f R_x. \qquad (7.2.8a)$$

A formula for the average skin friction $\bar{c}_f$ (averaged over the distance x) that makes use of the above equation was obtained by Schoenherr [2]:

$$\frac{1}{\sqrt{\bar{c}_f}} = 4.13 \log \bar{c}_f R_x. \qquad (7.2.8b)$$

Power-Law Velocity Profiles

By relating the profile parameter Π to the displacement thickness δ^* and to the momentum thickness θ as well as to the local skin-friction coefficient c_f, as is done in the analysis leading to Eq. (7.2.8a), we can obtain relations between δ, c_f, δ^*, θ, and H. Much simpler but less accurate relations can be obtained by assuming that the velocity profile can be represented by the "power law"

$$\frac{u}{u_e} = \left(\frac{y}{\delta}\right)^{1/n}. \tag{7.2.9}$$

Here n is about 7 in zero-pressure-gradient flow, increasing slowly with Reynolds number. Using Eq. (7.2.9) and the definitions of δ^*, θ, and H, we can show that

$$\frac{\delta^*}{\delta} = \frac{1}{1+n}. \tag{7.2.10a}$$

$$\frac{\theta}{\delta} = \frac{n}{(1+n)(2+n)}, \tag{7.2.10b}$$

$$H = \frac{2+n}{n} \tag{7.2.10c}$$

Other formulas obtained from power-law assumptions with $n = 7$, given by Schlichting [1] are the following equations valid only for Reynolds numbers R_x between 5×10^5 and 10^7:

$$c_f = 0.059 R_x^{-0.20}, \tag{7.2.11}$$

$$\bar{c}_f = 0.074 R_x^{-0.20}, \tag{7.2.12}$$

$$\frac{\delta}{x} = 0.37 R_x^{-0.20}, \tag{7.2.13}$$

$$\frac{\theta}{x} = 0.036 R_x^{-0.20}. \tag{7.2.14}$$

Equations (7.2.8b) and (7.2.12) assume that the boundary layer is turbulent from the leading edge onward, that is, the effective origin is at $x = 0$. If the flow is turbulent but the Reynolds number is moderate, we should consider the portion of the laminar flow that precedes the turbulent flow. There are several empirical formulas for $\bar{c}_f$ that account for this effect. One is the formula quoted by Schlichting [1]. It is given by

$$\bar{c}_f = \frac{0.455}{(\log R_x)^{2.58}} - \frac{A}{R_x}, \tag{7.2.15}$$

and another is

$$\bar{c}_f = 0.047 R_x^{-0.20} - \frac{A}{R_x}, \quad 5 \times 10^5 < R_x < 10^7. \tag{7.2.16}$$

Here A is a constant that depends on the transition Reynolds number $R_{x_{tr}}$. It is given by

$$A = R_{x_{tr}}\left(\bar{c}_{f_{tr}} - \bar{c}_{f_{tr}}\right), \tag{7.2.17}$$

where $\bar{c}_{f_t}$ and $\bar{c}_{f_l}$ correspond to the values of average skin-friction coefficient for turbulent and laminar flow at $R_{x_{tr}}$. We note that although Eq. (7.2.16) is restricted to the indicated R_x range, Eq. (7.2.15) is valid for a wide range of R_x and has given good results up $R_x = 10^9$.

Heat-Transfer Formulas on Smooth Surfaces with Specified Temperature

For a zero-pressure-gradient flow, the energy integral equation (3.6.26b) can be integrated to obtain the Stanton number St as a function of Reynolds number. This can be done by inserting the velocity profile expression given by Eq. (4.4.37) and the similar expression for the temperature profile given by Eq. (4.4.43) into the definition of θ_T. Since we already have an expression for c_f, however, a simpler procedure would be to evaluate Eq. (4.4.43) at $y = \delta$ and make use of the relation given by Eq. (7.2.6). For example, at $y = \delta$, Eq. (4.4.43) becomes

$$\frac{T_w - T_e}{T_\tau} = \frac{1}{\kappa_h}\ln\frac{R_\delta}{z} + c_h + \frac{2\Pi_h}{\kappa_h}. \tag{7.2.18}$$

Using the definition of T_τ and the local Stanton number St, the left-hand side of Eq. (7.2.18) becomes

$$\frac{T_w - T_e}{T_\tau} = \frac{1}{\mathrm{St}}\sqrt{\frac{c_f}{2}}. \tag{7.2.19}$$

If we equate the two expressions for $\ln(R_\delta/z)$ obtained from Eqs. (7.2.6) and (7.2.5) and substitute Eq. (7.2.19) in the left-hand side of Eq. (7.2.18), we get, after rearranging

$$\frac{\mathrm{St}}{c_f/2} = \frac{\kappa_h/\kappa}{1 - \left\{\left[c\kappa - c_h\kappa_h + 2\left(\Pi - \Pi_h\right)\right]/\kappa\right\}\sqrt{c_f/2}}. \tag{7.2.20}$$

Because of the scatter in temperature-profile data, it is simplest to choose the value of the quantity in braces as one that gives the best agreement with St data. For air, this is

$$\frac{\mathrm{St}}{c_f/2} = \frac{1.11}{1 - 1.20\sqrt{c_f/2}} \tag{7.2.21}$$

for $\kappa = 0.40$ and $\kappa_h = 0.44$. This equation for the Reynolds analogy factor $\mathrm{St}/(c_f/2)$ is quoted in the literature with a wide range of values for the empirical constants. For

$c_f = 3 \times 10^{-3}$, a typical value for a laboratory boundary layer, the constants quoted here give St/ $\frac{1}{1} c_f = 1.16$, whereas at very high Reynolds number, where c_f is small, St/ $\frac{1}{2} c_f$ asymptotes to 1.11 [3].

Substituting Eq. (7.2.11) into Eq. (7.2.21) yields

$$\mathrm{St} = \frac{0.0327 R_x^{-0.20}}{1 - 0.206 R_x^{-0.10}}. \tag{7.2.22}$$

According to an extensive study conducted by Kader and Yaglom [4], the empirical formula

$$\mathrm{St} = \frac{1}{\mathrm{Pr}} \frac{\sqrt{c_f}}{4.3 \ln R_x c_f + 3.8} \tag{7.2.23}$$

fits the existing data on air (Pr = 0.7) well. For fluids with Pr $\geq$ 0.7, they recommend

$$\mathrm{St} = \frac{\sqrt{c_f/2}}{2.12 \ln R_x c_f + 12.5\,\mathrm{Pr}^{2/3} + 2.12 \ln \mathrm{Pr} - 7.2}. \tag{7.2.24}$$

Equations (7.2.23) and (7.2.24) utilize Von Karman's equation (7.2.8a) for c_f with a slightly different constant ahead of the log $c_f R_x$ term:

$$\frac{1}{\sqrt{c_f}} = 1.7 + 4.07 \log c_f R_x .$$

For isothermal flat plates, Reynolds et al. [5] recommend the following empirical formula for Stanton number:

$$\mathrm{St}\,\mathrm{Pr}^{0.4} \left(\frac{T_w}{T_e}\right)^{0.4} = 0.0296 R_x^{-0.20} \tag{7.2.25}$$

for $5 \times 10^5 < R_x < 5 \times 10^6$ and $0.5 \leq \mathrm{Pr} \leq 1.0$.

An approximate expression for Stanton number on an isothermal flat plate with unheated starting length can be obtained for turbulent flows by making suitable assumptions for velocity, temperature, and shear-stress profiles and by using eddy viscosity and turbulent Prandtl number concepts [6]. For example, from the definition of Stanton number with power-law profiles for velocity and temperature,

$$\frac{u}{u_e} = \left(\frac{y}{\delta}\right)^{1/n}, \quad \frac{T_w - T}{T_w - T_e} = \left(\frac{y}{\delta_t}\right)^{1/n}, \tag{7.2.26a}$$

and with a shear-stress profile in the form

$$\frac{\tau}{\tau_w} = 1 - \left(\frac{y}{\delta}\right)^{(n+2)/n} \tag{7.2.26b}$$

and with $\tau/\varrho = \varepsilon_m(\partial u/\partial y)$, $\mathrm{Pr}_t = 1$, we can write

$$\mathrm{St} = \frac{c_f}{2}\left(\frac{\delta_t}{\delta}\right)^{-1/n}. \tag{7.2.27}$$

Substituting this equation into the energy integral equation (3.3.26b) and using the momentum integral equation for zero-pressure gradient-flow, the resulting expression can be written in the form

$$\int_{\delta_{x_0}}^{\delta}\frac{d\delta}{\delta} = \int_0^{\delta_t/\delta}\frac{n+1}{n}\frac{(\delta_t/\delta)^{2/n}}{1-(\delta_t/\delta)^{(2+n)/n}}d\left(\frac{\delta_t}{\delta}\right), \tag{7.2.28}$$

where δ_{x_0} denotes the hydrodynamic boundary-layer thickness at $x = x_0$ (see Fig. 7.2).

Integrating Eq. 7.2.28, we obtain

$$\frac{\delta}{\delta_{x_0}} = \left[1-\left(\frac{\delta_t}{\delta}\right)^{(2+n)/n}\right]^{-(n+1)/(n+2)}$$

or

$$\frac{\delta_t}{\delta} = \left[1-\left(\frac{\delta_{x_0}}{\delta}\right)^{(n+2)/(n+1)}\right]^{n/(2+n)}. \tag{7.2.29}$$

In the range $5\times10^5 \le R_x \le 10^7$, δ varies as $x^{4/5}$ [see Eq. (7.2.13)]. Thus Eq. (7.2.29) may be written as

$$\frac{\delta_t}{\delta} = \left[1-\left(\frac{x_0}{x}\right)^{4(n+2)/5(n+1)}\right]^{n/(2+n)}. \tag{7.2.30}$$

Substituting Eq. (7.2.30) into Eq. (7.2.27) and taking $n = 7$, we get

$$\mathrm{St} = \frac{c_f}{2}\left[1-\left(\frac{x_0}{x}\right)^{9/10}\right]^{-1/9}. \tag{7.2.31}$$

For a plate heated from the leading edge ($x_0 = 0$), Eq. (7.2.31) becomes

$$\mathrm{St} = \mathrm{St}_T = \frac{c_f}{2},$$

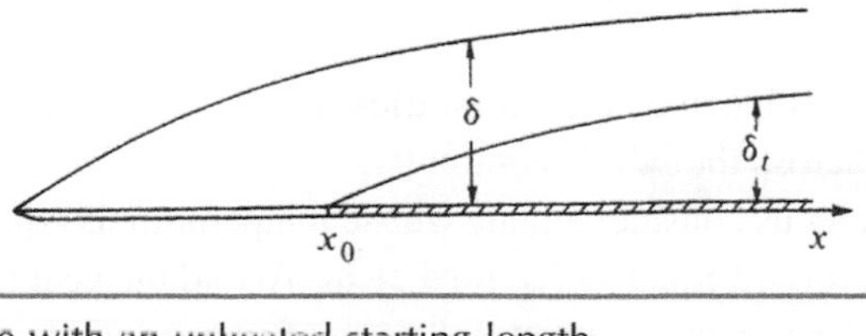

Fig. 7.2 Flat plate with an unheated starting length.

where St_T denotes the Stanton number of the isothermal flat plate. With this notation, Eq. (7.2.31) can be written as

$$\frac{\mathrm{St}}{\mathrm{St}_T} = \left[1 - \left(\frac{x_0}{x}\right)^{9/10}\right]^{-1/9}, x > x_0. \tag{7.2.32}$$

From the definition of heat-transfer coefficient $\hat{h}$ and Stanton number St, for an isothermal flat plate with unheated starting length,

$$\hat{h} = \varrho u_e c_p \mathrm{St}_T \left[1 - \left(\frac{x_0}{x}\right)^{9/10}\right]^{-1/9}. \tag{7.2.33}$$

Similarly, with St_T given by Eq. (7.2.25)

$$\mathrm{St}\,\mathrm{Pr}^{0.4}\left(\frac{T_w}{T_e}\right)^{0.4} = 0.0296 R_x^{-0.20}\left[1 - \left(\frac{x_0}{x}\right)^{9/10}\right]^{-1/9}. \tag{7.2.34}$$

For nonisothermal surfaces with arbitrary surface temperature, the heat flux at some distance x from the leading edge is, by superposition arguments [7],

$$\dot{q}_w = \int_{\xi=0}^{\xi=x} \hat{h}(\xi, x)\, dT_w(\xi), \tag{7.2.35}$$

where, with $\mathrm{St}_T(x)$ being evaluated at x,

$$\hat{h}(\xi, x) = \varrho u_e c_p \mathrm{St}_T(x)\left[1 - \left(\frac{\xi}{x}\right)^{9/10}\right]^{-1/9}. \tag{7.2.36}$$

The integration of Eq. (7.2.35) is performed in the "Stieltjes" sense rather than in the ordinary "Riemann" or "area" sense [8]. This must be done because specified surface temperature may have a finite discontinuity, so that dT_w is undefined at some point. The Stieltjes integral may, however, be expressed as the sum of an ordinary or Riemann integral and a term that accounts for the effect of the finite discontinuities [7]. The integral of Eq. (7.2.25) may be written as

$$\begin{aligned} \dot{q}_w(x) &= \int_{\xi=0}^{\xi=x} \hat{h}(\xi, x)\frac{dT_w(\xi)}{d\xi}\, d\xi \\ &+ \sum_{n=1}^{N} \hat{h}(x_{0n}, x)\left[T_w\left(x_{0n}^+\right) - T_w\left(x_{0n}^-\right)\right]. \end{aligned} \tag{7.2.37}$$

Here N denotes the number of discontinuities and $T_w(x_{0n}^+) - T_w(x_{0n}^-)$ denotes the temperature jump across the nth discontinuity.

As an example, let us consider a plate whose temperature is equal to T_{w_1} from the leading edge to $x = x_0$ and equal to T_{w_2} for $x > x_0$. To find the heat flux for $x > x_0$, we note that since $dT_w/d\xi$ is zero except at $x = x_0$, the first term on the right-hand side is zero. Therefore, we concentrate our attention on the second term. Since $N=2$, we can write

$$n = 1, \quad \hat{h}(0,x) = \varrho u_e c_p \mathrm{St}_T[1-0]^{-1/9},$$
$$T_w(0^+) - T_w(0^-) = T_{w_1} - T_e\,,$$

$$n = 2, \quad \hat{h}(x_0,x) = \varrho u_e c_p \mathrm{St}_T(x)\left[1-\left(\frac{x_0}{x}\right)^{9/10}\right]^{-1/9},$$
$$T_w(x_0^+) - T_w(x_0^-) = T_{w_2} - T_{w_1}\,.$$

Thus the heat transfer for $x > x_0$ is given by

$$\dot{q}_w(x) = \varrho u_e c_p \mathrm{St}_T(x)\left\{\left(T_{w_1} - T_e\right) + \left(T_{w_2} - T_{w_1}\right)\left[1-\left(\frac{x_0}{x}\right)^{9/10}\right]^{-1/9}\right\}. \quad (7.2.38)$$

Note that $\mathrm{St}_T(x)$ is computed with $T_w = T_{w_2}$.

Heat-Transfer Formulas on Smooth Surfaces with Specified Heat Flux

The analysis of thermal boundary layers on smooth surfaces with specified heat flux is similar to those with specified temperature. Based on the experiments of Reynolds et al., the following empirical formula is recommended in [9]:

$$\mathrm{St}\,\mathrm{Pr}^{0.4} = 0.030 R_x^{-0.2}, \quad (7.2.39)$$

which is nearly identical to the one for specified temperature, Eq. (7.2.25). Note that the difference in Stanton-number formulas between the constant wall heat-flux case and the constant wall temperature case is considerably more in laminar flows, where the difference is 36 percent.

When the plate has an arbitrary heat-flux distribution on the surface and also includes an unheated section, the difference between the surface temperature and edge temperature can be calculated from the following formula given by Reynolds et al. [7]:

$$T_w(x) - T_e = \int_{\xi=0}^{\xi=x} g(\xi,x)\dot{q}_w(\xi)\,d\xi, \quad (7.2.40)$$

where, with Γ denoting the gamma function (see Appendix 7A),

$$g(\xi,x) = \frac{\frac{9}{10}\mathrm{Pr}^{-0.6}R_x^{-0.8}}{\Gamma\left(\frac{1}{9}\right)\Gamma\left(\frac{8}{9}\right)(0.0287k)}\left[1-\left(\frac{\xi}{x}\right)^{9/10}\right]^{-8/9}.$$

The nature of the integrated in Eq. (7.2.40) is such that integration is always performed in the usual Riemann sense, including integration across discontinuities. To illustrate this point further, let us consider a plate that is unheated for a distance x_0 from the leading edge and is heated at a uniform rate $\dot{q}_w$ for $x > x_0$. To find the wall temperature for $x > x_0$, we write Eq. (7.2.40) as

$$T_w(x) - T_e = \frac{3.42\dot{q}_w}{\mathrm{Pr}^{0.6} R_x^{0.8} k} \int_{x_0}^{x} \left[1 - \left(\frac{\xi}{x}\right)^{9/10}\right]^{-8/9} d\xi. \tag{7.2.41}$$

The integral can be evaluated in terms of beta functions (see Appendix 7A), and the resulting expression can be written as

$$T_w(x) - T_e = \frac{33.61\dot{q}_w \mathrm{Pr}^{0.4} R_x^{0.2}}{\varrho c_p u_e} \frac{\beta_r(1/9, 10/9)}{\beta_1(1/9, 10/9)} \tag{7.2.42}$$

or, using the definition of Stanton number, as

$$\mathrm{St} = \frac{0.030\mathrm{Pr}^{-0.4} R_x^{-0.2}}{\left[\beta_r\left(\frac{1}{9}, \frac{10}{9}\right)\right] / \left[\beta_1\left(\frac{1}{9}, \frac{10}{9}\right)\right]}, \tag{7.2.43}$$

where $r = 1 - (x_0/x)^{9/10}$.

7.2.2 Incompressible Flow on a Rough Flat Plate

The discussion in the previous subsection is valid for smooth surfaces. In practice, many surfaces are "rough" in the hydraulic sense. It is often desirable to compute c_f, $\bar{c}_f$, R_θ, H, etc., on such surfaces. Here we show, as an example, how boundary-layer parameters can be obtained for sand-roughened plates by using an approach similar to that discussed in subsection 7.2.1.

It was shown in Section 4.5 that the law of the wall for flows with roughness is given by Eq. (4.5.1). By means of Eq. (4.5.2), it can be written as

$$u^+ = (1/\kappa)\ln y^+ - (1/\kappa)\ln k^+ + B_2. \tag{7.2.44}$$

The functional relationship, which can only be determined from experiments, also assumes that the slope of the velocity distribution on rough walls is the same as the slope on smooth surfaces. The best-known roughness configuration – one frequently used as standard roughness – is that of closely packed uniform sand grains. According to Nikuradse's experiments in sand-roughened pipes [10], the variation of B_2 with k^+ is that shown in Fig. 7.3. We see from the figure that B_2 varies differently in the three regions discussed in Section 4.5. For example, in the completely rough regime, B_2 is a constant equal to 8.48.

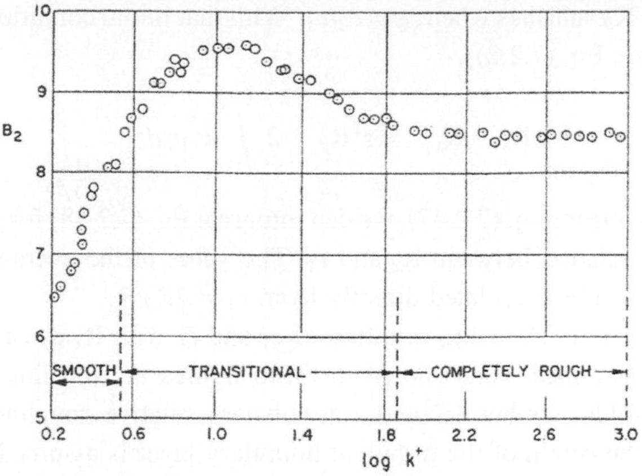

Fig. 7.3 Variation of B_2 with k^*.

Equation (7.2.44) applies only in the inner region of the boundary layer. For application to the entire boundary layer, it must be corrected for the wakelike behavior of the outer region. That can be done by using Coles' wake expression. With the correction, Eq. (7.2.44) becomes

$$u^+ = (1/\kappa) \ln (y/k) + B_2 + (\Pi/\kappa)w. \tag{7.2.45}$$

Evaluating that expression at the edge of the boundary layer and rearranging, we can write

$$\mathrm{R}_\delta = \mathrm{R}_k \exp[\kappa(z - \phi(1)], \tag{7.2.46}$$

where

$$\mathrm{R}_k = u_e k/\nu, \quad \phi(1) = B_2 + (2\Pi/\kappa).$$

Since the velocity-defect law is valid for both smooth and rough surfaces, we can still use the expression for R_θ as given by Eq. (7.2.5). From Eqs. (7.2.46) and (7.2.5), we can write

$$\mathrm{R}_\theta = (\mathrm{R}_k/z)[c_1 - (c_2/z)] \exp [\kappa(z - \phi(1)]. \tag{7.2.47}$$

As before, the constants c_1 and c_2 are 3.78 and 25, respectively.

A relation between c_f and R_x can now be obtained by using the momentum integral equation in the form given by Eq. (7.2.1). Before Eq. (7.2.1) is integrated, however, it is necessary to establish lower limits for R_x and R_θ. It is obvious from

Eq. (7.2.47) that R_θ vanishes when $z_0 = c_2/c_1$. With that initial condition we can write Eq. (7.2.1) as [see Eq. (7.2.2)],

$$R_x - R_{x_0} = z^2 R_\theta - 2 \int_{z_0}^{z} R_\theta z \, dz. \tag{7.2.48}$$

Since R_θ is known from Eq. (7.2.47), we can integrate Eq. (7.2.48) for given values of R_k and obtain a relation between R_x and c_f. The value of the average skin-friction coefficient can then be calculated directly from $\bar{c}_f = 2R_\theta/R_x$.

Figures 7.4 and 7.5 show the variation of c_f and $\bar{c}_f$ with R_x as calculated by the procedure just described. Also shown in these figures are the lines for constant-roughness Reynolds number R_k and for constant relative roughness x/k. As in previous cases, the origin of the turbulent boundary layer is assumed to be close to the leading edge of the plate, which means that the contribution of R_{x_0} can be neglected. In order to keep the calculation consistent with the empirical constants stipulated, it was necessary to modify the variation of B_2 shown in Fig. 7.3. Essentially, the adjustment consisted of making B_2 compatible with the chosen κ and B_1 values. Here, these values are 0.41 and 0.5, respectively, whereas Nikuradse's corresponding values are 0.40 and 5.5.

7.2.3 Compressible Flow on a Smooth Flat Plate

Skin-Friction Formulas

A number of empirical formulas for varying degrees of accuracy have been developed for calculating compressible turbulent boundary layers on flat plates. Those

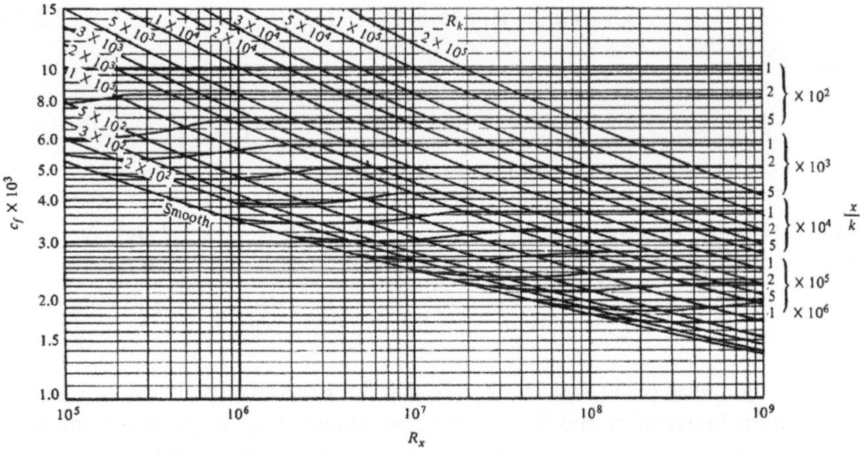

Fig. 7.4 Local skin-friction coefficient on a rough flat plate.

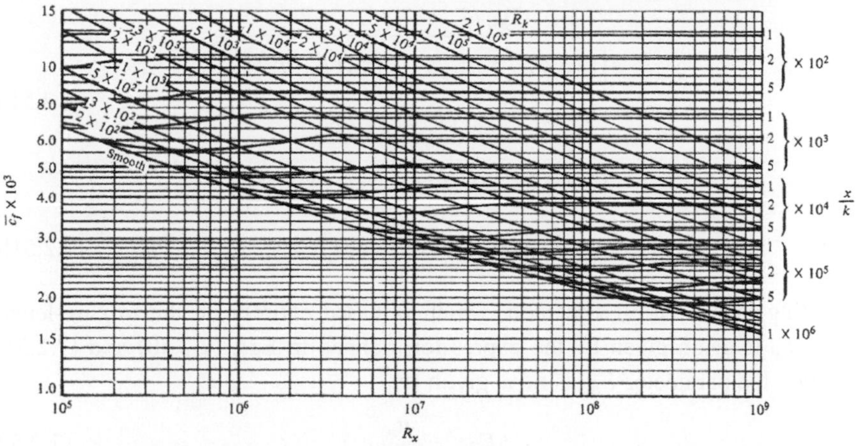

Fig. 7.5 Average skin-friction on a rough flat plate.

developed by Van Driest [11] and by Spalding and Chi [12] have higher accuracy than the rest and cover a wide range of Mach number and ratio of wall temperature to total temperature. These two methods have similar accuracy, although the approaches followed to obtain the formulas are somewhat different. Both methods define compressibility factors by the following relation between the compressible and incompressible values:

$$c_{f_i} = F_c c_f, \tag{7.2.49a}$$

$$R_{\theta_i} = F_{R_\theta} R_\theta, \tag{7.2.49b}$$

$$R_{x_i} = \int_0^{R_x} \frac{F_{R_\theta}}{F_c} dR_x = F_{R_x} R_x, \tag{7.2.49c}$$

Here the subscript i denotes the incompressible values, and the factors F_c, F_{R_θ}, and F_{R_x} ($\equiv F_{R_\theta}/F_c$) defined by Eq. (7.2.49) are functions of Mach number, ratio of wall temperature, and recovery factor. Spalding and Chi's method is based on the postulate that a unique relation exists between $c_f F_c$ and $F_{R_x} R_x$. The quantity F_c is obtained by means of mixing-length theory, and F_R is obtained semiempirically. According to Spalding and Chi,

$$F_c = \frac{T_{aw}/T_e - 1}{(\sin^{-1}\alpha + \sin^{-1}\beta)^2}, \quad F_{R_\theta} = \left(\frac{T_{aw}}{T_e}\right)^{0.772} \left(\frac{T_w}{T_e}\right)^{-1.474}, \tag{7.2.50}$$

where, with r denoting the recovery factor $(T_{aw} - T_e)/(T_{0e} - T_e)$,

$$\alpha = \frac{T_{aw}/T_e + T_w/T_e - 2}{\left[(T_{aw}/T_e + T_w/T_e)^2 - 4(T_w/T_e)\right]^{1/2}},$$
$$\beta = \frac{T_{aw}/T_e - T_w/T_e}{\left[(T_{aw}/T_e + T_w/T_e)^2 - 4(T_w/T_e)\right]^{1/2}}, \tag{7.2.51a}$$

$$\frac{T_{aw}}{Te} = 1 + \frac{\gamma - 1}{2} r M_e^2 . \tag{7.2.51b}$$

According to van Driest's method, which is based entirely on the mixing-length theory, F_c is again given by the expression defined in Eqs. (7.2.50) and (7.2.51). However, the parameter F_{R_θ} is now given by

$$F_{R_\theta} = \frac{\mu_e}{\mu_w}. \tag{7.2.52}$$

The development of Van Driest's formula for skin friction is analogous to the solution steps discussed for incompressible flows (see subsection 7.2.1) except that the derivation is more tedious. The solution requires the expansion of the integral into a series by means of integration by parts and a simple expression is again obtained when higher-order terms are neglected. With this procedure and with the power-law temperature-viscosity relation $\mu \propto T^\omega$, which implies $F_{R_\theta} = (T_e/T_w)^\omega$ the following relation for c_f and R_x is obtained for compressible turbulent boundary layers with and without heat transfer, with x measured from the effective origin of the turbulent flow:

$$\frac{0.242(\sin^{-1}\alpha + \sin^{-1}\beta)}{A\sqrt{c_f(T_w/T_c)}} = 0.41 + \log R_x c_f - \left(\frac{1}{2} + \omega\right)\log\frac{T_w}{T_e}, \tag{7.2.53}$$

where A

$$A^2 = \frac{\gamma - 1}{2}\frac{M_e^2}{T_w/T_e}.$$

This formula is based on Prandtl's mixing-length formula $l = \kappa y$. If the procedure leading to this equation is repeated with the mixing-length expression given by von Karman's similarity law

$$l = \kappa\left|\frac{\partial u/\partial y}{\partial^2 u/\partial y^2}\right|,$$

a formula similar to that given by Eq. (7.2.53) is obtained except that $\frac{1}{2} + \omega$ in Eq. (7.2.53) is replaced by ω. This formula is known as Van Driest II, in order to distinguish it from Eq. (7.2.53), which is known as Van Driest I, and may be written as

$$\frac{0.242(\sin^{-1}\alpha + \sin^{-1}\beta)}{A\sqrt{c_f(T_w/T_e)}} = 0.41 + \log R_x c_f - \omega \log \frac{T_w}{T_e}. \tag{7.2.54}$$

The predictions of Eq. (7.2.54) are in better agreement with experiment than those of Eq. (7.2.53) and Van Driest II should therefore be used in preference to Van Driest I.

Equations (7.2.53) and (7.2.54) constitute a compressible form of the von Karman equation, (7.2.8a). For an incompressible adiabatic flow, $T_w/T_e \to 1$ and $B = 0$, so that with Eq. (7.2.51a), we can write Eq. (7.2.54) as

$$\frac{0.242 \sin^{-1} A}{A\sqrt{c_f}} = 0.41 + \log R_x c_f .$$

In addition, A is of the order of M_e, and since it is small, $\sin^{-1} A = A$. The resulting equation is then identical to Eq. (7.2.8a).

According to Van Driest II, the average skin-friction coefficient $\bar{c}_f$ is obtained from the expression

$$\frac{0.242(\sin^{-1}\alpha + \sin^{-1}\beta)}{A\sqrt{\bar{c}_f(T_w/T_e)}} = \log R_x \bar{c}_f - \omega \log \frac{T_w}{T_e}. \tag{7.2.55}$$

Figures 7.6 and 7.7 show the variation of local and average skin-friction coefficients calculated from Eqs. (7.2.54) and (7.2.55), respectively, on an adiabatic flat plate for various Mach numbers. The recovery factor was assumed to be 0.88.

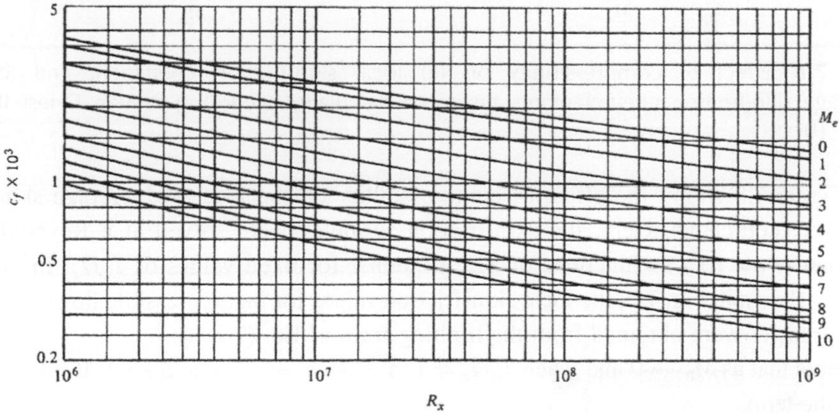

Fig. 7.6 Local skin-friction coefficient on a smooth adiabatic flat plate, according to Van Driest II.

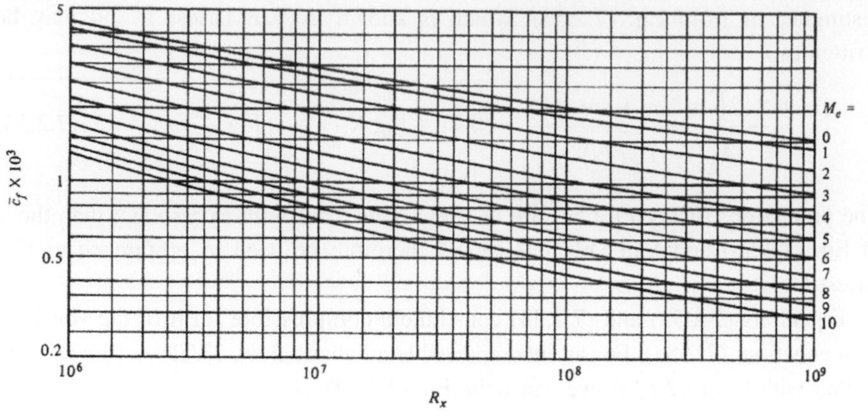

Fig. 7.7 Average skin-friction coefficient on a smooth adiabatic flat plate, according to Van Driest II.

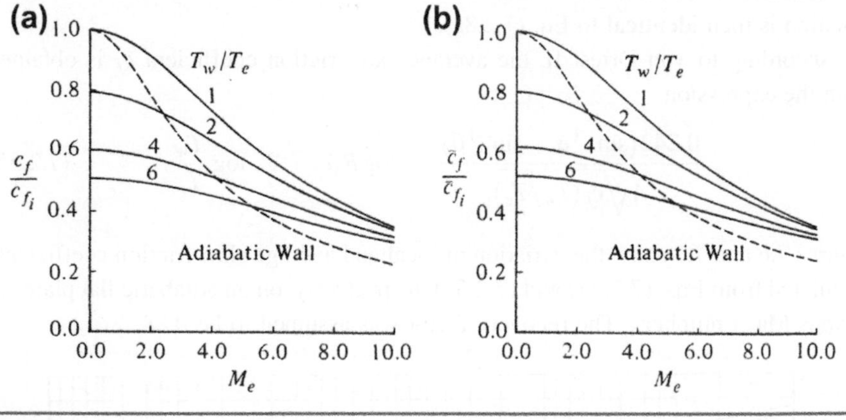

Fig. 7.8 Effect of compressibility on (a) local skin-friction coefficient and (b) average skin-friction coefficient on a smooth flat plate, according to Van Driest II. $R_x = 10^7$.

Figure 7.8 shows the effect of compressibility on the local and average skin-friction coefficients. Here, the skin-friction formulas were solved at a Reynolds number ($R_x = 10^7$) as functions of Mach number for fixed values of T_w/T_e. In the results shown in Fig. 7.6, the local skin-friction values for incompressible flows with heat transfer were obtained from the limiting form of Eq. (7.2.54).

We note that as $M_e \to 0$ and when $T_w/T_e = 1$, $A \to 0$, $\alpha \to -1$, and $\beta \to 1$. It follows that the term

$$\frac{\sin^{-1}\alpha + \sin^{-1}\beta}{A}$$

is indeterminate. Using L'Hospital's rule and recalling that $B = T_e/T_w - 1$, we can write Eq. (7.2.54) for an incompressible turbulent flow with heat transfer, after some algebraic manipulation, as

$$\frac{2}{\sqrt{T_w/T_e}+1}\frac{0.242}{\sqrt{c_f}} = 0.41 + \log R_x c_f - \omega \log \frac{T_w}{T_e}. \tag{7.2.56}$$

The average skin-friction formula, Eq. (7.2.55), can also be written for an incompressible flow by a similar procedure, yielding

$$\frac{2}{\sqrt{T_w/T_e}+1}\frac{0.242}{\sqrt{\bar{c}_f}} = \log R_x \bar{c}_f - \omega \log \frac{T_w}{T_e}. \tag{7.2.57}$$

Reynolds Analogy Factor

According to the studies conducted by Spalding and Chi [12] and Cary [13] it appears that for Mach numbers less than approximately 5 and near-adiabatic wall conditions, a Reynolds analogy factor of

$$\frac{\mathrm{St}}{c_f/2} = 1.16 \tag{7.2.58}$$

adequately represents the available experimental data. However, for turbulent flow with significant wall cooling and for Mach numbers greater than 5 at any ratio of wall temperature to total temperature, the Reynolds analogy factor is ill-defined. Data in [14] indicate that for local Mach numbers greater than 6 and T_w/T_0 less than approximately 0.3, the Reynolds analogy factor scatters around a value of 1.0. A sample of the results is presented in Fig. 7.9 for a Mach number of 11.3 and indicates that the Reynolds measured analogy factor is scattered from around 0.8 to 1.4 with no discernible trend for T_w/T_0.

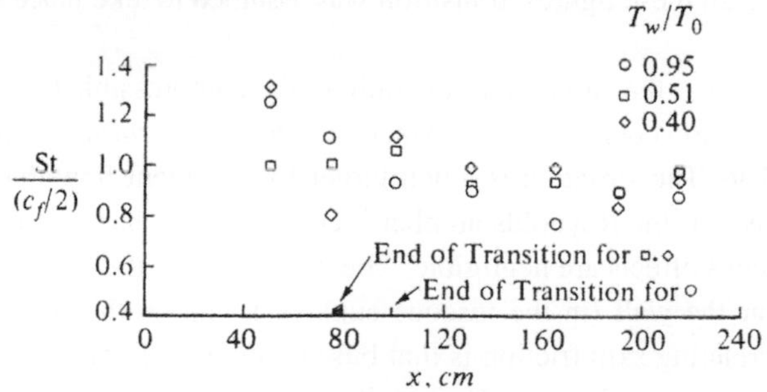

Fig. 7.9 Reynolds analogy factors at $M_e = 11.3$, $R_e/m = 54 \times 10^6$ [14].

7.2.4 Compressible Flow on a Rough Flat Plate

The skin-friction formulas for a smooth flat plate, Eqs. (7.2.55) and (7.2.56), can also be used to obtain formulas for sand-grain-roughened flat plates by assuming a relation between the compressible and incompressible values such as that given by Eq. (7.2.49a). According to the experiments of Goddard [15] on adiabatic fully rough flat plates,

$$F_c = \frac{T_{\text{aw}}}{T_e}\ , \tag{7.2.59}$$

and the experimental values of c_f verified the relation (7.2.59) for his chosen turbulent recovery factor, $r = 0.86$. It should be emphasized that this equation is for fully rough flow in which the flow on top of the roughness elements remains subsonic. It is consistent with the observation originally noted by Nikuradse for incompressible flow, namely, that the skin-friction drag for fully rough flow is the sum of the form drags of the individual roughnesses.

Fenter [16] also presented a theory for the effect of compressibility on the turbulent skin friction of rough plates with heat transfer. This gave results that agree with those of relation (7.2.59) only at Mach numbers close to unity and only for zero heat transfer. For $T_w = T_e$, the value of c_f given by this theory is 14 percent less than that given by Goddard's relation at $M_e = 2.0$ and 45 percent less at $M_e = 4.0$. Fenter presented experimental data for $M_e = 1.0$ and 2.0 that agreed well with this theory for the case of zero heat transfer. The difference in the experimental values of c_f of the two reports is probably within the accuracy to which the roughness heights were measured. The theory of Fenter is based on assumptions whose validity is questionable at high Mach numbers, and these assumptions may account for the difference in c_f predicted by Fenter and by Goddard for the case of $T_w = T_e$.

Figures 7.10 and 7.11 show the average skin friction distribution for a sand-roughened adiabatic plate, and Figs. 7.12 and 7.13 show the results for a sand-roughened plate with a wall temperature equal to the freestream temperature, all at $M_e = 1$ and 2. In all these figures, transition was assumed to take place at the leading edge.

Figure 7.14 shows the variation of the ratio of the compressible to incompressible values of skin-friction coefficient with Mach number for the various types of flow on an adiabatic plate. The variation is much larger for turbulent flow than for laminar flow and increases as the Reynolds number increases, being largest for a fully rough wall where viscous effects are negligible. The reason is that the effect of viscosity is felt mainly near the wall (in the viscous sublayer), and so the relevant Reynolds number for correlating skin friction is that based on the wall value of viscosity. The ratio of wall viscosity to freestream viscosity increases as M_e increases; so a given value of u_eL/ν_e corresponds to a smaller value of u_eL/ν_w and thus a larger c_f. The

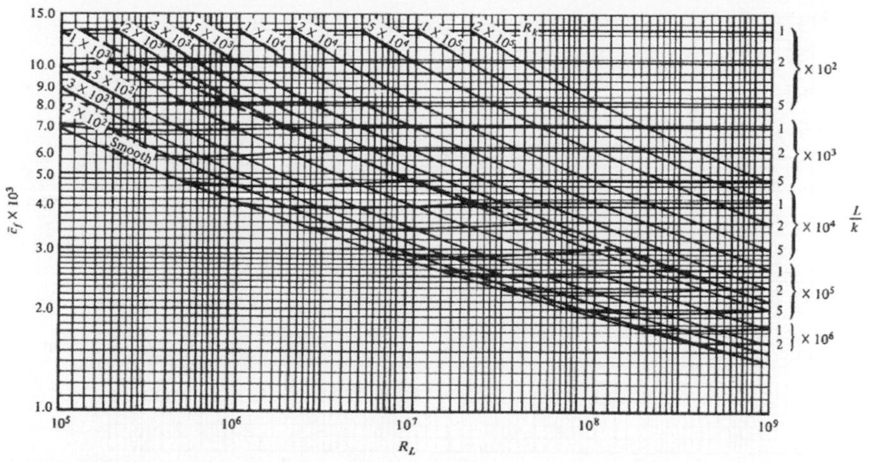

Fig. 7.10 Average skin-friction coefficient for a sand-roughened adiabatic flat plate at $M_e = 1$.

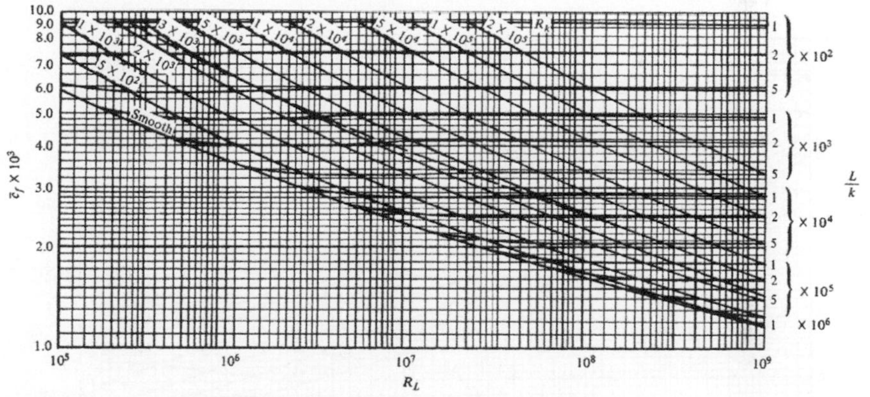

Fig. 7.11 Average skin-friction coefficient for a sand-roughened adiabatic flat plate at $M_e = 2$.

effect on c_f decreases as u_eL/ν_e increases because the change of c_f associated with R_L is smaller. The effect is absent on fully rough walls.

7.3 Flows with Pressure Gradient: Integral Methods

Integral methods are based on the solution of the integral equations of motion discussed in Section 3.6. They avoid the complexity of solving the differential form of the boundary layer equations, and they provide – with very short computation

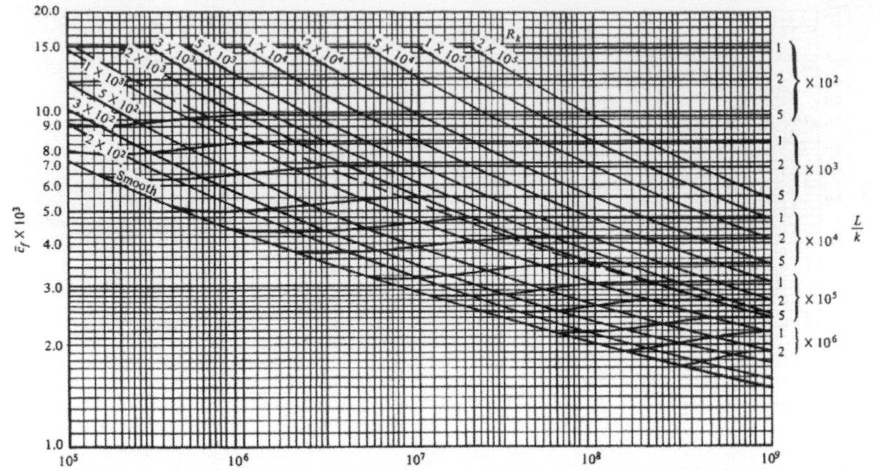

Fig. 7.12 Average skin-friction coefficient for a sand-roughened adiabatic flat plate with $T_w/T_e = 1$, $M_e = 1$.

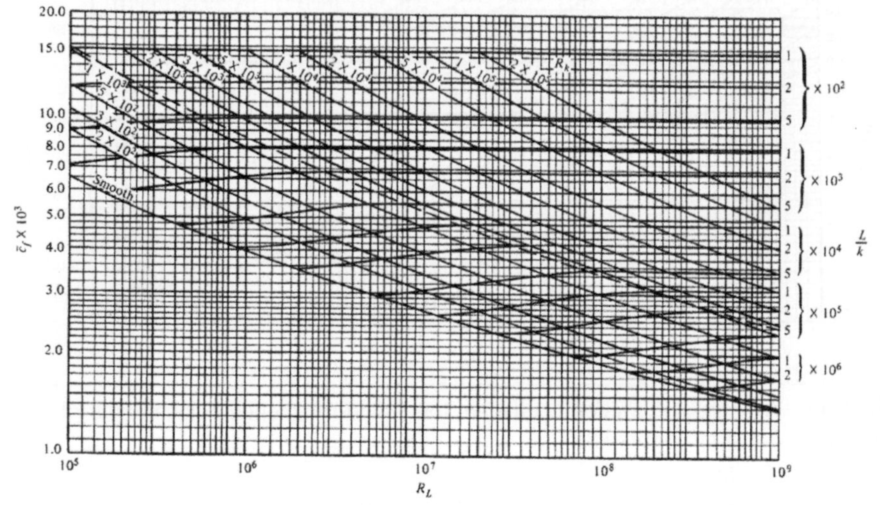

Fig. 7.13 Average skin-friction coefficient for a sand-roughened adiabatic flat plate with $T_w/T_e = 1$, $M_e = 2$.

times – a solution of the boundary layer equations. There are several integral methods for calculating momentum transfer in turbulent boundary layers and a more limited number for heat transfer. This disparity arises because of the difficulty of incorporating possible rapid changes in wall temperature or heat flux

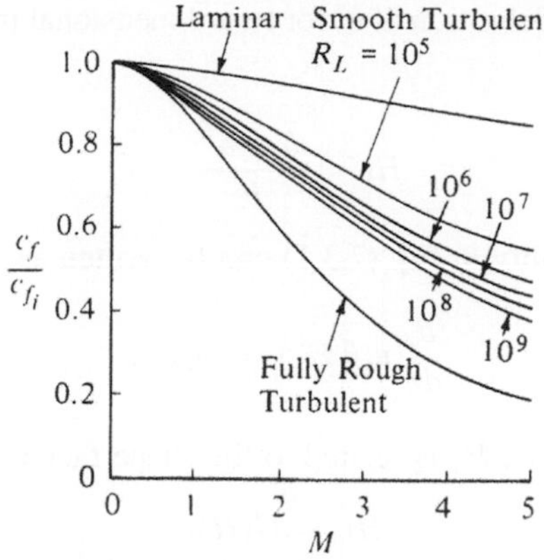

Fig. 7.14 Mach number variation of the ratio of the compressible to incompressible values of local skin-friction coefficient for the various types of air flow on an adiabatic flat plate, for given Reynolds number u_eL/ν_e.

into the temperature profiles used in the solution of the energy integral equation. In the following, we discuss integral methods, first for momentum transfer and then for heat transfer.

Head's Method

The momentum integral equation

$$\frac{d\theta}{dx}+\frac{\theta}{u_e}\frac{du_e}{dx}(H+2) = \frac{c_f}{2} \tag{3.6.6}$$

contains the three unknowns θ, H, and c_f, and assumed relationships between these integral parameters are required. There are several approaches to the achievement of this objective. One approach that we shall consider here adopts the notion that a turbulent boundary layer grows by a process of "entrainment" of nonturbulent fluid at the outer edge and into the turbulent region. It was first used by Head [17], who assumed that the mean velocity component normal to the edge of the boundary layer (which is known as the entrainment velocity v_E) depends only on the mean-velocity profile, specifically on H. He assumed that the dimensionless entrainment velocity v_E/u_e is given by

$$\frac{v_E}{u_e} \equiv \frac{1}{u_e}\frac{d}{dx}\int_0^\delta u\,dy = \frac{1}{u_e}\frac{d}{dx}u_e(\delta-\delta^*) = F(H_1)\ , \tag{7.3.1}$$

where we have used the definition of δ^* for two-dimensional incompressible flows. If we define

$$H_1 = \frac{\delta - \delta^*}{\theta} , \tag{7.3.2}$$

then the right-hand equality in Eq. (7.3.1) can be written as

$$\frac{d}{dx}(u_e \theta H_1) = u_e F . \tag{7.3.3}$$

Head also assumed that H_1 is related to the shape factor H by

$$H_1 = G(H). \tag{7.3.4}$$

The functions F and G were determined from experiment, and a best fit to several sets of experimental data showed that they can be approximated by

$$F = 0.0306(H_1 - 3.0)^{-0.6169}, \tag{7.3.5}$$

$$G = \begin{cases} 0.8234(H - 1.1)^{-1.287} + 3.3 & H \leq 1.6 , \\ 1.5501(H - 0.6778)^{-3.064} + 3.3 & H \geq 1.6 . \end{cases} \tag{7.3.6}$$

With F and G defined by Eqs. (7.3.5) and (7.3.6), Eq. (7.3.3) provides a relationship between θ and H. Another equation relating c_f to θ and/or H is needed, and Head used the semiempirical skin-friction law given by Ludwieg and Tillmann [18],

$$c_f = 0.246 \times 10^{-0.678H} R_\theta^{-0.268} , \tag{7.3.7}$$

where $R_\theta = u_e\theta/\nu$. The system [Eqs. (3.6.6) and (7.3.1)–(7.3.7)], which includes two ordinary differential equations, can be solved numerically for a specified external velocity distribution to obtain the boundary-layer development on a two-dimensional body with a smooth surface [19]. To start the calculations, say $x = x_0$, we note that initial values of two of the three quantities θ, H and c_f must be specified, with the third following from Eq. (7.3.7). When turbulent-flow calculations follow laminar calculations for a boundary layer on the same surface, Head's method is often started by assuming continuity of momentum thickness θ and taking the initial value of H to be 1.4, an approximate value corresponding to flat-plate flow.

This model, like most integral methods, uses a given value of the shape factor H as the criterion for separation. [Equation (7.3.7) predicts c_f to be zero only if H tends to infinity]. It is not possible to give an exact value of H corresponding to separation, and values between the lower and upper limits of H makes little difference in locating the separation point since the shape factor increases rapidly close to separation.

Green's Lag-Entrainment Method

A more refined integral method for computing momentum transfer in turbulent flows is Green's "lag-entrainment" method [20], which is an extension of Head's method in that the momentum integral equation and the entrainment equation are supplemented by an equation for the streamwise rate of change of entrainment coefficient F. This additional equation allows for more realistic calculations in rapidly changing flows and is a significant improvement over Head's method. In effect this is an "integral" version of the "differential" method of Bradshaw et al. discussed in Section 6.3. It requires the solution of Eqs. (3.6.6) and (7.3.1) as before and also considers the "rate of change of entrainment coefficient" equation given by

$$\theta(H_1 + H)\frac{dF}{dx} = \frac{F(F+0.02) + 0.2667c_{f_0}}{F+0.01}$$
$$\times\left\{2.8\left[\left(0.32c_{f_0} + 0.024\,F_{eq} + 1.2F_{eq}^2\right)^{1/2}\right.\right.$$
$$\left.\left. -\left(0.32c_{f_0} + 0.024F + 1.2F^2\right)^{1/2}\right] + \left(\frac{\delta}{u_e}\frac{du_e}{dx}\right)_{eq} - \frac{\delta}{u_e}\frac{du_e}{dx}\right\}, \tag{7.3.8}$$

where the numerical coefficients are from curve fits to experimental data and the empirical functions of Bradshaw et al. Here c_{f_0} is the flat-plate skin-friction coefficient calculated from the empirical formula

$$c_{f_0} = \frac{0.01013}{\log\,R_\theta - 1.02} - 0.00075\,. \tag{7.3.9}$$

The subscript eq in Eq. (7.3.8) refers to equilibrium flows, which are defined as flows in which the shape of the velocity and shear-stress profiles in the boundary layer do not vary with x. The xfunctional forms of the equilibrium values of F_{eq}, $[(\theta/u_e)\,(du_e/dx)]_{eq}$, and $[(\delta/u_e)\,(du_e/dx)]_{eq}$ are given by

$$F_{eq} = H_1\left[\frac{c_f}{2} - (H+1)\left(\frac{\theta}{u_e}\frac{du_e}{dx}\right)_{eq}\right], \tag{7.3.10}$$

$$\left(\frac{\theta}{u_e}\frac{du_e}{dx}\right)_{eq} = \frac{1.25}{H}\left[\frac{c_f}{2} - \left(\frac{H-1}{6.432H}\right)^2\right], \tag{7.3.11}$$

and an obvious consequence of the definitions of H and H_1,

$$\left(\frac{\delta}{u_e}\frac{du_e}{dx}\right)_{eq} = (H+H_1)\left(\frac{\theta}{u_e}\frac{du_e}{dx}\right)_{eq}. \tag{7.3.12}$$

The skin-friction formula and the relationship between the shape factors H and H_1 complete the number of equations needed to solve the system of ordinary differential equations (3.6.6), (7.3.1) and (7.3.8). The skin-friction equation is given by

$$\left(\frac{c_f}{c_{f_0}}+0.5\right)\left(\frac{H}{H_0}-0.4\right)=0.9\ , \tag{7.3.13a}$$

where

$$1-\frac{1}{H_0}=6.55\left(\frac{c_{f_0}}{2}\right)^{1/2}\ , \tag{7.3.13b}$$

so that Eqs. (7.3.9) and (7.3.13) give c_f as a function of H and R_θ with values close to Eq. (7.3.7).

The shape-factor relation is

$$H_1=3.15+\frac{1.72}{H-1}-0.01(H-1)^2 \tag{7.3.13c}$$

and gives values close to Eq. (7.3.6).

Comparisons with experiment show good accuracy in incompressible boundary layer flows and also in wakes. The method has also been extended to represent compressible flows [19].

Truckenbrodt's Method

Two dimensional turbulent boundary layers can also be computed by simple methods such as Thwaites' method for laminar flows [19]. Although these methods are limited and are not as accurate as the differential and integral methods, they are nevertheless useful to estimating boundary-layer parameters without the use of computers. According to Truckenbrodt's method, the momentum thickness is computed from

$$\Theta=\left(\frac{\theta}{c}\right)^{7/6}\left(\frac{u_e}{u_\infty}\right)^{7/2}=\frac{0.0076}{R_c^{1/6}}\int_{(x/c)_{\mathrm{tr}}}^{x/c}\left(\frac{u_e}{u_\infty}\right)^{10/2}d\left(\frac{x}{c}\right)+c_1\ . \tag{7.3.14}$$

Here c_1 is a constant determined by the initial values of u_e and θ at the transition point x_{tr}. The momentum thickness Reynolds number R_θ is defined by $u_e\theta/\nu$.

In order to calculate the development of the shape factor H, Truckenbrodt introduced a new shape factor L that can be calculated from the following expression:

$$L=\frac{\xi_{\mathrm{tr}}}{\xi}L_{\mathrm{tr}}+\ln\left(\frac{u_e(\xi)}{u_{e_{\mathrm{tr}}}}\right)$$
$$+\frac{1}{\xi}\int_{\xi_{\mathrm{tr}}}^{\xi}\left[0.0304\ \ln\ R_\theta-0.23-\ln\left(\frac{u_e(\xi)}{u_{e_{\mathrm{tr}}}}\right)\right]\ d\xi\ , \tag{7.3.15}$$

where $\xi=\Theta^4$.

The shape factor L is related to H by

$$L = \ln\left[\frac{0.775(H-0.379)^{1.61}}{H\,(H-1)^{0.61}}\right]. \tag{7.3.16}$$

Thus once the initial values of θ and H are known, one can calculate the initial value of L by Eq. (7.3.16) and consequently can calculate the development of θ, L, and H around the body.

The local skin friction can be calculated by means of the formula given by Ludwieg and Tillmann, Eq. (7.3.7).

Ambrok's Method

The use of integral procedures to predict heat transfer in turbulent boundary layers generally requires the solution of the integral forms of the energy and momentum equations, although solutions of the integral form of one equation and the differential form of the other have, on occasions, been used. Empirical information is, of course, required to allow the solution of the energy equation, and this usually involves a relationship between the wall heat flux and known integral quantities together with an equation to link the thickness of the temperature and velocity boundary layers. It is difficult to provide empirical relationships that can be used for more than the simplest flows; as a consequence, integral procedures are not widely used, and differential methods are generally to be preferred.

Where an approximate heat-transfer result is required in relation to a simple flow, expressions derived from integral procedures can be useful. The method of Ambrok [21], for example, assumes the Reynolds analogy and, with the integral energy equation, arrives at the approximate equation

$$\mathrm{St} = \frac{\dot{q}_w}{\varrho c_p u_e(T_w - T_e)} = \frac{\mathrm{Pr}^{-0.4} R_L^{-0.2}(T_w - T_e)^{0.25}}{\left[\int_0^{x^*} u_e^*(T_w - T_e)^{1.25} dx^*\right]^{0.2}}, \tag{7.3.17}$$

u_e^*, and R_L denote dimensionless quantities defined by

$$x^* = \frac{x}{L}, \qquad u_e^* = \frac{u_e}{u_{\mathrm{ref}}}, \qquad R_L = \frac{u_{\mathrm{ref}} L}{\nu}.$$

It is useful to note that Eq. (7.3.17) does represent, albeit approximately, the effect of variable surface temperature.

7.4 Prediction of Flow Separation in Incompressible Flows

In many problems it is necessary to know the boundary layer whether laminar or turbulent, will separate from the surface of a specific body and, if so, where the separation will occur. That is quite important, since in many design problems, such as those of the design of hydrofoils or airfoils, it is necessary to avoid flow separation in order to obtain low drag and high lift.

For two-dimensional steady flows, the separation point is defined as the point where the wall shear stree τ_w is equal to zero, that is,

$$(\partial u/\partial y)_w = 0 \ . \tag{7.4.1}$$

With high-speed computers, the boundary-layer equations for laminar flow can be solved exactly, and consequently the laminar separation point can be determined almost exactly. In addition, there are several "simple" methods that do not require the solution of the boundary layer equations in their differential form and that can be used to predict the separation point quite satisfactorily. Thwaites' method discussed in [19] and Stratford's method as cited in [22] are typical examples of two such methods. According to Thwaites' method, laminar separation is predicted when $\lambda(\equiv \frac{\theta^2}{\nu}\frac{du_e}{dx}) = -0.090$. Stratford's method does not even require the solution of the laminar boundary-layer equations. For a given pressure distribution, for example, $C_p(x)$, the expression

$$C_p^{1/2} x \left(dC_p/dx \right) \tag{7.4.2}$$

is calculated around the body. Separation is predicted when it reaches a value of 0.102. Here C_p is defined as

$$C_p = 1 - (u_e/u_o)^2 \ , \tag{7.4.3}$$

where u_o is the velocity at the beginning of the adverse pressure gradient.

The location of a separation point can also be calculated by using either a differential method or an integral method. In differential methods, the parameter used to predict the separation point is the zero-wall-shear stress. In integral methods, the shape factor H is usually used in locating the separation point. In integral methods separation is assumed to occur when H reaches a value between 1.8 and 2.4 for turbulent flows. In some cases, however, the value of H increases rapidly near separation and then begins to decrease. In each case[7] the point corresponding to the maximum value of H is taken as the separation point.

[7]Flows for which an experimental pressure distribution is used in the calculations.

Stratford's laminar method has also been extended to turbulent flows [23]. According to this method, for a given pressure distribution, the left-hand-side of the expression

$$C_p\left(x\frac{dC_p}{dx}\right)^{1/2}\left(10^{-6}R_x\right)^{-1/10}=\frac{2.5}{2}\kappa\equiv F(x),\quad C_p\leq\frac{4}{7}, \qquad (7.4.4)$$

is integrated as a function of x. Separation is predicted when it reaches its right-hand-side.

That analysis assumes an adverse pressure gradient starting from the leading edge, as well as fully turbulent flow everywhere. When there is a region of laminar flow or a region of turbulent flow with a favorable pressure gradient, Stratford defines a false origin x', replaces x by $(x-x')$ in Eq. (7.4.4) and takes the value of R_x as $u_m(x-x')/\nu$ with subscript m denoting the minimum pressure point. The appropriate value of x' is determined from

$$x_m - x' = 58\frac{\nu}{u_m}\left[\frac{u_{tr}}{\nu}\int_0^{x_{tr}}\left(\frac{u_e}{u_m}\right)^5 dx\right]^{3/5} - \int_{x_{tr}}^{x_m}\left(\frac{u_e}{u_m}\right)^4 dx. \qquad (7.4.5)$$

With the expression given by Eq. (7.4.5), the separation point in turbulent flows can be calculated from Eq. (7.4.4). In order to do this, however, it is necessary to assume a value for κ, which according to the mixing-length theory, is about 0.40. That means that the right-hand side of Eq. (7.4.4) should be of the order of 0.5, but a comparison with experiment, according to Stratford, suggests a smaller value of $F(x)$, about 0.35 or 0.40. For a typical turbulent boundary-layer flow with an adverse pressure gradient, it is found that $F(x)$ increases as separation is approached and decreases after separation. For that reason, after applying his method to several flows with turbulent separation, Stratford observed that if the maximum value of $F(x)$ is (a) greater than 0.40, separation is predicted when $F(x)=0.40$; (b) between 0.35 and 0.40, separation occurs at the maximum value; (c) less than 0.35, separation does not occur. On the other hand, in the study conducted by Cebeci et al. [24], Stratford's method gave better agreement with experiment, provided that the range of $F(x)$ was slightly changed from that given above, namely, if the maximum value of $F(x)$ is (a) greater than 0.50, separation is predicted when $F(x)=0.50$; (b) between 0.30 and 0.40, separation occurs at the maximum value; (c) less than 0.30, separation does not occur.

The accuracy of calculating the flow separation point in turbulent flows has been investigated by Cebeci et al. [23]. In that study several experimental pressure distributions that include observed or measured boundary-layer separation were considered. The CS method (the differential method of Cebeci-Smith, Chapter 8), Head's, Stratford's and Goldschmiedt's [25] methods were evaluated. Before we present a sample of results from that study, it is important to note that near separation the behavior of these methods with an experimental pressure distribution is quite different from that with an inviscid pressure distribution. The pressure distribution near the point of separation may be a characteristic of the phenomenon of separation, and inclusion of it in the specification of the flow is equivalent to being told the position of separation. For this reason, use of these separation-prediction methods with an experimental pressure distribution will only show their behavior close to separation and indicate whether the theoretical assumptions used in the methods are self-consistent. When one considers an experimental pressure distribution with separation and uses the CS method, it is quite possible that the wall shear stress at the experimental separation point may not reach zero. It may decrease as the separation point is approached and may then start to increase thereafter. Similarly, the shape factor H in Head's method may not show a continuous increase to the position of separation. Depending on the pressure distribution, which is distorted by the separated flow, the shape factor may even start to decrease after an increase. All that can be learned from a study is how these methods behave close to separation, and whether they predict an early separation or no separation at all.

Figure 7.15 shows the results for Schubauer's elliptic cylinder [26] which has a 3.98-in. minor axis. The experimental pressure distribution was given at a free-stream velocity of $u_\infty = 60$ ft/sec, corresponding to a Reynolds number of $R_D = 1.18 \times 10^5$. The transition region extended from $x/D = 1.25$ to $x/D = 2.27$, and experimental separation was indicated by $x/D = 2.91$.

In the calculations, the transition point was assumed to be at $x/D = 1.25$. It is interesting to note that while three methods predicted separation, the fourth method, Goldschmied's method [24], predicted no separation.

Figure 7.15b shows a comparison of calculated and experimental local skin-friction values. The calculations used the CS method. It is important to note that when the experimental pressure distribution was used, the local skin-friction

(a)

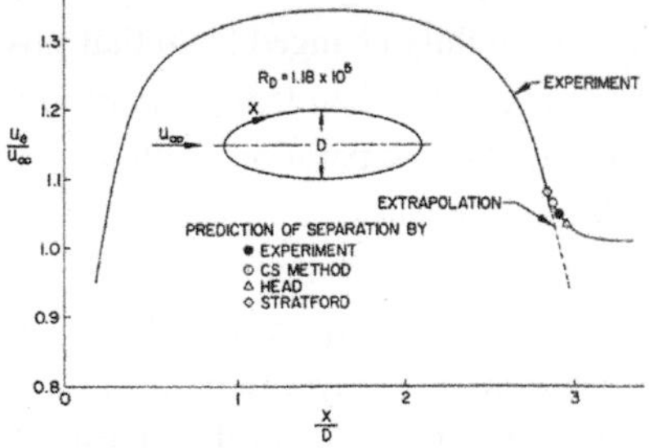

(b)

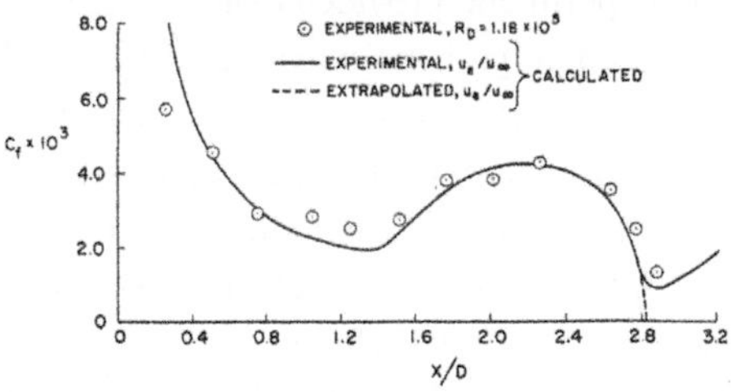

Fig. 7.15 Comparison of (a) predicted separation points with experiment and (b) calculated and experimental local skin-friction coefficients for Schubauer's elliptic cylinder [26].

coefficient began to increase near separation because the pressure distribution was distorted by the flow separation. However, when the calculations were repeated by using an extrapolated velocity distribution that could be obtained by an inviscid method, the skin friction went to zero at $x/D = 2.82$.

Figures 7.16 and 7.17 show the results for three airfoils where flow separation was observed. Figure 7.16 shows the results for the pressure distribution observed over an airfoil-like body at a Reynolds number per foot of 0.82×10^6. The experimental data, which are due to Schubauer and Klebanoff [27] gave the separation point at 25.7 ± 0.2 ft from the leading edge. The predictions of all methods are quite good.

As shown in Fig. 7.17a, agreement between the CS method and experiment is also very good for Newman's airfoil [28]. On the other hand, the other methods predict an early separation.

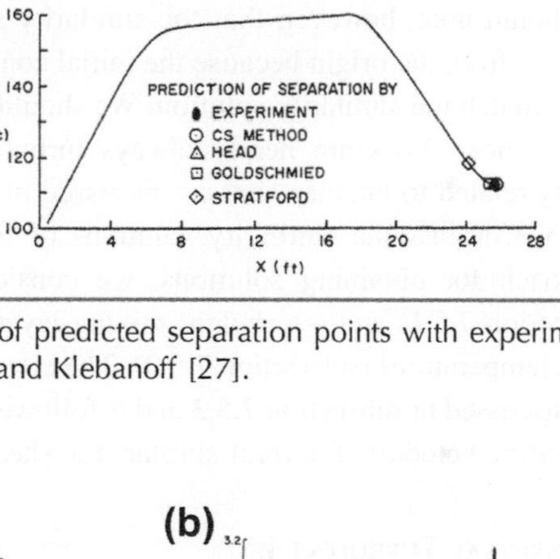

Fig. 7.16 Comparison of predicted separation points with experiment for the airfoil-like body of Schubauer and Klebanoff [27].

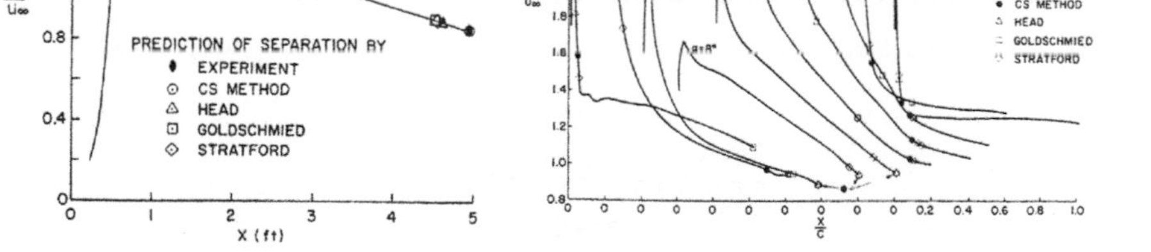

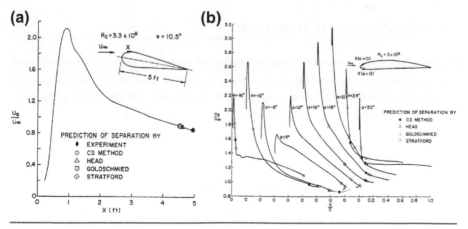

Fig. 7.17 Comparison of predicted separation points with experiment for (a) Newman's airfoil [28] and (b) the NASA 4412 airfoil section at various angles of attack.

For the pressure distribution of Fig. 7.17b, the experimental separation points were not given. The results show that, except at very high angles of attack, the CS method and Head's method predict separation at approximately the same locations, generally close to the characteristics "flattening" in the pressure distribution caused by separation. Stratford's method predicts a slightly earlier separation. Goldschmied's method shows results that are somewhat inconclusive, predicting early separation in some cases and late separation in others.

7.5 Free Shear Flows

As in the case of flow over walls, the boundary-layer equations admit similarity solutions for some laminar and turbulent free shear flows which are not adjacent to a solid surface [19]. Typical examples of such flows include (a) mixing layer between parallel streams, (b) boundary layer and wake of airfoil and (c) merging mixing layers in jet. We should note, however, that the similarity solutions become valid only at large distances from the origin because the initial conditions at, for example, a jet nozzle will not match the similarity solution. We should also note that while in practical cases free shear flows are nearly always turbulent, the turbulent-flow solutions are closely related to laminar ones as discussed in [19].

In this section we discuss the similarity solutions of free shear flows and to illustrate the approach for obtaining solutions, we consider a two-dimensional turbulent jet (subsection 7.5.1) and a turbulent mixing layer between two uniform streams at different temperatures (subsection 7.5.2). The effect of compressibility on free shear flows is discussed in subsection 7.5.3 and is followed by power laws for the width and the centerline velocity of several similar free shear layers.

7.5.1 Two-Dimensional Turbulent Jet

Figure 7.18 shows a two-dimensional heated jet emerging from a slot nozzle and mixing with the surrounding fluid, which is at rest and at another (uniform) temperature. Let the x direction coincide with the jet axis with the origin at the slot. Since the streamlines are nearly parallel within the jet, although the streamlines in

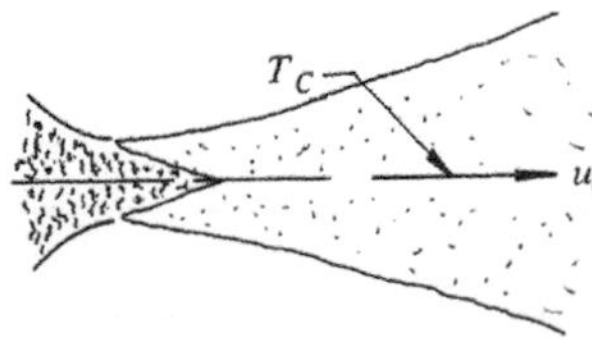

Fig. 7.18 The two-dimensional thermal jet.

the entraining flow are more nearly normal to the axis, the pressure variation in the jet is small and can be neglected. The boundary layer equations can be written as

$$\frac{\partial u}{\partial x} + \frac{\partial v}{\partial y} = 0, \tag{7.5.1}$$

$$u\frac{\partial u}{\partial x} + v\frac{\partial u}{\partial y} = \frac{1}{\varrho}\frac{\partial \tau}{\partial y}, \tag{7.5.2}$$

$$u\frac{\partial T}{\partial x} + \frac{\partial T}{\partial y} = -\frac{1}{\varrho c_p}\frac{\partial \dot{q}}{\partial y}. \tag{7.5.3}$$

Here in general

$$\tau = \mu\frac{\partial u}{\partial y} - \varrho\overline{u'v'}, \tag{7.5.4}$$

$$\dot{q} \equiv \dot{q}_y = -k\frac{\partial T}{\partial y} + \varrho c_p\overline{T'v'}. \tag{7.5.5}$$

These equations are subject to the symmetry and boundary conditions

$$y = 0, \quad v = 0, \quad \frac{\partial u}{\partial y} = 0, \quad \frac{\partial T}{\partial y} = 0, \tag{7.5.6a}$$

$$y = \infty, \quad u = 0, \quad T = T_e. \tag{7.5.6b}$$

Because the pressure is constant in the jet and the motion is steady, the total momentum in the x direction is constant; that is

$$J = \varrho\int_{-\infty}^{\infty} u^2\, dy \equiv 2\varrho\int_0^{\infty} u^2\, dy = \text{const.} \tag{7.5.7}$$

The heat flux (rate of transport of enthalpy of the mean flow) in the x direction is independent of x and equal to its value at the orifice; that is

$$K = 2\varrho c_p\int_0^{\infty} u(T - T_e)\, dy = \text{const}, \tag{7.5.8}$$

K being equal to the product of the initial mass flow rate and the mean enthalpy per unit mass.

To find the similarity solution for the above system, we define dimensionless velocity and temperature ratios by

$$f'(\eta) = \frac{u(x,\ y)}{u_c(x)}, \tag{7.5.9}$$

$$g(\eta) = \frac{T(x,\ y) - T_e}{T_c(x) - T_e}. \tag{7.5.10}$$

Here $u_c(x)$ and $T_c(x)$ denote the velocity and temperature, respectively, along the centerline $y = 0$, and η denotes the similarity variable defined by

$$\eta = \frac{y}{\delta(x)}, \tag{7.5.11}$$

where δ is the shear-layer thickness, to be defined quantitatively below. We assume that the stream function $\psi(x, y)$ is related to a dimensionless stream function $f(\eta)$, independent of x, by

$$\psi(x, y) = u_c(x)\delta(x)/f(\eta). \tag{7.5.12}$$

Note that since $\psi(x, y)$ has the units (length)^2/time, and since $f(\eta)$ is dimensionless, the product $u_c(x)\ \delta(x)$ has the same units as ψ. Our interest here is to find the functional form of $\delta(x)$.

Using Eqs. (7.5.9)–(7.5.11), we can write Eqs. (7.5.7) and (7.5.8) as

$$J = 2\varrho M \int_0^\infty (f')^2\, d\eta, \tag{7.5.13}$$

$$K = 2\varrho c_p N \int_0^\infty f'\, g\, d\eta, \tag{7.5.14}$$

where

$$M = u_c^2\delta\ , \ N = u_c\delta(T_c - T_e). \tag{7.5.15}$$

We note that since the total momentum J and the heat flux K are constant, then M and N must be constant, since the integrals in Eqs. (7.5.13) and (7.5.14) are pure numbers. By using Eqs. (7.5.9)–(7.5.12) and (7.5.15), together with the chain rule, we can write Eqs. (7.5.2) and (7.5.3) as

$$\frac{u_c^2}{2}\frac{d\delta}{dx}\left[(f')^2 + f f''\right] = -\frac{\tau'}{\varrho}, \tag{7.5.16}$$

$$\delta u_c \frac{dT_c}{dx}(fg)' = -\frac{1}{\varrho c_p}(\dot{q})', \tag{7.5.17}$$

$$\eta = 0,\ f = f'' = 0,\ g' = 0, \tag{7.5.18a}$$

$$\eta = \eta_e,\ f' = 0,\ g = 0. \tag{7.5.18b}$$

Equations (7.5.16) and (7.5.17) apply to both laminar and turbulent two-dimensional jets. For turbulent jets the contributions of the laminar shear stress and heat transfer to τ and $\dot{q}$, defined by Eqs. (7.5.4) and (7.5.5), respectively, are small, just as they are outside the sublayer in a wall flow, and can be neglected.

Assuming that the turbulent shear stress and the heat flux scale on similarity variables, so that

$$\frac{\tau}{\varrho} = -\overline{u'v'} = u_c^2 G(\eta), \tag{7.5.19}$$

$$-\frac{\dot{q}}{\varrho c_p} = -\overline{v'T'} = u_c(T_c - T_e)H(\eta), \tag{7.5.20}$$

where $\eta = y/\delta$, we can write Eqs. (7.5.16) and (7.5.17) as

$$\frac{1}{2}\frac{d\delta}{dx}\left[(f')^2 + f f''\right] + G' = 0, \tag{7.5.21}$$

$$\frac{\delta}{T_c - T_e}\frac{d}{dx}(T_c - T_e)\,(fg)' - H' = 0. \tag{7.5.22}$$

For similarity, the coefficients $d\delta/dx$ and

$$\frac{\delta}{T_c - T_e}\frac{d}{dx}(T_c - T_e)$$

must be constant so that

$$\delta \sim x, \qquad T_c - T_e \sim x^s, \tag{7.5.23a}$$

where s is a constant. From the definition of M given in Eq. (7.5.15)

$$u_c \sim x^{-1/2}. \tag{7.5.23b}$$

We have obtained the power laws for growth rate, centerline velocity, and temperature decay rate without introducing a turbulence model, but to integrate Eqs. (7.5.21) and (7.5.22) subject to the boundary conditions given by Eq. (7.5.6), relations between f' and G' and between g and H' are needed. If we use the eddy-viscosity and turbulent-Prandtl-number concepts and let

$$\frac{\tau}{\varrho} = \varepsilon_m \frac{\partial u}{\partial y} = \varepsilon_m \frac{u_c}{\delta} f'' = u_c^2 G(\eta) \tag{7.5.24}$$

and

$$-\frac{\dot{q}}{\varrho c_p} = \varepsilon_h \frac{\partial T}{\partial y} = \frac{\varepsilon_m}{\mathrm{Pr}_t}(T_c - T_e)\frac{g'}{\delta} = u_c(T_c - T_e)H(\eta) \tag{7.5.25}$$

and if we assume that it is accurate enough to take ε_m and Pr_t, to be independent of η, we can write Eqs. (7.5.21) and (7.5.22) as

$$\frac{u_c\delta}{2\varepsilon_m}\frac{d\delta}{dx}\left[(f')^2 + ff''\right] + f''' = 0, \tag{7.5.26}$$

$$\frac{\Pr_t}{\varepsilon_m}\frac{u_c\delta^2}{T_c - T_e}\frac{d}{dx}(T_c - T_e)\,(fg)' - g'' = 0. \tag{7.5.27}$$

If we define δ as the y distance where $u/uc = \frac{1}{2}$, then experimental data [1] suggest

$$\varepsilon_m = 0.037 u_c \delta. \tag{7.5.28}$$

If we write the first relation in Eq. (7.5.23a) as

$$\delta = Ax \tag{7.5.29a}$$

and use Eq. (7.5.28), the coefficient in Eq. (7.5.26) becomes

$$\frac{u_c\delta}{2\varepsilon_m}\frac{d\delta}{dx} = \text{const} = \frac{A}{2(0.037)} = c_1, \tag{7.5.29b}$$

as required for similar solution of Eq. (7.5.26) and Eq. (7.5.26) can be written as

$$f''' + c_1\left[(f')^2 + ff''\right] = 0. \tag{7.5.30}$$

After integrating it three times and using at first the boundary conditions that at $\eta = \eta_e$, $f' = f'' = 0$ and then the condition that at $\eta = 0$, $f' = 1$, $f = 0$, we find the solution to be

$$f = \sqrt{\frac{2}{c_1}}\tanh\sqrt{\frac{c_1}{2}}\eta. \tag{7.5.31}$$

Requiring that f' $(= u/u_c) = \frac{1}{2}$ at $y = \delta$, that is, $\eta = 1$, we find the value of c_1 to be 1.5523. Then it follows from Eq. (7.5.29b) that $A = 0.115$. As a result, the similarity solution for the dimensionless velocity profile of a two-dimensional turbulent jet can be written as

$$f' = \frac{u}{u_c} = \text{sech}^2 0.881\eta, \tag{7.5.32}$$

the dimensionless stream function f can be written as

$$f = 1.135\ \tanh 0.881\eta, \tag{7.5.33}$$

and Eq. (7.5.29a) for the width of the jet becomes

$$\delta = 0.115x. \tag{7.5.34}$$

We now insert Eq. (7.5.32) into Eq. (7.5.13), and upon integration we get

$$u_c = 2.40\sqrt{\frac{J/\varrho}{x}}. \tag{7.5.35}$$

The mass flow rate $\dot{m}$ is

$$\dot{m} = 0.625\sqrt{\varrho J x}. \tag{7.5.36}$$

To obtain the similarity solution of the energy equation (7.5.27), we denote

$$\frac{\Pr_t}{\varepsilon_m} \frac{u_c \delta^2}{T_c - T_e} \frac{d}{dx}\left(T_c - T_e\right) = \text{const} = -C\Pr_t, \tag{7.5.37}$$

and we write Eq. (7.5.27) as

$$g'' + C\Pr_t (fg)' = 0. \tag{7.5.38}$$

Letting $C = c_1$ ($\equiv$ 1.5523), we integrate Eq. (7.5.38) to get

$$g' + c_1 \Pr_t f\, g = c_2. \tag{7.5.39}$$

Noting that the constant of integration $c_2 = 0$ according to the centerline boundary condition imposed on g, and using the relation for f obtained from Eq. (7.5.33), we integrate Eq. (7.5.39) once more to get

$$g = \frac{T - T_e}{T_c - T_e} = \frac{c_3}{[\cosh 0.881\eta]^{2\Pr_t}} = c_3[\text{sech}\, 0.881\eta]^{2\Pr_t}, \tag{7.5.40}$$

where $c_3 = 1$ because $g(0) = 1$.

Clearly, if $\Pr_t = 1$, the velocity profile of Eq. (7.5.32) and the temperature profile of Eq. (7.5.40) are identical. The profile shapes are also identical with those given for a laminar jet [19] because the eddy viscosity is assumed to be independent of y. Since the eddy viscosity ε_m depends on x, the growth rate is different; the jet width varies linearly with x in turbulent flow and as $x^{2/3}$ in laminar flow.

7.5.2 Turbulent Mixing Layer Between Two Uniform Streams at Different Temperatures

Similarity solutions of the momentum and energy equations for a turbulent mixing layer between two uniform streams that move with velocities u_1 and u_2 and whose (uniform) temperatures are T_1 and T_2 (see Fig. 7.19) can be obtained by the method

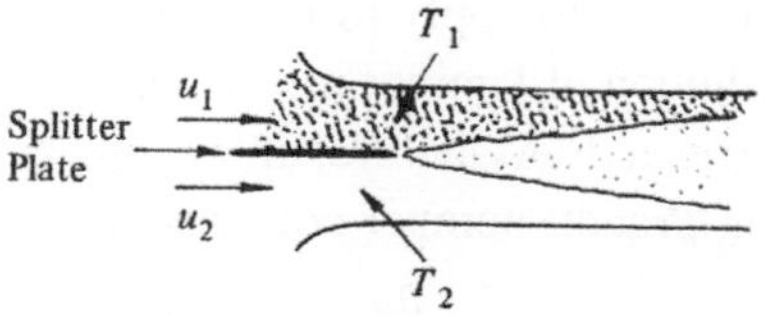

Fig. 7.19 The thermal mixing layer.

used for a two-dimensional jet but with different similarity variables. The governing equations are Eqs. (7.5.1)–(7.5.5). The boundary conditions in Eq. (7.5.6) are replaced by

$$y = \infty,\ u = u_1,\ T = T_1;\ y = -\infty,\ u = u_2,\ T = T_2. \tag{7.5.41}$$

Sometimes the velocity of one uniform stream may be zero. If we use the definition of stream function ψ and relate it to a dimensionless stream function f by

$$\psi(x,y) = u_1\delta(x)f(\eta) \tag{7.5.42}$$

then we can write

$$u = u_1 f',\ v = u_1\frac{d\delta}{dx}(f'\eta - f) \tag{7.5.43}$$

with $\eta = y/\delta(x)$. Here $y = 0$, defined as the line on which $v = 0$, is *not* in general parallel to the splitter plate dividing the two streams for $x < 0$. The lateral location of the profile is determined by the boundary conditions applied by the external flow. If there is a solid boundary, parallel to the splitter plate, at the upper edge of the high-velocity stream, then $v = 0$ for large positive y (where $f' = 1$, which requires $f = \eta$ for large η). If we now define a dimensionless temperature by

$$g(\eta) = \frac{T - T_2}{T_1 - T_2}, \tag{7.5.44}$$

then using the definition of η and the definition of dimensionless stream function given by Eq. (7.5.42), we can write the momentum and energy equations and their boundary conditions as

$$u_1^2\frac{d\delta}{dx}ff'' = -\frac{1}{\varrho}\tau', \tag{7.5.45}$$

$$(T_1 - T_2)\delta u_1\frac{d\delta}{dx}f\,g' = \frac{1}{\varrho c_p}\dot{q}', \tag{7.5.46}$$

$$\eta = \eta_e,\ f' = 1,\ g = 1;\ \eta = -\eta_e,\ f' = \frac{u_2}{u_1} \equiv \lambda,\ g = 0, \tag{7.5.47a}$$

$$\eta = 0,\ f = 0 \quad \text{or} \quad f' = \frac{1}{2}(1 + \lambda). \tag{7.5.47b}$$

Equations (7.5.45) and (7.5.46) apply to both laminar and turbulent flows. For turbulent flows, the contribution of laminar momentum and heat transfer to τ and $\dot{q}$ are small and can be neglected. As before, if we use the eddy-viscosity, eddy-conductivity, and turbulent-Prandtl-number concepts and let

$$-\overline{u'v'} = \varepsilon_m\frac{\partial u}{\partial y} = \varepsilon_m\frac{u_1}{\delta}f'' \tag{7.5.48a}$$

and

$$-\overline{T'v'} = \varepsilon_h \frac{\partial T}{\partial y} = \frac{\varepsilon_m}{\mathrm{Pr}_t}\frac{\partial T}{\partial y} = \frac{\varepsilon_m}{\mathrm{Pr}_t}\frac{T_0}{\delta} g'', \tag{7.5.48b}$$

where $T_0 = T_1 - T_2$, we can write Eqs. (7.5.45) and (7.5.46) as

$$f''' + \frac{u_1 \delta}{\varepsilon_m}\frac{d\delta}{dx} f f'' = 0 \tag{7.5.49}$$

and

$$g'' + \mathrm{Pr}_t \frac{u_1 \delta}{\varepsilon_m}\frac{d\delta}{dx} f g' = 0. \tag{7.5.50}$$

For similarity of the velocity field, we must have

$$\frac{u_1 \delta}{\varepsilon_m}\frac{d\delta}{dx} = \text{const.} \tag{7.5.51}$$

With Pr_t also assumed to be a constant, this requirement for similarity then applies to both velocity and temperature fields. If we take the constant in Eq. (7.5.51) to be $\frac{1}{2}$ then Eqs. (7.5.49) and (7.5.50) become

$$f''' + \frac{1}{2} f f'' = 0, \tag{7.5.52}$$

$$g'' + \frac{1}{2}\mathrm{Pr}_t f g' = 0. \tag{7.5.53}$$

As in the case of the jet, these equations, which are subject to the boundary conditions given by Eq. (7.5.47), are identical to those for laminar flows [19] if we replace Pr_t by Pr. In fact, if we assume Pr_t to be, say, 0.9, then the laminar-flow profile for $\mathrm{Pr} = 0.9$ will be the same as the turbulent-flow profile; as usual, if $\mathrm{Pr}_t = 1.0$, the velocity and temperature profiles will be identical.

The difference between the solutions of (7.5.51) for turbulent flows and the one for laminar flows is due to the definition of δ. For turbulent flows the solution of Eq. (7.5.51) requires an expression for ε_m. Several expressions can be used for this purpose. Here we use the one given by Prandtl. Assuming that $\varepsilon_m \sim \delta$, we expect that ε_m will be determined by the velocity and length scales of the mixing layer:

$$\varepsilon_m = \kappa_1 \delta (u_{\max} - u_{\min}) = \kappa_1 \delta (u_1 - u_2), \tag{7.5.54}$$

where κ_1 is an empirical factor, nominally dependent on y but usually taken as constant.

If we assume that $-\overline{u'v'} = u_1^2 H(\eta)$, then from similarity arguments it follows that δ is proportional to x; in laminar flow it is proportional to $x^{1/2}$. Denoting δ by cx, we can write Eq. (7.5.54), with $\kappa_1 c = C$, as

$$\varepsilon_m = C x u_1 (1 - \lambda). \tag{7.5.55}$$

For uniformity with the existing literature on turbulent mixing layers, we now introduce a parameter σ used by Görtler and defined by him as

$$\sigma = \frac{1}{2}\sqrt{\frac{1+\lambda}{(1-\lambda)C}}. \tag{7.5.56}$$

This can be written as

$$C = \frac{(1+\lambda)}{4\sigma^2(1-\lambda)}. \tag{7.5.57}$$

Substituting Eq. (7.5.57) into (7.5.55) and the resulting expression into Eq. (7.5.51) and taking the constant in Eq. (7.5.51) to be $\frac{1}{2}$, we get

$$\delta = \frac{x}{\sigma}\sqrt{\frac{1+\lambda}{8}}. \tag{7.5.58}$$

With δ given by this equation, we can now plot the solutions of Eqs. (7.5.52) and (7.5.53) in terms of f' ($\equiv u/u_1$) and g' [$\equiv (T - T_2)(T_1 - T_2)$] as a function of

$$\eta = \frac{y}{x}\sigma\sqrt{\frac{8}{1+\lambda}} \tag{7.5.59}$$

for a given value of λ. The Görtler parameter σ, a numerical constant, must be determined empirically. For a turbulent "half jet" (mixing layer in still air) for which $\lambda = 0$, experimental values are mostly between 11 and 13.5. For mixing layers with arbitrary velocity ratios λ, Abramovich [29] and Sabin [30] proposed that

$$\sigma = \sigma_0\left(\frac{1+\lambda}{1-\lambda}\right) \tag{7.5.60}$$

for flows with and without pressure gradient. In Eq. (7.5.60) σ_0 is, of course, the value of σ for the half jet, $\lambda = 0$. This relation was later confirmed by Pui and Gartshore [31] to be a good fit to data.

Figure 7.20 shows a comparison between the numerical solutions of Eq. (7.5.52) and the experimental data of Liepmann and Laufer [32] for a half jet, with $\sigma = 12.0$ and taking $y = 0$ where $u/u_1 = 0.5$.

Data for thermal mixing layers are rare but suggest a turbulent Prandtl number of the order of 0.5. This does not necessarily imply that the temperature profile is wider than the velocity profile but merely that the two shapes are different. However, good agreement with experiment near the edges would probably require Pr_t to depend on y.

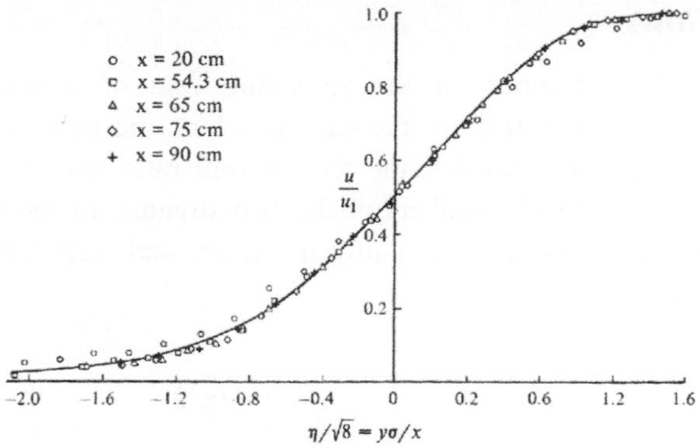

Fig. 7.20 A comparison between the numerical solutions of Eq. (7.5.52) for a turbulent mixing layer (shown by solid line) and the experimental data of Liepmann and Laufer [32], for a half jet with $\sigma \sim 12$.

Low-Speed Flows

The interpretation of experimental data obtained in free turbulent shear layers with large density differences is made difficult by the influence of initial conditions and the absence of flows that can be regarded as "fully developed". Indeed, in the case of the mixing layer between two streams of different densities and unequal (subsonic) speeds, even the direction of the change in spreading rate with density ratio is uncertain. The measurements covering the widest range of density ratio appear to be those of Brown and Roshko [33], who varied the ratio of low-speed stream density ϱ_2 to high-speed stream density ϱ_1 between 7 and $\frac{1}{7}$. In the former case the spreading rate was about 0.75 of that of a constant-density mixing layer, and in the latter case it was about 1.35 times as large as in the constant-density case. Other experiments over smaller ranges of density ratio are inconsistent, but it is clear that in most practical cases, such as the mixing of air and gaseous hydrocarbon fuel, the density ratio will be sufficiently near unity for the change in spreading rate to be negligible. Furthermore, the effect of density ratio on the percentage change of spreading rate with velocity ratio u_2/u_1 is also small.

In the case of a jet of one fluid emerging into another fluid of different density, the density ratio has inevitably fallen to a value fairly near unity at the location where the jet has become fully developed (say, $x/d = 20$), and the change of jet spreading rate with density ratio is effectively negligible. The case of low-speed wakes with significant density differences is not of great practical importance except for buoyant flows in the ocean, and there appear to be no data available.

High-Speed Flows

The effect of Mach number on the spreading rate of a mixing layer is extremely large. Most data refer to the case in which the total temperatures of both streams are the same, so that the temperature ratio and density ratio are uniquely related to the Mach numbers of the two streams. In the most common case, the mixing layer between a uniform stream and still air ($u_2 = 0$), the density ratio is given by

$$\frac{\varrho_2}{\varrho_1} = \frac{T_1}{T_2} = \frac{1}{1 + (\gamma - 1)M_1^2/2}, \tag{7.5.61}$$

where M_1 is the Mach number of the uniform stream. The usual measure of spreading rate is the Görtler parameter σ, related to the standard deviation of the "error function" that fits the velocity profiles at all Mach numbers to adequate accuracy. Figure 7.21 shows the data plotted by Birch and Morrisette [34] with a few later additions. Measurements at a Mach number of 19 are reported by Harvey and Hunter [37] and show a spreading parameter σ in the region of 50, which suggests that the trend of σ with Mach number flattens out considerably above the range of the data shown in Fig. 7.21. However, even the data in Fig. 7.21 show considerable scatter, mainly due to the effect of initial conditions (possibly including shock waves in the case where the pressure of the supersonic jet at exit was not adjusted to be accurately atmospheric). In cases where the exit

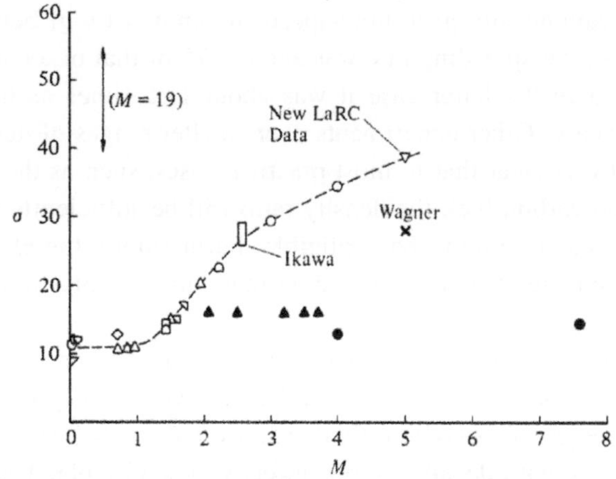

Fig. 7.21 Variation of σ with Mach number in single-stream turbulent mixing layers. Symbols □ and × denote data of Ikawa and Kubota [35] and Wagner [36], respectively; for other symbols, see [34].

pressure is significantly different from atmospheric pressure, the pattern of shock waves and/or expansions considerably affects the spreading rate of the mixing layer.

Equation (7.5.61) implies that the density ratio across a mixing layer at a Mach number of 5 is roughly equal to the factor of 7 investigated in a low-speed flow by Brown and Roshko, who found an increase in spreading rate of about 35 percent compared with the *decrease* of almost a factor of 3 indicated by Fig. 7.21. Clearly, the high Mach number implies an effect of compressibility on the turbulence, as well as on the mean density gradient. Indeed it is easy to show that the Mach-number *fluctuation* in a mixing layer is considerably higher than in a boundary layer at the same mean Mach number. A typical velocity fluctuation can be expressed in terms of the shear stress, so that a representative maximum root-mean-square (rms) velocity fluctuation can be written as

$$\sqrt{\frac{\tau_m}{\varrho}}, \tag{7.5.62}$$

where τ_m is the maximum shear stress within the layer. The square of the speed of sound, a^2, is $\gamma p/\varrho$, and we can see that the Mach number based on the above-mentioned representative velocity fluctuation and the local speed of sound can be written in terms of the external stream Mach number and a shear-stress coefficient based on external stream parameters, that is,

$$M_1\sqrt{\frac{\tau_m}{\varrho_1 u_1^2}}. \tag{7.5.63}$$

The quantity under the square root sign is of order 0.01 in a mixing layer at low speeds, whereas it is equal to $c_f/2$, which is of order 0.001, in a boundary layer in zero pressure gradient (where the maximum shear stress is equal to the wall value). Thus, the Mach-number fluctuation in a mixing layer at a given freestream Mach number is approximately 3 times as large as in a boundary layer at the same freestream Mach number. (This result refers to low Mach number; as the Mach number increases, the skin-friction coefficient in a boundary layer decreases and, as we have seen, the spreading rate and turbulence intensity in a mixing layer also decrease, so that the factor of 3 is at least roughly maintained). We can, therefore, argue that compressibility effects on the turbulence in a mixing layer at a Mach number of 1 are as strong as in a boundary layer at a Mach number of 3. The fact that the spreading rate of a mixing layer does not start to decrease until the Mach number is greater than unity, and that turbulence models with no explicit compressibility effects perform well in boundary layers at Mach numbers up to at least 3, supports this explanation. The implication that significant compressibility effects on turbulence may occur in boundary layers at Mach numbers in excess of 3 is overshadowed by the effects of the very large heat-transfer rates found in practice at hypersonic speeds and the fact that the viscous sublayer becomes extremely thick in hypersonic boundary layers.

No convincing explanation of the compressibility effects exists. Clearly, pressure fluctuations are in some way responsibly because large density differences at low speeds have very little effect. Pressure fluctuations within a turbulent flow are of the order of the density multiplied by the mean-square velocity fluctuation, which, we argued above, is in turn of the same order as the shear stress. In fact the ratio of the maximum shear stress to the absolute pressure is, except for a factor of γ, equal to the square of the Mach number fluctuation derived above. Since the root mean square of the Mach number fluctuation in a mixing layer is about 0.1 of the stream Mach number, this suggests that the ratio of the rms pressure fluctuation to the absolute pressure is of order $\frac{1}{4}$ at a Mach number of 5. It is not necessary that these large pressure fluctuations are caused by shock waves, although the latter may well occur, nor is it necessary to suppose that the main reason for the decrease in spreading rate with increase in Mach number is the increasing loss of turbulent kinetic energy by acoustic radiation ("eddy Mach waves"), although the latter may have some effect. It is known that pressure fluctuations play a large part in the generation and destruction of shear stress in turbulent flow, and this effect of pressure fluctuations is certain to alter if the pressure fluctuations become a significant fraction of the absolute pressure. However, this approach to the role of pressure fluctuations does not explain why the spreading rate should *decrease* with increasing Mach number.

In jets and wakes, the Mach number based on the maximum velocity difference between the shear layer and the external flow falls rapidly with increasing distance downstream, and the density ratio returns rapidly toward unity. As in the case of low-speed jets and wakes with significant density differences, it is difficult to establish general effects of compressibility on spreading rate, independent of the initial conditions. There is considerable interest in the wakes of axisymmetric bodies moving at high speeds, with reference to the detection of reentering missiles. In this case, the most important variables are the temperature and the electron density in the partly ionized gas. Wake data for moderate freestream Mach numbers are given by Demetriades [38,39].

7.5.3 Power Laws for the Width and the Centerline Velocity of Similar Free Shear Layers

The variation of the width, δ, and the centerline velocity, u_c or u_1, of several turbulent shear layers are summarized in Table 7.1.

TABLE 7.1 Power laws for width and centerline velocity of turbulent similar free shear layers.

Flow	Sketch	Width, δ	Centerline velocity, $u_c(x)$ and u_1
Two-dimensional jet	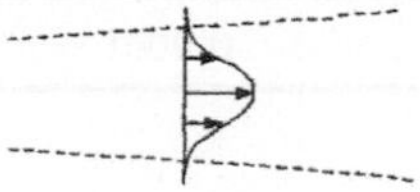	x	$x^{-1/2}$
Axisymmetric jet	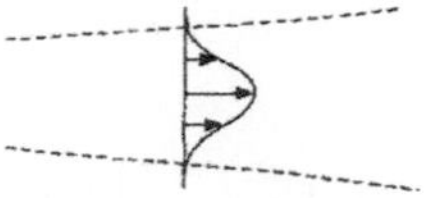	x	x^{-1}
Two-dimensional wake	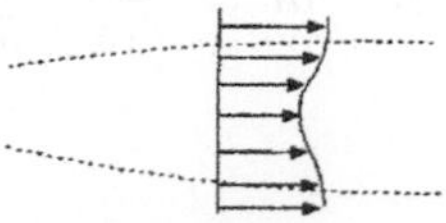	$x^{1/2}$	$x^{-1/2}$
Axisymmetric wake	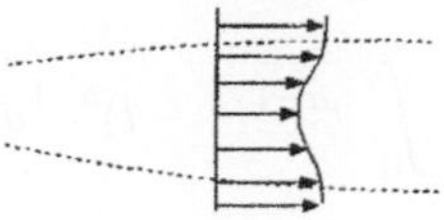	$x^{1/3}$	$x^{-2/3}$
Two uniform streams	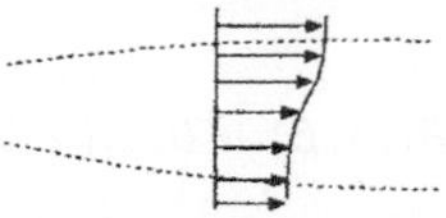	x	x^0

Appendix 7A Gamma, Beta and Incomplete Beta Functions

Gamma function definition

$$\Gamma(\alpha) = \int_0^\infty t^{\alpha-1}e^{t}dt$$

Recursion formula:

$$\Gamma(\alpha+1) = \alpha\Gamma(\alpha)$$

α	$\Gamma(\alpha)$	α	$\Gamma(\alpha)$	α	$\Gamma(\alpha)$
1.00	1.0000	1.35	0.8912	1.70	0.9086
1.05	0.9735	1.40	0.8873	1.75	0.9191
1.10	0.9514	1.45	0.8857	1.80	0.9314
1.15	0.9330	1.50	0.8862	1.85	0.9456

(Continued)

(Continued)

α	$\Gamma(\alpha)$	α	$\Gamma(\alpha)$	α	$\Gamma(\alpha)$
1.20	0.9182	1.55	0.8889	1.90	0.9618
1.25	0.9064	1.60	0.8935	1.95	0.9799
1.30	0.8975	1.65	0.9001	2.00	1.0000

Beta function definition:

$$B_1(\alpha,\beta) = \int_0^1 t^{\alpha-1}(1-t)^{\beta-1}dt = \frac{\Gamma(\alpha)\Gamma(\beta)}{\Gamma(\alpha+\beta)} = B_1(\beta,\alpha)$$

Incomplete Beta function definition:

$$B_x(\alpha,\beta) = \int_0^x t^{\alpha-1}(1-t)^{\beta-1}dt$$

Recursion formula:

$$B_x(\alpha,\beta) = B_1(\alpha,\beta) - B_{1-x}(\alpha,\beta)$$

The following table [40] gives the functional ratios $I_x(\alpha, \beta) = B_x(\alpha, \beta)/B_1(\alpha, \beta)$ for typical combinations of α and β:
Incomplete beta function ratios $I_x(\alpha, \beta)$

x	$\alpha = 1/3$ $\beta = 2/3$	$\alpha = 1/3$ $\beta = 4/3$	$\alpha = 1/3$ $\beta = 8/3$	$\alpha = 2/3$ $\beta = 4/3$	$\alpha = 1/9$ $\beta = 8/9$	$\alpha = 1/9$ $\beta = 10/9$	$\alpha = 1/9$ $\beta = 20/9$	$\alpha = 8/9$ $\beta = 10/9$
0	0	0	0	0	0	0	0	0
0.02	0.2249	0.3068	0.4007	0.0912	0.6346	0.6588	0.7281	0.0342
0.04	0.2838	0.3859	0.5007	0.1443	0.6856	0.7113	0.7845	0.0628
0.06	0.3254	0.4410	0.5684	0.1886	0.7173	0.7439	0.8186	0.0917
0.08	0.3588	0.4845	0.6204	0.2278	0.7407	0.7679	0.8431	0.1174
0.10	0.3872	0.5210	0.6627	0.2636	0.7595	0.7870	0.8622	0.1416
0.20	0.4924	0.6506	0.8008	0.4124	0.8213	0.8490	0.9199	0.2607
0.30	0.5694	0.7377	0.8793	0.5321	0.8603	0.8870	0.9506	0.3715

(Continued)

(Continued)

x	$\alpha = 1/3$ $\beta = 2/3$	$\alpha = 1/3$ $\beta = 4/3$	$\alpha = 1/3$ $\beta = 8/3$	$\alpha = 2/3$ $\beta = 4/3$	$\alpha = 1/9$ $\beta = 8/9$	$\alpha = 1/9$ $\beta = 10/9$	$\alpha = 1/9$ $\beta = 20/9$	$\alpha = 8/9$ $\beta = 10/9$
0.40	0.6337	0.8038	0.9284	0.6339	0.8895	0.9146	0.9696	0.4765
0.50	0.6911	0.8566	0.9599	0.7225	0.9133	0.9362	0.9820	0.5767
0.60	0.7448	0.8998	0.9796	0.7999	0.9335	0.9538	0.9901	0.6725
0.70	0.7970	0.9352	0.9912	0.8671	0.9515	0.9686	0.9952	0.7640
0.80	0.8501	0.9640	0.9972	0.9244	0.9679	0.9812	0.9982	0.8507
0.90	0.9084	0.9863	0.9996	0.9706	0.9835	0.9917	0.9996	0.9313
1.00	1.0000	1.0000	1.0000	1.0000	1.0000	1.0000	1.0000	1.0000
$B_1(\alpha, \beta)$	3.6275	2.6499	2.0153	1.2092	9.1853	8.8439	7.9839	1.0206

Problems

7.1 A thin flat plate is immersed in a stream of air at atmospheric pressure and at 25 °C moving at a velocity of 50 ms^{-1}. Calculate the momentum thickness, boundary-layer thickness, local skin-friction coefficient, and average skin-friction coefficient at $x = 3$ m. Assume that $\nu = 1.5 \times 10^{-5}$ m^{2}s^{-1} and $R_{x_{cr}} = 3 \times 10^6$.

7.2 Air at 70 °F and 1 atm flows at 100 ft s^{-1} past a flat plate of length 15 ft. Assume $R_{x_{tr}} = 3 \times 10^6$, take $\nu = 1.6 \times 10^{-4}$ ft^{2}s^{-1}.

(a) Find the effective origin x_0 of the turbulent boundary layer.
Hint: To estimate x_0 neglect the transitional region, assume that the momentum thickness is continuous at transition, and replace x in Eq. (7.2.14) by $x_{tr} - x_0$.

(b) With Reynolds number based on the effective origin, calculate the local and average skin-friction coefficients at $x = 15$ ft.

(c) At $x = 3$ ft, calculate the distances from the surface at which y^+ is equal to 5, 50, 100, 500 and 1000.

7.3 (a) If in Problem 7.2 the surface temperature of the plate is maintained at 80 °F, calculate the rate of cooling of the plate per unit width. Use the arithmetic-mean film temperature T_f to evaluate the fluid properties.

(b) What error is involved if the boundary layer is assumed to be turbulent from the leading edge?

(c) Repeat (a) and (b) for a velocity of 50 ft s^{-1} with all the other data remaining the same. Discuss the results.

7.4 Use Eq. (7.2.37) to obtain an expression for the heat transfer rate on a flat plate for $x > x_2$ and with $T_w = T_{w_1}$ for $0 < x < x_1$, $T_w = T_{w_2}$ for $x_1 < x < x_2$ and $T_w = T_{w_1}$ for $x > x_2$.

7.5 Air at $u_e/\nu = 3 \times 10^6$ m^{-1} flows past a 3m-long flat plate. Consider the plate: (a) heated at uniform wall temperature T_w, and (b) the heated portion preceded by an unheated portion x_0 of 1 m. Calculate the Stanton number distribution along the plate for both cases. What role does the term $(T_w/T_e)^{0.4}$ in Eq. (7.2.34) play in the results. Assume the flow to be turbulent from the leading edge with $T_w/T_e = 1.1$ and $\Pr = 0.7$.

7.6 Use Eq. (7.2.37) to derive an expression for wall heat flux on a flat plate for which the difference between wall temperature and freestream temperature varies linearly with x, that is,

$$T_w - T_e = A + Bx.$$

Hint: Note that there is a temperature jump at the leading edge of the plate where $T_w - T_e = A$.

7.7 Use Eq. (7.2.37) to obtain an expression for the heat transfer rate on a flat plate for $x > x_2$ and with $T_w = T_{w_1}$, for $0 < x < x_1$, $T_w = T_{w_2}$, for $x_1 < x < x_2$, and $T_w = T_{w_3}$, for $x > x_2$.

7.8 Air at $u_e/\nu = 10^7$ m s^{-1} flows past a 3 m long plate covered with spanwise square-bar roughness elements. Determine the local skin-friction coefficient at $x = 1$ m and the average skin-friction coefficient of the plate. As a simplification, assume that roughness causes the transition to be at the leading edge so that the contribution of laminar flow can be neglected, and take $k = 0.0005$ m.

Hint: First determine the equivalent sand-grain height of the square-bar roughness distribution tested by Moore (see Problem 4.11) and shown in Fig. 4.22.

7.9 Consider the flat-plate problem in Problem 7.1, but assume that (a) the plate surface is covered with camouflage paint (see Table P7.1) applied in mass production conditions and (b) the plate surface is a dip-galvanized metal surface. Calculate the momentum thickness, boundary-layer thickness, local skin-friction coefficient, and average skin-friction coefficient at $x = 3$ m. As a simplification assume that roughness causes the transition to be at the leading edge so that we can neglect the contribution of laminar flow.

7.10 Water at 20°C flows at a velocity of 3 ms^{-1} past a flat plate. Assume $R_{x_{tr}} = 3 \times 10^6$ and use Eq. (7.2.15) to determine the average skin-friction drag of the first 10 m of the plate. Check the contribution of the turbulent portion by Head's method, assuming that $H = 1.5$ at the end of the transition.

TABLE P7.1 Equivalent sand roughness for several types of surfaces.

Type of surface	k_s, cm
Aerodynamically smooth surface	0
Polished metal or wood	$0.05–0.2 \times 10^{-3}$
Natural sheet metal	0.4×10^{-3}
Smooth matte paint, carefully applied	0.6×10^{-3}
Standard camouflage paint, average application	1×10^{-3}
Camouflage paint, mass-production spray	3×10^{-3}
Dip-galvanized metal surface	15×10^{-3}
Natural surface of cast iron	25×10^{-3}

7.11 Consider flow over a NACA 0012 airfoil whose coordinates, $(x/c, y/c)$ are given in tabular form in the accompanying CD-ROM, and its external velocity distribution on the upper airfoil surface for $\alpha = 0°, 2°, 4°$.

(a) Using Head's method, compute the portion of the flow that is turbulent and free of separation. Plot the variation of θ/c, c_f with x/c for a chord Reynolds number of $R_c = 3 \times 10^6$.

(b) Repeat (a) using Truckenbrodt's method and compare the results with those obtained by Head's method.

Note: Since the transition location for this flow is not known, it is necessary to compute it. Also since integral methods require initial conditions, it is necessary to calculate the boundary-layer development on the airfoil starting at the stagnation point.

A practical integral method for calculating the laminar boundary layer development in an incompressible two-dimensional or axisymmetric flow is Thwaites' method described in [19]. According to this method, the momentum thickness for an axisymmetric flow, (θ_3/L), is calculated from

$$\left(\frac{\theta_3}{L}\right)^2 R_L = \frac{0.45}{(u_e^*)^6 (r_0^*)^{2k}} \int_0^{x_3^*} (u_e^*)^5 (r_0^*)^{2k} dx_3^* + \left(\frac{\theta_3}{L}\right)^2 R_L \left(\frac{u_{e0}^*}{u_e^*}\right)^6 \quad \text{(P7.11.1)}$$

Here L is a reference length, x_3^* is the dimensionless surface distance, r_0^* is dimensionless body radius, u_e^* is dimensionless velocity and R_L is a Reynolds number, all defined by

$$x_3^* = \frac{x_2}{L}, \quad r_0^* = \frac{r_0}{L}, \quad u_e^* = \frac{u_e}{u_{\text{ref}}}, \quad R_L = \frac{u_{\text{ref}} L}{\nu} \quad \text{(P7.11.2)}$$

Tha parameter k is flow index being equal to 0 for two-dimensional flows and 1 for axisymmetric flows.

For an axisymmetric stagnation point flow,

$$\left(\frac{\theta_3}{L}\right)^2 R_L = \frac{0.056}{(du_e^*/dx_3^*)_0} \tag{P7.11.3}$$

and for a two-dimensional flow

$$\left(\frac{\theta}{L}\right)_0^2 R_L = \frac{0.075}{(du_e^*/dx^*)_0} \tag{P7.11.4}$$

Once θ_3 is calculated from Eq. (P7.11.3), then the variables δ^*, H and c_f can be calculated from the following relations with λ, c_f and R_θ defined by

$$\lambda = \frac{\theta^2}{\nu}\frac{du_e}{dx}, \quad \frac{c_f}{2} = \frac{l}{R_\theta}, \quad R_\theta = \frac{u_e\theta}{\nu} \tag{P7.11.5}$$

For $0 \leq \lambda \leq 0.1$

$$\begin{aligned} \ell &= 0.22 + 1.57\lambda - 1.8\lambda^2 \\ H &= 2.61 - 3.75\lambda + 5.24\lambda^2 \end{aligned} \tag{P7.11.6a}$$

For $-0.1 \leq \lambda \leq 0$

$$\begin{aligned} \ell &= 0.22 + 1.402\lambda + \frac{0.018\lambda}{0.107 + \lambda} \\ H &= \frac{0.0731}{0.14 + \lambda} + 2.088 \end{aligned} \tag{P7.11.6b}$$

A useful method for predicting transition in two-dimensional incompressible flows is the expression based on Michel's method and Smith's e^9-correlation [19]. It is given by

$$R_{\theta_{\mathrm{tr}}} = 1.174\left(1 + \frac{22,400}{R_{x_{\mathrm{tr}}}}\right) R_{x_{\mathrm{tr}}}^{0.46}. \tag{P7.11.7}$$

According to this method, the boundary-layer development on the body is calculated for a laminar flow starting at the leading-edge of the flow so that both R_θ and Rx can be determined. Usually, the calculated Reynolds numbers are beneath the curve given by Eq. (P7.11.7). The location where the (R_θ, R_x) values intersect this curve corresponds to the onset of transition location. In some cases, however, before this happens, flow separation takes place; in those cases, the separation point is assumed to correspond to the onset of transition location.

In the accompanying CD-ROM, we include the FORTRAN programs for Thwaites' and Michel's method. Their input and output instructions are described in Section 10.2.

7.12 Consider Problem 7.11 and assume that the airfoil surface temperature, T_w, is at 80 °F and outside temperature, T_e, is at 50 °F. Taking $\Pr = 0.72$, we wish to calculate the Stanton number distribution on the airfoil for laminar and turbulent flows.

Note: The Stanton number for turbulent flow can be calculated by using Ambrok's method discussed in Sect. 7.3. The Stanton number for laminar flows can be calculated from the integral method of Smith and Spalding discussed in [21]. According to this method, Stanton number defined by

$$\mathrm{St} = \frac{\dot{q}_w}{\varrho c_p u_e (T_w - T_e)} = \frac{k}{\varrho c_p u_e \delta_c} = \frac{\mathrm{Nu}_x}{R_x \Pr} \tag{P7.12.1}$$

is calculated from

$$\mathrm{St} = \frac{c_1 \left(u_e^*\right)^{c_2}}{\left[\int_0^{x^*} \left(u_e^*\right)^{c_3} dx^*\right]^{1/2}} \frac{1}{\sqrt{R_L}}. \tag{P7.12.2}$$

Here $c_1 = \Pr^{-1} A^{-1/2}$, $c_2 = B/2 - 1$, $c_3 = B - 1$ (see Table P7.2).

TABLE P7.2 Constants in Eq. (P7.12.2) for various prandtl numbers

Pr	c_1	c_2	c_3
0.7	0.418	0.435	1.87
0.8	0.384	0.450	1.90
1.0	0.332	0.475	1.95
5.0	0.117	0.595	2.19
10.0	0.073	0.685	2.37

Smith-Spalding method can also be used for axisymmetric flows. As discussed in [21], using the Mangler transformation, Eq. (P7.12.2) can be written as

$$\mathrm{St} = \frac{c_1 \left(\overset{*}{r_0}\right)^K \left(\overset{*}{u_e}\right)^{c_2}}{\left[\int_0^{\overset{*}{x_3}} \left(\overset{*}{u_e}\right)^{c_3} \left(\overset{*}{r_0}\right)^{2K} d\overset{*}{x_3}\right]^{\frac{1}{2}}} R_L^{-\frac{1}{2}}. \tag{P7.12.3}$$

Here the constants c_1, c_2 and c_3 are the same as those given in Table P7.2.

The location of transition again can be calculated from Eq. (P7.11.7) if we assume that heat transfer has negligable effect on transition. Another practical method for predicting transition with heat transfer is the H-R_x method described in [21]. Here H and R_x are the shape factor ($\equiv \delta^*/\theta$) and the Reynolds number based on surface distance ($\equiv u_e x/\nu$), respectively. This method is simple to use for two-dimensional and axisymmetric flows with pressure gradient, suction and wall heating or cooling. It is given by

$$\begin{aligned} \log\left[R_x(e^9)\right] &= -40.4557 + 64.8066H - 26.7538H^2 + 3.3819H^3, \\ &\qquad 2.1 < H < 2.8. \end{aligned} \tag{P7.12.4}$$

This method is restricted to heating rates where the difference between surface temperature and freestream temperature, $T_w - T_\infty$ does not exceed about 23°C.

(a) First calculate the laminar boundary-layer development using Thwaites' method.
(b) Calculate the location of transition using Eqs. (P7.11.7) and (P7.12.3) and compare the results.
(c) Compute Stanton number distribution up to transition by using Smith-Spalding method.
(d) Compute Stanton number distribution for turbulent flow using Ambrok's method.

In the accompanying CD-ROM, we include the FORTRAN programs for Smith-Spalding, Ambrok methods and H-R_x method for predicting transition. Their input and output instructions are described in Section 10.2.

7.13 The boundary-layer equations and their boundary conditions for a heated laminar jet can be written as

$$\frac{\partial}{\partial x}(ur) + \frac{\partial}{\partial r}(vr) = 0, \tag{P7.13.1}$$

$$u\frac{\partial u}{\partial x} + v\frac{\partial u}{\partial r} = \frac{\nu}{r}\frac{\partial}{\partial r}\left(r\frac{\partial u}{\partial r}\right), \tag{P7.13.2}$$

$$u\frac{\partial T}{\partial x} + v\frac{\partial T}{\partial r} = \frac{\nu}{\Pr}\frac{1}{r}\frac{\partial}{\partial r}\left(r\frac{\partial T}{\partial r}\right), \tag{P7.13.3}$$

$$r = 0, \quad v = 0, \quad \frac{\partial u}{\partial r} = 0, \quad \frac{\partial T}{\partial r} = 0 \tag{P7.13.4a}$$

$$r \to \infty, \quad u \to 0, \quad T \to T_c. \tag{P7.13.4b}$$

In addition to the above equations, the total momentum denoted by J, and the heat flux denoted by K (both in the x-direction) remain constant and are independent of the distance x from the orifice. Hence

$$J = 2\pi\varrho \int_0^\infty u^2 r\, dr = \text{const.} \tag{P7.13.5}$$

$$K = 2\pi\varrho c_p \int_0^\infty ur(T - T_c)\, dr = \text{const.} \tag{P7.13.6}$$

In Eq. (P7.13.6), K is equal to the product of the initial mass flow rate and mean enthalpy at the orifice.

(a) Using the matrix-elimination procedure discussed in Problem 4.1 show that the similarity variable η and dimensionless strem function for continuity and momentum equations are

$$\eta = \frac{r}{x}, \quad f(\eta) = \frac{\psi}{x}, \tag{P7.13.7}$$

Note that the second expression in Eq. (P7.13.7) is dimensionally incorrect. It can easily be corrected by rewriting it as

$$\psi = \nu x f(\eta)\,. \tag{P7.13.8}$$

(b) From the definitions of η and stream function

$$ru = \frac{\partial \psi}{\partial r}, \quad rv = -\frac{\partial \psi}{\partial x}$$

and from Eq. (P7.13.8), show that

$$\frac{u}{u_c} = \frac{1}{(u_c x/\nu)}\frac{f'}{\eta}. \tag{P7.13.9}$$

Note that the right-hand side of Eq. (P7.13.9) is independent of x by virtue of

$$u_c x = \text{const.} \tag{P7.13.10}$$

As a result we can redefine η as

$$\eta = \left(\frac{u_c x}{\nu}\right)^{1/2} \frac{r}{x}\,. \tag{P7.13.11}$$

(c) Use the transformation defined by Eqs. (P7.13.8) and (P7.13.11), observe the chain-rule and show that Eqs. (P7.13.1) to (P7.13.4) can be written as

$$\left[\eta\left(\frac{f'}{\eta}\right)'\right]' + f\left(\frac{f'}{\eta}\right)' + \frac{(f')^2}{\eta} = 0, \tag{P7.13.12}$$

$$\left(\frac{\eta}{Pr}G' + fG\right)' = 0, \tag{P7.13.13}$$

$$\eta = 0,\quad f = G' = 0,\quad f'' = 0 \tag{P7.13.14a}$$

$$\eta = \eta_c,\quad f' = G = 0 \tag{P7.13.14b}$$

where

$$G(\eta) = \frac{T - T_e}{T_c - T_e}\,.$$

(d) Note that

$$\lim_{\eta\to\infty} \frac{f'(\eta)}{\eta} \to 0 \quad\text{and}\quad \lim_{\eta\to\infty} f''(\eta) \to 0 \tag{P7.13.15}$$

and show that the solutions of Eq. (P7.13.12) subject to $f(0) = 0$ are given by

$$f(\eta) = \frac{1/2\eta^2}{1 + 1/8\eta^2} \tag{P7.13.16}$$

and

$$\frac{f'(\eta)}{\eta} = \frac{1}{[1 + 1/8\eta^2]^2}\,. \tag{P7.13.17}$$

(e) With $f(\eta)$ given by Eq. (P7.13.15), show that the solution of Eq. (P7.13.13) subject to the boundary conditions given by Eq. (P7.13.14) is

$$G = \frac{1}{[1 + \eta^2/8]^{2\mathrm{Pr}}}\,. \tag{P7.13.18}$$

7.14 For the transition location determined in Problem 7.12, compute the location of flow separation on the NACA 0012 airfoil for $\alpha = 2°$ and $4°$ using Stratford's method and compare its predictions with Head's method.

7.15 Show that the velocity defect in the wake of a tall building, approximating a two-dimensional cylinder with a diameter of 100 ft (30 m), exceeds 10% of the wind velocity for a distance of 1 mile downstream of the building.

7.16 Trailing vortices from an airliner can endanger following aircraft. Do the jet exhausts significantly affect the decay of the trailing vortices by enhancing turbulent mixing? A simplified version of this question is to ask whether the jet velocity at the vortex position (say 20 nozzle diameters outboard of the jet axis) ever exceeds, say, 5%of the exhaust velocity. Answer the question, assuming that the velocity profile in a circular jet in still air can be approximated by $u/u_c = \frac{1}{2}(1 + \cos \pi r/2R)$, where R is the radius at which $u/u_c = 0.5$ and the approximation applies for $r < 2R$ only. State the main assumptions made in simplifying the question and any further assumptions you make.

7.17 Two tubes in a cross-flow heat exchanger can be idealized as parallel circular cylinders of 1 cm diameter, 10 cm apart. Find the distance downstream at which the two wakes meet, taking the sectional drag coefficients of each cylinder as 1.0.

References

[1] H. Schlichting, Boundary-Layer Theory, McGraw-Hill, New York, 1981.

[2] K.E. Schoenherr, Resistance of flat surface moving through a fluid, Trans. Soc. Nav. Architects Mar. Eng. 40 (1932) 279.

[3] J.C. Simonich, P. Bradshaw, Effect of freestream turbulence on heat transfer through a turbulent boundary layer, J. Heat Transfer 100 (1978) 671.

[4] B.A. Kader, A.M. Yaglom, Heat and mass transfer laws for fully turbulent wall flows, Int. J. Heat Mass Transfer 15 (1977) 2329.

[5] W.C. Reynolds, W.M. Kays, S.J. Klime, Heat transfer in the turbulent incompressible boundary layer, I. Constant wall temperature, NASA MEMO 12-1-58W (1958).

[6] W.C. Reynolds, W.M. Kays, S.J. Klime, Heat transfer in the turbulent incompressible boundary layer, I Step wall-temperature distribution, NASA MEMO 12-22-58W (1958).

[7] W.C. Reynolds, W.M. Kays, S.J. Klime, Heat transfer in the turbulent incompressible boundary layer, III. Arbitrary wall temperature and heat flux, NASA MEMO 12-3-58W (1958).

[8] F.B. Hildebrand, Advanced Calculus for Applications, Prentice-Hall, Englewood Cliffs, NJ, 1962.

[9] W.M. Kays, M.E. Crawford, Convective Heat and Mass Transfer, McGraw-Hill, New York, 1980.

[10] J. Nikuradse, Law of flow in rough pipes, Tech. Memo. NASA No 129 (1955).

[11] E.R. Van Driest, Turbulent boundary layer in compressible fluids, J. Aeronaut. Sci. 18 (1951) 145.

[12] D.B. Spalding, S.W. Chi, The drag of a compressible turbulent boundary layer on a smooth flat plate with and without heat transfer, J. Fluid Mech. 18 (1964) 117.

[13] A.M. Cary, Summary of available information on Reynolds analogy for zero-pressure gradient, compressible turbulent-boundary-layer flow, NASA TN D-5560 (1970).

[14] A.M. Cary, M.H. Bertram, Engineering prediction of turbulent skin friction and heat transfer in high-speed flow, NASA TN D-7507 (1974).

[15] F.E. Goddard Jr., Effect of uniformly distributed roughness on turbulent skin-friction drag at supersonic speeds, J. Aero/Space Sci. 26 (1959) 1–15.

[16] F.W. Fenter, The effect of heat transfer on the turbulent skin-friction of uniformly rough surfaces in compressible flow, The University of Texas, Defense Research Lab. Rept, April 1956. DLR-368, CM-839.

[17] M.R. Head, Entrainment in the turbulent boundary layers, ARC R&M 3643 (1969).

[18] H. Ludwieg, W. Tillmann, Investigation of the wall shearing stress in turbulent boundary layers, NACA Rept. TM 1285 (1949).

[19] T. Cebeci, J. Cousteix, Modeling and Computation of Boundary-Layer Flows, Horizons Pub., Long Beach, C. A. and Springer-Verlag, Germany, Heidelberg, 1998.

[20] J.E. Green, D.J. Weeks, J.W.F. Brooman, Prediction of turbulent boundary layers and wakes in incompressible flow by a lag-entrainment method, ARC R&M 3791 (1973).

[21] T. Cebeci, Convective Heat Transfer, Horizons Pub., Long Beach, C. A., 2002. and Springer-Verlag, Heidelberg, Germany.

[22] L. Rosenhead, Laminar Boundary Layers, Oxford Univ. Press (Clarendon), London and New York, 1963.

[23] B.S. Stratford, The prediction of separation of the turbulent boundary layer,, J. Fluid Mech. 5 (1959) 1.

[24] T. Cebeci, G.J. Mosinskis, A.M.O. Smith, Calculation of separation point in turbulent flows,, J. Aircraft 9 (1972) 618.

[25] F.R. Goldschmied, An approach to turbulent incompressible separation under adverse pressure gradients, J. Aircraft 2 (1965) 108.

[26] G.B. Schubauer, Air flow in the boundary layer of an elliptic cylinder, NACA Rep. No. 652 (1939).

[27] G.B. Schubauer, P.S. Klebanoff, Investigation of separation of the turbulent boundary layer, NACA Tech. Note No. 2133 (1950).

[28] B.G. Newman, Some contributions to the study of the turbulent boundary layer near separation, Austr, Dept. Supply Rep. ACA-53 (1951).

[29] G.N. Abramovich, The Theory of Turbulent Jets, M.I.T. Cambridge, Mass (1963).

[30] C.M. Sabin, An analytical and experimental study of the plane, incompressible, turbulent free-shear layer with arbitrary velocity ratio and pressure gradient, J. Basic Eng. 87 (1965) 421.

[31] N.K. Pui, I.S. Gartshore, Measurement of the growth rate and structure in plane turbulent mixing layers, J. Fluid Mech. 91 (1979) 111.

[32] H.W. Liepmann, J. Laufer, Investigations of free turbulent mixing, NACA TN 1257 (1947).

[33] G.L. Brown, A. Roshko, On density effects and large structure in turbulent mixing layers, J. Fluid Mech. 64 (1974) 775.

[34] S.F. Birch, J.M. Eggers, A critical review of the experimantal data for developed free turbulent shear layers, in Free Turbulent Shear Flows, NASA SP-321 11 (1972).

[35] H. Ikawa, T. Kubota, Investigation of supersonic turbulent mixing layer with zero pressure gradient, AIAA J. 13 (1975) 566.

[36] R.D. Wagner, Mean flow and turbulence measurements in a Mach 5 free shear layer, NASA TN D-7366 (1973).

[37] W.D. Harvey, W.W. Hunter, Experimental study of a free turbulent shear flow at Mach 19 with electron-beam and conventional probes, NASA TN D-7981 (1975).

[38] A. Demetriades, Turbulence measurements in a supersonic two-dimensional wake, Phys. Fluids 13 (1970) 1672.

[39] A. Demetriades, Turbulence correlations in a compressible wake, J. Fluid Mech. 74 (1976) 251.

[40] D.C. Baxter, W.C. Reynolds, Fundamental solutions for heat transfer from non-isothermal flat plates, J. Aero. Sci. 25 (1958) 403.

Chapter 8

Differential Methods with Algebraic Turbulence Models

Chapter Outline Head

Analysis of Turbulent Flows with Computer Programs. http://dx.doi.org/10.1016/B978-0-08-098335-6.00008-2

8.1 Introduction

Differential methods are based on the solution of the boundary-layer equations in their partial-differential equation form. They vary depending on the numerical method used to solve the equations and the turbulence model employed to model the Reynolds stresses. Unlike integral methods, they are general, accurate depending on the numerical method and turbulence model and can handle various initial and boundary conditions. The differential methods, which have largely superseded integral methods with the advent of modern computers, however, require more computer time than the integral methods.

An accurate and efficient differential method is the method developed by Cebeci and Smith [1]. It uses the Cebeci-Smith algebraic eddy viscosity formulation discussed in Chapter 5 to model the Reynolds shear stress term in the momentum equation. In this method, CS method, the boundary-layer equations are solved for both laminar and turbulent flows by specifying the onset of the transition location. The laminar flow calculations are performed up to this location, and the turbulent flow calculations, including the transition region are performed.

In Section 8.2 we describe the formulation and the numerical method used in the CS method. In Section 8.3 and the following sections up to 8.7, we discuss the prediction of incompressible and compressible, two-dimensional and axisymmetric flows with the CS method. Section 8.7 describes the so-called *standard* and *inverse* approaches for calculating boundary-layer flows with and without separation and Section 8.8 extends the standard approach of Section 8.2 to flows with separation. The results obtained with this approach are described in Sections 8.9 and 8.10 for two- and three-dimensional flows.

8.2 Numerical Solution of the Boundary-Layer Equations with Algebraic Turbulence Models

There are several numerical methods for solving the boundary-layer equations in differential form. The Crank-Nicolson and Keller's box methods are the most convenient ones as discussed in some detail in [1,2]. Of the two, Keller's method has significant advantages over the other, and in this section it will be used to solve the boundary-layer equations with algebraic turbulence models and in Sections 9.2 and 9.3 with transport-equation turbulence models for two-dimensional flows.

For two-dimensional incompressible flows, the continuity and momentum equations given by Eqs. (3.3.24) and (3.3.25) for compressible axisymmetric flows reduce to the equations given by Eqs. (5.2.8) and (5.2.9), that is,

$$\frac{\partial u}{\partial x}+\frac{\partial v}{\partial y}=0 \tag{8.2.8}$$

$$u\frac{\partial u}{\partial x}+v\frac{\partial u}{\partial y}=-\frac{1}{\varrho}\frac{dp}{dx}+\nu\frac{\partial^2 u}{\partial y^2}-\frac{\partial}{\partial y}\left(\overline{u'v'}\right) \tag{8.2.9}$$

With Bernoulli's equation, the above momentum equation can be written as

$$u\frac{\partial u}{\partial x}+v\frac{\partial u}{\partial y}=u_{\mathrm{e}}\frac{du_{\mathrm{e}}}{dx}+\nu\frac{\partial^2 u}{\partial y^2}-\frac{\partial}{\partial y}\left(\overline{u'v'}\right) \tag{8.2.1}$$

The boundary conditions for Eqs. (8.2.8) and (8.2.1) are

$$y=0\quad u=0\quad v=v_w(x) \tag{8.2.2a}$$

$$y=\delta\quad u=u_e \tag{8.2.2b}$$

The above equations can be solved in the form they are expressed or in the form after they are expressed as a third order equation by using the definition of stream function $\psi(x, y)$. Noting that

$$u=\frac{\partial\psi}{\partial y},\qquad v=-\frac{\partial\psi}{\partial x} \tag{8.2.3}$$

Eqs. (5.2.8) and (8.2.1), with a prime denoting differentiation with respect to y, and with an eddy viscosity ε_m defined by Eq. (5.2.9),

$$-\overline{u'v'}=\varepsilon_m\frac{\partial u}{\partial y} \tag{5.2.9}$$

can be written as

$$\left[(\nu+\varepsilon_m)\psi''\right]'+u_e\frac{du_e}{dx}=\psi'\frac{\partial\psi'}{\partial x}-\psi''\frac{\partial\psi}{\partial x} \tag{8.2.4}$$

In either form, for given initial conditions, say at $x = x_0$ and eddy viscosity distribution, these equations can be solved subject to their boundary conditions in the interval 0 to δ at each specified x-location greater than x_0. The boundary-layer thickness $\delta(x)$, however, increases with increasing downstream distance x for both laminar and turbulent flows; to maintain computational accuracy, it is necessary to take small steps in the streamwise direction.

Transformed coordinates employing similarity variables such as the one discussed in [1] provide another alternative to express the equations in a better form before solving. Such a choice can reduce the growth of transformed boundary-layer thickness and thus allow larger steps to be taken in the stream-wise direction. Furthermore, in some cases, they can also be used to generate the initial conditions needed in the solution of the boundary-layer equations.

We shall advocate the use of transformed coordinates employing similarity variables and for two-dimensional flows we will use the Falkner-Skan transformation discussed in [1]. With the similarity variable defined by

$$\eta = \sqrt{\frac{u_e}{\nu x}}y \tag{8.2.5a}$$

and the dimensionless stream function $f(x, \eta)$

$$\psi(x, y) = \sqrt{u_e \nu x} f(x, \eta), \tag{8.2.5b}$$

the continuity and momentum equations, Eq. (8.2.4) and their boundary conditions, Eqs. (8.2.2), can be written as

$$(bf'')' + \frac{m+1}{2} ff'' + m\left[1 - (f')^2\right] = x\left(f'\frac{\partial f'}{\partial x} - f''\frac{\partial f}{\partial x}\right) \tag{8.2.6}$$

$$\eta = 0, \quad f = f_w = -\frac{1}{\sqrt{u_e \nu x}} \int_0^x v_w dx, \quad f' = 0 \tag{8.2.7a}$$

$$\eta = \eta_e, \quad f' = 1 \tag{8.2.7b}$$

Here, a prime denotes differentiation with respect to η; the parameter b and pressure gradient parameter m are defined by

$$b = 1 + \varepsilon_m^+, \quad \varepsilon_m^+ = \frac{\varepsilon_m}{\nu}, \quad m = \frac{x}{u_e}\frac{du_e}{dx}. \tag{8.2.8}$$

To solve Eqs. (8.2.6) and (8.2.7) with Keller's box method, which is a two-point finite-difference scheme, we first express them as a first-order system by introducing new functions to represent the derivatives of f with respect to η

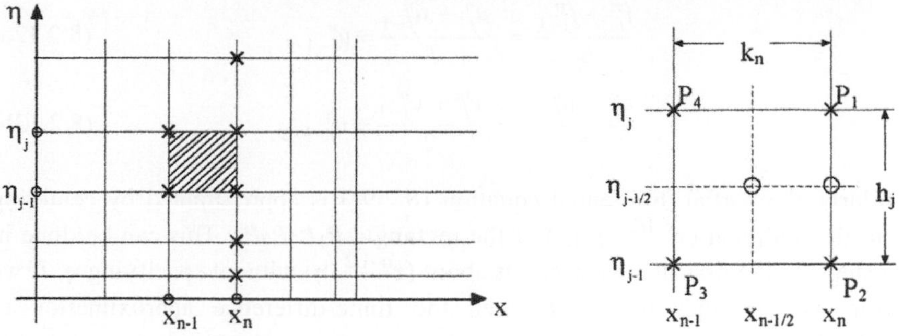

Fig. 8.1 Net rectangle for difference approximations.

(subsection 8.2.1). The first-order equations are approximated on an arbitrary rectangular net, Fig. 8.1, with "centered-difference" derivatives and averages at the midpoints of the net rectangle difference equations. The resulting system of equations which is implicit and nonlinear is linearized by Newton's method (subsection 8.2.2) and solved by the block-elimination method discussed in subsection 8.2.3.

8.2.1 Numerical Formulation

In order to express Eqs. (8.2.6) and (8.2.7) as a system of first-order equations, we define new variables $u(x, \eta)$ and $v(x, \eta)$ by

$$f' = u \tag{8.2.9a}$$

$$u' = v \tag{8.2.9b}$$

and write them as

$$(bv)' + \frac{m+1}{2} fv + m(1 - u^2) = x\left(u\frac{\partial u}{\partial x} - v\frac{\partial f}{\partial x}\right) \tag{8.2.9c}$$

$$\eta = 0, \quad u = 0, \quad f = f_w(x); \quad \eta = \eta_e, \quad u = 1 \tag{8.2.10}$$

We denote the net points of the net rectangle shown in Fig. 8.1 by

$$\begin{aligned} x_0 &= 0, \quad x_n = x_{n-1} + k_n, \quad n = 1, 2, \ldots, N \\ \eta_0 &= 0, \quad \eta_j = \eta_{j-1} + h_j, \quad j = 1, 2, \ldots, J \end{aligned} \tag{8.2.11}$$

and write the difference equations that are to approximate Eqs. (8.2.9) by considering one mesh rectangle as in Fig. 8.1. We start by writing the finite-difference approximations of the ordinary differential equations (8.2.9a,b) for the midpoint $(x^n, \eta_{j-1/2})$ of the segment P_1P_2, using centered-difference derivatives,

$$\frac{f_j^n - f_{j-1}^n}{h_j} = \frac{u_j^n + u_{j-1}^n}{2} \equiv u_{j-1/2}^n \tag{8.2.12a}$$

$$\frac{u_j^n - u_{j-1}^n}{h_j} = \frac{v_j^n + v_{j-1}^n}{2} \equiv v_{j-1/2}^n \tag{8.2.12b}$$

Similarly, the partial differential equation (8.2.9c) is approximated by centering about the midpoint ($x^{n-1/2}$, $\eta_{j-1/2}$) of the rectangle $P_1P_2P_3P_4$. This can be done in two steps. In the first step we center it about ($x^{n-1/2}$, η) without specifying η. If we denote its left-hand side by L, then the finite-difference approximation to Eq. (8.2.9c) is

$$\frac{1}{2}\left(L^n + L^{n-1}\right) = x^{n-1/2}\left[u^{n-1/2}\left(\frac{u^n - u^{n-1}}{k_n}\right) - v^{n-1/2}\left(\frac{f^n - f^{n-1}}{k_n}\right)\right] \tag{8.2.13}$$

$$\alpha^n = \frac{x^{n-1/2}}{k_n}, \quad \alpha_1 = \frac{m^n + 1}{2} + \alpha^n, \quad \alpha_2 = m^n + \alpha^n \tag{8.2.14a}$$

$$R^{n-1} = -L^{n-1} + \alpha^n\left[(fv)^{n-1} - (u^2)^{n-1}\right] - m^n \tag{8.2.14b}$$

$$L^{n-1} \equiv \left[(bv)' + \frac{m+1}{2}fv + m(1-u^2)\right]^{n-1} \tag{8.2.14c}$$

Eq. (8.2.12) can be written as

$$\left[(bv)'\right]^n + \alpha_1(fv)^n - \alpha_2\left(u^2\right)^n + \alpha^n\left(v^{n-1}f^n - f^{n-1}v^n\right) = R^{n-1} \tag{8.2.15}$$

The identity sign introduces a useful shorthand: $[\ ]^{n-1}$ means that the quantity in square brackets is evaluated at $x = x^{n-1}$.

We next center Eq. (8.2.15) about the point ($x^{n-1/2}$, $\eta_{j-1/2}$), that is, we choose $\eta = \eta_{j-1/2}$ and obtain

$$\begin{aligned} &h_j^{-1}\left(b_j^n v_j^n - b_{j-1}^n v_{j-1}^n\right) + \alpha_1(fv)_{j-1/2}^n - \alpha_2\left(u^2\right)_{j-1/2}^n \\ &+\alpha^n\left(v_{j-1/2}^{n-1}f_{j-1/2}^n - f_{j-1/2}^{n-1}v_{j-1/2}^n\right) = R_{j-1/2}^{n-1} \end{aligned} \tag{8.2.16}$$

where

$$R_{j-1/2}^{n-1} = -L_{j-1/2}^{n-1} + \alpha^n\left[(fv)_{j-1/2}^{n-1} - \left(u^2\right)_{j-1/2}^{n-1}\right] - m^n \tag{8.2.17a}$$

$$L_{j-1/2}^{n-1} = \left\{h_j^{-1}\left(b_j v_j - b_{j-1}v_{j-1}\right) + \frac{m+1}{2}(fv)_{j-1/2} + m\left[1 - \left(u^2\right)_{j-1/2}\right]\right\}^{n-1} \tag{8.2.17b}$$

Eqs. (8.2.12) and (8.2.16) are imposed for $j = 1, 2, \ldots, J-1$ at given η and the transformed boundary-layer thickness, η_e, is to be sufficiently large so that $u \to 1$ asymptotically. The latter is usually satisfied when $v(\eta_e)$ is less than approximately 10^{-3}.

The boundary conditions [Eq. (8.2.10)] yield, at $x = x^n$,

$$f_0^n = f_w, \quad u_0^n = 0, \quad u_J^n = 1 \tag{8.2.18}$$

8.2.2 Newton's Method

If we assume f_j^{n-1}, u_j^{n-1}, and, v_j^{n-1} to be known for $0 \le j \le J$, then Eqs. (8.2.12), (8.2.16) and (8.2.18) form a system of $3J+3$ equations for the solution of $3J+3$ unknowns $(f_j^n, u_j^n, v_j^n), j = 0, 1, \ldots, J$. To solve this nonlinear system, we use Newton's method; we introduce the iterates $[f_j^{(\nu)}, u_j^{(\nu)}, v_j^{(\nu)}]$, $\nu = 0, 1, 2, \ldots$, with initial value ($\nu = 0$) equal to those at the previous x-station x^{n-1} (which is usually the best initial guess available). For the higher iterates we set

$$f_j^{(\nu+1)} = f_j^{(\nu)} + \delta f_j^{(\nu)}, \quad u_j^{(\nu+1)} = u_j^{(\nu)} + \delta u_j^{(\nu)}, \quad v_j^{(\nu+1)} = v_j^{(\nu)} + \delta v_j^{(\nu)} \tag{8.2.19}$$

We then insert the right-hand sides of these expressions in place of f_j^n, u_j^n, and v_j^n in Eqs. (8.2.12) and (8.2.16) and drop the terms that are quadratic in $\delta f_j^{(\nu)}$, $\delta u_j^{(\nu)}$ and $\delta v_j^{(\nu)}$. This procedure yields the following *linear* system (the superscript n is dropped from f_j, u_j, v_j and ν from δ quantities for simplicity).

$$\delta f_j - \delta f_{j-1} - \frac{h_j}{2}\left(\delta u_j + \delta u_{j-1}\right) = (r_1)_j \tag{8.2.20a}$$

$$\delta u_j - \delta u_{j-1} - \frac{h_j}{2}\left(\delta v_j + \delta v_{j-1}\right) = (r_3)_{j-1} \tag{8.2.20b}$$

$$(s_1)_j \delta v_j + (s_2)_j \delta v_{j-1} + (s_3)_j \delta f_j + (s_4)_j \delta f_{j-1} + (s_5)_j \delta u_j + (s_6)_j \delta u_{j-1} = (r_2)_j \tag{8.2.20c}$$

where

$$(r_1)_j = f_{j-1}^{(\nu)} - f_j^{(\nu)} + h_j u_{j-1/2}^{(\nu)} \tag{8.2.21a}$$

$$(r_3)_{j-1} = u_{j-1}^{(\nu)} - u_j^{(\nu)} + h_j v_{j-1/2}^{(\nu)} \tag{8.2.21b}$$

$$(r_2)_j = R_{j-1/2}^{n-1} - \left[\begin{array}{l} h_j^{-1}\left(b_j^{(\nu)} v_j^{(\nu)} - b_{j-1}^{(\nu)} v_{j-1}^{(\nu)}\right) + \alpha_1 (fv)_{j-1/2}^{(\nu)} \\ -\alpha_2 (u^2)_{j-1/2}^{(\nu)} + \alpha^n\left(v_{j-1/2}^{n-1} f_{j-1/2}^{(\nu)} - f_{j-1/2}^{n-1} v_{j-1/2}^{(\nu)}\right) \end{array}\right] \tag{8.2.21c}$$

In writing the system given by Eqs. (8.2.20) we have used a certain order for them. The reason for this choice, as we shall see later, is to ensure that the A_0 matrix in Eq. (8.2.27a) is not singular.

The coefficients of the linearized momentum equation are

$$(s_1)_j = h_j^{-1} b_j^{(v)} + \frac{\alpha_1}{2} f_j^{(v)} - \frac{\alpha^n}{2} f_{j-1/2}^{n-1} \tag{8.2.22a}$$

$$(s_2)_j = -h_j^{-1} b_{j-1}^{(v)} + \frac{\alpha_1}{2} f_{j-1}^{(v)} - \frac{\alpha^n}{2} f_{j-1/2}^{n-1} \tag{8.2.22b}$$

$$(s_3)_j = \frac{\alpha_1}{2} v_j^{(v)} + \frac{\alpha^n}{2} v_{j-1/2}^{n-1} \tag{8.2.22c}$$

$$(s_4)_j = \frac{\alpha_1}{2} v_{j-1}^{(v)} + \frac{\alpha^n}{2} v_{j-1/2}^{n-1} \tag{8.2.22d}$$

$$(s_5)_j = -\alpha_2 u_j^{(v)} \tag{8.2.22e}$$

$$(s_6)_j = -\alpha_2 u_{j-1}^{(v)} \tag{8.2.22f}$$

The boundary conditions, Eq. (8.2.18) become

$$\delta f_0 = 0, \quad \delta u_0 = 0, \quad \delta u_J = 0 \tag{8.2.23}$$

As discussed in [1], the linear system given by Eqs. (8.2.20) and (8.2.23) has a block tridiagonal structure and can be written in matrix-vector form as

$$A \vec{\vec{\delta}} = \vec{\vec{r}} \tag{8.2.24}$$

where

$$A = \begin{vmatrix} A_0 & C_0 & & & & \\ B_1 & A_1 & C_1 & & & \\ & \cdot & \cdot & \cdot & & \\ & & \cdot & \cdot & \cdot & \\ & & B_j & A_j & C_j & \\ & & & \cdot & \cdot & \cdot \\ & & & B_{J-1} & A_{J-1} & C_{J-1} \\ & & & & B_J & A_J \end{vmatrix} \quad \vec{\vec{\delta}} = \begin{vmatrix} \vec{\delta}_0 \\ \vec{\delta}_1 \\ \cdot \\ \vdots \\ \vec{\delta}_j \\ \cdot \\ \vdots \\ \vec{\delta}_J \end{vmatrix} \quad \vec{\vec{r}} = \begin{vmatrix} \vec{r}_0 \\ \vec{r}_1 \\ \cdot \\ \vdots \\ \vec{r}_j \\ \cdot \\ \vdots \\ \vec{r}_J \end{vmatrix} \tag{8.2.25}$$

$$\vec{\delta}_j = \begin{vmatrix} \delta f_j \\ \delta u_j \\ \delta v_j \end{vmatrix} \quad \vec{r}_j = \begin{vmatrix} (r_1)_j \\ (r_2)_j \\ (r_3)_j \end{vmatrix} \quad 0 \le j \le J \tag{8.2.26}$$

and A_j, B_j, C_j are 3×3 matrices defined as

$$A_0 \equiv \begin{vmatrix} 1 & 0 & 0 \\ 0 & 1 & 0 \\ 0 & -1 & -h_1/2 \end{vmatrix} \quad A_j \equiv \begin{vmatrix} 1 & -h_j/2 & 0 \\ (s_3)_j & (s_5)_j & (s_1)_j \\ 0 & -1 & -h_{j+1}/2 \end{vmatrix} \quad 1 \le j \le J-1 \tag{8.2.27a}$$

$$A_J \equiv \begin{vmatrix} 1 & -h_j/2 & 0 \\ (s_3)_J & (s_5)_J & (s_1)_J \\ 0 & 1 & 0 \end{vmatrix} \quad B_j \equiv \begin{vmatrix} -1 & -h_j/2 & 0 \\ (s_4)_j & (s_6)_j & (s_1)_j \\ 0 & 0 & 0 \end{vmatrix} \quad 1 \le j \le J \tag{8.2.27b}$$

$$C_j \equiv \begin{vmatrix} 0 & 0 & 0 \\ 0 & 0 & 0 \\ 0 & 1 & -h_{j+1}/2 \end{vmatrix} \quad 0 \le j \le J-1 \tag{8.2.27c}$$

Note that the first two rows of A_0 and C_0 and the last row of A_J and B_J correspond to the boundary conditions [Eq. (8.2.23)]. To solve the continuity and momentum equations for different boundary conditions, only the rows mentioned above need altering.

8.2.3 Block-Elimination Method

The solution of Eq. (8.2.24) can be obtained efficiently and effectively by using the block-elimination method described by Cebeci and Cousteix [1]. According to this method, the solution procedure consists of two sweeps. In the first part of the so-called *forward* sweep, we compute Γ_j, Δ_j from the recursion formulas given by

$$\Delta_0 = A_0 \tag{8.2.28a}$$

$$\Gamma_j \Delta_{j-1} = B_j \qquad j = 1,\ 2, \ldots, J \tag{8.2.28b}$$

$$\Delta_j = A_j - \Gamma_j C_{j-1} \qquad j = 1,\ 2, \ldots, J \tag{8.2.28c}$$

where the Γ_j matrix has the same structure as B_j. In the second part of the forward sweep, we compute $\tilde{w}_j$ from the following relations

$$\tilde{w}_0 = \tilde{r}_0 \tag{8.2.29a}$$

$$\tilde{w}_j = \tilde{r}_j - \Gamma_j \tilde{w}_{j-1} \quad 1 \le j \le J \tag{8.2.29b}$$

In the so-called *backward* sweep, we compute $\tilde{\delta}_j$ from the recursion formulas given by

$$\Delta_J \vec{\delta}\ J = \vec{w}_J \tag{8.2.30a}$$

$$\Delta_j \vec{\delta}_j = \vec{w}_j - C_j \vec{\delta}_{j+1} \qquad j = J-1,\ J-2, \ldots, 0 \tag{8.2.30b}$$

The block elimination method is a general one and can be used to solve any system of first-order equations. The amount of algebra in solving the recursion formulas given by Eqs. (8.2.28) to (8.2.30), however, depends on the order of the matrices A_j, B_j, C_j. When it is small, the matrices Γ_j, Δ_j and the vector $\vec{w}_j$ can be obtained by relatively simple expressions, as discussed in subsection 8.2.4. However, this procedure, though very efficient, becomes increasingly tedious as the order of matrices increases and requires the use of an algorithm that reduces the algebra internally. A general algorithm, called the "matrix solver" discussed by Cebeci and Cousteix [1] and also in subsection 10.9.7 can be used for this purpose.

8.2.4 Subroutine SOLV3

The solution of Eq. (8.2.24) by the block-elimination method can be obtained by using the recursion formulas given by Eqs. (8.2.28) to (8.2.30), and determining the expressions such as Δ_j, Γ_j and $\vec{w}_j$ and $\vec{\delta}_j$. To describe the procedure let us first consider Eq. (8.2.28). Noting that the Γ_j matrix has the same structure as B_j and denoting the elements of Γ_j by γ_{ik} $(i,\ k = 1,\ 2,\ 3)$, we can write Γ_j as

$$\Gamma_j \equiv \begin{vmatrix} (\gamma_{11})_j & (\gamma_{12})_j & (\gamma_{13})_j \\ (\gamma_{21})_j & (\gamma_{22})_j & (\gamma_{23})_j \\ 0 & 0 & 0 \end{vmatrix} \tag{8.2.31a}$$

Similarly, if the elements of Δ_j are denoted by α_{ik} we can write Δ_j as [note that the third row of Δ_j follows from the third row of A_j according to Eq. (8.2.28c)]

$$\Delta_j \equiv \begin{vmatrix} (\alpha_{11})_j & (\alpha_{12})_j & (\alpha_{13})_j \\ (\alpha_{21})_j & (\alpha_{22})_j & (\alpha_{23})_j \\ 0 & -1 & -h_{j+1/2} \end{vmatrix} \qquad 0 \le j \le J-1 \tag{8.2.31b}$$

and for $j = J$, the first two rows are the same as the first two rows in Eq. (8.2.31b), but the elements of the third row, which correspond to the boundary conditions at $j = J$, are (0, 1, 0).

For $j = 0$, $\Delta_0 = A_0$; therefore the values of $(\alpha_{ik})_0$ are

$$\begin{aligned} (\alpha_{11})_0 &= 1 & (\alpha_{12})_0 &= 0 & (\alpha_{13})_0 &= 0 \\ (\alpha_{21})_0 &= 0 & (\alpha_{22})_0 &= 1 & (\alpha_{23})_0 &= 0 \end{aligned} \tag{8.2.32a}$$

and the values of (γ_{ik}) are

$$\begin{aligned} (\gamma_{11})_1 &= -1 & (\gamma_{12})_1 &= -\frac{1}{2}h_1 & (\gamma_{13})_1 &= 0 \\ (\gamma_{21})_1 &= (s_4)_1 & (\gamma_{23})_1 &= -2\left[\frac{(s_2)_1}{h_1}\right] & (\gamma_{22})_1 &= (s_6)_1 + (\gamma_{23})_1 \end{aligned} \tag{8.2.32b}$$

The elements of the Δ_j matrices are calculated from Eq. (8.2.28c). Using the definitions of A_j, Γ_j and C_{j-1}, we find from Eq. (8.2.28c) that for $j = 1, 2,\ldots, J$,

$$(\alpha_{11})_j = 1 \qquad (\alpha_{12})_j = -\frac{h_j}{2} - (\gamma_{13})_j \qquad (\alpha_{13})_j = \frac{h_j}{2}(\gamma_{13})_j$$

$$(\alpha_{21})_j = (s_3)_j \qquad (\alpha_{22})_j = (s_5)_j - (\gamma_{23})_j \qquad (\alpha_{23})_j = (s_1)_j + \frac{h_j}{2}(\gamma_{23})_j \tag{8.2.33a}$$

To find the elements of the Γ_j matrices, we use Eq. (8.2.28b). With Δ_j defined by Eq. (8.2.31b) and B_j by Eq. (8.2.27b), it follows that for $2 \le j \le J$,

$$(\gamma_{11})_j = \left\{(\alpha_{23})_{j-1} + \frac{h_j}{2}\left[\left(\frac{h_j}{2}\right)(\alpha_{21})_{j-1} - (\alpha_{22})_{j-1}\right]\right\}\Big/\Delta_0$$

$$(\gamma_{12})_j = -\left\{\frac{h_j}{2}\frac{h_j}{2} + (\gamma_{11})_j\left[(\alpha_{12})_{j-1}\frac{h_j}{2} - (\alpha_{13})_{j-1}\right]\right\}\Big/\Delta_0$$

$$(\gamma_{13})_j = \left[(\gamma_{11})_j(\alpha_{13})_{j-1} + \left(\gamma_{12}\right)(\alpha_{23})_{j-1}\right]\Big/\frac{h_j}{2}$$

$$(\gamma_{21})_j = \left\{(s_2)_j(\alpha_{21})_{j-1} - (s_4)_j(\alpha_{23})_{j-1} + \frac{h_j}{2}\left[(s_4)_j(\alpha_{22})_{j-1} - (s_6)_j(\alpha_{21})_{j-1}\right]\right\}\Big/\Delta_0$$

$$(\gamma_{22})_j = \left\{(s_6)_j\frac{h_j}{2} - (s_2)_j + (\gamma_{21})_j\left[(\alpha_{13})_{j-1} - (\alpha_{12})_{j-1}\frac{h_j}{2}\right]\right\}\Big/\Delta_1 \tag{8.2.33b}$$

$$(\gamma_{23})_j = (\gamma_{21})_j(\alpha_{12})_{j-1} + (\gamma_{22})_j(\alpha_{22})_{j-1} - (s_6)_j$$

$$\Delta_0 = (\alpha_{13})_{j-1}(\alpha_{21})_{j-1} - (\alpha_{23})_{j-1}(\alpha_{11})_{j-1} - \frac{h_j}{2}\left[(\alpha_{12})_{j-1}(\alpha_{21})_{j-1} - (\alpha_{22})_{j-1}(\alpha_{11})_{j-1}\right]$$

$$\Delta_1 = (\alpha_{22})_{j-1}\frac{h_j}{2} - (\alpha_{23})_{j-1}$$

To summarize the calculation of Γ_j and Δ_j matrices, we first calculate α_{ik} from Eq. (8.2.31b) for $j = 0$, γ_{ik} from Eq. (8.2.32b) for $j = 1$, α_{ik} from Eq. (8.2.33a) for $j = 1$, then γ_{ik} from Eq. (8.2.33b) for $j = 2$, α_{ik} from Eq. (8.2.33a) for $j = 2$, then γ_{ik} from Eq. (8.2.33b), α_{ik} from Eq. (8.2.33a) for $j = 3$, etc.

In the second part of the forward sweep we compute $\vec{w}_j$ from the relations given by Eq. (8.2.29). If we denote the components of the vector $\vec{w}_j$ by

$$\vec{w}_j \equiv \begin{vmatrix} (w_1)_j \\ (w_2)_j \\ (w_3)_j \end{vmatrix} \qquad 0 \le j \le J \tag{8.2.34}$$

Then it follows from Eq. (8.2.29a) that for $j = 0$,

$$(w_1)_0 = (r_1)_0 \; (w_2)_0 = (r_2)_0 \; (w_3)_0 = (r_3)_0 \tag{8.2.35a}$$

and from Eq. (8.2.29b) for $1 \le j \le J$,

$$\begin{aligned} (w_1)_j &= (r_1)_j - (\gamma_{11})_j (w_1)_{j-1} - (\gamma_{12})_j (w_2)_{j-1} - (\gamma_{13})_j (w_3)_{j-1} \\ (w_2)_j &= (r_2)_j - (\gamma_{21})_j (w_1)_{j-1} - (\gamma_{22})_j (w_2)_{j-1} - (\gamma_{23})_j (w_3)_{j-1} \\ (w_3)_j &= (r_3)_j \end{aligned} \tag{8.2.35b}$$

In the backward sweep, $\vec{\delta}_j$ is computed from the formulas given by Eq. (8.2.30). With the definitions of $\vec{\delta}_j$, Δ_j and $\vec{w}_j$, it follows from Eq. (8.2.30a) that

$$\delta u_J = (w_3)_J \tag{8.2.36a}$$

$$\delta v_J = \frac{e_2(\alpha_{11})_J - e_1(\alpha_{21})_J}{(\alpha_{23})_J(\alpha_{11})_J - (\alpha_{13})_J(\alpha_{21})_J} \tag{8.2.36b}$$

$$\delta f_J = \frac{e_1 - (\alpha_{13})_J \delta v_J}{(\alpha_{11})_J} \tag{8.2.36c}$$

where

$$\begin{aligned} e_1 &= (w_1)_J - (\alpha_{12})_J \delta u_J \\ e_2 &= (w_2)_J - (\alpha_{22})_J \delta u_J \end{aligned}$$

The components of $\vec{\delta}$, for $j = J-1, J-2, \ldots, 0$, follow from Eq. (8.2.30b)

$$\delta v_j = \frac{(\alpha_{11})_j \left[(w_2)_j + e_3 (\alpha_{22})_j \right] - (\alpha_{21})_j (w_1)_j - e_3 (\alpha_{21})_j (\alpha_{12})_j}{\Delta_2} \tag{8.2.37a}$$

$$\delta u_j = -\frac{h_{j+1}}{2} \delta v_j - e_3 \tag{8.2.37b}$$

$$\delta f_j = \frac{(w_1)_j - (\alpha_{12})_j \delta u_j - (\alpha_{13})_j \delta v_j}{(\alpha_{11})_j} \tag{8.2.37c}$$

where

$$\begin{aligned} e_3 &= (w_3)_j - \delta u_{j+1} + \frac{h_{j+1}}{2} \delta v_{j+1} \\ \Delta_2 &= (\alpha_{21})_j (\alpha_{12})_j \frac{h_{j+1}}{2} - (\alpha_{21})_j (\alpha_{13})_j \\ &\quad - \frac{h_{j+1}}{2} (\alpha_{22})_j (\alpha_{11})_j + (\alpha_{23})_j (\alpha_{11})_j \end{aligned} \tag{8.2.37d}$$

To summarize, one iteration of Newton's method is carried out as follows. The vectors $\vec{r}_j$ defined in Eq. (8.2.25) are computed from Eq. (8.2.21) by using the latest iterate. The matrix elements of A_j, B_j and C_j defined in Eq. (8.2.24) are next determined by Eq. (8.2.22a) to (8.2.22f). Using the relations in Eqs. (8.2.27) and (8.2.28), the matrices Γ_j and Δ_j and vectors $\vec{w}_j$ are calculated. The matrix elements for Γ_j defined in Eq. (8.2.31a) are determined from Eq. (8.2.32b) and (8.2.33b). The components of the vector $\vec{w}_j$ defined in Eq. (8.2.34) are determined from Eq. (8.2.35). In the backward sweep, the components of $\vec{\delta}_j$ are computed from Eqs. (8.2.36) and (8.2.37). A subroutine which makes use of these formulas and called SOLV3 is given on the companion site, store.elsevier.com/components/9780080983356.

8.3 Prediction of Two-Dimensional Incompressible Flows

In this section and the following sections up to Section 8.7, we discuss the prediction of incompressible and compressible, two-dimensional and axisymmetric flows with the numerical method described in the previous section and the CS algebraic eddy-viscosity formulation in Chapter 5. These results were all obtained previously without improvements to the CS model proposed by Cebeci and Chang for strong pressure-gradient flows as discussed in Section 5.4. The predictions of the CS model with the Cebeci-Chang improvement are described in Sections 8.7 and 8.8 for two-dimensional and three-dimensional flows with separation.

8.3.1 Impermeable Surface with Zero Pressure Gradient

We first examine the accuracy of the momentum equation and compute the local skin-friction coefficients for a Reynolds number of $10^6 \leq R_x \leq 10^9$. We choose a unit Reynolds number of 1×10^6 ($u_e = 160$ ft/sec, $\nu = 1.6 \times 10^{-4}$ ft^2/sec) and specify the flow as turbulent at $x = 0.01$ ft. The boundary-layer thickness η_e is calculated in the program as the calculations proceed downstream.

Figure 8.2 shows the results with two different $\Delta\eta$ and Δx spacings (see Section 12.3). Figure 8.2 shows the results with fixed $\Delta\eta$ spacing ($h_1 = 0.002$, $K = 1.226$) and with variable Δx spacing. The latter was chosen to be such that starting from $R_x = 10^6$, the ΔR_x spacing was $2^{n/2} \times 10^6$, $2^{n/4} \times 10^6$, yielding approximately 20 and 40 x-stations, respectively, at $R_x = 10^9$. The c_f values shown in Fig. 8.2a do not seem to be very sensitive to Δx spacing. On the other hand, the computed values of transformed boundary-layer thickness (Fig. 8.2b) show appreciable irregularity. The number of η points J remains approximately constant.

Figure 8.3 shows a comparison between the calculations obtained with the CS method (present method) and experimental data for incompressible flows with zero pressure gradient. Figure 8.3a presents the local skin friction results for the data of

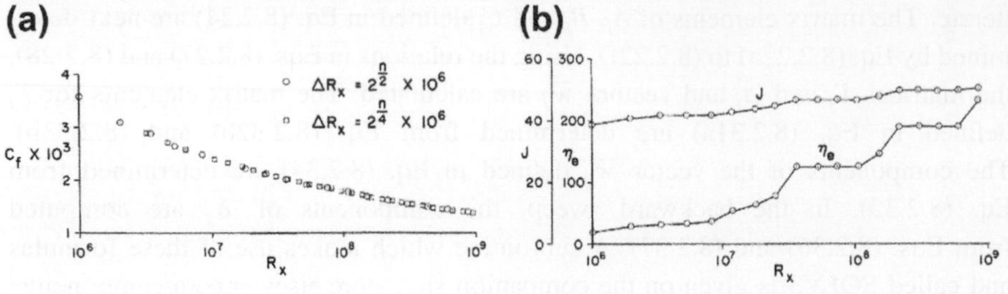

Fig. 8.2 Flat-plate flow: effect of Δx spacing on the computed results. Calculations were made for a fixed spacing. (a) c_f values, (b) η_e values and J, the number of points across the boundary layer.

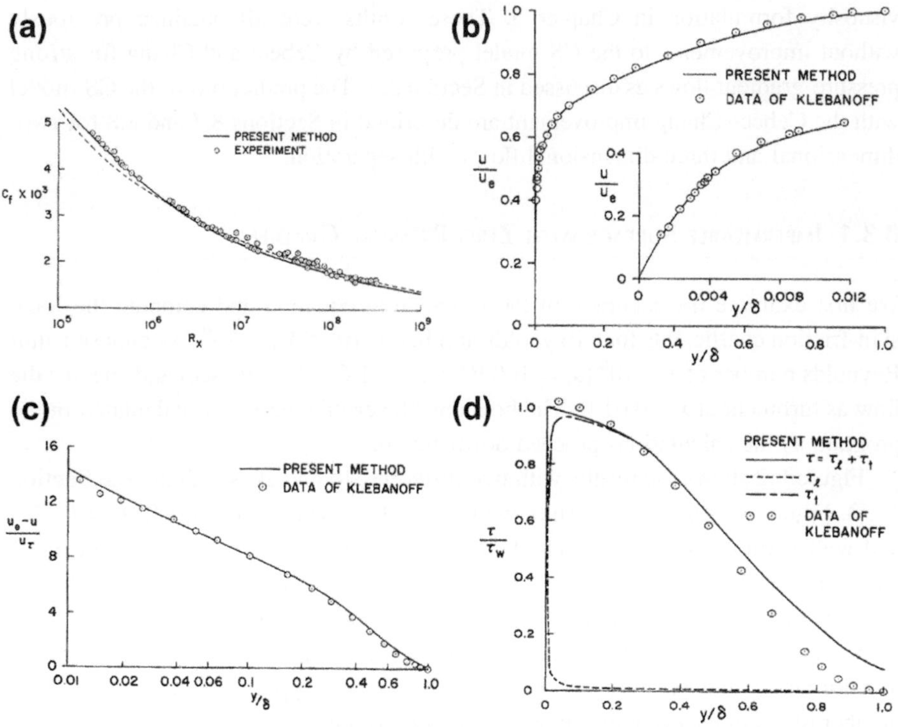

Fig. 8.3 Comparison of calculated and experimental (a) local skin-friction coefficients, (b, c) velocity profiles and (d) shear-stress distributions with experiment. The dashed line in Fig. 8.3a is the solid line of Fig. 7.1.

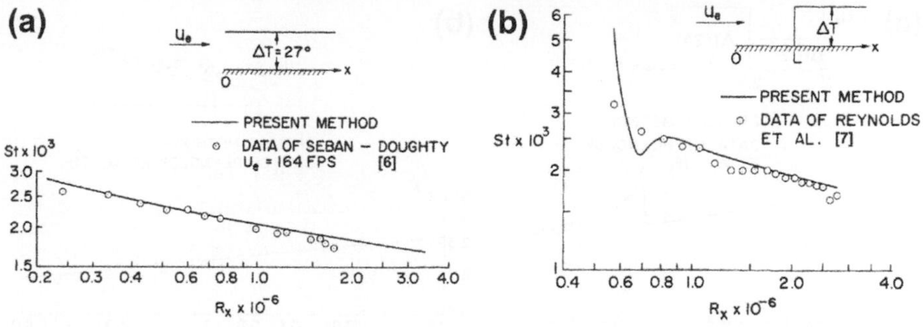

Fig. 8.4 Comparison of calculated and experimental Stanton numbers (a) uniform and (b) step variation of wall temperature.

Coles [3], Figs. 8.3b and 8.3c the velocity profiles and Fig. 8.3d the shear stress profiles, all for the data of Klebanoff [4].

In plotting the velocity profiles in Figs. 8.3b and 8.3c, we have used the definition of $\delta \left(\equiv \frac{\kappa(\delta^* u_e)}{u_\tau(1+\Pi)} \right)$, which can also be written as (with $\Pi = 0.55$)

$$\delta = (2/c_f)^{1/2}\left(\delta^*/3.78\right) = 0.375\left(\delta^*/c_f^{1/2}\right) .$$

That was necessary because Klebanoff's experimental data are for a unit Reynolds number of approximately 3×10^5. Furthermore, the above relation has the advantage that it eliminates the difficulty of dealing with an ill-defined quantity δ.

We now study the accuracy of the energy equation for smooth impermeable walls with different wall-temperature distributions, and compare the computed local Stanton number with experiment. For more comparisons, see [5].

Figure 8.4 shows the results for (a) an isothermal heated plate and (b) step variation of wall temperature. Similarly Fig. 8.5 shows the results for (a) double-step and (b) step-ramp wall temperatures.

8.3.2 Permeable Surface with Zero Pressure Gradient

Again we first study the accuracy of the momentum equation, this time for flows with suction and blowing for a wide range of mass-transfer parameters F defined as $F = (\varrho v)_w/\varrho_e u_e$, which for incompressible flows is simply $F = v_w/u_e$.

Figure 8.6 shows the computed velocity profiles for the boundary layer measured by (a) McQuaid [9] for $F = 0.0046$ and (b) Simpson et al. [10] for $F = 0.00784$. Figure 8.6b also shows a comparison between the calculated local skin-friction values and the values given by Simpson et al. Simpson's values were obtained by

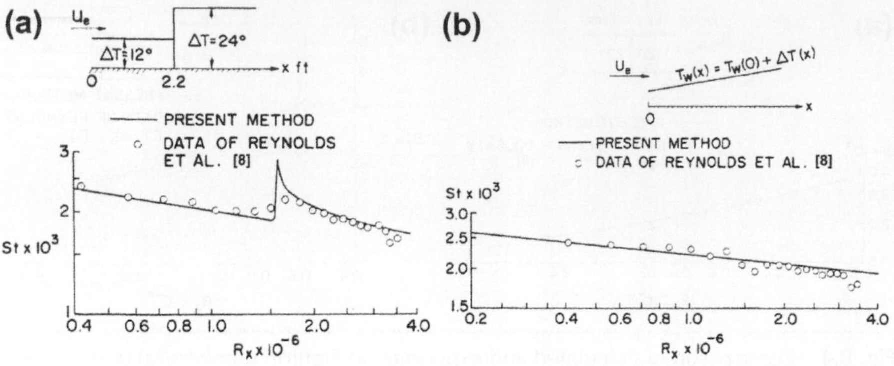

Fig. 8.5 Comparison of calculated and experimental Stanton numbers for (a) double-step and (b) step-ramp variation of wall temperature.

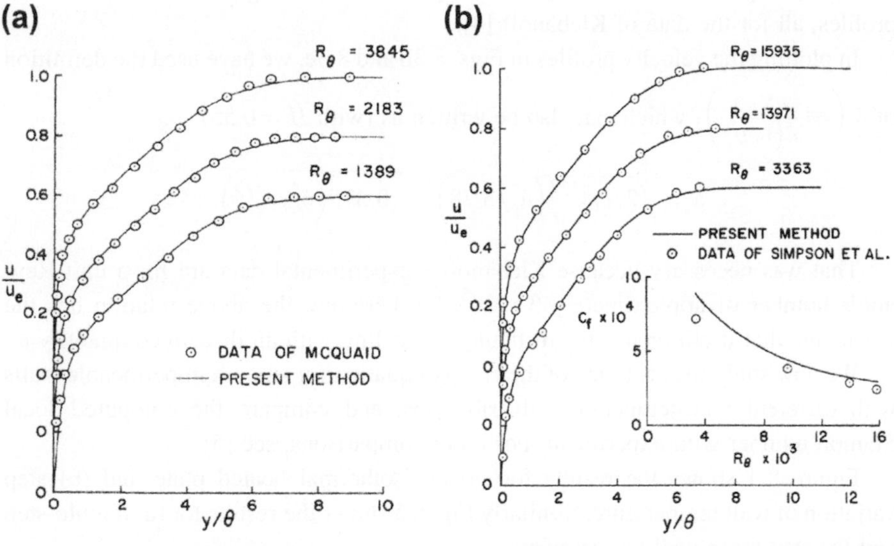

Fig. 8.6 Comparison of calculated velocity profiles for (a) $F = 0.0046$, $u_e = 50$ ft/sec and (b) $F = 0.00784$ with experiment.

using the momentum integral equation and by a method based on a viscous-sublayer model.

Figure 8.7 shows the results for a flow involving discontinuous injection. The experimental data are due to McQuaid [9]. The calculations were started by using the experimental velocity profile at $x = 0.958$ ft and were continued downstream with a uniform injection rate, $F = 0.0034$, up to and including $x = 1.460$ ft. The blowing rate was set at zero at $x = 1.460$ ft and at all subsequent downstream locations.

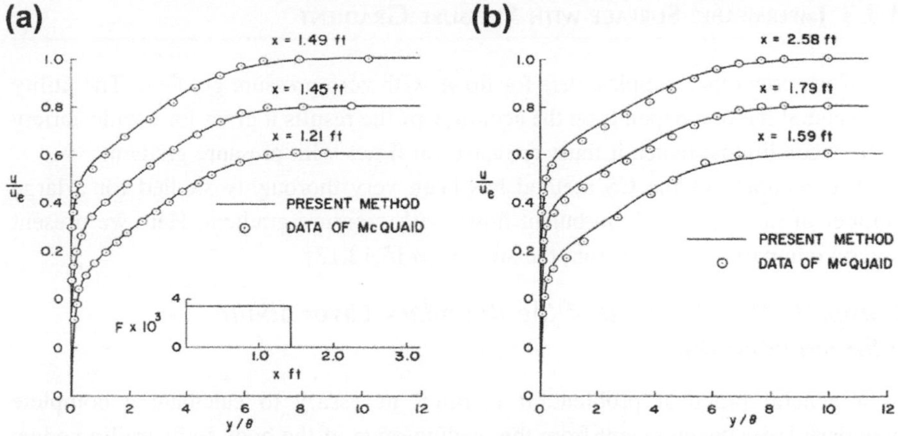

Fig. 8.7 Comparison of calculated velocity profiles for the discontinuous-injection flow measured by McQuaid [9]. Profiles (a) upstream of discontinuity, (b) downstream.

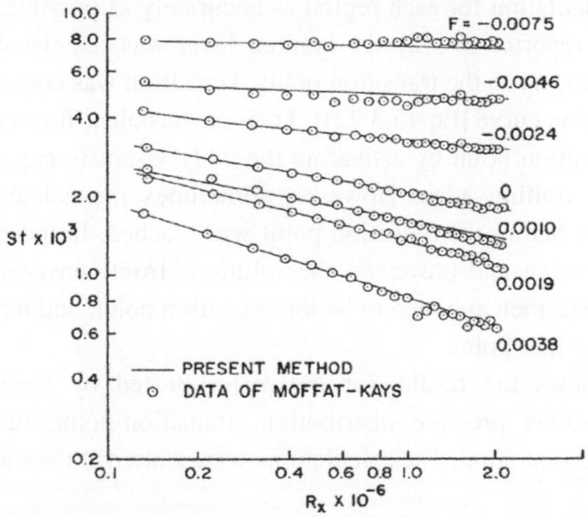

Fig. 8.8 Comparison of calculated and experimental Stanton numbers for a turbulent boundary layer with suction and blowing.

Results show that the experimental trends are closely followed by the calculations, including the results for the region after which the mass transfer is zero.

We now investigate the accuracy of calculating local Stanton number for flows with heat and mass transfer. Figure 8.8 shows the computed results for a wide range of the mass-transfer parameter *F*. The experimental data are due to Moffat and Kays [11]. As the results show, the agreement with experiment is good for all mass-transfer rates.

8.3.3 Impermeable Surface with Pressure Gradient

All of the previous examples were for flows with zero pressure gradient. The utility of a general method depends on the accuracy of the results it gives for a wide variety of flow conditions; hence it must be tested on flows with pressure gradient.

The accuracy of the CS method has been very thoroughly studied for a large number of incompressible turbulent flows with pressure gradient. Here we present several comparisons taken from the studies in [5,12,13].

Complete Development of the Boundary Layer about a Streamlined Body

In a general practical problem, it is often necessary to calculate a complete boundary-layer development from the leading edge of the body to its trailing edge, which means that it is necessary to calculate the laminar layer, locate the transition point, and then calculate the turbulent layer. Thus, for example, if one is interested in calculating the total skin-friction drag of the body, the accuracy of the result depends on doing each calculation for each region as accurately as possible.

In the studies reported in [12], the laminar layer was calculated by solving the governing equations up to the transition point. Transition was computed by Smith's transition-correlation curve [Eq. (5.3.22)]. Then the turbulent-flow calculations were started at the transition point by activating the eddy-viscosity expression and were continued to the trailing edge. However, sometimes the calculations indicated laminar separation before the transition point was reached. In those cases, the wall shear became negative and prevented the solutions from converging; the laminar separation point was then assumed to be tbe transition point, and turbulent flow was assumed to start at that point.

Figure 8.9 shows the results for the airfoil tested by Newman [14]. The measurements include pressure distribution, transition-point, turbulent velocity profiles, and separation point. The calculations were started at the stagnation point of

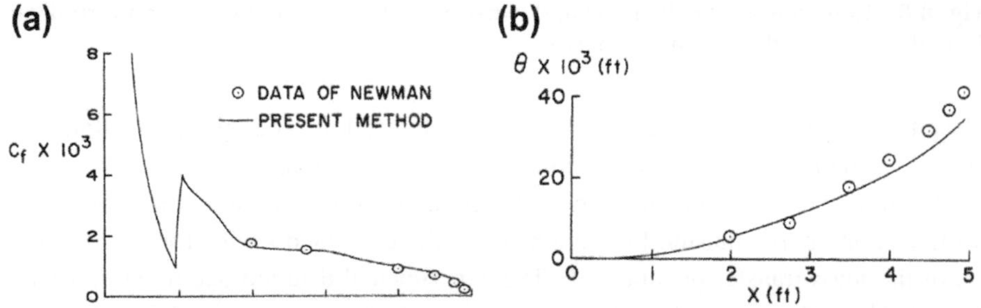

Fig. 8.9 Comparison of calculated and experimental results for Newman's airfoil.

the airfoil. The experimental transition point was at $x = 1.169$ ft, but at $x = 1.009$ ft the calculations predicted laminar-flow separation. Consequently, the transition point was shifted to $x = 1.009$ ft, at which point the turbulent-flow calculations were started without the γ_{tr} term in the eddy-viscosity formulas and were continued until $x = 4.926$ ft. At this point, the calculations predicted turbulent-flow separation, which agreed with the experimental separation point, within the accuracy of the measurement. It is also interesting to note that the calculated result obtained by starting the turbulent flow calculations at $x = 2.009$ ft with the experimental velocity profiles agreed extremely well with those obtained by starting them at the stagnation point.

Equilibrium Flows

Figure 8.10 shows a comparison of calculated and experimental results for an equilibrium flow in a (a) favorable and (b) adverse pressure gradient flow. It can be seen that in general there is a good agreement with data.

Nonequilibrium and Separating Flows

From a practical standpoint, nonequilibrium and separating flows are perhaps the most important flows, since they are so often encountered in the design of diffusers and lifting surfaces. We now present the results for a flow of this type and consider an airfoil-like body that has both favorable and adverse pressure gradients. The body is

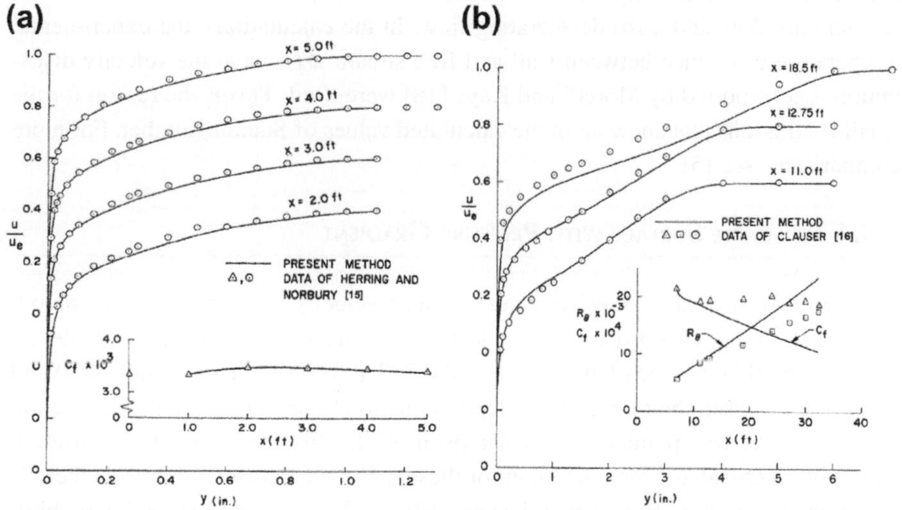

Fig. 8.10 Comparison of calculated results for an equilibrium flow in (a) a favorable pressure gradient $\beta = -0.35$ and (b) an adverse pressure gradient $\beta = 1.8$.

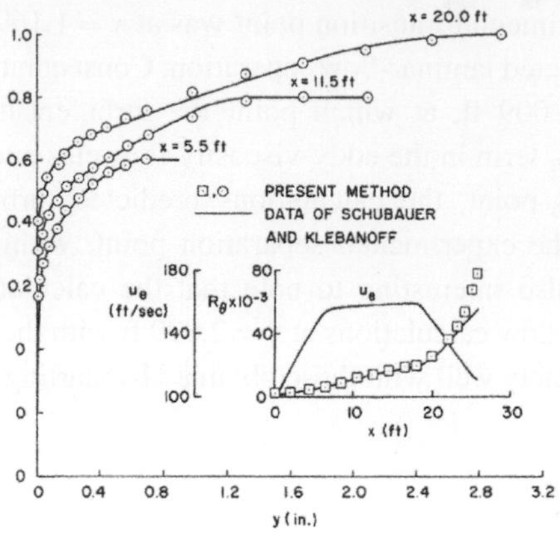

Fig. 8.11 Comparison of calculated and experimental results on a large airfoil-like body [17].

two-dimensional and has a sharp nose. It is at a slight angle of attack, which produces a pressure peak at the leading edge that causes transition. Separation is reported to have taken place at 25.7±0.2 ft from the leading edge. Figure 8.11 shows some computed and experimental velocity profiles. The experimental data are due to Schubauer and Klebanoff [17].

Thermal Boundary Layers

Figure 8.12 shows a comparison of calculated and experimental results for an (a) accelerating flow and a (b) decelerating flow. In the calculations, the experimental temperature difference between wall and free stream $\Delta T(x)$ and the velocity distribution $u_e(x)$ reported by Moretti and Kays [18] were used. This is the reason for the small oscillations that show up in the calculated values of Stanton number. For more comparisons, see [5].

8.3.4 Permeable Surface with Pressure Gradient

McQuaid [9] made an extensive series of mean-velocity measurements on smooth permeable surfaces with distributed injection. He measured boundary-layer developments for blowing rates F between 0 and 0.008 at free-stream velocities of 50 and 150 ft/sec. He used the momentum integral equation to obtain the local skin-friction coefficient. As was pointed out by Simpson et al. [19] the reported skin-friction values of McQuaid are very uncertain for these data. The reasons are that (1) there is variation in injection velocity over the test surface, (2) the usable test section is short, and (3) the fact that F is subtracted from the momentum-thickness gradient to obtain local skin-friction coefficient.

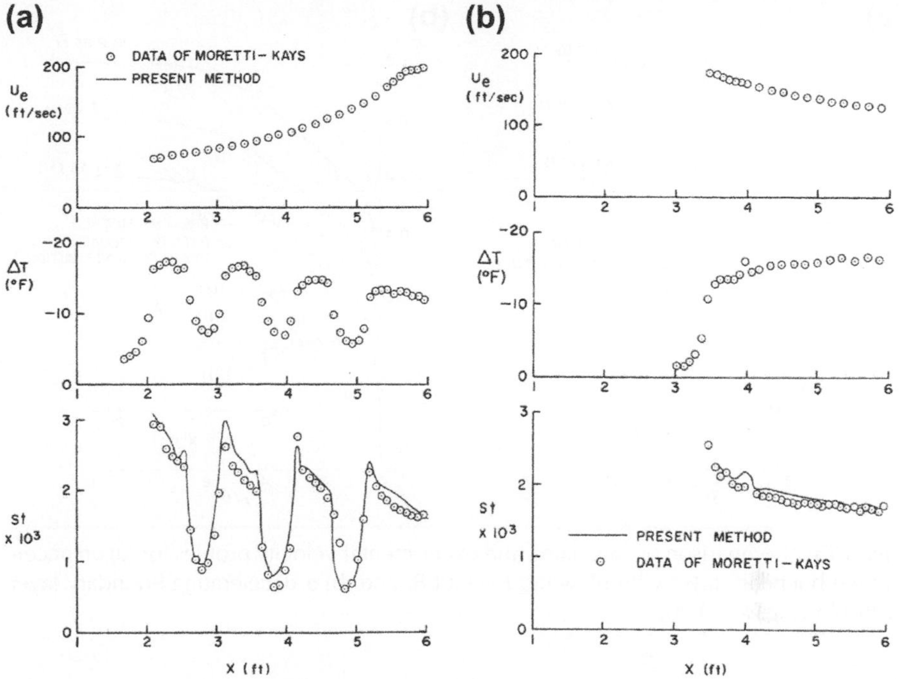

Fig. 8.12 Comparison of calculated and experimental Stanton numbers for (a) an accelerating and (b) a decelerating boundary layer, Runs 24 and 36, respectively.

Figure 8.13a shows experimental and calculated velocity-profile comparisons for a boundary layer in a favorable pressure gradient with $F = 0.008$. The calculations were started by using the experimental velocity profile at $x = 0.958$ ft and were continued downstream with the given blowing rate. Figure 8.13b shows the results for an adverse pressure gradient with $F = 0.002$. The calculations were started by initially matching zero-pressure-gradient data for the given blowing rate at $x = 0.958$ ft and were continued downstream with the experimental velocity distribution. In both of those figures, skin-friction comparisons were omitted because of the uncertainty in experimental c_f data.

Figure 8.14 shows the results for two highly accelerating flows with mass transfer. In these cases, the flow starts in zero pressure gradient, then accelerates, and later ends in zero pressure gradient, all with mass transfer. In both cases, the acceleration parameter K $\left(\equiv \frac{\mu_e}{\varrho_e u_e^2}\frac{du_e}{dx}\right)$ is not very large. The results in Fig. 8.14a show that the CS method computes the accelerating boundary layers with blowing ($F = 0.006$) quite well. Calculated velocity profiles and skin-friction values agree well with experiment. On the other hand, the results in Fig. 8.14b show that the CS method does not compute the highly accelerating boundary layers with suction

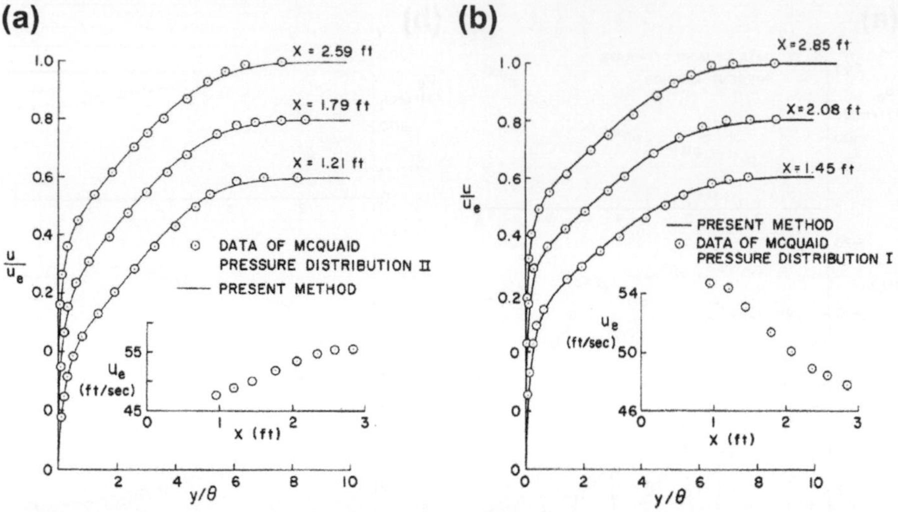

Fig. 8.13 Comparison of calculated and experimental velocity profiles for (a) an accelerating boundary layer with blowing, $F=0.008$, and (b) a decelerating boundary layer with blowing, $F=0.002$.

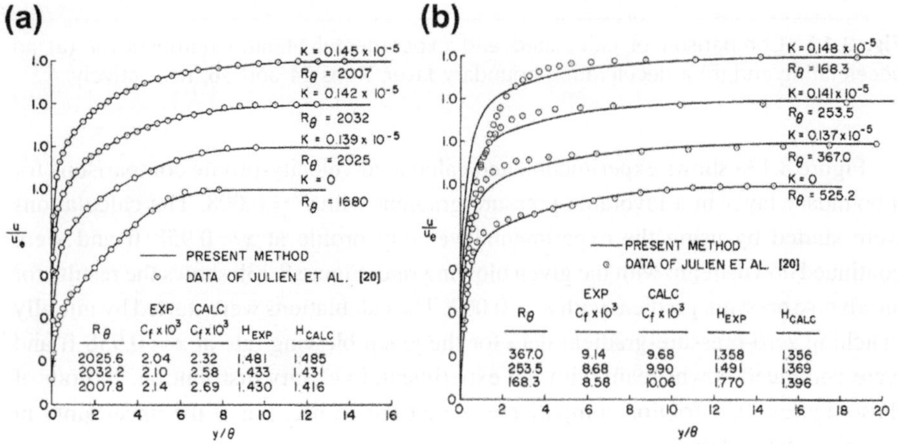

Fig. 8.14 Comparison of calculated and experimental results for two highly accelerating boundary layers with mass transfer: (a) blowing, $F=0.006$; (b) suction, $F=-0.004$.

($F=-0.004$) well. It should be noted, however, that for that suction case the Reynolds number is quite low and that with acceleration and suction, the boundary layer, which initially had an already low Reynolds number, has no doubt laminarized.

8.4 Axisymmetric Incompressible Flows

We first consider flows with zero pressure gradient, namely, flows past slender cylinders. As was discussed in subsection 5.3.2, in such flows the transverse-curvature effect plays an important role and causes the boundary-layer development to be significantly different from those in flat-plate flows. The difference increases with the slenderness of the cylinder.

Figure 5.4 in subsection 5.3.2 shows the velocity profiles and Table 8.1 presents the local skin-friction values for two cylinders of diameters $d = 1$ in. and 0.024 in. The experimental data are due to Richmond [21]. The calculations were made by Cebeci [22,23] and by White [24]. The values of Richmond, which were estimated by using the "streamline hypothesis" for the 0.024-in. cylinder, have been pointed out by Rao [22] to be in error by a factor of 2. White's values were obtained by solving the momentum integral equation. They are given by

$$c_f = \begin{cases} 0.0015 + \left[0.20 + 0.016(x/r_0)^{0.4}\right] R_x^{-1/3}, & 10^6 \le R_x \le 10^9 \\ (4/R_{r_0})\left[(1/G) + (0.5772/G^2)\right], & (x/r_0)\langle 5 \times 10^5,\ R_{r_0} \le 25 \end{cases},$$

where $G = \ln(4R_x/R_{r_0}^2)$ and $R_{r0} = u_e r_0/\nu$. White also gives the following equations for the average skin-friction coefficient $\bar{c}_f$:

$$\bar{c}_f = \begin{cases} 0.0015 + \left[0.34 + 0.07(L/r_0)^{0.4}\right] R_L^{-1/3}, & 10^6 \le R_L \le 10^9 \\ (4/R_{r_0})\left[(1/G) + (0.5772/G^2)\right], & (L/r_0) \le 10^6,\ G \ge 6,\quad R_{r_0} \le 20, \end{cases}$$

where $G = \ln(4R_L/R_{r_0}^2)$ and $R_L = u_e/\nu$. Here L is the length of the cylinder.

The values of Cebeci [22,23] were obtained by using the CS method with a two-dimensional mixing-length expression, Eq. (5.2.12), instead of the expression given by Eq. (5.2.13). As may be seen from the results in Table 8.1, the Cebeci [23] values agree closely with White's values. Furthermore, the value of the local skin friction for the 1-in. cylinder remains unchanged from that obtained earlier by

TABLE P8.1 Comparison of calculated c_f values with other reported values for incompressible flows on slender cylinders.

			$c_f \times 10^3$			
d (in.)	$R_{\theta_{3-d}}$	δ/r_0	Richmond [21]	White [24]	Cebeci [22]	Cebeci [23]
0.024	2100	75	4.95	7.71	10.73	8.21
1.00	8750	2	2.90	3.18	3.03	3.02

Cebeci [22]. However, for the 0.024-in. cylinder, the calculated c_f differs from the earlier one by 30%.

Figure 8.15 shows the results for axisymmetric flows with pressure gradient. Figure 8.15a shows the results for a 285-ft-long airship with fineness ratio of 4.2. The experimental data are due to Cornish and Boatwright [25]; the measurement was along the top, where the flow should be nearly axisymmetric. The pressure distribution and boundary-layer measurements were made in flight at speeds from 35 to 70 mph. No transition data were given, but from the configuration of the airship it was inferred that the boundary layer was tripped at approximately $x/L = 0.05$. The example is of great importance because of the very large Reynolds number. The good agreement establishes validity of the CS method at large scale.

Figure 8.15b shows the results for the axisymmetric bodies measured by Murphy [26]. The experimental data include pressure distributions, skin-friction coefficients, velocity profiles, and very carefully determined separation-point locations.

The calculations were made for three different shapes that represented a combination of one basic nose shape (A-2), a constant-area section, and different tail shapes (Tails A-2, C-2, and C-4). Transition was tripped at an axial location 31 in. from the nose of the body by a 2-in.-wide porous strip, which was used for mass-transfer measurements (sealed for zero mass transfer). The skin-friction coefficients were obtained by Preston tube, and experimental total-drag coefficients were obtained from the wake profile. As can be seen from the results in Fig. 8.15b the

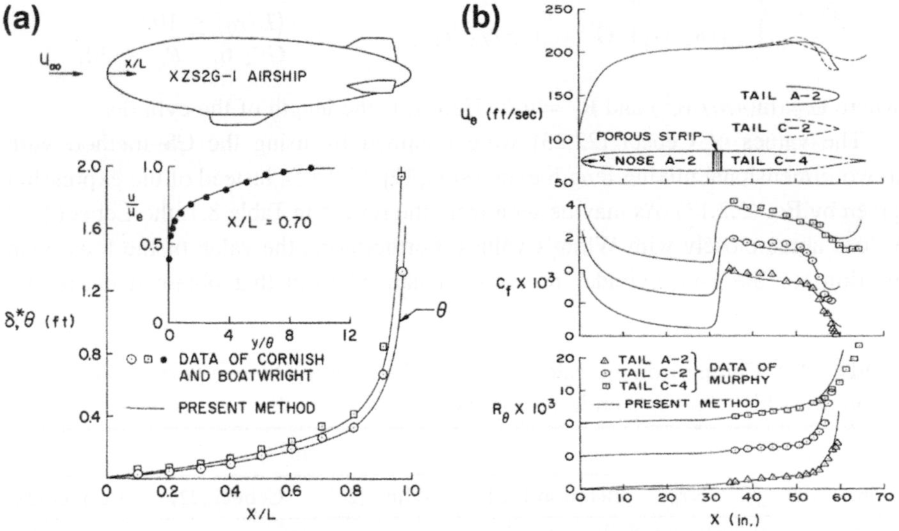

Fig. 8.15 Comparison of calculated and experimental results for the axisymmetric bodies measured by (a) Cornish and Boatwright and (b) Murphy.

agreement is quite good, considering the fact that the calculations were started at the stagnation point and transition was specified at 31 in. The values calculated in this manner match the experimental values; a slight discrepancy in skin friction may be attributed to the effect of the porous strip.

8.5 Two-Dimensional Compressible Flows

8.5.1 Impermeable Surface with Zero Pressure Gradient

The accuracy of the CS method has been studied for several compressible turbulent boundary layers with heat and mass transfer for Mach numbers up to 7 [5,22,27]. Because of the scarcity of experimental data with pressure gradients, most of the data considered in the studies have been restricted to zero-pressure-gradient flows.

Adiabatic Flows

A considerable amount of data exists on adiabatic turbulent boundary layers with zero pressure gradient. The data consist of accurate velocity profiles, Mach profiles, and local skin-friction values, mostly for Mach numbers up to 5. Here we shall present several comparisons of calculated and experimental results taken from the study of such flows reported in [22].

Figure 8.16a shows a comparison of calculated and experimental velocity and Mach-number profiles and local skin-friction coefficients for the boundary layer measured by Coles [28]. Skin-friction coefficients were measured by floating element. Computations for that flow and for the flows to be discussed next were made by starting the flow as compressible laminar at $x=0$ and specifying that the flow become turbulent at the next x station, which was arbitrarily taken to be at $x=0.001$ ft. The computations were than continued on downstream until the experimental R_θ was obtained. Where experimental values of R_θ were not reported, the same procedure was used in matching the experimental R_x. Then calculated results at that x location were compared with the experimental data.

Figure 8.16b shows a comparison of calculated and experimental velocity and Mach profiles for the boundary layer measured by Matting et al. [29].

Figure 8.17a shows a comparison of skin-friction values for the boundary layer measured by Moore and Harkness [30] at a nominal $M_e=2.8$: The agreement is good, even at very high Reynolds numbers. The experimental skin friction was obtained by floating element.

Figure 8.17b shows a summary of calculated and experimental skin-friction coefficients for compressible adiabatic turbulent zero-pressure-gradient flows

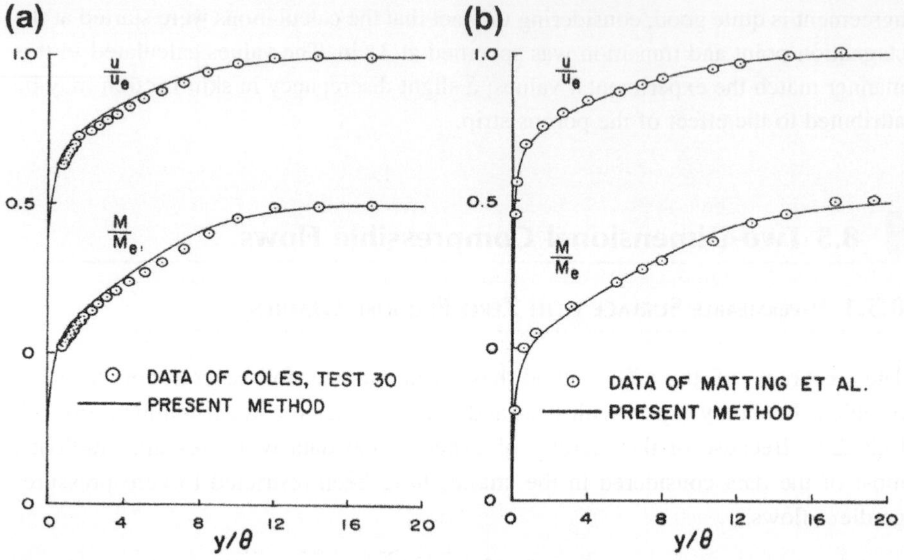

Fig. 8.16 Comparison of calculated and experimental results for the data of (a) Coles and (b) Matting et al.

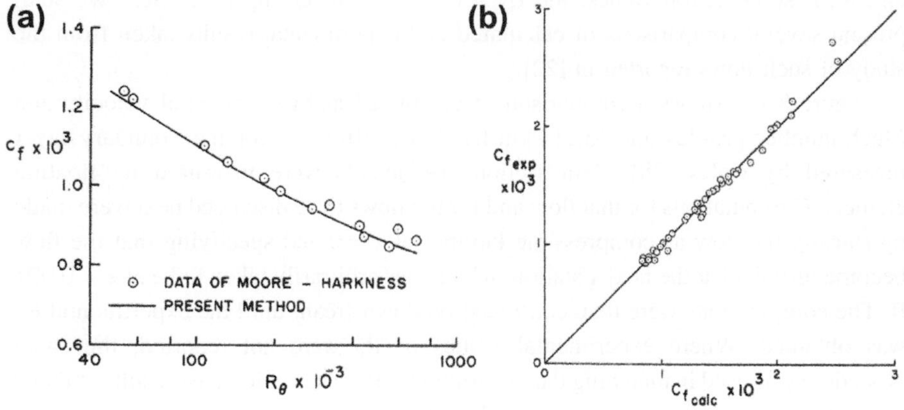

Fig. 8.17 Comparison of calculated and experimental local skin-friction values for adiabatic zero-pressure gradient flows (a) data of Moore and Harkness and (b) 43 experimental values.

studied by Cebeci et al. [22]. The experimental values of skin friction were all measured by the floating-element technique. The calculated values cover a Mach number range of 0.40 to 5 and a momentum-thickness Reynolds number range of 1.6×10^3 to 702×10^3. The rms error based on 43 experimental values, all obtained by the floating-element technique, is 3.5%, which is within the experimental scatter.

Flows with Heat Transfer

We now present several comparisons of calculated and experimental results for compressible turbulent flows with heat transfer taken from the studies of Cebeci [5,27]. Figure 8.18a shows a comparison of calculated and experimental Mach profiles for the boundary layer measured by Michel [31] at a Mach number of 2.57. The calculations were made by starting the flow as compressible laminar at $x = 0$ and specifying that the flow become turbulent at the next x station, which was arbitrarily taken to be at $x = 0.001$ ft for $T_w/T_e = 1.95$, which was assumed constant along the plate. The computations were then continued downstream until the experimental R_x was obtained.

Figure 8.18b shows a comparison of calculated and experimental results for the boundary layer measured by Pappas [32] for $M_e = 2.27$ and $T_w/T_e = 2.16$. Again the calculations were made by assuming the flow to be compressible laminar at $x = 0$ and specifying that the flow become turbulent at the next x station ($x = 0.01$ ft). The experimental Reynolds number based on momentum thickness varied between 3500 and 9500. For that reason, the calculated Mach profiles shown in Fig. 8.18b were compared with the experimental data for $R_\theta = 3500$ and 9500. The agreement is good, and the calculations account for the R_θ effect. The Stanton number was calculated from the formula

$$\mathrm{St} = -q_w/\varrho_e u_e (H_{aw} - H_w).$$

The adiabatic wall enthalpy H_{aw} was obtained by repeating the calculations for adiabatic flow.

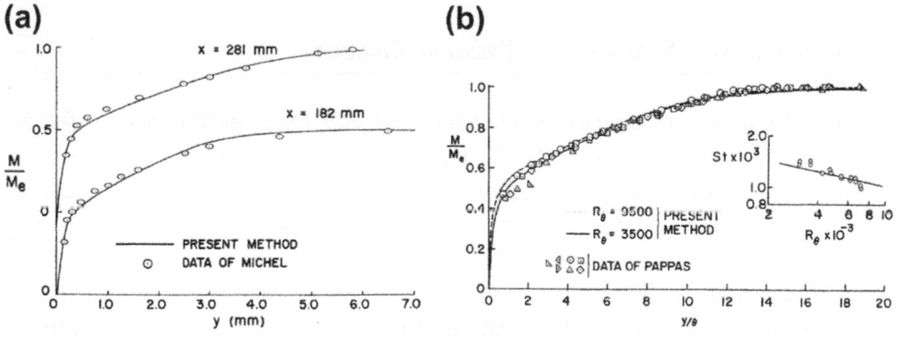

Fig. 8.18 Comparison of calculated and experimental results for a zero-pressure-gradient flow with heat transfer. (a) $T_w/T_e = 1.95$; $M_e = 2.57$ and (b) $T_w/T_e = 2.16$; $M_e = 2.27$.

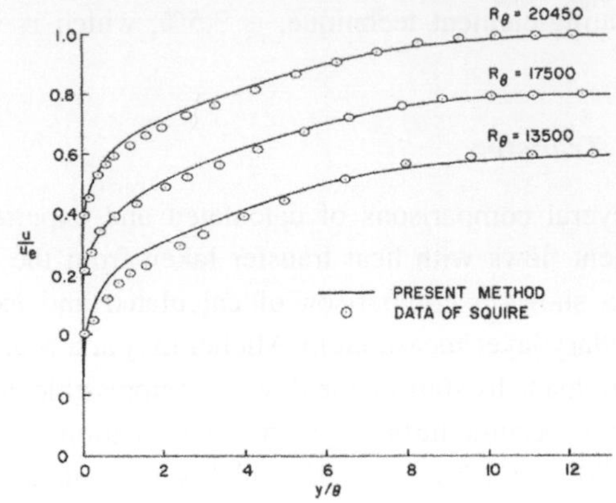

Fig. 8.19 Comparison of calculated and experimental results for an adiabatic flat-plate flow with mass transfer. $M_e = 1.8$; $F = 0.0013$.

8.5.2 Permeable Surface with Zero Pressure Gradient

Figure 8.19 shows a comparison of calculated and experimental velocity profiles for the adiabatic boundary layer measured by Squire [33]. The calculations were made for a blowing rate $F = 0.0013$ with $M_e = 1.8$, and they were started by initially matching the experimental momentum thickness at the first measuring station, $x = 8.6$ in. downstream of the leading edge of the porous plate. Comparison of calculated and experimental skin-friction values are omitted in Fig. 8.19 because the experimental c_f values, which were derived from a momentum balance, are subject to large errors, see [34]. As can be seen, the agreement of the velocity profiles with experiment is quite good. Similar good agreement with experiment was also reported for the higher blowing rates studied by Thomas et al. [34].

8.5.3 Impermeable Surface with Pressure Gradient

Figure 8.20 shows a comparison of calculated and experimental results for an accelerating adiabatic flow measured by Pasiuk et al. [35]. Calculations were started by assuming an adiabatic flat-plate flow that matched the experimental momentum thickness value at $x = 0.94$ ft. Then the experimental Mach-number distribution was used to compute the rest of the flow. The edge Mach number varied from $M_e = 1.69$ at $x = 0.94$ ft to $M_e = 2.97$ at $x = 3.03$ ft. Figures 8.20a and 8.20b show a comparison of calculated and experimental velocity and temperature profiles together with c_f values, respectively, for three x stations.

Figure 8.21 shows a comparison of calculated and experimental results for an accelerating flow with constant heat flux. Again, the measurements are due to

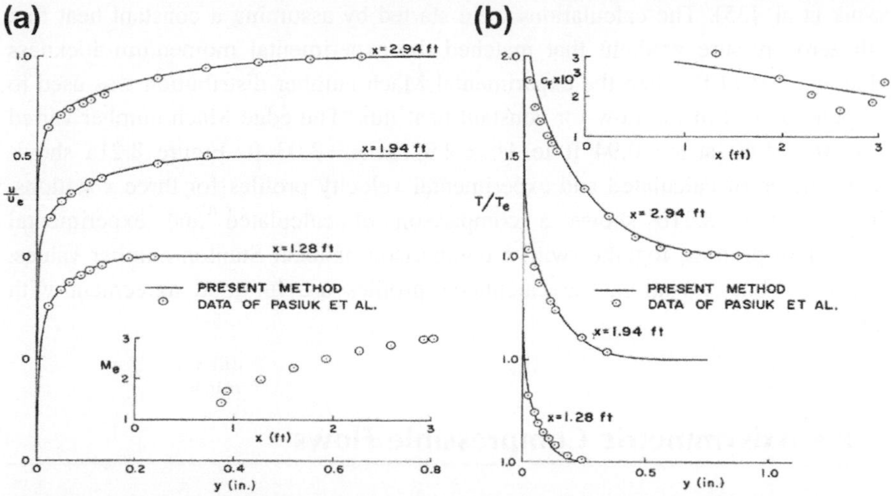

Fig. 8.20 Comparison of calculated and experimental results for an accelerating adiabatic flow. The experimental skin-friction values were not measured but were deduced by means of the momentum integral equation.

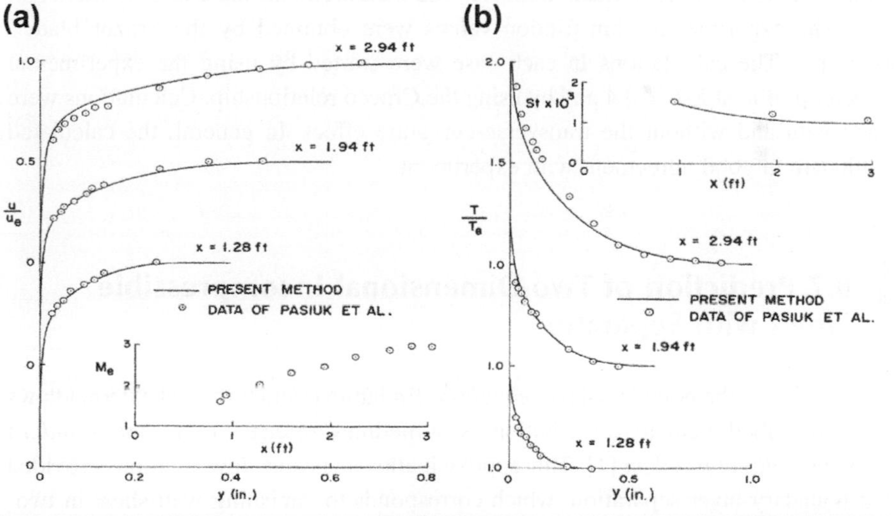

Fig. 8.21 Comparison of calculated and experimental results for an accelerating flow with constant heat flux. (a) Velocity profiles; (b) temperature profiles and Stanton-number distribution.

Pasiuk et al. [35]. The calculations were started by assuming a constant heat flux with zero pressure gradient that matched the experimental momentum-thickness value at $x = 0.94$ ft. Then the experimental Mach-number distribution was used to compute the rest of the flow for constant heat flux. The edge Mach number varied from $M_e = 1.69$ at $x = 0.94$ ft to $M_e = 2.97$ at $x = 3.03$ ft. Figure 8.21a shows a comparison of calculated and experimental velocity profiles for three x stations. Similarly, Fig. 8.21b shows a comparison of calculated and experimental temperature profiles, together with a comparison of local Stanton-number values. Except for one x station, the calculated profiles are in good agreement with experiment.

8.6 Axisymmetric Compressible Flows

Figure 5.4b in subsection 5.3.2 presented a comparison of calculated and experimental velocity profiles in adiabatic compressible turbulent boundary layers on slender cylinder. As we discussed in Sects. 5.3.2 and 8.4, in flows past such bodies, the transverse-curvature effect causes the boundary-layer development to be significantly different from those in two-dimensional zero-pressure gradient flows. A modification of the two-dimensional eddy-viscosity distribution for thick axisymmetric boundary layers improves the calculations.

Figure 8.22 shows the results for a waisted body of revolution for an adiabatic compressible flow at two Mach numbers. The measurements are due to Winter et al. [36]. The experimental skin-friction values were obtained by the "razor blade" technique. The calculations in each case were started by using the experimental velocity profile at $X/L = 0.4$ and by using the Crocco relationship. Calculations were made with and without the transverse-curvature effect. In general, the calculated results are in good agreement with experiment.

8.7 Prediction of Two-Dimensional Incompressible Flows with Separation

The solution of the boundary-layer equations for laminar and turbulent external flows with prescribed velocity distribution is sometimes referred to as the *standard problem* or *direct problem* [1]. This approach allows viscous flow solutions provided that boundary-layer separation, which corresponds to vanishing wall shear in two-dimensional steady flows, does not occur. If the wall shear vanishes at some x-location, solutions breakdown and convergence cannot be obtained. This is referred to as the singular behaviour of the boundary-layer equations at separation.

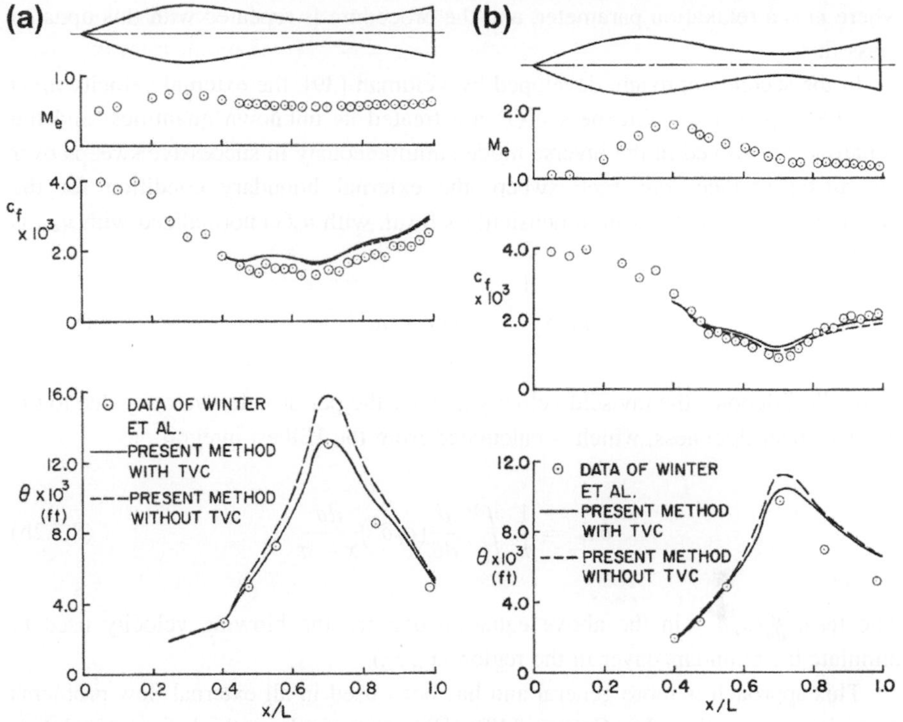

Fig. 8.22 Comparison of calculated and experimental results for a waisted body of revolution for an adiabatic flow at two Mach numbers: (a) $M_\infty = 0.6$, $R_L = 10^7$; (b) $M_\infty = 1.4$, $R_L = 10^7$ (TVC stands for transverse curvature).

The boundary-layer equations are not singular at separation if the external velocity or pressure is computed as part of the solution. This procedure is known as the *inverse problem* and has been extensively used for airfoil flows. In general two procedures have been pursued. In the first procedure, developed by Le Balleur [37] and Carter and Wornom [38], the solution of the boundary-layer equations is obtained by the standard method, and a displacement-thickness, $\delta^{*0}(x)$, distribution is determined. If this initial calculation encounters separation, $\delta^{*0}(x)$ is extrapolated to the trailing edge of the airfoil. For the given $\delta^{*0}(x)$ distribution, the boundary-layer equations are then solved in the inverse mode to obtain an external velocity $u_{\text{ev}}(x)$. An updated inviscid velocity distribution, $u_{\text{ei}}(x)$, is then obtained from the inviscid flow method with the added displacement thickness. A relaxation formula is introduced to define an updated displacement-thickness distribution,

$$\delta^*(x) = \delta^{*0}(x)\left\{1 + \omega\left[\frac{u_{\text{ev}}(x)}{u_{\text{ei}}(x)} - 1\right]\right\} \quad (8.7.1\text{d})$$

where ω is a relaxation parameter, and the procedure is repeated with this updated mass flux.

In the second approach, developed by Veldman [39], the external velocity $u_e(x)$ and the displacement thickness $\delta^*(x)$ are treated as unknown quantities, and the equations are solved in the inverse mode simultaneously in successive sweeps over the airfoil surface. For each sweep, the external boundary condition for the boundary-layer equations in dimensionless form, with $u_e(x)$ normalized with u_∞, is written as

$$u_e(x) = u_e^0(x) + \delta u_e(x) \tag{8.7.2a}$$

Here $u_e^0(x)$ denotes the inviscid velocity and δu_e the perturbation velocity due to the displacement thickness, which is calculated from the Hilbert integral

$$\delta u_e = \frac{1}{\pi}\int_{x_a}^{x_b} \frac{d}{d\sigma}(u_e\delta^*)\frac{d\sigma}{x-\sigma} \tag{8.7.2b}$$

The term $\frac{d}{d\sigma}(u_e\delta^*)$ in the above equation denotes the blowing velocity used to simulate the boundary-layer in the region (x_a, x_b).

This approach is more general and has been used in all external flow problems requiring interaction by Cebeci [40]. His numerical method for calculating two-dimensional incompressible flows is briefly described in Sections 8.8 and 8.9 and in detail in [40]. A sample of results are given in Section 8.10 for airfoil flows and in Section 8.11 for wing flows, following a brief description of the interaction problem in subsection 8.7.1.

8.7.1 Interaction Problem

Predicting the flowfield by solutions based on inviscid-flow theory is usually adequate as long as the viscous effects are negligible. A boundary layer that forms on the surface causes the irrotational flow outside it to be on a surface displaced into the fluid by a distance equal to the displacement thickness δ^*, which represents the deficiency of mass within the boundary layer. Thus, a new boundary for the inviscid flow, taking the boundary-layer effects into consideration, can be formed by adding δ^* to the body surface. The new surface is called the displacement surface and, if its deviation from the original surface is not negligible, the inviscid flow solutions can be improved by incorporating viscous effects into the inviscid flow equations [40].

A convenient and popular approach described in detail in [40,42] for aerodynamic flows, is based on the concept that the displacement surface can also be formed by distributing a blowing or suction velocity on the body surface. The strength of the blowing or suction velocity v_b is determined from the boundary-layer solutions according to

$$v_b = \frac{d}{dx}\left(u_e\delta^*\right) \tag{8.7.3}$$

where x is the surface distance of the body, and the variation of v_b on the body surface simulates the viscous effects in the potential flow solution. This approach, which can be used for both incompressible and compressible flows [40], is used in this section to address the interaction problem for an airfoil in subsonic flows.

The approach to simulate a turbulent wake with the transpiration model is similar to the approach discussed for an airfoil surface. A dividing streamline is chosen in the wake to separate the upper and lower parts of the inviscid flow, and on this line discontinuities are required in the normal components of velocity, so that it can be thought of as a source sheet.

At points C and D on the upper and lower sided of the dividing streamline (Fig. 8.23), the components of transpiration velocity, v_{iu} v_{il} are, respectively, see Eq. (8.7.4)

$$v_{iu} = \frac{1}{\varrho_{iu}}\frac{d}{dx}(\varrho_{iu}u_{iu}\delta_u{}^*) \tag{8.7.4}$$

and

$$v_{il} = \frac{1}{\varrho_{il}}\frac{d}{dx}(\varrho_{il}u_{il}\delta_l{}^*) \tag{8.7.5}$$

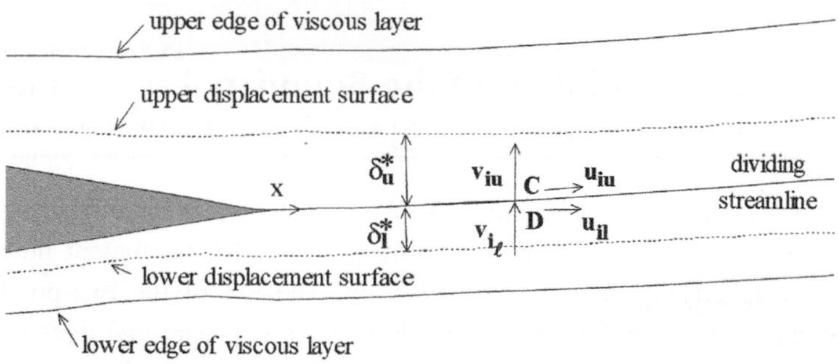

Fig. 8.23 Notation for the airfoil trailing-edge region.

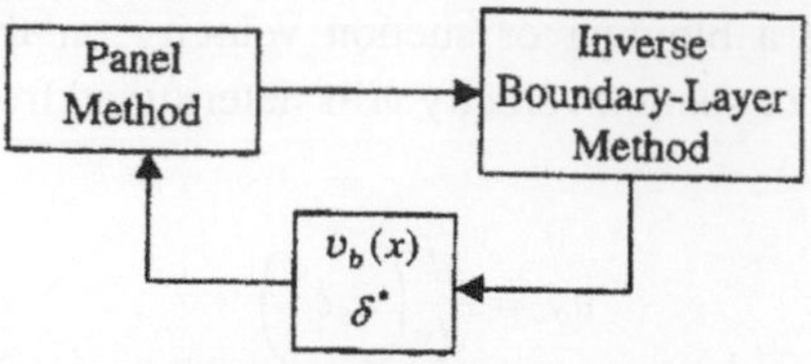

Fig. 8.24 Interactive boundary-layer scheme.

Here the sign convention has been used the v is measured positive in the direction of the upward normal to the wake. Hence a jump Δv in the component of velocity normal to the wake is required; it is given by

$$\Delta_{vi} \equiv v_{iu} - v_{il} = \frac{1}{\varrho_{iu}} \frac{d}{dx}(\varrho_{iu} u_{iu} \delta_u^*) + \frac{1}{\varrho_{il}} \frac{d}{dx}(\varrho_{il} u_{il} \delta_l^*) \tag{8.7.6}$$

For a given airfoil geometry and freestream flow conditions, the inviscid velocity distribution is usually obtained with a panel method, then the boundary-layer equations are solved in the inverse mode as described in Section 8.8. The blowing velocity distribution, $v_b(x)$, is computed from Eqs. (8.7.3–8.7.6) and the displacement thickness distribution $\delta^*(x)$ on the airfoil and in the wake are then used in the panel method to obtain an improved inviscid velocity distribution with viscous effects (See Section 8.9). The δ^*_{te} is used to satisfy the Kutta condition in the panel method at a distance equal to δ^*_{te}; this is known as the off-body Kutta condition (Fig. 8.24). In the first iteration between the inviscid and the inverse boundary-layer methods, $v_b(x)$ is used to replace the zero blowing velocity at the surface. At the next and following iterations, a new value of $v_b(x)$ in each iteration is used as a boundary condition in the panel method. This procedure is repeated for several cycles until convergence is obtained, which is usually based on the lift and total drag coefficients of the airfoil. Studies discussed in [40] show that with three boundary-layer sweeps for one cycle, convergence is obtained in less than 10 cycles.

8.8 Numerical Solution of the Boundary-Layer Equations in the Inverse Mode with Algebraic Turbulence Models

We consider a laminar and turbulent flow. We assume the calculations start at the leading edge, $x = 0$, for laminar flow and are performed for turbulent flow at any x-location by specifying the transition location. The use of the two-point finite-difference approximations for streamwise derivatives is proper and does not cause numerical difficulties if there is no flow separation. If there is one, then it is necessary to use backward difference formulas [40].

We employ two separate but closely related transformations. The first one is the Falkner-Skan transformation in which the dimensionless similarity variable η and a dimensionless stream function $f(x, \eta)$ are defined by Eqs. (8.2.5).

The resulting equations with this transformation are given by Eqs. (8.2.6) and (8.2.7).

In the inverse mode, since $u_e(x)$ is also an unknown, slight changes are made to the transformation given by Eq. (8.2.5), replacing $u_e(x)$ by u_∞ and redefining new variables Y and F by

$$Y = \sqrt{\frac{u_\infty}{\nu x}}y, \quad \psi(x, y) = \sqrt{u_\infty \nu x}F(\xi, Y), \quad \xi = \frac{x}{L} \tag{8.8.1}$$

The resulting equation and its wall boundary equations can be written as

$$(bF'')' + \frac{1}{2}FF'' = \xi\left(F'\frac{\partial F'}{\partial \xi} - F''\frac{\partial F}{\partial \xi}\right) - \xi w\frac{dw}{d\xi} \tag{8.8.2}$$

$$Y = 0, \quad F' = 0, \quad F = 0 \tag{8.8.3}$$

Here primes denote differentiations with respect to Y and $w = u_e/u_\infty$.

The edge boundary condition is obtained from Eq. (8.7.2). By applying a discretization approximation to the Hilbert integral, Eq. (8.7.2b), we can write

$$u_e(x_i) = u_e^0(\xi_i) + C_{ii}D_i + \sum_{j=1}^{i-1} C_{ij}D_j + \sum_{j=i+1}^{N} C_{ij}D_j \tag{8.8.4}$$

where the subscript i denotes the ξ-station where the inverse calculations are to be performed, C_{ij} is a matrix of interaction coefficients obtained by the procedure described in subroutine HIC, and D is given by $D = u_e\delta^*$. In terms of transformed variables, the parameter D becomes

$$\overline{D} = \frac{D}{Lu_\infty} = \sqrt{\frac{\xi}{R_L}}(Y_e w - F_e) \tag{8.8.5}$$

and the relation between the external velocity u_e and displacement thickness δ^* provided by the Hilbert integral can then be written in dimensionless form as

$$Y = Y_e, \quad F_e'(\xi^i) - \lambda\left[Y_e F_e'(\xi^i) - F_e(\xi^i)\right] = g_i \tag{8.8.6}$$

where

$$\lambda = C_{ii}\sqrt{\xi^i/R_L} \tag{8.8.7a}$$

$$g_i = \overline{u_e^0}(\xi^i) + \sum_{j=1}^{i-1} C_{ij}\overline{D}_j + \sum_{j=i+1}^{N} C_{ij}\overline{D}_j \tag{8.8.7b}$$

8.8.1 Numerical Formulation

The numerical method for the inverse problem is similar to the numerical method described for the standard problem in Section 8.2. Since $u_e(\xi)$ must be computed as part of the solution procedure, we treat it as an unknown. Remembering that the external velocity w is a function of ξ only, we write

$$w' = 0 \tag{8.8.8}$$

As in the case of the standard problem, new variables $U(\xi, Y)$, $V(\xi, Y)$ are introduced and Eq. (8.8.2) and its boundary conditions, Eq. (8.8.3) and (8.8.6), are expressed as a first-order system,

$$F' = U \tag{8.8.9a}$$

$$U' = V \tag{8.8.9b}$$

$$(bV)' + \frac{1}{2}FV = \xi\left(U\frac{\partial U}{\partial \xi} - V\frac{\partial F}{\partial \xi}\right) - \xi w\frac{dw}{d\xi} \tag{8.8.9c}$$

$$Y = 0, \quad F = U = 0 \tag{8.8.10a}$$

$$Y = Y_e, \quad U = w, \quad \lambda F + (1 - \lambda Y_e)w = g_i \tag{8.8.10b}$$

Finite-difference approximations to Eqs. (8.8.8) and (8.8.9) are written in a similar fashion to those expressed in the original Falkner-Skan variables, yielding

$$h_j^{-1}\left(w_j^n - w_{j-1}^n\right) = 0 \tag{8.8.11a}$$

$$h_j^{-1}\left(F_j^n - F_{j-1}^n\right) = U_{j-1/2}^n \tag{8.8.11b}$$

$$h_j^{-1}\left(U_j^n - U_{j-1}^n\right) = V_{j-1/2}^n \tag{8.8.11c}$$

$$\begin{aligned} h_j^{-1}\left(b_j^n V_j^n - b_j^n V_{j-1}^n\right) &+ \left(\frac{1}{2} + \alpha^n\right)(FV)_{j-1/2}^n \\ &+ \alpha^n\left[\left(w^2\right)_{j-1/2}^n - FLARE\left(U^2\right)_{j-1/2}^n\right] \\ &+ \alpha^n\left(V_{j-1/2}^{n-1}F_{j-1/2}^n - F_{j-1/2}^{n-1}V_{j-1/2}^n\right) = R_{j-1/2}^{n-1} \end{aligned} \tag{8.8.11d}$$

where

$$R_{j-1/2}^{n-1} = -L_{j-1/2}^{n-1} + \alpha^n\left[(FV)_{j-1/2}^{n-1} - FLARE\left(U^2\right)_{j-1/2}^{n-1}\right] \tag{8.8.12a}$$

$$L_{j-1/2}^{n-1} = \left[h_j^{-1}\left(b_j v_j - b_{j-1} v_{j-1}\right) + \frac{1}{2}(FV)_{j-1/2} - \alpha^n \left(w^2\right)_{j-1/2}\right]^{n-1} \tag{8.8.12b}$$

In Eq. (8.8.11d), the parameter FLARE refers to the Flügge-Lotz-Reyhner approximation [1] used to set $u\dfrac{\partial u}{\partial x}$ equal to zero in the momentum equation whenever $u < 0$. As a result, the numerical instabilities that plague attempts to integrate the boundary-layer equations against the local directions of flow are avoided. In regions of positive streamwise velocity ($u_j > 0$), it is taken as unity and as zero in regions of negative streamwise velocity ($u_j \leq 0$).

The linearized form of Eqs. (8.8.11) and (8.8.10) can be expressed in the form given by Eq. (8.2.24) or

$$\begin{array}{r|cccc|cccc|cccc|}
 & \delta F_o & \delta U_o & \delta V_o & \delta w_o & \delta F_j & \delta U_j & \delta V_j & \delta w_j & \delta F_J & \delta U_J & \delta V_J & \delta w_J \\
\text{b.c.} & 1 & 0 & 0 & 0 & 0 & 0 & 0 & 0 & & & & \\
\text{b.c.} & 0 & 1 & 0 & 0 & 0 & 0 & 0 & 0 & & & & \\
 & 0 & -1 & \frac{-h_1}{2} & 0 & 0 & -1 & \frac{-h_1}{2} & 0 & & & & \\
 & 0 & 0 & 0 & -1 & 0 & 0 & 0 & 1 & & & & \\
\hline
 & -1 & \frac{-h_j}{2} & 0 & 0 & -1 & \frac{-h_j}{2} & 0 & 0 & 0 & 0 & 0 & 0 \\
 & (s_4)_j & (s_6)_j & (s_2)_j & (s_8)_j & (s_3)_j & (s_5)_j & (s_1)_j & (s_7)_j & 0 & 0 & 0 & 0 \\
 & 0 & 0 & 0 & 0 & 0 & -1 & \frac{h_j+1}{2} & 0 & 0 & 1 & \frac{h_j+1}{2} & 0 \\
 & 0 & 0 & 0 & 0 & 0 & 0 & 0 & -1 & 0 & 0 & 0 & 1 \\
\hline
 & & & & & -1 & \frac{-h_J}{2} & 0 & 0 & 1 & \frac{-h_J}{2} & 0 & 0 \\
 & & & & & (s_4)_J & (s_6)_J & (s_2)_J & (s_8)_J & (s_3)_J & (s_5)_J & (s_1)_J & (s_7)_J \\
\text{b.c.} & & & & & 0 & 0 & 0 & 0 & 0 & 0 & 0 & 0 \\
\text{b.c.} & & & & & 0 & 0 & 0 & 0 & 0 & 1 & 0 & -1
\end{array}
\begin{vmatrix} \delta F_o \\ \delta U_o \\ \delta V_o \\ \delta w_o \\ \delta F_j \\ \delta U_j \\ \delta V_j \\ \delta w_j \\ \delta F_J \\ \delta U_J \\ \delta V_J \\ \delta w_J \end{vmatrix}
\begin{vmatrix} (r_1)_o \\ (r_2)_o \\ (r_3)_o \\ (r_4)_o \\ (r_1)_j \\ (r_2)_j \\ (r_3)_j \\ (r_4)_j \\ (r_1)_J \\ (r_2)_J \\ (r_3)_J \\ (r_4)_J \end{vmatrix} \tag{8.8.13}$$

With $\underset{\sim}{\delta}_j$ and $\underset{\sim}{r}_j$ now defined by

$$\underset{\sim}{\delta}_j = \begin{vmatrix} \delta F_j \\ \delta U_j \\ \delta V_j \\ \delta w_j \end{vmatrix}, \quad \underset{\sim}{r}_j = \begin{vmatrix} (r_1)_j \\ (r_2)_j \\ (r_3)_j \\ (r_4)_j \end{vmatrix} \tag{8.8.14}$$

and A_j, B_j, C_j becoming 4×4 matrices defined by

$$A_0 = \begin{vmatrix} 1 & 0 & 0 & 0 \\ 0 & 1 & 0 & 0 \\ 0 & -1 & -\dfrac{h_1}{2} & 0 \\ 0 & 0 & 0 & -1 \end{vmatrix}, \quad A_j = \begin{vmatrix} 1 & -\dfrac{h_j}{2} & 0 & 0 \\ (s_3)_j & (s_5)_j & (s_1)_j & (s_7)_j \\ 0 & -1 & -\dfrac{h_{j+1}}{2} & 0 \\ 0 & 0 & 0 & -1 \end{vmatrix}, \quad 1 \leq j \leq J-1 \tag{8.8.15a}$$

$$A_J = \begin{vmatrix} 1 & -\frac{h_J}{2} & 0 & 0 \\ (s_3)_J & (s_5)_J & (s_1)_J & (s_7)_J \\ \gamma_1 & 0 & 0 & \gamma_2 \\ 0 & 1 & 0 & -1 \end{vmatrix}, \quad B_j = \begin{vmatrix} -1 & -\frac{h_j}{2} & 0 & 0 \\ (s_4)_j & (s_6)_j & (s_2)_j & (s_8)_j \\ 0 & 0 & 0 & 0 \\ 0 & 0 & 0 & 0 \end{vmatrix}, \quad 1 \le j \le J$$

(8.8.15b)

$$C_j = \begin{vmatrix} 0 & 0 & 0 & 0 \\ 0 & 0 & 0 & 0 \\ 0 & 1 & -\frac{h_{j+1}}{2} & 0 \\ 0 & 0 & 0 & 1 \end{vmatrix}, \quad 0 \le j \le J-1 \tag{8.8.15c}$$

Here the first two rows of A_0 and C_0 and the last two rows of B_J and A_J correspond to the linearized boundary conditions,

$$\delta F_0 = \delta U_0 = 0; \quad \delta U_J - \delta w_J = w_J - U_J, \quad \gamma_1 \delta F_J + \gamma_2 \delta w_J = \gamma_3 \tag{8.8.16}$$

where

$$\gamma_1 = \lambda, \quad \gamma_2 = 1 - \lambda Y_J, \quad \gamma_3 = g_i - (\gamma_1 F_J + \gamma_2 w_J) \tag{8.8.17}$$

As a result

$$(r_1)_0 = (r_2)_0 = 0 \tag{8.8.18a}$$

$$(r_3)_J = \gamma_3, \quad (r_4)_J = w_J - U_J \tag{8.8.18b}$$

The third and fourth rows of A_0 and C_0 correspond to Eq. (8.2.20b) and the linearized form of Eq. (9.2.18a), that is,

$$\delta w_j - \delta w_{j-1} = w_{j-1} - w_j = (r_4)_{j-1} \tag{8.8.19}$$

if the unknows f, u, v are replaced by F, U and V. Similarly, the first and second rows of A_j and B_j correspond to Eq. (8.2.20a) and (9.2.18a) with two terms added to its left-hand side,

$$(s_7)_j \delta w_j + (s_8)_j \delta w_{j-1} \tag{8.8.20a}$$

with $(s_7)_j$ and $(s_8)_j$ defined by

$$(s_7)_j = \alpha^n w_j, \quad (s_8)_j = \alpha^n w_{j-1} \tag{8.8.20b}$$

The coefficients $(s_1)_j$ to $(s_6)_j$ defined by Eqs. (8.2.22) remain unchanged provided we set

$$\alpha_1 = \frac{1}{2} + \alpha^n, \quad \alpha_2 = \alpha^n \tag{8.8.21}$$

and define $(r_2)_j$ by

$$\begin{aligned}(r_2)_j = R^{n-1}_{j-1/2} - \Big[& h_j^{-1}(b_j V_j - b_{j-1} V_{j-1}) + \left(\frac{1}{2} + \alpha^n\right)(FV)_{j-1/2} \\ & + \alpha^n \left[(w^2)_{j-1/2} - FLARE(U^2)_{j-1/2} \right] \\ & + \alpha^n \left(V^{n-1}_{j-1/2} F_{j-1/2} - F^{n-1}_{j-1/2} V_{j-1/2} \right) \Big]\end{aligned} \tag{8.8.22}$$

The remaining elements of the $\underset{\sim}{r}_j$ vector follow from Eqs. (8.2.21), and (8.8.19) and (8.8.22) so that, for $l \le j \le J$, $(r_1)_j$, $(r_2)_j$, $(r_3)_{j-1}$ are given by Eqs. (8.2.21a), (8.8.22) and (8.2.21b), respectively. For the same j-values, $(r_4)_{j-1}$ is given by the right-hand side of Eq. (8.8.19).

The parameters γ_1, γ_2 and γ_3 in Eq. (8.8.16) determine whether the system given by the linearized form of Eqs. (8.8.11) and their boundary conditions is to be solved in standard or inverse form. For an inverse problem, they are represented by the expressions given in Eq. (8.8.16) and for a standard problem by $\gamma_1 = 0$, $\gamma_2 = 1.0$ and $\gamma_3 = 0$.

It should be noted that for flows with separation, it is necessary to use backward differences as discussed for the CS and k-ε models in Sections 10.7 to 10.10.

In that case, the coefficients $(s_1)_j$ to $(s_6)_j$ are given by Eq. (9.2.25), and $(r_2)_j$ by Eq. (9.2.26) with the relations given by Eq. (8.8.22).

The solution of Eq. (8.2.24), with $\underset{\sim}{\delta}_j$ and $\underset{\sim}{r}_j$ defined by Eq. (8.8.14) and with A_j, B_j and C_j matrices given by Eqs. (8.8.15), can again be obtained by the block-elimination of subsection 8.2.3. The resulting algorithm, is similar to SOLV3, and is called SOLV4.

Numerical Method for Wake Flows

In interaction problems involving airfoils, it is usually sufficient to neglect the wake effect and perform calculations on the airfoil only, provided that there is no or little flow separation on the airfoil. With flow separation, the relative importance of including the wake effect in the calculations depends on the flow separation as shown in Fig. 8.25 taken from [40]. Figure (8.25a) shows the computed separation locations on a NACA 0012 airfoil at a chord Reynolds number, R_c of 3×10^6. When the wake effect is included, separation is encountered for angles of attack greater than 10°, and attempts to obtain results without consideration of the wake effect lead to erroneously large regions of recirculation that increases with angle of attack, as discussed in [40]. Figure (8.25b) shows that the difference in displacement thickness at the trailing edge is negligible for $\alpha = 10°$ but more than 30% for $\alpha = 16°$ [40].

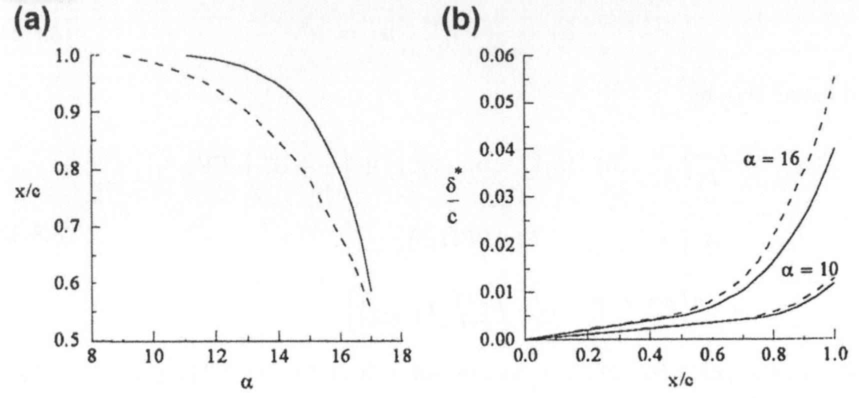

Fig. 8.25 Wake effect on (a) flow separation and (b) displacement thickness – NACA 0012 airfoil. —, with wake:– – – , without wake.

As discussed in [40], the inverse boundary layer method described here can also be extended to include wake flows. This requires the specification of a turbulence model for wake flows and minor modifications to the numerical method.

The extension of the CS model for wall boundary layers to wake flows is given by the following expressions described in [41]:

$$\varepsilon_m = (\varepsilon_m)_w + \left[(\varepsilon_m)_{\text{t.e.}} - (\varepsilon_m)_w\right] \exp\frac{-(x - x_{\text{t.e.}})}{\lambda\delta_{\text{t.e.}}} \tag{8.8.23}$$

where $\delta_{\text{t.e.}}$ is the boundary layer thickness at the trailing edge, λ is an empirical parameters, $(\varepsilon_m)_{\text{t.e.}}$ is the eddy viscosity at the trailing edge, and $(\varepsilon_m)_w$ is the eddy-viscosity in the far wake given by the larger of

$$(\varepsilon_m)_w^l = 0.064 \int_{-\infty}^{y_{\min}} (u_e - u)dy \tag{8.8.24}$$

and

$$(\varepsilon_m)_w^u = 0.064 \int_{y_{\min}}^{\infty} (u_e - u)dy \tag{8.8.25}$$

with $y_{\min}$ denoting the location where the velocity is a minimum.

The studies conducted in [41] indicate that a choice of $\lambda = 20$ is satisfactory for single airfoils. Calculations with different values of λ essentially produced similar results, indicating that the modeling of wake flows with Eq. (8.8.23) was not too sensitive to the choice of λ. The application of the above model to wake flows with strong adverse pressure gradient, however, indicated that this was not the case and the value of the parameter is an important one. On the basis of that study, a value of $\lambda = 50$ was found to produce best results and is used in the computer program discussed in Section 10.14.

A modification to the numerical method of the previous section arises due to the boundary conditions along the wake dividing streamline. The new "wall" boundary conditions on f and u now become

$$\eta = 0, \quad f_0 = 0, \quad v_0 = 0 \tag{8.8.26}$$

so that the second row of Eq. (8.8.23) can de written as

$$0 \quad 0 \quad 1 \quad 0 \tag{8.8.27}$$

Before the boundary-layer equations can be solved for wake flows, the initial velocity profiles must satisfy the wall and edge boundary conditions. When the calculations are first performed for wall boundary layer flows and are then to be extended to wake flows, it is necessary to modify the velocity profiles computed for wall boundary layers. This is done in subroutine WAKEPR of the computer program.

8.9 Hess-Smith (HS) Panel Method

For incompressible flows, a panel method is an ideal inviscid method for interactive boundary layer approach. Of the several panel methods, here we choose the one due to Hess and Smith [40]. The procedure for incorporating the viscous effects into the panel method is discussed in Section 8.9.1. Changes required in an inviscid method to extend the viscous flow calculations into the wake of an airfoil are discussed in Section 8.9.2. A brief description of the computer program for the HS method with viscous effects is given in Section 10.4.

We consider an airfoil at rest in an onset flow of velocity V_∞. We assume that the airfoil is at an angle of attack, α (the angle between its chord line and the onset velocity), and that the upper and lower surfaces are given by functions $Y_u(x)$ and $Y_l(x)$, respectively. These functions can be defined analytically, or (as is often the case) by a set of (x, y) values of the airfoil coordinates. We denote the distance of any field point (x, y) from an arbitrary point, b, on the airfoil surface by r, as shown in Fig. 8.25. Let $\vec{n}$ also denote the unit vector normal to the airfoil surface and directed from the body into the fluid, and $\vec{t}$ a unit vector tangential to the surface, and assume that the inclination of $\vec{t}$ to the x-axis is given by θ. It follows from Fig. 8.25 that with $\vec{i}$ and $\vec{j}$ denoting unit vectors in the x- and y-directions, respectively,

$$\begin{aligned} \vec{n} &= -\sin\theta\,\vec{i} + \cos\theta\,\vec{j} \\ \vec{t} &= \cos\theta\,\vec{i} + \sin\theta\,\vec{j} \end{aligned} \tag{8.9.1}$$

If the airfoil contour is divided into a large number of small segments, ds, then we can write

$$\begin{aligned} dx &= \cos\theta \, ds \\ dy &= \sin\theta \, ds \end{aligned} \tag{8.9.2}$$

with ds calculated from $ds = \sqrt{(dx)^2 + (dy)^2}$.

We next assume that the airfoil geometry is represented by a finite number (N) of short straight-line elements called panels, defined by $(N+1)(x_j,\ y_j)$ pairs called boundary points. It is customary to input the $(x,\ y)$ coordinates starting at the lower surface trailing edge, proceeding clockwise around the airfoil, and ending back at the upper surface trailing edge. If we denote the boundary points by

$$(x_1,\ y_1),\ (x_2,\ y_2),\ \ldots,\ (x_N,\ y_N),\ (x_{N+1},\ y_{N+1}) \tag{8.9.3}$$

then the pairs $(x_1,\ y_1)$ and $(x_{N+1},\ y_{N+1})$ are identical for a closed trailing edge (but not for an open trailing edge) and represent the trailing edge. It is customary to refer to the element between $(x_j,\ y_j)$ and $(x_{j+1},\ y_{j+1})$ as the j-th panel, and to the midpoints of the panels as the *control points*. Note from Fig. 8.26 that as one traverses from the i-th boundary point to the (i+1)-th boundary point, the airfoil body is on the right-hand side. This numbering sequence is consistent with the common definition of the unit normal vector $\vec{n_i}$ and unit tangential vector $\vec{t_i}$ for all panel surfaces, i.e., $\vec{n_i}$ is directed from the body into the fluid and $\vec{t_i}$ from the i-th boundary point to the $(i + 1)$-th boundary point with its inclination to the x-axis given by θ_i.

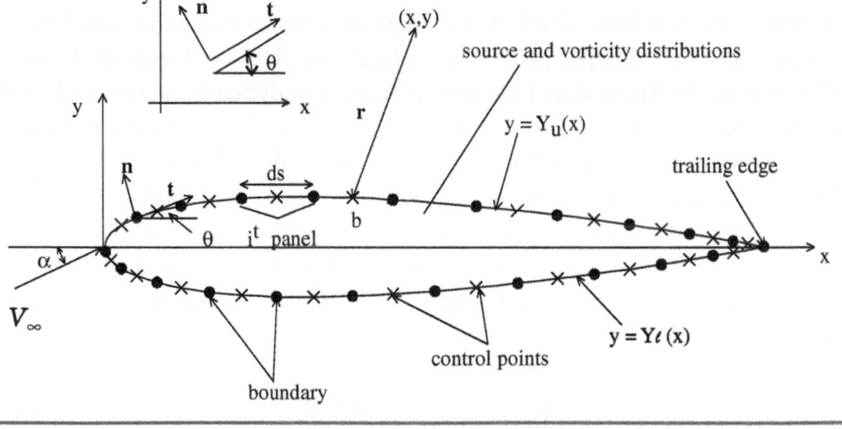

Fig. 8.26 Panel representation of airfoil surface and notation for an airfoil at incidence α.

In the HS panel method, the velocity $\vec{V}$ at any point (x, y) is represented by

$$\vec{V} = \vec{U} + \vec{v} \tag{8.9.4}$$

where $\vec{U}$ is the velocity of the uniform flow at infinity

$$\vec{U} = V_\infty \left(\cos\alpha\, \vec{i} + \sin\alpha\, \vec{j}\right) \tag{8.9.5}$$

and $\vec{v}$ is the disturbance field due to the body which is represented by two elementary flows corresponding to source and vortex flows. A source or vortex on the j-th panel causes an induced source velocity $\vec{v_s}$ at (x, y) or an induced vortex velocity $\vec{v}_v$ at (x, y), respectively, and these are obtained by taking radients of a potential source

$$\phi_s = \frac{q}{2\pi} \ln r \tag{8.9.6}$$

and a potential vortex

$$\phi_r = \frac{I'}{2\pi}\theta, \tag{8.9.7}$$

both centered at the origin, so that, with integrals applied to the airfoil surface,

$$\vec{v}(x, y) = \int \vec{v_s} q_j(s) ds_j + \int \vec{v_v} \tau_j(s) ds_j \tag{8.9.8}$$

Here $q_j ds_j$ is the source strength for the element ds_j on the j-th panel. Similarly, $\tau_j ds_j$ is the vorticity strength for the element ds_j on the same panel.

Each of the N panels is represented by similar sources and vortices distributed on the airfoil surface. The induced velocities in Eq. (8.9.8) satisfy the irrotationality condition and the boundary condition at infinity

$$u = \frac{\partial\phi}{\partial x} = \frac{\partial\psi}{\partial y} = V_\infty \cos\alpha \tag{8.9.9a}$$

$$v = \frac{\partial\phi}{\partial y} = -\frac{\partial\psi}{\partial x} = V_\infty \sin\alpha \tag{8.9.9b}$$

For uniqueness of the solutions, it is also necessary to specify the magnitude of the circulation around the body. To satisfy the boundary conditions on the body, which correspond to the requirement that the surface of the body is a streamline of the flow, that is,

$$\psi = \text{constant} \quad \text{or} \quad \frac{\partial\phi}{\partial n} = 0 \tag{8.9.10}$$

at the surface on which n is the direction of the normal, the sum of the source induced and vorticity-induced velocities and freestream velocity is set to zero in the direction normal to the surface of each of the N panels. It is customary to choose the *control points* to numerically satisfy the requirement that the resultant flow is tangent to the surface. If the tangential and normal components of the total velocity at the control point of the i-th panel are denoted by $(V^t)_i$ and $(V^n)_i$, respectively, the flow tangency conditions are then satisfied at panel control points by requiring that the resultant velocity at each control point has only $(V^t)_i$, and

$$(V^n)_i = 0 \quad i = 1, 2, ..., N \tag{8.9.11}$$

Thus, to solve the Laplace equation with this approach, at the i-th panel control point we compute the normal $(V^n)_i$ and tangential $(V^t)_i$, $(i = 1, 2, \ldots, N)$ velocity components induced by the source and vorticity distributions on all panels, j $(j = 1, 2, \ldots, N)$, including the i-th panel itself, and separately sum all the induced velocities for the normal and tangential components together with the freestream velocity components. The resulting expressions, which satisfy the irrotationality condition, must also satisfy the boundary conditions discussed above. Before discussing this aspect of the problem, it is convenient to write Eq. (8.9.4) expressed in terms of its velocity components $(V^n)_i$ and $(V^t)_i$ by

$$(V^n)_i = \sum_{j=1}^{N} A_{ij}^n q_j + \sum_{j=1}^{N} B_{ij}^n \tau_j + V_\infty \sin(\alpha - \theta_i) \tag{8.9.12a}$$

$$(V^t)_i = \sum_{j=1}^{N} A_{ij}^t q_j + \sum_{j=1}^{N} B_{ij}^t \tau_j + V_\infty \cos(\alpha - \theta_i) \tag{8.9.12b}$$

where A^n_{ij}, B^n_{ij}, A^t_{ij}, B^t_{ij} are known as influence coefficients, defined as the velocities induced at a control point (x_{m_i}, y_{m_i}); more specifically, A^n_{ij} and A^t_{ij} denote the normal and tangential velocity components, respectively, induced at the i-th panel control point by a unit strength source distribution on the j-th panel, and B^n_{ij} and B^t_{ij} are those induced by unit strength vorticity distribution on the j-th panel. The influence coefficients are related to the airfoil geometry and the panel arrangement; they are given by the following expressions:

$$A_{ij}^n = \begin{cases} \dfrac{1}{2\pi}\left[\sin(\theta_i - \theta_j)\ \ln\dfrac{r_{i,j+1}}{r_{i,j}} + \cos(\theta_i - \theta_j)\beta_{ij}\right] & i \neq j \\ \dfrac{1}{2} & i = j \end{cases} \tag{8.9.13}$$

$$A_{ij}^t = \begin{cases} \dfrac{1}{2\pi}\left[\sin(\theta_i - \theta_j)\ \beta_{ij} - \cos(\theta_i - \theta_j)\ \ln\dfrac{r_{i,j+1}}{r_{i,j}}\right] & i \neq j \\ 0 & i = j \end{cases} \tag{8.9.14}$$

$$B_{ij}^n = -A_{ij}^t \quad B_{ij}^t = A_{ij}^n \tag{8.9.15}$$

Here

$$
\begin{aligned}
r_{i,j+1} &= \left[(x_{m_i} - x_{j+1})^2 + (y_{m_i} - y_{j+1})^2\right]^{1/2} \\
r_{i,j} &= \left[(x_{m_i} - x_j)^2 + (y_{m_i} - y_j)^2\right]^{1/2} \\
x_{m_i} &= \frac{1}{2}(x_i + x_{j+1}), \quad y_{m_i} = \frac{1}{2}(y_i + y_{j+1}) \\
\theta_i &= \tan^{-1}\left(\frac{y_{i+1} - y_i}{x_{i+1} - x_i}\right), \quad \theta_j = \tan^{-1}\left(\frac{y_{j+1} - y_j}{x_{j+1} - x_j}\right) \\
\beta_{ij} &= \tan^{-1}\left(\frac{y_{m_i} - y_{j+1}}{x_{m_i} - x_{j+1}}\right) - \tan^{-1}\left(\frac{y_{m_i} - y_j}{x_{m_i} - x_j}\right)
\end{aligned}
\tag{8.9.16}
$$

Regardless of the nature of $q_j(s)$ and $\tau_j(s)$, Eq. (8.9.12) satisfies the irrotationality condition and the boundary condition at infinity, Eq. (8.9.9). To satisfy the requirements given by Eq. (8.9.11) and the condition related to the circulation, it is necessary to adjust these functions. In the approach adopted by Hess and Smith [40], the source strength $q_j(s)$ is assumed to be constant over the j-th panel and is adjusted to give zero normal velocity over the airfoil, and the vorticity strength τ_j is taken to be constant on all panels ($\tau_j = \tau$) and its single value is adjusted to satisfy the condition associated with the specification of circulation. Since the specification of the circulation renders the solution unique, a rational way to determine the solution is required.

The best approach is to adjust the circulation to give the correct force on the body as determined by experiment. However, this requires advance knowledge of that force, and one of the principal aims of a flow calculation method is to calculate the force and not to take it as given. Thus, another criterion for determining circulation is needed.

For smooth bodies such as ellipses, the problem of rationally determining the circulation has yet to be solved. Such bodies have circulation associated with them, and resulting lift forces, but there is no rule for calculating these forces. If, on the other hand, we deal with an airfoil havinga sharp trailing edge, we can apply the Kutta condition [40]. It turns out that for every value of circulation except one, the inviscid velocity is infinite at the trailing edge. The Kutta condition states that the particular value of circulation that gives a finite velocity at the trailing edge is the proper one to choose. This condition does not include bodies with nonsharp trailing edges and bodies on which the viscous effects have been simulated by, for example, surface blowing, as discussed [4]. Thus, the classical Kutta condition is of strictly limited validity. It is customary to apply a "Kutta condition" to bodies outside its narrow definition, but this is an approximation; nevertheless the calculations are often in close accord with experiment.

In the panel method, the Kutta condition is indirectly applied by deducing another property of the flow at the trailing edge that is a direct consequence of the finiteness of velocity; this property is used as "the Kutta condition". Properties that have been used in lieu of "the Kutta condition" in panel methods include the following:

(a) A streamline of the flow leaves the trailing edge along the bisector of the trailing-edge angle.
(b) Upper and lower displacement total velocities approach a common limit at the trailing edge. The limiting value is zero if the trailing-edge angle is nonzero.
(c) Source and/or vorticity strengths at the trailing edge must satisfy conditions to allow finite velocity.

Of the above, property (b) is more widely used. At first it may be thought that this property requires setting both the upper and lower surface velocities equal to zero. This gives two conditions, which cannot be satisfied by adjusting a single parameter. The most reasonable choice is to make these two total velocities in the downstream direction at the 1st and N-th panel control points equal so that the flow leaves the trailing edge smoothly. Since the normal velocity on the surface is zero according to Eq. (8.9.11), the magnitudes of the two tangential velocities at the trailing edge must be equal to each other, that is,

$$(V^t)_N = -(V^t)_1 \tag{8.9.17}$$

Introducingthe flow tangency condition, Eq. (8.9.11), into Eq. (8.9.12a) and noting that $\tau_j = \tau$, we get

$$\sum_{j=1}^{N} A_{ij}^n q_j + \tau \sum_{j=1}^{N} B_{ij}^n + V_\infty \sin(\alpha - \theta_i) = 0, \quad i = 1, 2, \ldots, N \tag{8.9.18}$$

In terms of the unknowns, q_j (j = 1, 2, . . . , N) and τ, the Kutta condition of Eq. (8.9.17) and Eq. (8.9.18) for a system of algebraic equations which can be written in the following form,

$$A\underset{\sim}{x} = \underset{\sim}{b} \tag{8.9.19}$$

Here A is a square matrix of order (N + 1), that is

$$A \equiv \begin{vmatrix} a_{11} & a_{12} & \cdots & a_{1j} & \cdots & a_{1N} & a_{1,N+1} \\ a_{21} & a_{22} & \cdots & a_{2j} & \cdots & a_{2N} & a_{2,N+1} \\ \vdots & \vdots & \vdots & \vdots & \vdots & \vdots & \vdots \\ a_{i1} & a_{i2} & \cdots & a_{ij} & \cdots & a_{iN} & a_{i,N+1} \\ \vdots & \vdots & \vdots & \vdots & \vdots & \vdots & \vdots \\ a_{N1} & a_{N2} & \cdots & a_{Nj} & \cdots & a_{NN} & a_{N,N+1} \\ a_{N+1,1} & a_{N+1,2} & \cdots & a_{N+1,j} & \cdots & a_{N+1,N} & a_{N+1,N+1} \end{vmatrix} \tag{8.9.20}$$

and $\overrightarrow{x} = (q_1, \ldots, q_i, \ldots, q_N, \tau)^T$ and $b = (b_1, \ldots, b_i, \ldots, b_N, b_{N+1})^T$ with denoting the transpose. The elements of the coefficient matrix A follow from Eq. (8.9.18)

$$a_{ij} = A^n_{ij}, \quad \begin{matrix} i = 1, 2, \ldots, N \\ j = 1, 2, \ldots, N \end{matrix} \tag{8.9.21a}$$

$$a_{i,N+1} = \sum_{j=1}^{N} B^n_{ij}, \quad i = 1, 2, \ldots, N \tag{8.9.21b}$$

$A^n{}_{ij}$ are given by Eq. (8.9.13) and $B^n{}_{ij}$ by Eq. (8.9.15). The relation in Eq. (8.9.20) follows from the definition of $\overrightarrow{x}$ where τ is essentially x_{N+1}.

To find $a_{N+1,j}$ ($J = 1, \ldots, N$) and $a_{N+1,N+1}$ in the coefficient matrix A, we use the Kutta condition and apply Eq. (8.9.17) to Eq. (8.9.12b) and, with τ as a constant, we write the resulting expression as

$$\sum_{j=1}^{N} A^t_{1j} q_j + \tau \sum_{j=1}^{N} B^t_{1j} + V_\infty \cos(\alpha - \theta_1)$$
$$= -\left[\sum_{j=1}^{N} A^t_{Nj} q_j + \tau \sum_{j=1}^{N} B^t_{Nj} + V_\infty \cos(\alpha - \theta_N) \right]$$

or as

$$\sum_{j=1}^{N} (A^t_{1j} + A^t_{Nj}) q_j + \tau \sum_{j=1}^{N} (B^t_{1j} + B^t_{Nj}) = -V_\infty \cos(\alpha - \theta_1) - V_\infty \cos(\alpha - \theta_N) \tag{8.9.22b}$$

so that,

$$a_{N+1,j} = A^t_{1j} + A^t_{Nj}, \quad j = 1, 2, \ldots, N \tag{8.9.23a}$$

$$a_{N+1,N+1} = \sum_{j=1}^{N} (B^t_{1j} + B^t_{Nj}) \tag{8.9.23b}$$

where now $A^t{}_{1j}$ and $A^t{}_{Nj}$ are computed from Eq. (8.9.14) and $B^t{}_{1j}$ and $B^t{}_{Nj}$ from Eq. (8.9.15).

The components of $\overrightarrow{b}$ again follow from Eqs. (8.9.18) and (8.9.21). From Eq. (8.9.18),

$$b_i = -V_\infty \sin(\alpha - \theta_i), \quad i = 1, \ldots, N \tag{8.9.24a}$$

and from Eq. (8.9.22),

$$b_{N+1} = -V_\infty \cos(\alpha - \theta_1) - V_\infty \cos(\alpha - \theta_N) \tag{8.9.24b}$$

With all the elements of a_{ij} determined from Eqs. (8.9.21) and (8.9.23) and the elements of $\overrightarrow{b}$ from Eq. (8.9.24), the solution of Eq. (8.9.19) can be obtained by the Gaussian elimination method [42]. The elements of $\overrightarrow{x}$ are given by

$$x_i = \frac{1}{a_{ii}^{(i-1)}} - \left[b_i^{(i-1)} - \sum_{j=i+1}^{N+1} a_{ij}^{(i-1)} x_j \right] \quad i = N+1, \ldots, 1 \tag{8.9.25}$$

where

$$a_{ij}^{(k)} = a_{ij}^{(k-1)} - \frac{a_{ik}^{(k-1)}}{a_{kk}^{(k-1)}} a_{kj}^{(k-1)}, \quad \begin{array}{l} k = 1, \ldots, N \\ j = k+1, \ldots N+1 \\ i = k+1, \ldots, N+1 \\ a_{ij}^{(0)} = a_{ij} \end{array} \tag{8.9.26a}$$

$$b_i^{(k)} = b_i^{(k-1)} - \frac{a_{ik}^{(k-1)}}{a_{kk}^{(k-1)}} b_k^{(k-1)}, \quad \begin{array}{l} k = 1, \ldots, N \\ i = k+1, \ldots N+1 \\ b_i^{(0)} = b_i \end{array} \tag{8.9.26b}$$

8.9.1 Viscous Effects

The viscous effects can be introduced into the panel method by (1) replacing the zero normal-velocity condition, Eq. (8.9.11), by a nonzero normal-velocity condition $V_{iw}(x)$ and by (2) satisfying the Kutta condition, Eq. (8.9.17), not on the surface of the airfoil trailing edge but at some distance away from the surface.

Here it will be assumed that the nonzero normal-velocity distribution $V_{iw}(x)$ along the surface of the airfoil and in its wake is known, together with the distance from the surface, say displacement thickness δ^*, where the Kutta condition is to be satisfied. We now describe how these two new conditions can be incorporated into the panel method.

To include the nonzero normal-velocity condition into the solution procedure, we write Eq. (8.9.18) as

$$\sum_{j=1}^{N} A_{ij}^n q_j + \tau \sum_{j=1}^{N} B_{ij}^n q_j + V_\infty \sin(\alpha - \theta_i) = v_{iw}(x_{m_i}) \tag{8.9.27}$$

To satisfy the Kutta condition at the normal distance δ^* from the surface of the trailing edge, called the "off-body" Kutta condition, the total velocities at the N-th and first off-body control points are again required to be equal. Since the normal velocity component is not zero, we write the off-body Kutta condition at distance δ^* as

$$(V)_N = -(V)_1 \tag{8.9.28}$$

where V is the total velocity at the two control points. The off-body total velocities are computed from

$$V = \frac{(V^n)^2 + (V^t)^2}{V} = V^n\frac{V^n}{V} + V^t\frac{V^t}{V} = V^n\sin\phi + V^t\cos\phi \qquad (8.9.29)$$

where V^n and V^t are computed by expressions identical to those given by Eqs. (8.9.12) at the two off-body control points, $I = 1, I = N$, that is,

$$(V^n)_I = \sum_{j=1}^{N} A^n_{Ij}q_j + \tau\sum_{j=1}^{N} B^n_{Ij} + V_\infty \sin(\alpha - \theta_I) \qquad (8.9.30a)$$

$$(V^t)_I = \sum_{j=1}^{N} A^t_{Ij}q_j + \tau\sum_{j=1}^{N} B^t_{Ij} + V_\infty \cos(\alpha - \theta_I) \qquad (8.9.30b)$$

and where

$$\phi = \tan^{-1}[(V^n)_I/(V^t)_I] \qquad (8.9.31)$$

With Eqs. (8.9.30), the expression for the total velocity given by Eq. (8.9.29) can be written as

$$V = \sum_{j=1}^{N}(A^n_{Ij}\cdot\sin\phi + A^t_{Ij}\cdot\cos\phi)q_j + \tau\sum_{j=1}^{N}(B^n_{Ij}\cdot\sin\phi + B^t_{Ij}\cdot\cos\phi)$$
$$+ V_\infty \sin(\alpha - \theta_I)\sin\phi + V_\infty \cos(\alpha - \theta_I)\cos\phi \qquad (8.9.32a)$$

or as

$$V = \sum_{j=1}^{N} A'_{Ij}q_j + \tau\sum_{j=1}^{N} B'_{Ij} + V_\infty \cos(\alpha - \theta_I - \phi) \qquad (8.9.32b)$$

where

$$A'_{Ij} = A^n_{Ij}\cdot\sin\phi + A^t_{Ij}\cdot\cos\phi, \quad B'_{Ij} = B^n_{Ij}\cdot\sin\phi + B^t_{Ij}\cdot\cos\phi \qquad (8.9.33a)$$

$$A^n_{Ij} = \frac{1}{2\pi}\left[\sin(\theta_I - \theta_j)\ \ln\frac{r_{I,j+1}}{r_{I,j}} + \cos(\theta_I - \theta_j)\beta_{Ij}\right] \qquad (8.9.33b)$$

$$A^t_{Ij} = \frac{1}{2\pi}\left[\sin(\theta_I - \theta_j)\ \beta_{Ij} - \cos(\theta_I - \theta_j)\ \ln\frac{r_{I,j+1}}{r_{I,j}}\right] \qquad (8.9.33c)$$

$$B^n_{Ij} = -A^t_{Ij}, \quad B^t_{Ij} = A^n_{Ij} \qquad (8.9.33d)$$

If we define

$$\theta'_I = \theta_I + \phi \qquad (8.9.34)$$

then it can be shown that Eq. (8.9.32b) can be written as

$$V = \sum_{j=1}^{N} A'_{Ij} q_j + \tau \sum_{j=1}^{N} B'_{Ij} + V_\infty \cos(\alpha - \theta'_I) \tag{8.9.35}$$

where

$$A'_{Ij} = \frac{1}{2\pi}\left[\sin(\theta'_I - \theta_j)\,\beta_{Ij} - \cos(\theta'_I - \theta_j)\ln\frac{r_{I,j+1}}{r_{I,j}}\right] \tag{8.9.36a}$$

$$B'_{Ij} = \frac{1}{2\pi}\left[\sin(\theta'_I - \theta_j)\ln\frac{r_{I,j+1}}{r_{I,j}} + \cos(\theta'_I - \theta_j)\beta_{Ij}\right] \tag{8.9.36b}$$

The off-body Kutta condition can now be expressed in a form similar to that of Eq. (8.9.22). Applying Eq. (8.9.28) to Eq. (8.9.35), we write the resulting expression as

$$\sum_{j=1}^{N} A'_{Nj} q_j + \tau \sum_{j=1}^{N} B'_{Nj} + V_\infty \cos(\alpha - \theta'_N)$$
$$= -\left[\sum_{j=1}^{N} A'_{1j} q_j + \tau \sum_{j=1}^{N} B'_{1j} + V_\infty \cos(\alpha - \theta'_1)\right] \tag{8.9.37a}$$

or as

$$\sum_{j=1}^{N} (A'_{1j} + A'_{Nj}) q_j + \tau \sum_{j=1}^{N} (B'_{1j} + B'_{Nj}) + V_\infty \cos(\alpha - \theta'_1) + V_\infty \cos(\alpha - \theta'_N) = 0 \tag{8.9.37b}$$

8.9.2 Flowfield Calculation in the Wake

The calculation of airfoils in incompressible viscous flows can be accomplished without taking into account the wake effect; that is, the viscous flow calculations are performed up to the trailing edge only and are not extended into the wake. This procedure, which may be sufficient at low to moderate angles of attack without flow separation, is not sufficient at higher angles of attack, including post-stall flows. Additional changes are required in the panel method (and in the boundary-layer method), as discussed in this section.

The viscous wake calculations usually include a streamline issuing from the trailing edge of the airfoil. The computation of the location of this streamline is relatively simple if conformal mapping methods are used to determine the velocity field. In this case, the stream function ψ is usually known, and because the airfoil surface is represented by $\psi(x, y) = \text{const}$, the calculation of the wake streamline amounts to tracing the curve after it leaves the airfoil. When the flowfield is computed by a panel method or by a finite-difference method, however, the results are known only at discrete points in the field in terms of the velocity components. In this case, the wake streamline is determined from the numerical integration of

$$\frac{dy}{dx} = \frac{v}{u} \tag{8.9.38}$$

aft of the trailing edge with known initial conditions. However, some care is necessary in selecting the initial conditions, especially when the trailing edge is blunt. As a general rule, the initial direction of the streamline is given to a good approximation by the bisector of the trailing-edge angle of the airfoil.

The panel method, which was modified only for an airfoil flow, now requires similar modifications to include the viscous effects in the wake which behaves as a distribution of sinks. It is divided into nwp panels along the dividing streamline with suction velocities or sink strengths $q_i = \Delta_{vi}$ $(N + 1 \leq i \leq N + nwp)$, distributed on the wake panels and determined from boundary-layer solutions in the wake by Eq. (8.9.12). As before, off-body boundary points and "control" points are introduced at the intersections of the δ^* surface with the normals through panel boundary points and panel control points, respectively. Summation of all the induced velocities, separately for the normal and tangential components and together with the freestream velocity components, produces $(V^n)_I$ and $(V^t)_I$ at $I = 1, 2, \ldots, N + nwp$. The wake velocity distribution, as the airfoil velocity distribution, is computed on the δ^*-surface, rather than on the dividing streamline.

The total velocities are again computed from Eq. (8.9.29), with $(V^n)_I$ and $(V^t)_I$ from Eqs. (8.9.30), except that now

$$(V^n)_I = \sum_{j=1}^{N+nwp} A^n_{Ij} q_j + \tau \sum_{j=1}^{N} B^n_{Ij} + V_\infty \sin(\alpha - \theta_I) \tag{8.9.39a}$$

$$(V^t)_I = \sum_{j=1}^{N+nwp} A^t_{Ij} q_j + \tau \sum_{j=1}^{N} B^t_{Ij} + V_\infty \cos(\alpha - \theta_I) \tag{8.9.39b}$$

As before, the expression for the total velocity is written in the same form as Eq. (8.9.32a), except that now

$$\begin{aligned} V = & \sum_{j=1}^{N+nwp} (A^n_{Ij} \cdot \sin\phi + A^t_{Ij} \cdot \cos\phi) q_j + \tau \sum_{j=1}^{N} (B^n_{Ij} \cdot \sin\phi + B^t_{Ij} \cdot \cos\phi) \\ & + V_\infty \sin(\alpha - \theta_I) \sin\phi + V_\infty (\alpha - \theta_I) \cos\phi \end{aligned} \tag{8.9.40}$$

where A''_{Ij}, A'_{Ij}, B''_{Ij} and B'_{Ij}, are identical to those given by Eq. (8.9.33). Similarly, Eq. (8.9.35) with A_{Ij} and B_{Ij} given by Eq. (8.9.36) is

$$V = \sum_{j=1}^{N+nwp} A'_{Ijqj} + \tau \sum_{j=1}^{N} B'_{Ij} + V_\infty \cos(\alpha - \theta'_I) \tag{8.9.41}$$

and the Kutta condition given by Eqs. (8.9.37a) becomes

$$\sum_{j=1}^{N+nwp} A'_{Njqj} + \tau \sum_{j=1}^{N} B'_{Nj} + V_\infty \cos(\alpha - \theta'_N) = -\left[\sum_{j=1}^{N+nwp} A'_{1jqj} + \tau \sum_{j=1}^{N} B'_{1j} + V_\infty \cos(\alpha - \theta'_1)\right] \tag{8.9.42a}$$

or

$$\sum_{j=1}^{N+nwp} (A'_{1j} + A'_{Nj})qj + \tau \sum_{j=1}^{N} (B'_{1j} + B'_{Nj}) + V_\infty \cos(\alpha - \theta'_1) + V_\infty \cos(\alpha - \theta'_N) = 0 \tag{8.9.42b}$$

In computing the wake velocity distribution at distances δ^* from the wake dividing streamline, the velocities in the upper wake are equal to those in the lower wake for a symmetrical airfoil at zero angle of attack. This is, however, not the case if the airfoil is asymmetric or if the airfoil is at an angle of incidence. While the external velocities on the upper and lower surfaces at the trailing edge are equal to each other, they are not equal to each other in the wake region since the δ^*-distribution in the upper wake is different from the δ^*-distribution in the lower wake.

8.10 Results for Airfoil Flows

The interactive boundary-layer method dicussed in subsection 8.8 employing the improved CS model (subsection 5.4.2) has been extensively tested for single and multielement airfoils with extensive flow separation. A sample of results were presented in Fig. 5.8b for an airfoil at low Reynolds number (see subsection 5.3.4) and in Figs. 5.12 and 5.13 for airfoils at high Reynolds number (see subsection 5.4.2). Here we present more results for single airfoils and also include multielement airfoils. For additional results, see [40].

Figure 8.27 shows the variation of the lift and drag coefficients of the NACA 0012 airfoil for a chord Reynolds number of 3×10^6. As can be seen from Fig. 8.27a, viscous effects have a considerable effect on the maximum lift coefficient, $(c_l)_{max}$,

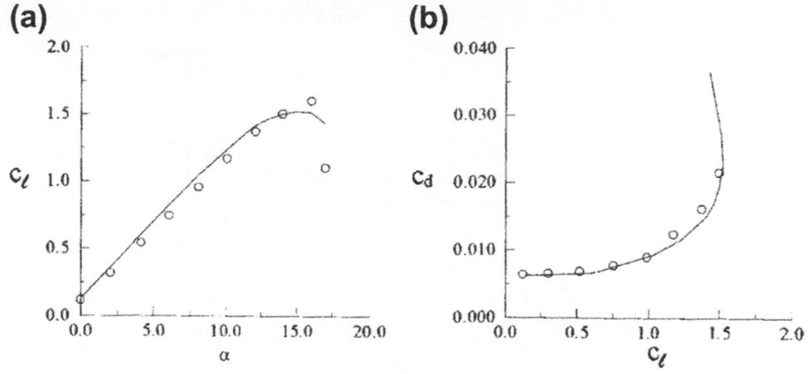

Fig. 8.27 Comparison between calculated (*solid lines*) and experimental values (*symbols*) of: (**a**) c_l vs α, and (**b**) c_d vs c_l. NACA 0012 airfoil at $R_c = 3 \times 10^6$.

of the airfoil, which occurs at a stall angle of around 16° and the calculated results agree well with measurements [40,42].

Figure 8.27b shows the variation of the drag coefficient with lift coefficient. As can be seen, the measurements of drag coefficients do not extend beyond an angle of attack of 12 degrees and at smaller angles agree well with the calculations. The nature of the lift-drag curve is interesting at higher angles of attack with the expected increase in drag coefficient and reduction in lift coefficient for post-stall angles.

Figure 8.28 shows the variation of the local skin-friction coefficient c_f for the same airfoil at the same Reynolds number. As can be seen, flow separation occurs around $\alpha = 10°$ and its extent increases with increasing angle of attack. At an angle of attack $\alpha = 18°$, the flow separation on the airfoils is 50% of the chord length.

Figure 8.29 shows a comparison of calculated and experimental velocity profiles for the NACA 66_3-018 airfoil. The transition location was at $\frac{x}{c} = 0.81$. The

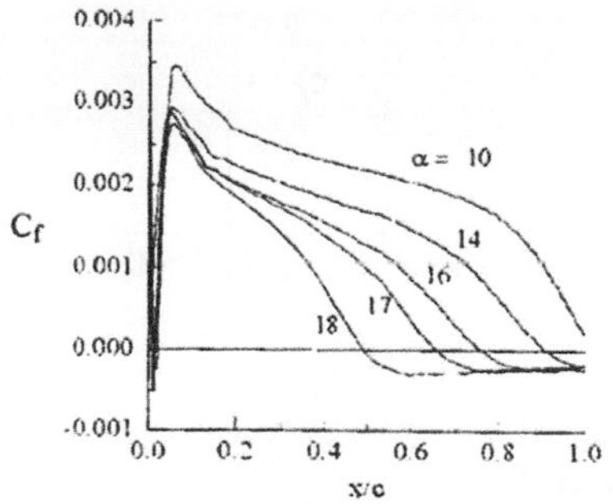

Fig. 8.28 Variation of local skin-friction coefficient distribution. NACA 0012 airfoil at $R_c = 3 \times 10^6$.

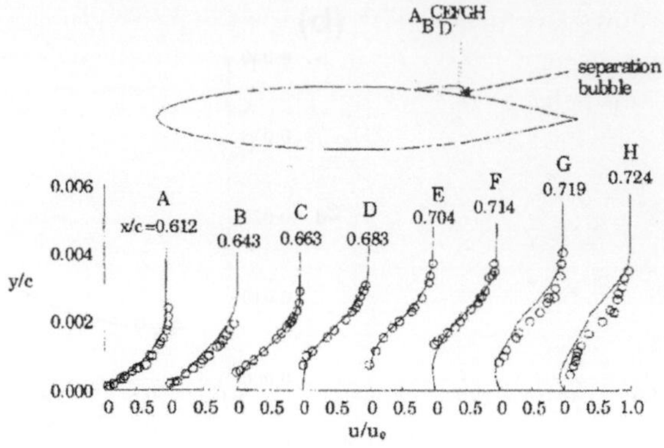

Fig. 8.29 Comparison of calculated (*solid lines*) and measured (*symbols*) velocity profiles for the NACA 66_3-018 airfoil for $\alpha=0°$, $R_c=2\times10^6$.

calculations with transition location at $\frac{x}{c}=0.81$ made use of the CS model with low Reynolds number and transitional flow effects (subsection 5.3.4). As in the results for Fig. 5.8b, the agreement between the calculated and experimental results is very good.

The accuracy of the calculation method employing the CS model has also been investigated extensively for several multielement airfoil configurations. Here we show the results for the airfoil/flap configuration of Van den Berg and Oskam, see [40], which corresponds to a supercritical main airfoil (NLR 7301) with a flap of 32% of the main chord at a deflection angle of 20 degrees and with a gap of 2.6% chord. Measurements of surface pressure and velocity profiles were obtained at a chord Reynolds number of 2.51×10^6 and for angles of attack of 6 and 13.1 degrees.

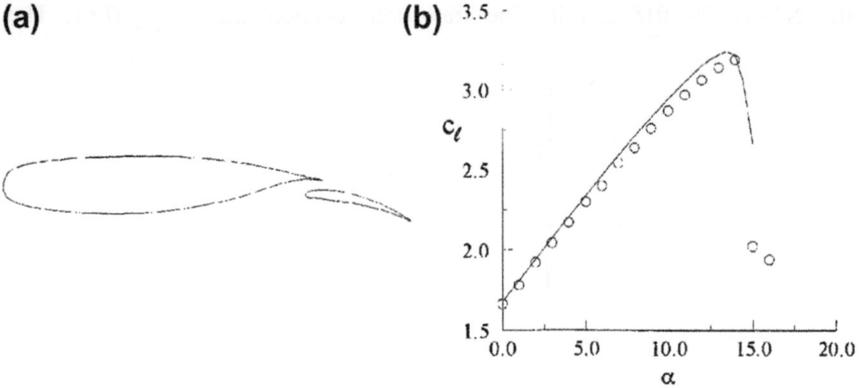

Fig. 8.30 (**a**) NLR 7301 airfoil with flap and (**b**) calculated and measured lift coefficients.

Figure 8.30b shows a comparison between calculated and measured lift coefficients. While some discrepancies exist at higher angles of attack, the stall is predicted accurately. It is believed that these discrepancies are due to the merging of the airfoil shear layer with the boundary-layer on the upper surface of the flap which was not considered in the calculation method described in [40].

8.11 Prediction of Three-Dimensional Flows with Separation

The calculation method described in the previous section for airfoils and multielement airfoils has also been extended and evaluated for wing and multielement wings as described in [40]. Here again we present results for one wing and two slat-wing-flap configurations.

Figure 8.31a shows the lift coefficient variation with angle of attack for the RAE wing tested by Lovell, see [40]. This wing has an airfoil section having a considerable rear loading with the maximum thickness of 11.7% occuring at 37.5% chord and the maximum camber occuring at 75% chord. It has no twist nor dihedral, but has a quarter-chord sweep angle of 28°, a taper ratio of 1/4 and an aspect ratio of 8.35. The experiments were conducted at a test Reynolds number of 1.35×10^6 with one set of measurements corresponding to free transition and with another to fixed transition for a freestream Mach number of 0.223. The wing has a semispan of 1.07 m and a mean aerodynamic chord of 0.26 m.

The calculations for this wing were performed with the Hess panel method [40] and the inverse boundary-layer method of Cebeci [40] described in Section 8.8. Initially, the calculations were done with angle of attack increments of 2° until 10°.

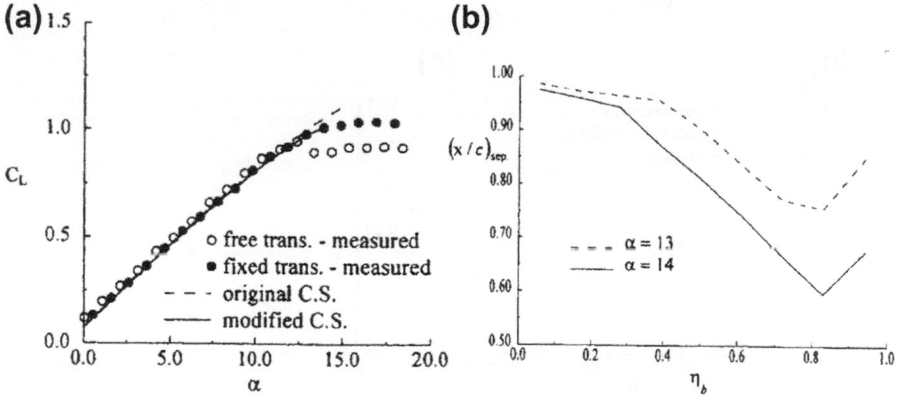

Fig. 8.31 (**a**) Effect of turbulence model on the lift coefficient of the RAE wing and (**b**) distribution of flow separation along the span at two angles of attack.

The increment in angle of attack is reduced to 1° beyond that point. Results show that with the original CS turbulence model (see Section 5) the lift coefficient keeps increasing past the measured stall angle (around 12° with free transition). With the modified CS turbulence model, on the other hand, the agreement between measured and calculated lift coefficient is excellent up to 14°. At 15°, the boundary layer calculations did not converge near the trailing edge due to the large separated flow region.

Figure 8.31b shows flow separation along the span at angles of attack, $\alpha = 13°$ and 14°. As can be seen, there is a significant increase in the amount of flow separation with one degree increase in α.

Figures 8.32 and 8.33 present results for the RAE slat-wing-flap configurations with the slat deflected at 25°, and the flap deflected at 10° and 25°, respectively. Again, the inviscid lift coefficient is included to show how the introduction of the viscous effects allows obtaining reasonable predictions of lift and drag coefficients. The discrepancies may be due to the merging of shear layers which was not accounted for. In addition, the large recirculating flow region in the slat cove – larger at low angles of incidence – was removed with the fairing and may contribute to the disagreement at low angles of attack.

Stall is not captured for the configurations tested. However, it is worthwhile to note that, at the present time, the reliable prediction of stall for slat-wing-flap configurations still offers significant challenges for two-dimensional flows. Unlike for single element and wing-flap configurations, stall can occur without flow separation on the body but may be due instead to a sudden increase of the wake thickness thus reducing the circulation on the entire configuration. Therefore, the results of the calculation method of [40] should be viewed as quite satisfactory.

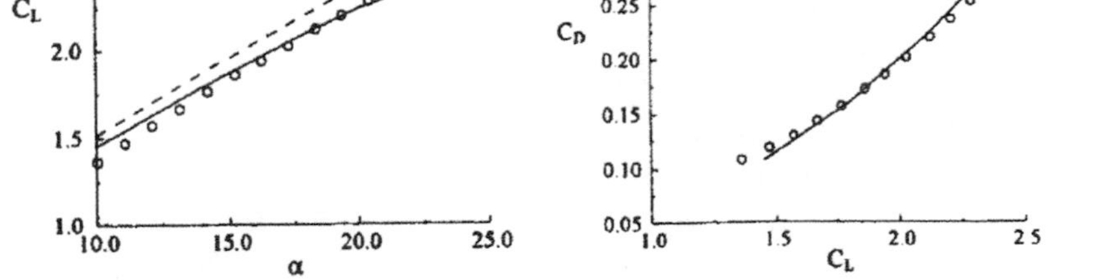

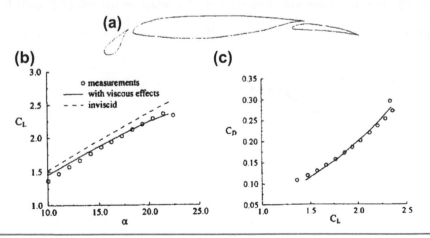

Fig. 8.32 RAE wing with slat deflected at 25° and flap deflected at 10°, (**a**) wing cross-section, (**b**) lift coefficient, and (**c**) drag coefficient.

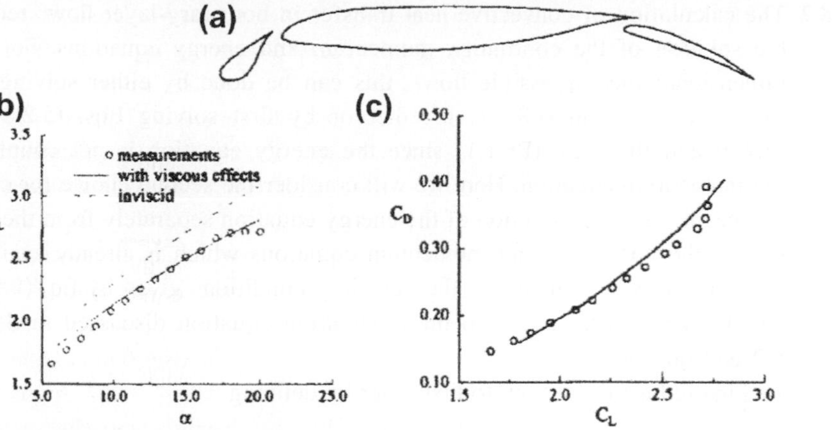

Fig. 8.33 RAE wing with both slat and flap deflected at 25°, (**a**) wing cross-section, (**b**) lift coefficient, and (**c**) drag coefficient.

Problems

8.1 For a two-dimensional steady incompressible laminar and turbulent flow, the energy equation Eq. (3.3.23), and its boundary conditions for specified wall temperature can be written as

$$u\frac{\partial T}{\partial x} + v\frac{\partial T}{\partial y} = \alpha\frac{\partial^2 T}{\partial y^2} - \frac{\partial}{\partial y}(\overline{T'v'}) \tag{P8.1.1}$$

$$y = 0, \quad T = T_w(x) \tag{P8.1.2a}$$

$$y = \delta, \quad T = T_e \tag{P8.1.2b}$$

Using the turbulent Prandtl number concept, $\mathrm{Pr_t}$, and the Falkner-Skan transformation given by Eq. (8.2.5), show that Eqs. (P8.1.1) and (P8.1.2) can be written as

$$(eg')' + \frac{m+1}{2}fg' + n(1-g)f' = x\left(f'\frac{\partial g}{\partial x} - g'\frac{\partial f}{\partial x}\right) \tag{P8.1.3}$$

$$\eta = 0, \quad g = 0 \tag{P8.1.4a}$$

$$\eta = \eta_e, \quad g = 1. \tag{P8.1.4b}$$

Here a prime denotes differentiation with respect to η and

$$g = \frac{T_w - T}{T_w - T_e}, \quad n = \frac{x}{T_w - T_e}\frac{d}{dx}(T_w - T_e) \tag{P8.1.5a}$$

$$e = \frac{1}{\mathrm{Pr}}\left(1 + \varepsilon_m^+\frac{\mathrm{Pr}}{\mathrm{Pr}_t}\right) \tag{P8.1.5b}$$

8.2 The calculation of convective heat transfer in boundary-layer flows requires the solution of the continuity, momentum and energy equations. For two-dimensional incompressible flows, this can be done by either solving Eqs. (5.2.8), (5.2.9) and (P8.1.1) together or by first solving Eqs. (5.2.8) and (5.2.9) and then Eq. (P8.1.1) since the energy equation is not coupled to the momentum equation. Here we will consider the second choice for convenience and seek the solution of the energy equation separately from the solution of the continuity and momentum equations which is already available. Solve Eq. (P8.1.3) subject to the boundary conditions given by Eq. (P8.1.4). Use the procedure similar to the momentum equation discussed in Section 8.2 and outlined below.

1. First reduce the system to first order by defining

$$g' = p \tag{P8.2.1}$$

and using the same definitions of u, v for f' and f'' in the momentum equation, Eq. (8.2.6).

2. Write difference approximations for the two first-order equations and express the first-order energy equation in the following form

$$(s_1)_j p_j + (s_2)_j p_{j-1} + (s_3)_j \left(g_j + g_{j-1}\right) = (r_1)_j, \quad 1 \le j \le J \tag{P8.2.2}$$

3. Write the resulting system in the following form

$$\begin{array}{l} \text{b.c.} \\ \\ \\ \\ \\ \text{b.c.} \end{array}
\begin{bmatrix}
\overset{g_0}{1} & \overset{p_0}{0} & \overset{g_j}{0} & \overset{p_j}{0} & \overset{g_J}{} & \overset{p_J}{} \\
-1 & \frac{-h_1}{2} & 1 & \frac{-h_1}{2} & & \\
(s_3)_j & (s_2)_j & (s_3)_j & (s_1)_j & 0 & 0 \\
0 & 0 & -1 & \frac{-h_{j+1}}{2} & 1 & \frac{-h_{j+1}}{2} \\
 & & (s_3)_J & (s_2)_J & (s_3)_J & (s_2)_J \\
 & & 0 & 0 & 1 & 0
\end{bmatrix} \tag{P8.2.3}$$

$$\times \begin{bmatrix} \begin{pmatrix} g_0 \\ p_0 \end{pmatrix} \\ \begin{pmatrix} g_j \\ p_j \end{pmatrix} \\ \begin{pmatrix} g_J \\ p_j \end{pmatrix} \end{bmatrix} = \begin{bmatrix} \begin{pmatrix} (r_1)_0 \\ (r_2)_0 \end{pmatrix} \\ \begin{pmatrix} (r_1)_j \\ (r_2)_j \end{pmatrix} \\ \begin{pmatrix} (r_1)_J \\ (r_2)_J \end{pmatrix} \end{bmatrix}$$

Note that there is no need for Newton's method since the energy equation is linear.

Noting that Eq. (P8.2.3) is of the matrix-vector form given by Eqs. (8.2.24), use the block elimination method to solve the linear system. Check your algorithm, which we shall call SOLV2 (subsection 10.13.2). Note that this algorithm is written for wall and edge boundary conditions in the form

Wall:

$$\alpha_1 g_0 + \alpha_1 p_0 = \gamma_0 \tag{P8.2.4a}$$

Edge:

$$\beta_0 g_J + \beta_1 p_J = \gamma_1 \tag{P8.2.4b}$$

When g_0 is specified, $\alpha_0 = 1$, $\alpha_1 = 0$ and γ_0 is known. When p_0 is specified, $\alpha_0 = 0$, $\alpha_1 = 1$ and γ_1 is known. Similarly, when g_J is specified $\beta_0 = 1$ and $\beta_1 = 0$. When p_J is specified $\beta_0 = 0$ and $\beta_1 = 1.0$

In our problem, $\alpha_0 = 1$, $\alpha_1 = 0$, $\beta_0 = 1$ and $\beta_1 = 0$.

8.3 Using the computer program discussed in Problem 8.2 and the computer program (BLP2) for solving the continuity and momentum equations (Section 10.3), obtain solution of the energy equation for similar laminar flows with uniform wall temperature ($n = 0$). Take $m = 0$, with $\Pr = 0.72$. Compare the wall heat transfer parameter g'_w with the values given below.

m	g'_w
0	0.2957
1	0.5017

8.4 Repeat Problem 8.3 for a laminar and turbulent flow over a uniformly heated flat plate of length $L = 3$ ft, $u_\infty = 160$ ft/sec, $\nu = 1.6 \times 10^{-4}$ ft^2/cm. Assume transition at $x = 1$ ft and take $\Pr_t = 1.0$.

8.5 For an incompressible laminar and turbulent flow over an infinite swept wing, the boundary-layer equations are given by the continuity equation, Eq. (5.2.8), x-momentum equation, Eq. (5.2.9), and the z-momentum equations given by

$$u\frac{\partial w}{\partial x} + v\frac{\partial w}{\partial y} = \nu\frac{\partial^2 w}{\partial y^2} - \frac{\partial}{\partial y}\left(\overline{w'v'}\right). \tag{P8.5.1}$$

(a) Using the eddy viscosity concept, ε_m, and the Falkner-Skan transformation given by Eq. (8.2.5), show that Eq. (P8.5.1) and its boundary conditions

$$y = 0, \quad w = 0 \tag{P8.5.2a}$$

$$y = \delta, \quad w = w_e \tag{P8.5.2b}$$

can be written as

$$(bg'')' + \frac{m+1}{2} f'g'' = x\left(f'\frac{\partial g'}{\partial x} - g''\frac{\partial f}{\partial x}\right) \tag{P8.5.3}$$

$$\eta = 0, \quad g' = 0 \tag{P8.5.4a}$$

$$\eta = \eta_e, \quad g' = 1. \tag{P8.5.4b}$$

Here a prime again denotes differentiation with respect to η and

$$b = 1 + \varepsilon_m^+, \quad g' = \frac{w}{w_e} \tag{P8.5.5}$$

(b) Express the eddy-viscosity formulation given by Eqs. (5.7.4) for three-dimensional flows in terms of Falkner-Skan variables.

8.6 Using the boundary-layer program BLP2 discussed in Section 10.3, develop a new program to solve the infinite swept wing equations, which, in transformed variables, are given by Eqs. (8.2.7), (P8.2.3) and (P8.2.4). Follow the steps below.

1. Reduce Eq. (P8.5.3) to second order by defining

$$g' = G$$

and to a system of two first order equations by defining

$$G' = P$$

Use the same definitions of u, v for f' and f'' in the momentum equation, Eq. (8.2.6).

2. Write difference approximations similar to the procedure used in Problem 8.2.

3. Solve the resulting linear system with SOLV2.

8.7 As discussed in [1,41], for incompressible flows the external velocity distribution for an infinite swept wing can be obtained from a panel method for two-dimensional flows. The streamwise external velocity u_e/v_∞ can be calculated from

$$\frac{u_e}{V_\infty} = \left(\frac{u_e}{u_\infty}\right)_{2D} \cos\lambda \tag{P8.7.1a}$$

and the spanwise velocity w_e/v_∞ from

$$\frac{w_e}{V_\infty} = \sin\lambda \tag{P8.7.1b}$$

Here λ is the sweep angle and v_∞ is the total velocity,

$$v_\infty = \sqrt{u_\infty^2 + w_e^2} \tag{P8.7.2}$$

(a) Use the panel method (HSPM) given in Section 10.4 and calculate u_e/V_∞ and u_e/V_∞ for the upper surface of an infinite swept wing having the NACA 0012 airfoil cross-section (given on the companion website: store.elsevier.com/companions/9780080983356) and $\lambda = 30°$, $\alpha = 0°$.

Note: The identification of upper and lower surface requires the location of the airfoil stagnation point.

(b) Using the computer program discussed in Problem 8.4, obtain laminar flow solutions up to flow separation for the external velocity given in (a).

8.8 Repeat Problem 8.7(b) for a laminar and turbulent flow with transition at $x/c = 0.20$.

8.9 In some problems, it is desirable to start turbulent flow calculations by specifying the initial velocity profiles. A convenient formula for this purpose is to use Eq. (4.4.41) for $y^+ \leq 50$ and Eq. (4.4.38) for $y^+ \geq 50$.

In terms of the Falkner-Skan variables, Eq. (4.4.41) can be written as

$$f'\sqrt{\frac{2}{c_f}} = \begin{cases} e_1\eta & \eta \leq \dfrac{4}{e_1} \\ c_1 + c_2 \ln\left(e_1\eta\right) + c_3 \ln\left(e_1\eta\right)^2 + c_4(e_1\eta)^3 & \dfrac{4}{e_1} \leq \eta \leq \dfrac{50}{e_1} \end{cases} \quad \text{(P8.9.1)}$$

where c_1, c_2, c_3 and c_4 are the coefficients of Eq. (P8.9.1) and

$$e_1 = \sqrt{R_x}\sqrt{\frac{c_f}{2}}, \quad R_x = \frac{u_e x}{\nu} = R_L \bar{u}_e \xi, \quad R_L = \frac{u_\infty L}{\nu} \quad \text{(P8.9.2)}$$

Similarly, for $\eta \geq 50/e_1$, Eq. (4.4.38) can be written as

$$f'\sqrt{\frac{2}{c_f}} = \frac{1}{\kappa}\ln\left(e_1\eta\right) + c + \frac{1}{\kappa}\left[\Pi\left(1 - \cos\pi\frac{\eta}{\eta_e}\right) + \left(\frac{\eta}{\eta_e}\right)^2 - \left(\frac{\eta}{\eta_e}\right)^3\right] \quad \text{(P8.9.3)}$$

where

$$\eta_e = \sqrt{R_x}\,\frac{\delta}{x} \quad \text{(P8.9.4)}$$

It is clear that a complete velocity profile for a turbulent boundary-layer can be obtained from Eqs. (P8.9.1) and (P8.9.3) provided that the boundary-layer thickness δ and the profile parameter Π are known. Since they are not known at first, they must be calculated in a manner they are compatible with R_θ and c_f.

A convenient procedure is to assume δ^ν ($\nu = 0$) and, calculate Π from Eq. (4.4.38) evaluated at the boundary-layer edge, $\eta = \eta_e$. The initial estimate of δ is obtained from the power-law relation, Eq. (7.2.10b), which for $n = 7$

$$\frac{\theta}{\delta} \sim 0.10$$

Here θ is calculated from the specified value of R_θ,

$$\theta = \frac{\nu}{u_e} R_\theta \tag{P8.9.5}$$

The next values of δ^ν ($\nu = 1, 2, \ldots, n$) are obtained from

$$\delta^{(\nu+1)} = \delta^{(\nu)} - \frac{\phi}{\left(\frac{d\phi}{d\delta}\right)} \tag{P8.9.6}$$

Where, with $\phi_1 (\equiv R_\theta / R_\delta)$ given by Eq. (4.4.39c) and $\tilde{\theta}$ denoting the momentum thickness calculated from Eq. (P8.9.3),

$$\phi = \tilde{\theta} - \delta\phi_1 \tag{P8.9.7a}$$

$$\frac{d\phi}{d\delta} = -\phi_1 - \delta \frac{d\phi_1}{d\Pi} \frac{d\Pi}{d\delta} \tag{P8.9.7b}$$

and with

$$\frac{d\Pi}{d\delta} = -\frac{1}{2\delta} \tag{P8.9.7c}$$

obtained by differentiating Eq. (4.4.40) with respect to δ.

Using the computer program BLP2 (Section 10.3) and the subroutine IVPT, see subsection 10.13.1, both given on the companion website: store.elsevier.com/companions/9780080983356, we can perform turbulent flow calculations for a given external velocity distribution with initial values of R_θ and H given at $\xi = \xi_0$.

(a) Compute turbulent flow on a flat plate of length 20 ft for a Reynolds number per foot, u_e/ν, equal to 10^6. Take uniform steps in ξ ($\Delta\xi$) equal to 1, with $h_1 = 0.01$, $k = 1.14$, $\xi_0 = 5$ ft with R_θ and H at $\xi = \xi_0$ equal to 6000 and 1.4, respectively.

(b) Repeat (a) with $\Delta x = 2$ ft.

Note: Experience shows that the calculations which begin with velocity profiles generated in this way show oscillations in wall shear for x-stations greater than x_0. A convenient procedure is to perform calculations, say for the first two x-stations x_1 and x_2, then average the solutions in the midpoint of x_0 and x_1, $x\frac{1}{2}$, and x_1 and x_2, $x_1\frac{1}{2}$ and then average the solutions $x\frac{1}{2}$ and $x_1\frac{1}{2}$ at x_1 to define a new solution and start the calculations at $x = x_2$. Another useful procedure which is effective in avoiding oscillations is the use of first-order backward differences for the streamwise derivatives.

References

[1] T. Cebeci, J. Cousteix, Modeling and Computation of Boundary-Layer Flows, Horizons Publishing, Long Beach, CA, and Springer-Verlag, Heidelberg, Germany, 1998.

[2] P. Bradshaw, T. Cebeci, J.H. Whitelaw, Engineering Calculation Methods for Turbulent Flows, Academic Press, New York, NY, 1981.

[3] D. Coles, Measurements in the boundary layer on a smooth flat plate in supersonic flow. I. The problem of the turbulent boundary layer. Rep. A20–69, Jet Propulsion Lab., California Inst. Technol., Pasadena, CA, 1953.

[4] P.S. Klebanoff, Characteristics of turbulence in a boundary layer with zero pressure gradient, NACA Tech. Note 3178 (1954).

[5] T. Cebeci, Application of exact turbulent boundary-layer equations as a means of calculating heat and mass transfer in incompressible and compressible flows. Proc. Int. Heat Transfer Conf. 4th vol. 2, Elsevier, Amsterdam, 1970.

[6] R.A. Seban, D.L. Doughty, Heat transfer to turbulent boundary layers with variable free-stream velocity, J. heat Transfer 78 (1956) 217.

[7] W.C. Reynolds, W.M. Kays, S.J. Kline, Heat transfer in the turbulent incompressible boundary layer – II: step wall-temperature distribution, NASA Memo 12-2-58W (1958).

[8] W.C. Reynolds, W.M. Kays, S.J. Kline, Heat transfer in the turbulent incompressible boundary layer – III: arbitrary wall temperature and heat flux, NASA Memo 12-3-58W (1958).

[9] J. McQuaid, Experiments on incompressible turbulent boundary layers with distributed injection. British ARC 28735, HMT 135. Also N69–21334 (1967).

[10] R.L. Simpson, Characteristics of turbulent layers at low Reynolds numbers with and without transpiration, J. Fluid Mech. 42 (1970) 769.

[11] R.J. Moffat, W.M. Kays, The turbulent boundary layer on a porous plate: experimental heat transfer with uniform blowing and suction, Int. J. Heat Mass Transfer 11 (1968) 1547.

[12] T. Cebeci, Behavior of turbulent flow near a porous wall with pressure gradient, AIAA J. 8 (1970) 2152.

[13] T. Cebeci, G.J. Mosinskis, Prediction of turbulent boundary layers with mass addition, including accelerating flows, J. heat Transfer 93 (1971) 271.

[14] B.G. Newman, Some contributions to the study of the turbulent boundary layer near separation, Austr. Dept. Supply Rep. ACA-53 (1951).

[15] T.E. Herring, J. Norbury, Some experiments on equilibrium turbulent boundary layers in favorable pressure gradients, J. Fluid Mech. 27 (1967) 541.

[16] F.H. Clauser, Turbulent boundary layers in adverse pressure gradient, J. Aero. Sci. 21 (1954) 91.

[17] G.B. Schubauer, P.S. Klebanoff, Investigation of separation of the turbulent boundary layer, NACA Tech. Note No. 2133 (1950).

[18] P.M. Moretti, W.M. Kays, Heat transfer to a turbulent boundary layer with varying free-stream velocity and varying surface temperature, an experimental study, Int. J. Heat Mass Transfer 8 (1965) 1187.

[19] R.L. Simpson, W.M. Kays, R.J. Moffat, The turbulent boundary layer on a porous plate: An experimental study of the fluid dynamics with injection and suction. Rep. No. HMT-2, Stanford Univ., Mech. Eng. Dept., Stanford, California, 1967.

[20] H.L. Julien, W.M. Kays, R.J. Moffat, The turbulent boundary layer on a porous plate: experimental study of the effects of a favorable pressure gradient. Rep. No. HMT-4, Stanford Univ. Also N69-34951, NASA-CR-104140, 1969.

[21] R. Richmond, Experimental investigation of thick axially symmetric boundary layers on cylinders at subsonic and hypersonic speeds. Ph.D. Thesis, California Inst. Technol., Pasadena, California, 1957.

[22] T. Cebeci, Laminar and turbulent incompressible boundary layers on slender bodies of revolution in axial flow, J. Basic Eng. 92 (1970) 545.

[23] T. Cebeci, Eddy viscosity distribution in thick axisymmetric turbulent boundary layers, J. Fluids Eng. 95 (1973) 319.

[24] F.M. White, An analysis of axisymmetric turbulent flow past a long cylinder, J. Basic Eng. 94 (1972) 200.

[25] J.J. Cornish III, D.W. Boatwright, Application of full scale boundary layer measurements to drag reduction of airships. Rep. No. 28, Aerophys. Dept. Mississippi State Univ., Jackson, Mississippi, 1960.

[26] J.S. Murphy, The separation of axially-symmetric turbulent boundary layers. Part I. Preliminary experimental results on several bodies in incompressible flows. Rep. No. ES 17513, Douglas Aircraft Co., Long Beach, CA. Also AD 432–666. Part II. Detailed measurements in the boundary layers on several slender bodies in incompressible flow. Rep. No. ES 17513, Douglas Aircraft Co., Long Beach, CA., 1954. Also AD 432–491.

[27] T. Cebeci, Calculation of compressible turbulent boundary layers with heat and mass transfer, AIAA J. 9 (1971) 1091.

[28] D. Coles, Measurements in the boundary layer on a smooth flat plate in supersonic flow. III. Measurements in a flat-plate boundary layer at the Jet Propulsion Lab. Rep. No. 20–71, Jet Propulsion Lab., California Inst. Technol., Pasadena, CA, 1953.

[29] F.W. Matting, D.R. Chapman, J.R. Nyholm, A.G. Thomas, Turbulent skin friction at high Mach numbers and Reynolds numbers in air and helium. NASA TR R-82. Also N62–70506 (1961).

[30] D.R. Moore, J. Harkness, Experimental investigation of the compressible turbulent boundary layer at very high Reynolds numbers. $M = 2.8$, Rep. No. 0.71000/4R-9, LTV Res. Center. Also N64–29671 (1964).

[31] R. Michel, Resultants sur la couche limite turbulent aux grandes vitesses, Memo Tech 22 (1961). Office Nat. d'Ütudes et de Rech, Aeronaut.

[32] C.S. Pappas, Measurement of heat transfer in the turbulent boundary layer on a flat plate in supersonic flow and comparison with skin-friction results, NACA Tech. Note No. 3222 (1954).

[33] L.C. Squire, Further experimental investigation of compressible turbulent boundary layers with an injection. ARC R&M 3627. Also N71–15789 (1970).

[34] G.D. Thomas, V.K. Verma, L.C. Squire, A comparison of two prediction methods with experiment for compressible turbulent boundary layers with air injection, Aeronaut. Quart. XXIII (4) (1972) 301.

[35] L. Pasiuk, S.M. Hastings, R. Chatham, Experimental Reynolds analogy factor for a compressible turbulent boundary layer with a pressure gradient. Naval Ordinance Rep. NOLTR 64–200, White Oak, Maryland. Also AD610-663 (1965).

[36] K.G. Winter, L. Gaudet, Turbulent boundary-layer studies at high Reynolds numbers at Mach numbers between 0.2 and 2.8. RAE Tech. Rep. 70251. Also AD890–133 (1970).

[37] J.C. LeBalleur, Couplage visqueux-non visqueux: méthode numérique et applications aux ecoulements bidimensionnels transsoniques et supersoniques, La Recherche Aérospatiale (March-April 1978).

[38] J. Carter, S.F. Wornom, Solutions for incompressible separated boundary layers including viscous-inviscid interaction, Aerodynamic Analysis Requiring Advanced Computers (1975) 125. NASA SP-347.

[39] A.E.P. Veldman, A numerical method for the calculation of laminar incompressible boundary layers with strong inviscid interaction, NLRTR 79023L (1979).

[40] T. Cebeci, An Engineering Approach to the Calculation of Aerodynamic Flows, Horizons Publishing, Long Beach, CA, and Springer-Verlag, Heidelberg, Germany, 1999.

[41] T. Cebeci, K.C. Chang, Compressibility and Wake Effects on the Calculations of Flow over High Lift Configuration, 48th Annual Conference, Canadian Aeronautics and Space Institute, Toronto, Canada, April 2001.

[42] T. Cebeci, Turbulence Models and Their Application, Horizons Publishing, Long Beach, CA, and Springer-Verlag, Heidelberg, Germany, 2003.

[43] C. Gleyzes, J. Cousteix, J.L. Bonnet, A calculation method of leading-edge separation bubbles, in: T. Cebeci (Ed.), Numerical and Physical Aspects of Aerodynamic Flow, II, Springer-Verlag, Heidelberg, Germany, 1984.

[44] J.C. LeBalleur, New possibilities of viscous-inviscid numerical techniques for solving viscous flow equations with massive separation, in: T. Cebeci (Ed.), Numerical and Physical Aspects of Aerodynamic Flows, IV, Springer-Verlag, Heidelberg, Germany, 1990.

Chapter 9

Differential Methods with Transport-Equation Turbulence Models

Analysis of Turbulent Flows with Computer Programs. http://dx.doi.org/10.1016/B978-0-08-098335-6.00009-4

9.1 Introduction

In Sections 9.2 and 9.3 we discuss the numerical solution of the boundary-layer equations employing transport-equation turbulence models. There are several models that can be used for this purpose, and there are several approaches that can be pursued. For example, in one approach the solution of the k-ε model equations can be obtained with and without the wall functions as discussed in subsection 6.2.1. In the case without wall functions, the usual boundary conditions are replaced by boundary conditions specified at some distance $y = y_0$. In the case with wall functions, the boundary conditions are specified at $y = 0$. Another approach which we shall refer to it as the zonal approach, the boundary-layer equations are solved in two regions with each region employing different turbulence models. In effect this approach may be regarded as the use of the k-ε model with wall functions. In Section 9.2 we discuss the numerical solution of the k-ε model equations with this zonal method; in Section 9.3, the numerical solution of the k-ε model equations with and without wall functions; and in Section 9.4, the numerical solution of the k-ω and SST model equations.

In Section 9.5 we consider four turbulence models discussed in Sections 6.2 and 6.3 and evaluate their relative performance for free-shear flows and attached and separated boundary-layers flows. In discussing the performance of these transport-equation turbulence models, it would be more consistent with this book to present results obtained from the solution of the boundary-layer equations. However, such a study is yet to be conducted for a range of flows including free-shear layer and wall boundary layers with and without separation. For this reason, we present results obtained from the solution of the Navier-Stokes equations.

9.2 Zonal Method for k-ε Model

In the zonal method considered here, the boundary-layer is divided into two zones. The inner zone is identified by $y \leq y_0$, $y_0^+ = (y_0 u_\tau / \nu) \approx 100$, where the continuity and momentum equations, Eqs. (5.2.8) and (8.2.1), are solved subject to the wall boundary conditions given by Eq. (8.2.2a), with eddy viscosity ε_m given by the inner region of the CS model. In the outer zone, $y > y_0$, the continuity Eq. (5.2.8), momentum Eq. (8.2.1), turbulent kinetic energy Eq. (6.2.7) and rate of dissipation Eq. (6.2.8) are solved subject to the inner boundary condition given by Eqs. (6.2.13) and (6.2.14) and edge boundary conditions given by Eqs. (6.2.16) and (6.2.18), with eddy viscosity ε_m computed from Eq. (6.2.6).

The turbulence equations and boundary conditions for this zonal method are provided in the following subsection; the finite-difference equations together with Newton's method are discussed in the subsequent subsection.

9.2.1 Turbulence Equations and Boundary Conditions

As with the differential method with algebraic turbulence models, we again express the turbulence equations in terms of transformed variables. The mean-flow equations remain the same as those considered in Chapter 8, namely, Eqs. (8.2.6) and (8.2.7). The turbulent kinetic energy and rate of dissipation equations and their boundary conditions are also expressed in transformed variables. With the transformation

$$\psi = \sqrt{u_e \nu x} f, \quad \eta = \sqrt{\frac{u_e}{\nu x}} y, \quad \tilde{k} = \frac{k}{u_e^2}, \quad \tilde{\varepsilon} = \frac{\varepsilon x}{u_e^3} \tag{9.2.1}$$

and with the definition of stream function, Eq. (8.2.3), they can be written as

$$\left(b_2 k'\right)' + m_1 f k' + \varepsilon_m^+ (f'')^2 - \varepsilon - 2mf'k = x\left(f' \frac{\partial k}{\partial x} - k' \frac{\partial f}{\partial x}\right) \tag{9.2.2}$$

$$\left(b_3 \varepsilon'\right)' + m_1 f \varepsilon' + C_{\varepsilon 1} f_1 \frac{\varepsilon}{k} \varepsilon_m^+ (f'')^2 - C_{\varepsilon 2} f_2 \frac{\varepsilon^2}{k} - (3m-1) f' \varepsilon = x\left(f' \frac{\partial \varepsilon}{\partial x} - \varepsilon' \frac{\partial f}{\partial x}\right) \tag{9.2.3}$$

where the tilde has been dropped from the equations and

$$b_2 = 1 + \frac{\varepsilon_m^+}{\sigma_k}, \quad b_2 = 1 + \frac{\varepsilon_m^+}{\sigma_\varepsilon}, \quad \varepsilon_m^+ = \frac{\varepsilon_m}{\nu}, \quad m_1 = \frac{m+1}{2} \tag{9.2.4}$$

With the introduction of new variables

$$f' = u, \quad u' = v, \quad k' = s, \quad \varepsilon' = q \tag{9.2.5}$$

Eqs. (8.2.6), (9.2.2) and (9.2.3) can be written as

$$\left(b_1 v\right)' + m_1 f v + m(1 - u^2) = x\left(u \frac{\partial u}{\partial x} - v \frac{\partial f}{\partial x}\right) \tag{9.2.6}$$

$$\left(b_2 s\right)' + m_1 f s + \varepsilon_m^+ v^2 - \varepsilon - 2muk = x\left(u \frac{\partial k}{\partial x} - s \frac{\partial f}{\partial x}\right) \tag{9.2.7}$$

$$\left(b_3 q\right)' + m_1 f q + C_{\varepsilon 1} f_1 \frac{\varepsilon}{k} \varepsilon_m^+ v^2 - C_{\varepsilon 2} f_2 \frac{\varepsilon^2}{k} - (3m-1) u \varepsilon = x\left(u \frac{\partial \varepsilon}{\partial x} - q \frac{\partial f}{\partial x}\right) \tag{9.2.8}$$

These equations form a system of seven first-order differential equations with seven dependent variables (f, u, v, k, s, ε, and q) for the outer zone and three for the inner zone. The definitions, $k' = 0$, $s' = 0$, $\varepsilon' = 0$, and $q' = 0$, respectively, replace the last two expressions in Eq. (9.2.5) and the equations (9.2.7) and (9.2.8) in the solution algorithm to represent the equations for the inner zone, so there are seven first-order equations from the wall to the edge of the boundary layer that require solution subject to the seven boundary conditions given by Eqs. (8.2.2), (6.2.14)–(6.2.16) and (6.2.18), which in terms of transformed variables can be written as

$$\eta = 0, \quad f = u = 0 \tag{9.2.9a}$$

$$\eta = \eta_0, \quad (\varepsilon_m^+)_{\mathrm{CS}} = \mathrm{R}_x f_\mu c_\mu \frac{k^2}{\varepsilon}, \quad \varepsilon = (\varepsilon_m^+)_{\mathrm{CS}} v^2 \tag{9.2.9b}$$

$$\begin{gathered}\eta = \eta_e, \quad u = 1.0, \quad x\frac{\partial k}{\partial x} + \varepsilon + 2m_2 k = 0, \\ x\frac{\partial \varepsilon}{\partial x} + C_{\varepsilon 2} f_2 \frac{\varepsilon^2}{k} + (3m - 1)\varepsilon = 0\end{gathered} \tag{9.2.9c}$$

9.2.2 Solution Procedure

In general, differential methods for turbulent flows require the specification of initial profiles at $x = x_0$. With methods employing algebraic viscosity models, the initial profiles correspond to streamwise u and normal v velocity profiles. However, when the calculations are performed for both laminar and turbulent flows, as was discussed in Chapter 8, the initial velocity profiles may be assumed to correspond to those at the transition location. With methods employing transport-equation models, since the calculations are for turbulent flows, it is often necessary to specify initial profiles, which in the case of k-ε model, correspond not only to u and v profiles, but also to k, ε_m and ε profiles.

Experience with the box method discussed in Chapter 8 has shown that when profiles are used to start the turbulent flow calculations, the solutions at the subsequent x-locations oscillate. A common cure to this problem is to compute the first two x-stations equally spaced and take an average of the solutions at the midpoint of x_0 and x_1, say x_m and x_1 and x_2, say x_e. Then another average of the solutions is taken at x_m and x_e defining a new solution at x_1. When new calculations begin at x_2 with averaged profiles at $x = x_1$, the solutions at $x \geq x_2$ do not exhibit oscillations.

While this cure is relatively easy to incorporate into a computer program and in most cases provides stable solutions in adverse pressure gradient flows, sometimes the solutions may breakdown due to oscillations. On the other hand, the author and his colleagues observed that if one uses backward difference approximations for the x-derivates in the boundary-layer equations, rather than central differences as used in the box method, the solutions do not oscillate and are more stable. For this reason, when the box method is used for turbulent flow calculations with initial profiles, we will represent the x-derivates with backward finite difference approximations.

As discussed in Section 6.2, the k-ε model equations without wall functions use "wall" boundary conditions specified at some distance y_0 outside the viscous sublayer. The boundary conditions on u and v are usually represented by Eqs. (6.2.10) and those for k and ε by Eqs. (6.2.13) and (6.2.14), although in the latter case, there are other choices. In such cases, the friction velocity u_τ appearing in u and v equations is unknown and must be determined as part of the solution. One approach is to assume u_τ, τ_0 (say from the initial profiles at the previous x-station), solve the governing equations subject to the "wall" and edge boundary conditions. From the solution determine τ_0 at y_0,

$$\tau_0 = \varrho(\varepsilon_m)_0\left(\frac{\partial u}{\partial y}\right)_0 \tag{9.2.10}$$

and compute u_τ from Eq. (6.2.11). If the calculated value of u_τ does not agree with the estimated value within a specified tolerance parameter δ_1,

$$\left|u_\tau^{\nu+1} - u_\tau^{\nu}\right| < \delta_1 \tag{9.2.11}$$

then a new solution is obtained with the updated values of u_τ and τ_0. This procedure is repeated until convergence.

This iterative procedure can be replaced with a more efficient one by treating u_τ as an unknown. Since u_τ is a function of x only, we can write

$$u_\tau' = 0 \tag{9.2.12}$$

thus increasing the number of first-order equations from seven to eight. Although the A_j, B_j, C_j matrices now become 8×8, rather than 7×7, this procedure does not increase the storage much and allows the solutions to converge faster, especially for flows with strong adverse pressure gradient.

In the solution procedure described here, the numerical method is formulated for eight unknowns, not only for the k-ε model equations with the zonal method but also for the k-ε model equations with and without wall functions and with the zonal method. This choice does not increase the complexity of the solution procedure, and as we shall discuss in Chapter 10, it paves the way to solve the k-ε model equations or others in an inverse mode if the solution procedure is to be extended to flows with separation (see Section 10.9).

Inner Region

The numerical solution of the k-ε model equations with the zonal method requires that in the inner region Eq. (8.2.6) is solved subject to the true wall boundary conditions $f = 0$, $u = 0$. Since, however, the solution procedure is being formulated for the general case which includes the solution of the k-ε model equations without wall functions, it is nescessary to specify a boundary condition for u_τ. This can be done as described below.

From the definition of $u_\tau \left(\equiv u_e\sqrt{\frac{c_f}{2}}\right)$, we can write

$$\frac{u_\tau}{u_e} \equiv w = \sqrt{\frac{c_f}{2}} \tag{9.2.13a}$$

or in transformed variables,

$$w = \frac{\sqrt{f''_w}}{R_x^{1/4}} \tag{9.2.13b}$$

The boundary condition for w is

$$w_0 = \frac{\sqrt{v_0}}{R_x^{1/4}} \tag{9.2.14}$$

Next the eight first-order equations can be written by letting $u' = v$, $k' = 0$, $s' = 0$, $\varepsilon' = 0$, $w' = 0$, $q' = 0$, $f' = u$ and the momentum equation (9.2.6). For $j = 0$, with the first three equations corresponding to boundary conditions, *the equations for the inner region* are ordered as

$$f_0 = 0 \tag{9.2.15a}$$

$$u_0 = 0 \tag{9.2.15b}$$

$$w_0 = \frac{\sqrt{v_0}}{R_x^{1/4}} \tag{9.2.15c}$$

$$u' = v \tag{9.2.15d}$$

$$k' = 0 \tag{9.2.15e}$$

$$s' = 0 \tag{9.2.15f}$$

$$\varepsilon' = 0 \tag{9.2.15g}$$

$$q' = 0 \tag{9.2.15h}$$

With finite-difference approximations and linearization, they become

$$\delta f_0 = (r_1)_0 = 0 \tag{9.2.16a}$$

$$\delta u_0 = (r_2)_0 = 0 \tag{9.2.16b}$$

$$\delta v_0 - 2\sqrt{R_x}w_0\delta w_0 = (r_3)_0 = \sqrt{R_x}w_0^2 - v_0 \tag{9.2.16c}$$

$$\delta u_j - \delta u_{j-1} - \frac{h_j}{2}(\delta v_j + \delta v_{j-1}) = (r_4)_j = u_{j-1} - u_j + h_j v_{j-1/2} \tag{9.2.16d}$$

$$\delta k_j - \delta k_{j-1} = (r_5)_j = 0 \tag{9.2.16e}$$

$$\delta s_j - \delta s_{j-1} = (r_6)_j = 0 \tag{9.2.16f}$$

$$\delta \varepsilon_j - \delta \varepsilon_{j-1} = (r_7)_j = 0 \tag{9.2.16g}$$

$$\delta q_j - \delta q_{j-1} = (r_8)_j = 0 \tag{9.2.16h}$$

For $1 \leq j \leq j_s$, the order of the equations is the same as those above except that the first three equations are replaced by

$$w' = 0 \tag{9.2.17a}$$

$$f' = u \tag{9.2.17b}$$

$$\text{momentum Eq.(9.2.6)} \tag{9.2.17c}$$

In linearized form they can be written as

$$\delta w_j - \delta w_{j-1} = (r_1)_j = 0 \tag{9.2.18a}$$

$$\delta f_j - \delta f_{j-1} - \frac{h_j}{2}(\delta u_j + \delta u_{j-1}) = (r_2)_j = f_{j-1} - f_j + h_j u_{j-1/2} \tag{9.2.18b}$$

$$(s_1)_j\delta f_j + (s_2)_j\delta f_{j-1} + (s_3)_j\delta u_j + (s_4)_j\delta u_{j-1} + (s_5)_j\delta v_j + (s_6)_j\delta v_{j-1} = (r_3)_j \tag{9.2.18c}$$

The finite-difference procedure for Eq. (9.2.6) is identical to the procedure described in subsection 8.2.1. The only difference occurs in the solution of Eq. (9.2.6) where we use three-point or two-point backward finite-difference formulas for the x-wise derivatives rather than central differences as we did in subsection 8.2.1. For this purpose, for any variable V, the derivative of $\frac{\partial V}{\partial x}$ is defined by

$$\left(\frac{\partial V}{\partial x}\right)^n = A_1 V^{n-2} + A_2 V^{n-1} + A_3 V^n \tag{9.2.19}$$

where for first-order

$$A_1 = 0, \quad A_2 = -\frac{1}{x_n - x_{n-1}}, \quad A_3 = -\frac{1}{x_n - x_{n-1}} \tag{9.2.20}$$

and second-order

$$
\begin{aligned}
A_1 &= \frac{(x_n - x_{n-1})}{(x_{n-2} - x_{n-1})\,(x_{n-2} - x_n)} \\
A_2 &= \frac{(x_n - x_{n-2})}{(x_{n-1} - x_{n-2})\,(x_{n-1} - x_n)} \qquad (9.2.21) \\
A_3 &= \frac{2x_n - x_{n-1} - x_{n-2}}{(x_n - x_{n-2})\,(x_n - x_{n-1})}
\end{aligned}
$$

Representing the x-derivatives in Eq. (9.2.6) with either *two-point or three-point backward difference approximations at* $x = x^n$ and using central differences in the η-direction, we can write Eq. (9.2.6) as

$$
\begin{aligned}
&h_j^{-1}\left[(bv)_j^n - (bv)_{j-1}^n\right] + m_1^n (fv)_{j-1/2}^n + m^n\left[1 - (u^2)_{j-1/2}^n\right] \\
&= \frac{1}{2}x^n\left[\frac{\partial}{\partial x}(u^2)\right]_{j-1/2}^n - \frac{x^n}{2}\left[\left(v\frac{\partial f}{\partial x}\right)_j^n + \left(v\frac{\partial f}{\partial x}\right)_{j-1}^n\right] \qquad (9.2.22)
\end{aligned}
$$

Linearizing we get

$$
\begin{aligned}
&h_j^{-1}\left(b_j^n \delta v_j - b_{j-1}^n \delta v_{j-1}\right) \\
&+\frac{m_1^n}{2}\left(f_j^n \delta v_j + v_j^n \delta f_j + f_{j-1}^n \delta v_{j-1} + v_{j-1}^n \delta f_{j-1}\right) \\
&-m^n\left(u_j \delta u_j + u_{j-1}\delta u_{j-1}\right) \\
&= \frac{x^n}{4}\left[\frac{\partial}{\partial u}\left(\frac{\partial u^2}{\partial x}\right)_j^n \delta u_j + \frac{\partial}{\partial u}\left(\frac{\partial u^2}{\partial x}\right)_{j-1}^n \delta u_{j-1}\right] \qquad (9.2.23) \\
&-\frac{x^n}{2}\left[\left(\frac{\partial f}{\partial x}\right)_j^n \delta v_j + v_j^n \frac{\partial}{\partial f}\left(\frac{\partial f}{\partial x}\right)_j^n \delta f_j + \left(\frac{\partial f}{\partial x}\right)_{j-1}^n \delta v_{j-1}\right. \\
&\left. + v_{j-1}^n \frac{\partial}{\partial f}\left(\frac{\partial f}{\partial x}\right)_{j-1}^n \delta f_{j-1}\right]
\end{aligned}
$$

From Eq. (9.2.19), it follows that

$$
\begin{aligned}
&\frac{\partial}{\partial u}\left(\frac{\partial u^2}{\partial x}\right)_j^n = 2A_3 u_j^n, \quad \frac{\partial}{\partial u}\left(\frac{\partial u^2}{\partial x}\right)_{j-1}^n = 2A_3 u_{j-1}^n \\
&\frac{\partial}{\partial f}\left(\frac{\partial f}{\partial x}\right)_j^n = A_3, \quad \frac{\partial}{\partial f}\left(\frac{\partial f}{\partial x}\right)_{j-1}^n = A_3 \qquad (9.2.24)
\end{aligned}
$$

The linearized expression can be written in the form given by Eq. (9.2.18c). The coefficients $(s_1)_j$ to $(s_6)_j$ and $(r_2)_j$ are given by

$$(s_1)_j = \frac{1}{2}\left(m_1^n + x^n\right)v_j^n \tag{9.2.25a}$$

$$(s_2)_j = \frac{1}{2}\left(m_1^n + x^n\right)v_{j-1}^n \tag{9.2.25b}$$

$$(s_3)_j = -\left(m^n + \frac{x^n}{2}A_3\right)u_j^n \tag{9.2.25c}$$

$$(s_4)_j = -\left(m^n + \frac{x^n}{2}A_3\right)u_{j-1}^n \tag{9.2.25d}$$

$$(s_5)_j = h_j^{-1}b_j^n + \frac{m_1^n}{2}f_j^n + \frac{x^2}{2}\left(\frac{\partial f}{\partial x}\right)_j^n \tag{9.2.25e}$$

$$(s_6)_j = -h_j^{-1}b_{j-1}^n + \frac{m_1^n}{2}f_{j-1}^n + \frac{x^2}{2}\left(\frac{\partial f}{\partial x}\right)_{j-1}^n \tag{9.2.25f}$$

$$\begin{aligned}(r_3)_j = &-\left[h_j^{-1}\left[(bv)_j^n - (bv)_{j-1}^n\right] + m_1^n(fv)_{j-1/2}^n + m^n\left[1 - (u^2)_{j-1/2}^n\right]\right]\\ &+ \frac{1}{2}x^n\left[\frac{\partial}{\partial x}(u^2)\right]_{j-1/2}^n - \frac{x^n}{2}\left[\left(v\frac{\partial f}{\partial x}\right)_j^n + \left(v\frac{\partial f}{\partial x}\right)_{j-1}^n\right]\end{aligned} \tag{9.2.26}$$

The linearized finite-difference equations and their boundary conditions, Eqs. (9.2.16) and (9.2.17), are again written in matrix-vector form, with eight dimensional vectors $\vec{\delta}_j$ and $\vec{r}_j$ for each value of j defined by

$$\vec{\delta}_j = \begin{vmatrix} \delta f_j \\ \delta u_j \\ \delta v_j \\ \delta k_j \\ \delta s_j \\ \delta \varepsilon_j \\ \delta q_j \\ \delta w_j \end{vmatrix}, \quad \vec{r}_j = \begin{vmatrix} (r_1)_j \\ (r_2)_j \\ (r_3)_j \\ (r_4)_j \\ (r_5)_j \\ (r_6)_j \\ (r_7)_j \\ (r_8)_j \end{vmatrix} \tag{9.2.27}$$

leading to the following definitions of 8×8 matrices A_j, B_j and C_j in the inner region, $0 \le j \le j_s$

$$A_0 = \begin{vmatrix} 1 & 0 & 0 & 0 & 0 & 0 & 0 & 0 \\ 0 & 1 & 0 & 0 & 0 & 0 & 0 & 0 \\ 0 & 0 & 1 & 0 & 0 & 0 & 0 & -2\sqrt{R_x}w_0 \\ 0 & -1 & -\frac{h_1}{2} & 0 & 0 & 0 & 0 & 0 \\ 0 & 0 & 0 & -1 & 0 & 0 & 0 & 0 \\ 0 & 0 & 0 & 0 & -1 & 0 & 0 & 0 \\ 0 & 0 & 0 & 0 & 0 & -1 & 0 & 0 \\ 0 & 0 & 0 & 0 & 0 & 0 & -1 & 0 \end{vmatrix} \tag{9.2.28a}$$

$$C_j = \begin{vmatrix} 0 & 0 & 0 & 0 & 0 & 0 & 0 & 0 \\ 0 & 0 & 0 & 0 & 0 & 0 & 0 & 0 \\ 0 & 0 & 0 & 0 & 0 & 0 & 0 & 0 \\ 0 & 1 & -\frac{h_{j+1}}{2} & 0 & 0 & 0 & 0 & 0 \\ 0 & 0 & 0 & 1 & 0 & 0 & 0 & 0 \\ 0 & 0 & 0 & 0 & 1 & 0 & 0 & 0 \\ 0 & 0 & 0 & 0 & 0 & 1 & 0 & 0 \\ 0 & 0 & 0 & 0 & 0 & 0 & 1 & 0 \end{vmatrix} \quad 0 \le j \le j_s - 1 \tag{9.2.28b}$$

$$A_j = \begin{vmatrix} 0 & & 0 & 0 & 0 & 0 & 0 & 1 \\ 1 & -\frac{h_j}{2} & 0 & 0 & 0 & 0 & 0 & 0 \\ (s_1)_j & (s_3)_j & (s_5)_j & 0 & 0 & 0 & 0 & 0 \\ 0 & -1 & -\frac{h_{j+1}}{2} & 0 & 0 & 0 & 0 & 0 \\ 0 & 0 & 0 & -1 & 0 & 0 & 0 & 0 \\ 0 & 0 & 0 & 0 & -1 & 0 & 0 & 0 \\ 0 & 0 & 0 & 0 & 0 & -1 & 0 & 0 \\ 0 & 0 & 0 & 0 & 0 & 0 & -1 & 0 \end{vmatrix} \quad 1 \le j \le j_s - 1 \tag{9.2.28c}$$

$$B_j = \begin{vmatrix} 0 & 0 & 0 & 0 & 0 & 0 & 0 & -1 \\ -1 & -\frac{h_j}{2} & 0 & 0 & 0 & 0 & 0 & 0 \\ (s_2)_j & (s_4)_j & (s_6)_j & 0 & 0 & 0 & 0 & 0 \\ 0 & 0 & 0 & 0 & 0 & 0 & 0 & 0 \\ 0 & 0 & 0 & 0 & 0 & 0 & 0 & 0 \\ 0 & 0 & 0 & 0 & 0 & 0 & 0 & 0 \\ 0 & 0 & 0 & 0 & 0 & 0 & 0 & 0 \\ 0 & 0 & 0 & 0 & 0 & 0 & 0 & 0 \end{vmatrix} \quad 1 < j \leq j_s \tag{9.2.28d}$$

Interface between Inner and Outer Regions

The first-order system of equations are now ordered as

$$w' = 0 \tag{9.2.29a}$$

$$f' = u \tag{9.2.29b}$$

$$\text{momentum Eq. (9.2.6)} \tag{9.2.29c}$$

$$\text{b.c. Eq. (9.2.9b)} \tag{9.2.29d}$$

$$\text{b.c. Eq. (9.2.9b)} \tag{9.2.29e}$$

$$u' = v \tag{9.2.29f}$$

$$k' = s \tag{9.2.29g}$$

$$\varepsilon' = q \tag{9.2.29h}$$

The resulting A_j and C_j matrices from the linearized equations, with B_j given by Eq. (9.2.28d) and $(s_1)_j$ to $(s_6)_j$ by. Eq. (9.2.25) at $j = j_s$ are

$$A_{j_s} = \begin{vmatrix} 0 & 0 & 0 & 0 & 0 & 0 & 0 & 1 \\ 1 & -\frac{h_{js}}{2} & 0 & 0 & 0 & 0 & 0 & 0 \\ (s_1)_{j_s} & (s_3)_{j_s} & (s_5)_{j_s} & 0 & 0 & 0 & 0 & 0 \\ 0 & 0 & D_1 & D_2 & 0 & D_3 & 0 & 0 \\ 0 & 0 & D_4 & D_5 & 0 & D_6 & 0 & 0 \\ 0 & -1 & -\frac{h_{js+1}}{2} & 0 & 0 & 0 & 0 & 0 \\ 0 & 0 & 0 & -1 & -\frac{h_{js+1}}{2} & 0 & 0 & 0 \\ 0 & 0 & 0 & 0 & 0 & -1 & -\frac{h_{js+1}}{2} & 0 \end{vmatrix} \tag{9.2.30a}$$

$$C_{js} = \begin{vmatrix} 0 & 0 & 0 & 0 & 0 & 0 & 0 & 0 \\ 0 & 0 & 0 & 0 & 0 & 0 & 0 & 0 \\ 0 & 0 & 0 & 0 & 0 & 0 & 0 & 0 \\ 0 & 0 & 0 & 0 & 0 & 0 & 0 & 0 \\ 0 & 0 & 0 & 0 & 0 & 0 & 0 & 0 \\ 0 & 1 & -\frac{h_{js+1}}{2} & 0 & 0 & 0 & 0 & 0 \\ 0 & 0 & 0 & 1 & -\frac{h_{js+1}}{2} & 0 & 0 & 0 \\ 0 & 0 & 0 & 0 & 0 & 1 & -\frac{h_{js+1}}{2} & 0 \end{vmatrix} \tag{9.2.30b}$$

Here the fourth and fifth rows of A_{js} follow from the boundary conditions, Eq. (9.2.9b), at $\eta = \eta_0$. After the application of Newton's method to the finite-difference form of these equations, D_1 to D_6 are given by the following expressions.

$$D_1 = \varepsilon_{j_s}\frac{\partial}{\partial v}\left(\varepsilon_m^+\right)_{\mathrm{CS}}, \quad D_2 = -2\mathrm{R}_x c_\mu k_{j_s}, \quad D_3 = \left(\varepsilon_m^+\right)_{\mathrm{CS}} \tag{9.2.31a}$$

$$D_4 = 2\mathrm{R}_x c_\mu k_{j_s}^2 v_{j_s}, \quad D_5 = -2\mathrm{R}_x c_\mu k_{j_s} v_{j_s}^2, \quad D_6 = -2\varepsilon_{j_s} \tag{9.2.31b}$$

The associated $(r_4)_{js}$ and $(r_5)_{js}$ are

$$(r_4)_{j_s} = \mathrm{R}_x c_\mu k_{j_s}^2 - \left(\varepsilon_m^+\right)_{\mathrm{CS}}\varepsilon_{j_s} \tag{9.2.32a}$$

$$(r_5)_{j_s} = \varepsilon_{j_s}^2 - \mathrm{R}_x c_\mu \left(k_{j_s} v_{j_s}\right)^2 \tag{9.2.32b}$$

Outer Region

The finite-difference approximations for the outer region defined for $j_s + 1 \leq j \leq J$ are written by using a similar procedure described for the inner region equations. The system of first-order equations are arranged similar to those given by Eqs. (9.2.29) except that Eqs. (9.2.29d) and (9.2.29e) are replaced by Eqs. (9.2.7) and (9.2.8). The resulting matrices from the linearized equations, with C_j matrix remaining the same as that given by Eq. (9.2.30b) for $j_s < j \leq J - 1$, are

$$B_j = \begin{vmatrix} 0 & 0 & 0 & 0 & 0 & 0 & 0 & -1 \\ -1 & -\frac{h_j}{2} & 0 & 0 & 0 & 0 & 0 & 0 \\ (s_2)_j & (s_4)_j & (s_6)_j & (s_8)_j & 0 & (s_{12})_j & 0 & 0 \\ (\alpha_2)_j & (\alpha_4)_j & (\alpha_6)_j & (\alpha_8)_j & (\alpha_{10})_j & (\alpha_{12})_j & 0 & 0 \\ (\beta_2)_j & (\beta_4)_j & (\beta_6)_j & (\beta_8)_j & 0 & (\beta_{12})_j & (\beta_{14})_j & 0 \\ 0 & 0 & 0 & 0 & 0 & 0 & 0 & 0 \\ 0 & 0 & 0 & 0 & 0 & 0 & 0 & 0 \\ 0 & 0 & 0 & 0 & 0 & 0 & 0 & 0 \end{vmatrix} \quad j_s + 1 < j \leq J \tag{9.2.33a}$$

$$A_j = \begin{vmatrix} 0 & 0 & 0 & 0 & 0 & 0 & 0 & 1 \\ 1 & -\frac{h_j}{2} & 0 & 0 & 0 & 0 & 0 & 0 \\ (s_1)_j & (s_3)_j & (s_5)_j & (s_7)_j & 0 & (s_{11})_j & 0 & 0 \\ (\alpha_1)_j & (\alpha_3)_j & (\alpha_5)_j & (\alpha_7)_j & (\alpha_9)_j & (\alpha_{11})_j & 0 & 0 \\ (\beta_1)_j & (\beta_3)_j & (\beta_5)_j & (\beta_7)_j & 0 & (\beta_{11})_j & (\beta_{13})_j & 0 \\ 0 & -1 & -\frac{h_{j+1}}{2} & 0 & 0 & 0 & 0 & 0 \\ 0 & 0 & 0 & -1 & -\frac{h_{j+1}}{2} & 0 & 0 & 0 \\ 0 & 0 & 0 & 0 & 0 & -1 & -\frac{h_{j+1}}{2} & 0 \end{vmatrix} \quad j_s + 1 < j \leq J - 1 \tag{9.2.33b}$$

$$A_J = \begin{vmatrix} 0 & 0 & 0 & 0 & 0 & 0 & 0 & 1 \\ 1 & -\frac{h_J}{2} & 0 & 0 & 0 & 0 & 0 & 0 \\ (s_1)_J & (s_3)_J & (s_5)_J & (s_7)_J & 0 & (s_{11})_J & 0 & 0 \\ (\alpha_1)_J & (\alpha_3)_J & (\alpha_5)_J & (\alpha_7)_J & (\alpha_9)_J & (\alpha_{11})_J & 0 & 0 \\ (\beta_1)_J & (\beta_3)_J & (\beta_5)_J & (\beta_7)_J & 0 & (\beta_{11})_J & (\beta_{13})_J & 0 \\ 0 & 1 & 0 & 0 & 0 & 0 & 0 & 0 \\ 0 & 0 & 0 & E_1 & 0 & E_2 & 0 & 0 \\ 0 & 0 & 0 & E_3 & 0 & E_4 & 0 & 0 \end{vmatrix} \tag{9.2.33c}$$

Here $(s_1)_j$ to $(s_{12})_j$, $(\alpha_1)_j$, to $(\alpha_{12})_j$ and $(\beta_1)_j$ to $(\beta_{14})_j$ given in Appendix 9A correspond to the coefficients of the linearized momentum (9.2.6), kinetic energy of turbulence (9.2.7), and rate of dissipation (9.2.8) equations written in the following forms, respectively,

$$\begin{aligned} &(s_1)_j \delta f_j + (s_2)_j \delta f_{j-1} + (s_3)_j \delta u_j + (s_4)_j \delta u_{j-1} + (s_5)_j \delta v_j \\ &\quad + (s_6)_j \delta v_{j-1} + (s_7)_j \delta k_j + (s_8)_j \delta k_{j-1} + (s_{11})_j \delta \varepsilon_j \\ &\quad + (s_{12})_j \delta \varepsilon_{j-1} = (r_3)_j \end{aligned} \tag{9.2.34}$$

$$\begin{aligned} &(\alpha_1)_j \delta f_j + (\alpha_2)_j \delta f_{j-1} + (\alpha_3)_j \delta u_j + (\alpha_4)_j \delta u_{j-1} + (\alpha_5)_j \delta v_j \\ &\quad + (\alpha_6)_j \delta v_{j-1} + (\alpha_7)_j \delta k_j + (\alpha_8)_j \delta k_{j-1} + (\alpha_9)_j \delta s_j \\ &\quad + (\alpha_{10})_j \delta s_{j-1} + (\alpha_{11})_j \delta \varepsilon_j + (\alpha_{12})_j \delta \varepsilon_{j-1} = (r_4)_j \end{aligned} \tag{9.2.35}$$

$$\begin{aligned}(\beta_1)_j\delta f_j + (\beta_2)_j\delta f_{j-1} + (\beta_3)_j\delta u_j + (\beta_4)_j\delta u_{j-1} + (\beta_5)_j\delta v_j \\ +(\beta_6)_j\delta v_{j-1} + (\beta_7)_j\delta k_j + (\beta_8)_j\delta k_{j-1} + (\beta_{11})_j\delta\varepsilon_j \\ +(\beta_{12})_j\delta\varepsilon_{j-1} + (\beta_{13})_j\delta q_j + (\beta_{14})_j\delta q_{j-1} = (r_5)_j\end{aligned} \tag{9.2.36}$$

The last three rows of the A_J matrix correspond to the edge boundary conditions and follow from the linearized forms of Eq. (9.2.9c). They are given by

$$\begin{aligned} E_1 &= 2m^n + x^n \frac{\partial}{\partial k}\left(\frac{\partial k}{\partial x}\right)_J^n, \\ E_2 &= 1, \quad E_3 = -c_{\varepsilon_2} f_2^n \frac{(\varepsilon^2)_J^n}{(k^2)_J^n}, \\ E_4 &= 3m^n - 1 + x^n \frac{\partial}{\partial \varepsilon}\left(\frac{\partial \varepsilon}{\partial x}\right)_J^n + c_{\varepsilon_2} f_2^n \frac{2\varepsilon_J^n}{k_J^n} \end{aligned} \tag{9.2.37}$$

where

$$\frac{\partial}{\partial k}\left(\frac{\partial k}{\partial x}\right)_J^n = A_3, \quad \frac{\partial}{\partial \varepsilon}\left(\frac{\partial \varepsilon}{\partial x}\right)_J^n = A_3 \tag{9.2.38}$$

The coefficients $(r_7)_J$ and $(r_8)_J$ are given by

$$(r_7)_J = -\left[x^n\left(\frac{\partial k}{\partial x}\right)_J^n + \varepsilon_J^n + 2m^n k_J^n\right] \tag{9.2.39a}$$

$$(r_8)_J = -\left[x^n\left(\frac{\partial \varepsilon}{\partial x}\right)_J^n + c_{\varepsilon_2} f_2^n \frac{(\varepsilon^2)_J^n}{k_J^n} + (3m^n - 1)\varepsilon_J^n\right] \tag{9.2.39b}$$

As before, the linear system expressed in the form of Eq. (8.2.24) can be solved by the block-elimination method discussed in subsection 8.2.3. The solution procedure, however, is somewhat more involved than that used to solve the boundary-layer equations with an algebraic eddy-viscosity formulation since the formulation of the zonal method requires that the linearized inner boundary conditions resulting from Eq. (9.2.9b) also be satisfied as well as the usual boundary conditions at the surface and the boundary-layer edge. Subsection 10.9.5 presents an algorithm called KESOLV for this purpose. It employs the block-elimination method and follows the structure of the solution procedure used in the zonal method as well as the procedure used in the solution of the k-ε model equations with and without wall functions discussed in the following section. Sections 10.7 to 10.11 describe a computer program for the zonal method discussed in Section 9.2 and the method described in Section 9.3. A computer program is given on the companion website.

9.3 Solution of the k-ε Model Equations with and without Wall Functions

The solution of the k-ε model equations with and without wall functions is similar to the solution of the k-ε model equations with the zonal method. Their solution with either one can be accomplished with minor changes to the solution algorithm described in the previous section. In both cases changes are made to the A_{js} matrix, Eq. (9.2.30a), by modifying or redefining the elements of the first five rows which in this case correspond to the boundary conditions at $y = y_0$ or $y = 0$. In either case, for $j = 0$, after the five boundary conditions are specified, the next three equations correspond to those given by Eqs. (9.2.29f) to (9.2.29h). For $j \geq 1$, the ordering of the first-order equations is identical to that used for the outer region, that is, the equations are ordered according to those given by Eqs. (9.2.29) except that Eqs. (9.2.29d) and (9.2.29e) are replaced by Eqs. (9.2.7) and (9.2.8), respectively. In addition of course, the coefficients of the momentum, kinetic energy and rate of dissipation are different.

9.3.1 Solution of the k-ε Model Equations without Wall Functions

The k-ε model equations without wall functions given by Eqs. (6.2.7) and (6.2.8) for high Reynolds number together with the continuity and momentum equations are subject to the four boundary conditions given by Eqs. (6.2.10), (6.2.13) and (6.2.15) at $y = y_0$ and to those at the edge, $\eta = \delta$, given by Eqs. (6.2.16) and (6.2.18) together with the relation given by Eq. (6.2.12). To discuss the solution procedure in terms of transformed variables, let us consider first the two boundary conditions at $\eta = \eta_0$. In terms of transformed variables, Eqs. (6.2.10) become

$$f_0' = w_0\left[\frac{1}{\kappa}\ln\left(\sqrt{R_x}w_0\eta_0\right) + c\right] \tag{9.3.1a}$$

$$x\frac{\partial f_0}{\partial x} + m_1 f_0 = f'\eta\left[m_1 + \frac{x}{w_0}\frac{dw_0}{dx}\right] \tag{9.3.1b}$$

where

$$R_x = \frac{u_e x}{\nu}, \quad c = 5.2, \quad \kappa = 0.41, \quad m_1 = \frac{m+1}{2} \tag{9.3.2}$$

with $f' = u$, $u_\tau/u_e = w_0$, Eqs. (9.3.1) in linearized form can be written as

$$\delta u_0 + \alpha_8 \delta w_0 = (r_1)_0 \tag{9.3.3}$$

$$\beta_1 \delta f_0 + \beta_2 \delta u_0 + \beta_8 \delta w_0 = (r_2)_0 \tag{9.3.4}$$

where

$$\alpha_8 = -\frac{1}{\kappa} - \left[\frac{1}{\kappa}\ln\left(\sqrt{R_x}w_0\eta_0\right) + c\right], \quad \beta_1 = \alpha + m, \quad \alpha = \frac{x_n}{k_n} \tag{9.3.5a}$$

$$\beta_2 = -\eta_0\left[m_1 + \alpha\left(1 - \frac{w_0^{n-1}}{w_0^n}\right)\right], \quad \beta_8 = -u_0\eta_0\alpha\frac{w_0^{n-1}}{\left(w_0^n\right)^2} \tag{9.3.5b}$$

$$(r_1)_0 = w_0\left[\frac{1}{\kappa}\ln\left(\sqrt{R_x}w_0\eta_0\right) + c\right] - u_0 \tag{9.3.5c}$$

$$(r_2)_0 = u_0\eta_0\left[m_1 + \alpha\left(1 - \frac{w_0^{n-1}}{w_0^n}\right)\right] - \alpha\left(f_0^n - f_0^{n-1}\right) - m_1 f_0^n \tag{9.3.5d}$$

The third boundary condition in Eq. (6.2.13), which makes use of Bradshaw's relation in Eq. (6.3.2), and with

$$-\overline{u'v'} = \varepsilon_m\frac{\partial u}{\partial y}$$

and with ε_m defined by Eq. (6.2.6), can be written as

$$a_1 = c_\mu\frac{k}{\varepsilon}\frac{\partial u}{\partial y}$$

or in terms of dimensionless and transformed variables, as

$$a_1 = c_\mu\frac{\tilde{k}}{\tilde{\varepsilon}}\sqrt{R_x}v \tag{9.3.6}$$

all evaluated at $\eta = \eta_0$ with $\tilde{k} = k/u_e^2$ and $\tilde{\varepsilon} = \frac{\varepsilon x}{u_e^3}$ as defined before. Linearization gives

$$\gamma_3\delta v_0 + \gamma_4\delta k_0 + \gamma_6\delta\varepsilon_0 = r_3 \tag{9.3.7}$$

Here we have dropped the tilde ($\tilde{\ }$) from $\tilde{k}$ and $\tilde{\varepsilon}$ and defined

$$\tilde{\gamma}_3 = c_\mu\sqrt{R_x}k_0, \quad \tilde{\gamma}_4 = c_\mu\sqrt{R_x}v_0, \quad \tilde{\gamma}_6 = -\sqrt{c_\mu} \tag{9.3.8}$$

$$r_3 = \sqrt{c_\mu}\varepsilon_0 - c_\mu\sqrt{R_x}k_0v_0. \tag{9.3.9}$$

The fourth boundary condition at $\eta = \eta_0$ assumes

$$(\varepsilon_m)_{CS} = (\varepsilon_m)_{k-\varepsilon} \tag{9.3.10a}$$

that is,

$$l^2\frac{\partial u}{\partial y} = c_\mu\frac{k^2}{\varepsilon} \tag{9.3.10b}$$

In terms of dimensionless quantities and transformed variables, in linearized form, Eq. (9.3.10b) can be written as

$$\theta_3 \delta v_0 + \theta_4 \delta k_0 + \theta_6 \delta \varepsilon_0 = r_4 \tag{9.3.11}$$

where

$$\theta_3 = -c_v \varepsilon_0, \quad \theta_4 = 2c_\mu \sqrt{\mathrm{R}_x} k_0, \quad \theta_6 = -c_v v_0, \tag{9.3.12}$$

$$c_v = (\kappa\eta \cdot \text{damping})^2, \quad r_4 = c_v v_0 \varepsilon_0 - c_\mu \sqrt{\mathrm{R}_x} k_0^2 \tag{9.3.13}$$

If y_0 is sufficiently away from the wall, i.e. $y_0^+ \geq 60$, then the damping term, such as the one used in the CS model, is equal to 1.0.

The fifth boundary condition which connects τ_0 at $y = y_0$ and τ_w at $y = 0$, is obtained from Eq. (6.2.12). With Thompson's and log law velocity profiles, it can be written as

$$\tau_0 = \tau_w + \alpha^* y_0 \frac{d\tau_w}{dx} + y_0 \frac{dp}{dx} \tag{9.3.14}$$

Here α^* is given by

$$\alpha^* = 0.5 \left[c_1 \ln\left(y_0^+\right)^2 + c_2 \ln y_0^+ + c_3 + \frac{c_4}{y_0^+} \right] \tag{9.3.15}$$

where

$$c_1 = 5.9488, \quad c_2 = 13.4682, \quad c_3 = 13.5718, \quad c_4 = -785.20$$

$$y_0^+ = \sqrt{\mathrm{R}_x \bar{u}_\tau \eta_0}, \quad y_0 = \frac{x\eta_0}{\sqrt{\mathrm{R}_x}} \tag{9.3.16}$$

In terms of transformed variables, Eq. (9.3.14), after linearization, can be expressed in the form

$$\delta_3 \delta v_0 + \delta_4 \delta k_0 + \delta_6 \delta \varepsilon_0 + \delta_8 \delta w_0 = (r_5)_0 \tag{9.3.17}$$

where

$$\delta_3 = c_\mu \mathrm{R}_x \frac{k_0^2}{\varepsilon_0} \tag{9.3.18a}$$

$$\delta_4 = 2c_\mu \mathrm{R}_x \frac{k_0}{\varepsilon_0} v_0 \tag{9.3.18b}$$

$$\delta_6 = -c_\mu \mathrm{R}_x \frac{k_0^2}{\varepsilon_0^2} v_0 \tag{9.3.18c}$$

$$\delta_8 = -2w_0\{\sqrt{R_x} + \alpha^* \eta_0 (\alpha + 2m_1)\}$$
$$-\left(\frac{\partial \alpha^*}{\partial w}\right)_0 \left\{ \alpha \left[(w_0^n)^2 - (w_0^{n-1})^2 \right] + 2m_1 (w_0^n)^2 \right\} \tag{9.3.18d}$$

and

$$(r_5)_0 = \sqrt{R_x} w^2 + \alpha^* \eta_0 \left\{ \alpha \left[(w_0^n)^2 - (w_0^{n-1})^2 \right] + 2m_1 (w_0^n)^2 \right\}$$
$$-\eta_0 m_2 - c_\mu R_x \frac{k_0^2}{\varepsilon_0} v_0^2 \tag{9.3.19}$$

With the five boundary conditions defined, the A_{j_s} matrix, which is essentially the A_0 matrix in this case, becomes

$$A_0 = \begin{vmatrix} 0 & 1 & 0 & 0 & 0 & 0 & 0 & \alpha_8 \\ \beta_1 & \beta_2 & 0 & 0 & 0 & 0 & 0 & \beta_8 \\ 0 & 0 & \gamma_3 & \gamma_4 & 0 & \gamma_6 & 0 & 0 \\ 0 & 0 & \theta_3 & \theta_4 & 0 & \theta_6 & 0 & 0 \\ 0 & 0 & \delta_3 & \delta_4 & 0 & \delta_6 & 0 & \delta_8 \\ 0 & -1 & -\frac{h_1}{2} & 0 & 0 & 0 & 0 & 0 \\ 0 & 0 & 0 & -1 & -\frac{h_1}{2} & 0 & 0 & 0 \\ 0 & 0 & 0 & 0 & 0 & -1 & -\frac{h_1}{2} & 0 \end{vmatrix} \tag{9.3.20}$$

9.3.2 Solution of the *k*-*ε* Model Equations with Wall Functions

The solution of the k-ε model equations with wall functions is similar to the procedure described for the case without wall functions. Again the only changes occur in the first five rows of the A_{j_s} matrix, Eq. (9.2.30a). Of the five boundary conditions at the wall, the first three are written in the order given by Eqs. (9.2.15a,b,c) and the fourth and fifth ones are given by

$$k_0 = 0 \tag{9.3.21a}$$

$$\varepsilon_0 = 0 \tag{9.3.21b}$$

or in linearized form

$$\delta k_0 = (r_4)_0 = 0 \tag{9.3.22a}$$

$$\delta \varepsilon_0 = (r_5)_0 = 0 \tag{9.3.22b}$$

The structure of the other matrices remain the same, but of course, the coefficients of the linearized momentum, kinetic energy and dissipation equations, Eqs. (9.2.34), (9.2.35) and (9.2.36), respectively are different than those for k-ε model equations without wall functions. These coefficients naturally vary depending on the wall functions used.

The A_0 matrix for the k-ε model equations with wall functions, with the last three rows identical to those in Eq. (9.3.20), is

$$A_0 = \begin{vmatrix} 1 & 0 & 0 & 0 & 0 & 0 & 0 & 0 \\ 0 & 1 & 0 & 0 & 0 & 0 & 0 & 0 \\ 0 & 0 & 1 & 0 & 0 & 0 & 0 & -2\sqrt{R_x}\omega_0 \\ 0 & 0 & 0 & 1 & 0 & 0 & 0 & 0 \\ 0 & 0 & 0 & 0 & 0 & 1 & 0 & 0 \\ 0 & -1 & -\frac{h_1}{2} & 0 & 0 & 0 & 0 & 0 \\ 0 & 0 & 0 & -1 & -\frac{h_1}{2} & 0 & 0 & 0 \\ 0 & 0 & 0 & 0 & 0 & -1 & -\frac{h_1}{2} & 0 \end{vmatrix} \tag{9.3.23}$$

In some model equations, the boundary conditions on $\varepsilon = 0$ is replaced by $\frac{\partial \varepsilon}{\partial y} = 0$; in that case, the fifth row of A_0-matrix becomes

$$0\,0\,0\,0\,0\,0\,1\,0 \tag{9.3.24}$$

9.4 Solution of the k-ω and SST Model Equations

The solution of the k-ω model equations is similar to the solution of the k-ε model equations with wall functions. Again the k-ω model equations, Eqs. (6.2.19), (6.2.23) and (6.2.28), are expressed in terms of Falkner-Skan variables.

Since the SST model equations make use of the k-ε model equations in the inner region and the k-ε model equations in the outer region we express them, for the sake of compactness, in the following form in transformed variables.

$$\left[(1 + \sigma_k \varepsilon_m^+)k'\right]' - 2mf'k + m_1 fk' + \varepsilon_m^+ (f'')^2 - \beta^* \omega k = x\left(f' \frac{\partial k}{\partial x} - k' \frac{\partial f}{\partial x}\right) \tag{9.4.1}$$

$$
\begin{aligned}
&[(1+\sigma_\omega\varepsilon_m^+)\omega']' + 2(1-F_1)\sigma_{\omega_2}\frac{R_x}{\omega}k'\omega' + m_1\omega' f - (m-1)f'\omega \\
&\quad -\beta\omega^2 + R_x(f'')^2 = x\left(f'\frac{\partial\omega}{\partial x} - \omega'\frac{\partial f}{\partial x}\right)
\end{aligned}
\tag{9.4.2}
$$

where ω and k are dimensions, normalized by x/u_e and $1/u_e^2$, respectively. Equations (9.4.1) and (9.4.2) are the equations used in the SST model. To recover Wilcox's k-ω model equations expressed in transformed variables, we let $F_1 = 1$ and take

$$
\begin{aligned}
\sigma_k &= 0.5, \quad \sigma_\omega = 0.5, \quad \beta = 0.075, \\
\beta^* &= 0.09, \quad \kappa = 0.41, \quad \gamma = \frac{\beta}{\beta^*} - \sigma_\omega\frac{\kappa^2}{\sqrt{\beta^*}}
\end{aligned}
\tag{9.4.3}
$$

In the SST model, the above constants are determined from the relation, Eq. (6.2.45)

$$
\phi = F_1\phi_1 + (1-F_1)\phi_2 \tag{6.2.45}
$$

where the constant ϕ_1 is determined from Eq. (6.2.46) and the constant ϕ_2 from Eq. (6.2.47). F_1 is determined from Eq. (6.2.41), where its $\arg_1$ given by Eq. (6.2.42) can be written us

$$
\arg_1 = \min[\max(\lambda_1, \lambda_2), \lambda_3] \tag{9.4.4}
$$

In terms of transformed quantities, λ_1 to λ_3 are

$$
\lambda_1 = \frac{\sqrt{k}}{0.09\omega y} = \frac{\sqrt{k}\sqrt{R_x}}{\omega\eta\, 0.09} \tag{9.4.5a}
$$

$$
\lambda_2 = \frac{500\nu}{y^2\omega} = \frac{500}{\eta^2\omega} \tag{9.4.5b}
$$

$$
\lambda_3 = \frac{4\varrho\sigma_{\omega_2}k}{CD_{c\omega}y^2} \tag{9.4.5c}
$$

$$
CD_{c\omega} = \max\left(2\varrho\sigma_{\omega_2}\frac{1}{\omega}\frac{\partial k}{\partial y}\frac{\partial\omega}{\partial y}, 10^{-20}\right) = \frac{2k}{\max\left(\frac{1}{\omega}k'\omega', 10^{-20}\right)\eta^2}
$$

We first find the maximum of λ_1 and λ_2 (say λ_4), then calculate the minimum of λ_4 and λ_3 and thus determine $\arg_1$ and F_1. Once F_1 is calculated, then the constants in Eqs. (9.4.1) and (9.4.2) are determined from the relation given by Eq. (6.2.45). For example,

$$
\begin{aligned}
\sigma_\omega &= 0.5F_1 + 0.856(1-F_1) \\
\beta &- 0.0750F_1 + 0.0828(1-F_1)\text{etc.}
\end{aligned}
$$

Next we determine the eddy viscosity distribution across the boundary layer. In terms of transformed variables, Eq. (6.2.36) can be written as ($v = f''$. $a_1 = 0.31$)

$$\varepsilon_m^+ = \frac{\varepsilon_m}{\nu} = \begin{cases} \mathrm{R}_x \dfrac{k}{\omega} & a_1\omega > \Omega F_2 \\ \dfrac{\sqrt{\mathrm{R}_x} a_1 k}{|v| F_2} & a_1\omega < \Omega F_2 \end{cases} \tag{9.4.6}$$

where

$$\Omega = \left|\frac{\partial u}{\partial y}\right| = u_e |f''| \sqrt{\frac{u_e}{\nu_x}}$$

and F_2 is determined from Eq. (6.2.37a) where $\arg_2$ is

$$\arg_2 = \max(2\lambda_1, \lambda_2) \tag{9.4.7}$$

In the SST model, once the constants are determined and the distribution of eddy viscosity is calculated, then Eqs. (9.4.1) and (9.4.2) are solved together with the continuity and momentum equations; a new $\arg_1$, $\arg_2$, F_1 and F_2, new constants and eddy viscosity distribution are determined. This procedure is repeated until convergence.

It should be noted that, for $F_1 = 1$, the whole region is the inner region governed by the k-ω model equations.When $F_1 = 0$, the whole region is governed by the k-ε model equations.

Before we discuss the solution procedure for the SST model equations, it is useful to point out that the structure of the solution algorithm for the k-ε model equations with wall functions is almost identical to the one for the SST model equations. This means all the A_j, B_j, C_j matrices have the same structure; the difference occurs in the definitions of the coefficients of the linearized momentum, kinetic energy and rate-of-dissipation equations and in the definition of the boundary condition for ω which occurs in the fourth row of A_0-matrix.

To describe the numerical method for the k-ω model equations, we start with the kinetic energy equation, Eq. (9.4.1), and write it in the same form as Eq. (9.4.2) by defining Q and F by

$$Q = \beta^* \omega k, \quad F = 0 \tag{9.4.8}$$

The definition of P remains the same. Next we write Eq. (9.4.2) in the form

$$(b_3\omega')' + P_1 - Q_1 + E = x\left(f' \frac{\partial \omega}{\partial x} - \omega' \frac{\partial f}{\partial x}\right) + (m-1) f'\omega - m_1 \omega' f \tag{9.4.9}$$

where

$$E = 2(1 - F_1)\sigma_{\omega_2}\frac{R_x}{\omega}k'\omega' \tag{9.4.10}$$

$$Q_1 = \beta\omega^2$$

$$P_1 = \gamma R_x (f'')^2$$

With

$$\omega' = q$$

Equation (9.4.9) can be written as

$$(b_3 q)' + P_1 - Q_1 + E = x\left(u\frac{\partial \omega}{\partial x} - q\frac{\partial f}{\partial x}\right) + (m-1)u\omega - m_1 q f \tag{9.4.11}$$

A comparison of Eq. (9.4.11) with Eq. (9A.10) shows that if we let $\varepsilon = \omega$ for notation purposes, then the coefficients of linearized specific dissipation rate equation are very similar to those given by Eqs. (9A.14a) and (9A.15a). Except for the definitions of Q and F in the kinetic energy-equation, Eq. (9.4.1), the coefficients of the linearized kinetic energy equation are identical to those given by Eqs. (9A.7) and (9A.8). Appropriate changes then can be easily made to subroutine KECOEF (see Section 11.9) in order to adopt the computer program of Sections 11.7 to 11.10 to solve the kinetic energy and specific dissipation rate kinetic energy equations in the SST model. Of course, other changes also should be made, but these are not discussed here. A good understanding of the computer program for the k-ε model equations is needed to make the necessary changes.

9.5 Evaluation of Four Turbulence Models

In Sections 8.3 to 8.6 and 8.9, 8.10 we discussed the evaluation of the *CS* model with a differential method based on the solution of the boundary-layer equations. In this section we present a similar discussion for transport-equation turbulence models with a differential method based on the solution of the Navier-Stokes equations. The discussion is based on the study conducted in [2] where Bardina et al. evaluated the performance of four higher-order turbulence models. The models were: 1) the k-ω model of Wilcox (subsection 6.2.2), 2) k-ε model of Launder and Sharma (subsection 6.2.1), 3) the *SST* model of Menter (subsection 6.2.3) and 4) the *SA* model (subsection 6.3.2). The flows investigated were five free shear flows and five boundary-layer flows consisting of an incompressible and compressible flat plate, a separated boundary layer, an axisymmetric shock-wave/boundary-layer interaction, and an RAE 2822 transonic airfoil.

In this section we present a sample of results for some of these flows obtained from this study and discuss a summary of the conclusions regarding the relative performance of the various models tested. For additional results and discussion, the reader is referred to [2].

9.5.1 Free-Shear Flows

Five free-shear flows corresponding to a mixing layer, plane jet, round jet, plane wake and a compressible mixing layer were considered in [2], and four eddy viscosity models were validated for the prediction of these flows. The validation of each model was mainly based on the ability of the models to predict the mean velocity profile and spreading rate of each one of these fully developed free-shear flows. Sensitivity analyses, the validation results to freestream turbulence, grid resolution and initial profiles were also included in their study. Here, however, we only present the mean velocity profiles and spreading rate of each flow.

Mixing Layer

Figure 9.1 shows a comparison of the predictions of the mean velocity profile u/u_1 against $\eta = y/x$ and the experimental data of Liepmann and Laufer [3] for a half jet. See also Fig. 7.20. The dimensionless coordinate η was defined with its origin located where the mean velocity ratio was 1/2.

According to the calculations in [2], the results of the k-ε and SST models are insensitive to freestream turbulence and show good agreement in the middle of the mixing zone and sharp edge profiles at the boundaries. The small difference between the predictions of these two models near the edge of the freestream at rest is due to the different value of their diffusion model constant, σ_ε. The results of the SA model

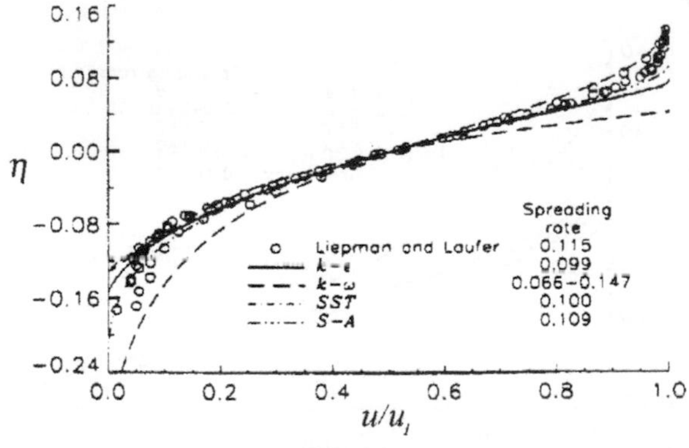

Fig. 9.1 Predictions of four turbulence models for the mixing layer of Liepmann and Laufer, after [3].

show very good agreement with the experimental data, but show a wider mean velocity profile with very large values of freestream eddy viscosity ($10^{-3} \leq N_1 \equiv \frac{\varepsilon_m}{u_1 x}$). In practice, these large values of dimensionless eddy viscosity are much larger than the molecular viscosity, and the errors can be controlled by limiting the eddy viscosity in the freestream, ($\varepsilon_m^+ \leq 10^{-3} R_L = u_\infty L/\nu$). The k-ω model shows two different results of mean velocity profiles, one for low values and another for high values of freestream ω, and a range of profiles in between these two values, ($10^{-2} \leq W \equiv \omega x/u_1 \leq 10$). The profiles show significant underprediction in the low-speed side and overprediction in the high-speed side of the mixing layer with low freestream ω, ($W \leq 10^{-2}$), and underprediction in the higher speed side with high freestream ω, ($W \geq 10$).

Figure 9.1 also compares the calculated and measured spreading rates for the mixing layer. Considering that the experimental value of 0.115 also shows an uncertainty of about ± 10%, the predictions of all four models are very good. The range of values reported for the k – ω model is due to the effects of low and high freestream ω values [2].

Plane Jet

Figure 9.2 shows a comparison of the predictions of the mean velocity profile u/u_1 against $\eta \equiv y/x$ and the experimental data of Bradbury for a plane jet [4]. See also subsection 7.5.1. As discussed in [2], the *SST* model gives excellent agreement with the experimental data and is also insensitive to low freestream values of ω. The profile of the k-ε model is similar, except near the freestream at rest, and is insensitive to low freestream values of ε. The small difference between the predictions of these two models is due to the different value of their diffusion model constant, σ_ε. The *SA* model overpredicts the mean velocity profile thickness; the results are insensitive to

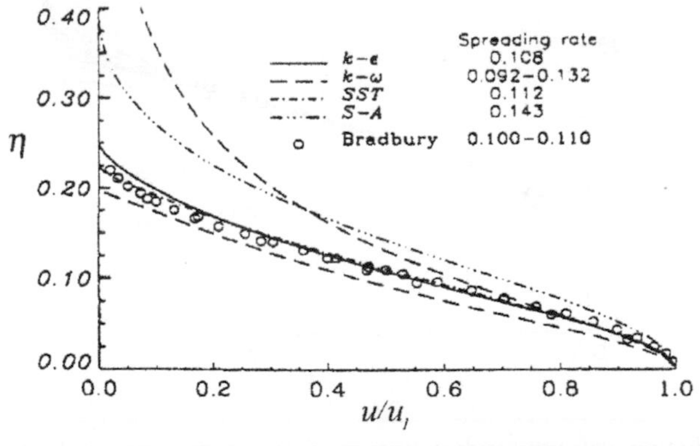

Fig. 9.2 Predictions of four turbulence models for the plane jet of Bradbury, after [ ].

freestream eddy viscosity for $N \equiv \varepsilon_m / u_1 x \leq 10^{-3}$. Results with larger freestream eddy viscosities give much larger overpredictions and are not shown in Fig. 9.2. In practice, these errors can be controlled by limiting the values of the eddy viscosity in the freestream, $\varepsilon_m^+ \leq 10^{-3} R_L$. The results of the k-ω model show two predictions; one largely overpredicts and the other underpredicts the thickness of the mean velocity profile, corresponding to low and high freestream $W \equiv \frac{\omega x}{u_1}$ values, ($W \leq 10^{-4}$ and $W \geq 10^3$), respectively. This model gives a set of intermediate solutions (not shown in Fig. 9.2) depending on the values of freestream ω, ($10^{-4} \leq W \leq 10^3$).

Figure 9.2 also compares the calculated and measured spreading rates for the plane jet. The range of experimental values is reported between 0.10 and 0.11 and is given only as reference values.

The k-ε and the *SST* models give close predictions of the experimental spreading rate, while the *SA* model overpredicts the spreading rate. The k-ω model predicts a range of values due to the effects of low and high freestream ω.

Round Jet

Figure 9.3 shows the comparison of the predictions of the mean velocity profile u/u_1 against η - y/x and the experimental data of Wygnanski and Fiedler [7]. As discussed in [2], all models overpredict the thickness of the experimental mean velocity profile. This classical anomaly is well known in these models that have been fine-tuned with empirical data of mixing layer, plane jet, and/or far wake experiments.

The results of the k-ε and *SST* models are closer to the experimental data and are also insensitive to low freestream values of ε or ω, respectively. The small difference between the predictions of these two models near the edge of the freestream is also due to the different value of their diffusion model constant, σ_ε. The *SA* model gives

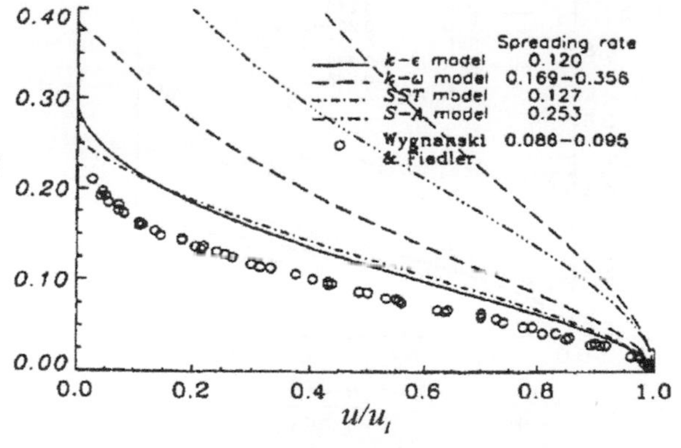

Fig. 9.3 Prediction of four turbulence models for the round jet of Wygnanski and Fiedler, after [2].

a considerably larger overprediction of the mean velocity profile thickness, and the results are insensitive to freestream eddy viscosity for $N \equiv \varepsilon_m/u_1\,x \leq 10^{-3}$. Results with larger freestream eddy viscosities give much larger overpredictions and are not shown in Fig. 9.3. In practice, these large values of dimensionless eddy viscosity are much larger than the molecular viscosity, and the errors can be controlled by limiting the eddy viscosity in the freestream $\varepsilon_m^+ \leq 10^{-3} R_L$. The results of the k-ω model show two overpredictions of the thickness of the mean velocity profile, corresponding to low and high freestream $W \equiv \omega x/u_1$ values, ($W \leq 10^{-4}$ and $W \geq 10^4$), and a set of intermediate solutions (not shown in Fig. 9.3) depending on the values of freestream ω, ($10^{-4} \leq W \leq 10^4$) [2].

The spreading rate is defined as the value of the nondimensional jet radius, $S = y/(x - x_0)$, where the mean speed is half its centerline value. This definition of spreading rate is one of several formulations that have been proposed. The spreading rate provides an estimate of the thickness of the round jet and is widely used in turbulence modeling. However, it is only one parameter and it does not provide information about the shape of the velocity profile.

Figure 9.3 also compares the spreading rates obtained with the turbulence models and the recommended experimental value. The range of experimental values is between 0.086 and 0.095. All models overpredict the spreading rate. The range of values reported for the k-ω model is due to the effects of low and high freestream ω values.

Plane Wake

Figure 9.4 shows a comparison of the predictions of the mean velocity profiles $\frac{u-u_{\min}}{u_e-u_{\min}}$ against the dimensionless coordinate $\eta = y(\varrho u_\infty/\mu x)^{1/2}$ compared with the

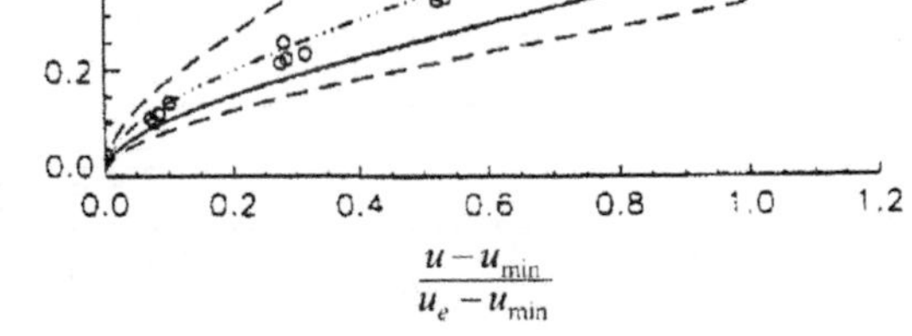

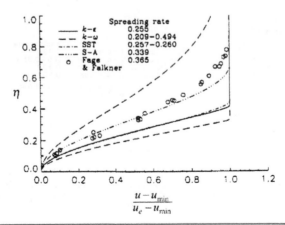

Fig. 9.4 Predictions of four turbulence models for the plane wake of Fage and Falkner, after [2].

experimental data of Fage and Falkner [8]. The k-ε and *SST* models give thinner profiles than the experiment and are insensitive to low freestream values of ε or ω, respectively. The small difference between the predictions of these two models near the freestream is due to the different value of their diffusion model constant, σ_ε. The *SA* model gives the best agreement with the experimental profile, and the results are insensitive to freestream eddy viscosity for $N \leq 10^{-4}$. Results with much larger freestream eddy viscosities give overpredictions and are not shown in Fig. 9.4; in practice, these errors can be controlled by limiting the eddy viscosity in the freestream, ($\varepsilon_m^+ \leq 10^{-3} R_L$). The results of the k-ω model show two predictions; one largely overpredicts and the other underpredicts the thickness of the mean velocity profile, corresponding to low and high freestream W values, ($W \leq 10^{-4}$ and $W \geq 10^3$), respectively. This model gives a set of intermediate solutions (not shown in Fig. 9.4) depending on the intermediate values of freestream ω, ($10^{-4} \leq W \leq 10^3$).

The spreading rate S is defined as the difference $S = \eta_{0.5} - \eta_0$ of the nondimensional coordinate η between the points where the nondimensional mean speed is one half and zero, respectively. The definition of spreading rate is one of several formulations that have been proposed and it is widely used in turbulence modeling.

Figure 9.4 also compares the spreading rates obtained with the turbulence models and the recommended experimental value 0.365 of Fage and Falkner [8]. As can be seen, the k-ε and the *SST* models underpredict the experimental spreading rate by 30%, while the *SA* model gives a value much closer to the experimental spreading rate (7%). The k-ω model predicts a range of values due to the effects of low and high freestream ω, within an underprediction of 43% and an overprediction of 35%.

Compressible Mixing Layer

Figure 9.5 shows a comparison between the predictions of the dimensionless mean velocity profile $\frac{u-u_2}{u_1-u_2}$ of the mixing layer, using the standard turbulence models and the experimental data of Samimy and Elliot against the coordinate $((\eta - \eta_{0.5})/\delta_\omega$ [9]. This particular coordinate system was used in order to show all the data in a simpler plot. The nondimensional coordinate, $\eta \equiv y/x$, the coordinate $\eta_{0.5}$ represents the point where the nondimensional speed is 0.5, and δ_ω is the vorticity thickness of the mixing layer where $\delta_\omega = (u_1 - u_2)/(\frac{du}{d\eta})_{\max}$. The experimental data are shown with convective Mach numbers of $M_c = 0.51$, 0.64, and 0.86. The experimental data of Liepmann and Laufer [3] for the incompressible mixing layer, $M_c = 0$, is also shown in this figure as a reference. The numerical predictions with the four different turbulence models were obtained over a wide range of convective Mach numbers, and Fig. 9.5 shows the predictions for $M_c = 0$, 0.8, and 1.6. The vertical arrows indicate the trend of the predictions with increasing convective Mach numbers. The convective Mach number, $M_c = (u_1 - u_2)/(a_1 + a_2)$ is defined in terms of the mean velocity, u_1, and the sound speed, a, in each freestream.

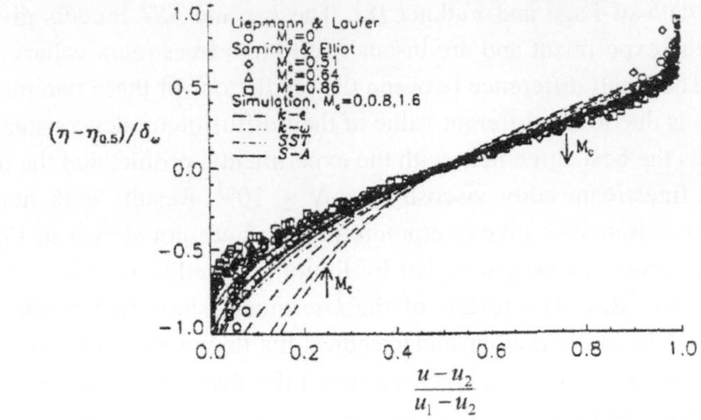

Fig. 9.5 Comparison of velocity profiles for compressible mixing layer, after [2].

All velocity profiles show some degree of agreement with the experimental data due to the particular coordinate system of the plot. These plot coordinates bound the range and collapse all data at the midpoint. The results of the *k-ε* and *SST* models show good agreement with the nondimensional shape of the experimental profile. The results of the *SA* model also show good agreement with the experimental data, except for very high freestream eddy viscosity. The mean velocity profiles of the *k-ω* model show sensitivity to low freestream ω values. The relative good agreement of prediction with data is due to the use of $\delta_{\overline{\omega}}$ in the nondimensional plots. The dimensional profiles and spreading rates show a much stronger dependence on Mach number, as discussed in detail in [2].

Figure 9.6 shows a comparison of spreading rates predicted with the four turbulence models with no compressibility corrections and the experimental Langley data [10]. Here the spreading rate S is defined as $d\delta/dx$ where $\delta(x)$ is the thickness of the mixing layer. The predicted results are shown with lines and the experimental data are shown with symbols. The most significant result is that all models fail to predict the experimental data on the decrease of spreading rate with increasing convective Mach number. This is a well-known weakness of present turbulence models. For additional details, see [2].

9.5.2 Attached and Separated Turbulent Boundary Layers

Studies in [2] for attached and separated turbulent boundary layers included the flow over an adiabatic flat plate of an incompressible flow and a compressible flow at Mach 5, an adverse pressure gradient flow on an axisymmetric cylinder, a shock/boundary layer flow on an axisymmetric bump, and a transonic flow on the RAE 2822 airfoil. A brief description of the performance of the four turbulence models for

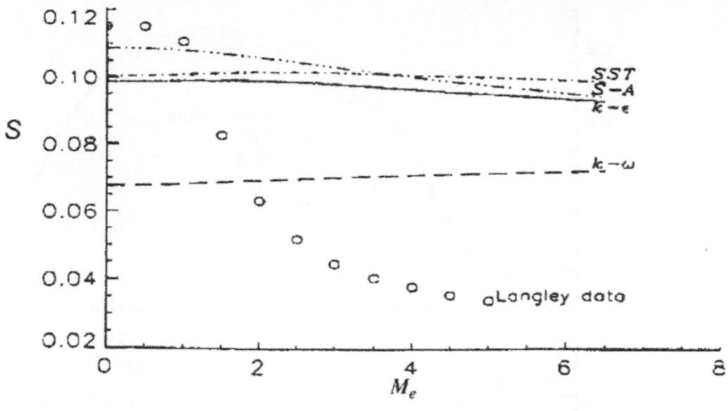

Fig. 9.6 Comparison of spreading rate of the mean velocity profile for compressible mixing layer, after [2].

these flows except the last one are given below. For additional discussion, the reader is referred to [2].

Flat Plate Flows

The studies for flat plate flows were conducted in order to investigate the predictions of the turbulence models with the well established correlations such as the velocity profile expression of Coles, Eq. (4.4.34), for incompressible flows and the local skin-friction coefficient expression of Van Driest, Eq. (7.2.54), for compressible flows. Studies were also conducted to investigate the sensitivity of the solutions to inlet conditions and to grid. Overall, all turbulence models performed well, as they should, for these zero-pressure gradient flows. The predicted boundary-layer parameters such as c_f, θ, H and velocity profiles agreed well with data and with correlations [2].

Axisymmetric Flow with Adverse Pressure Gradient

This flow corresponds to an axial flow along a cylinder with superimposed adverse pressure gradient. The experiment was performed by Driver [11]. Boundary layer suction was applied through slots on the wind tunnel walls, and this mass flow removal (about 10% of the incoming mass flow through the tunnel) allowed the flow to remain attached along the tunnel walls in the presence of the strong pressure gradient. Experimental data, including velocity and Reynolds stress profiles, have been measured in several locations. Since flow separation was observed experimentally, a full Navier–Stokes prediction method was performed and is recommended. The solution procedure requires the specification of an outer boundary such as a streamline. The experimental velocity profiles have been integrated to obtain the stream function and corresponding outer streamline. From a computational point of

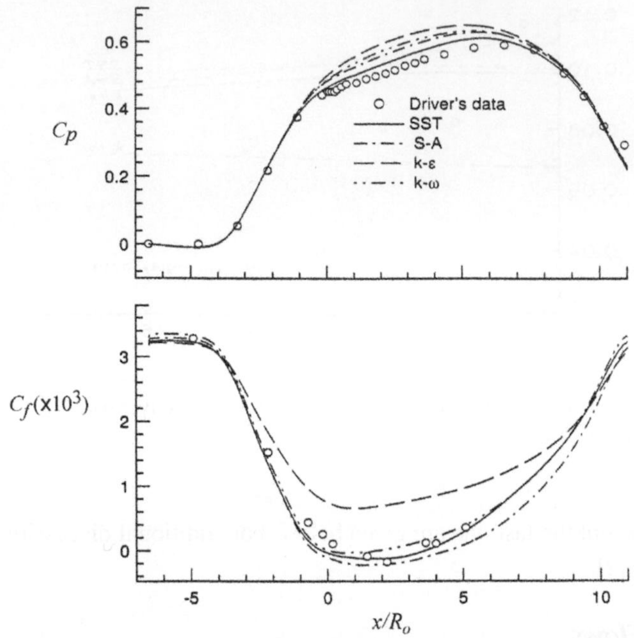

Fig. 9.7 Comparison of surface pressure and skin friction coefficients, after [2].

view, this method allows the flow to be treated as flow in an annular duct with one boundary defined with the surface of the cylinder (no-slip condition) and other boundary defined with an outer streamline (slip condition). The recommended outer streamline distance h as a function of the coordinate distance is given in [2].

Figure 9.7 shows comparisons of the pressure and skin friction coefficients. With the exception of the k-ε model, all models predict flow separation. Overall, the *SST* model gives the best performance.

Comparisons of a sample of the velocity, turbulent kinetic energy, and shear stresses at some specific measured locations are displayed in Figs. 9.8, 9.9, and 9.10 respectively. Additional results are given in [2]. Again, the figures show that the *SST* model gives the best overall performance, the k-ε model the worst, and the other two models are in between.

Transonic Flow with Separation over an Axisymmetric Body

The experiment [12] was conducted in the Ames 2- by 2-Foot Transonic Wind Tunnel with total temperature and total pressure of 302 K and 9.5×10^4 N/m^2, respectively. The axisymmetric flow model consisted of an annular bump on a circular cylinder aligned with the flow direction. The longitudinal section of the bump was a circular arc. The axisymmetric configuration was chosen to circumvent

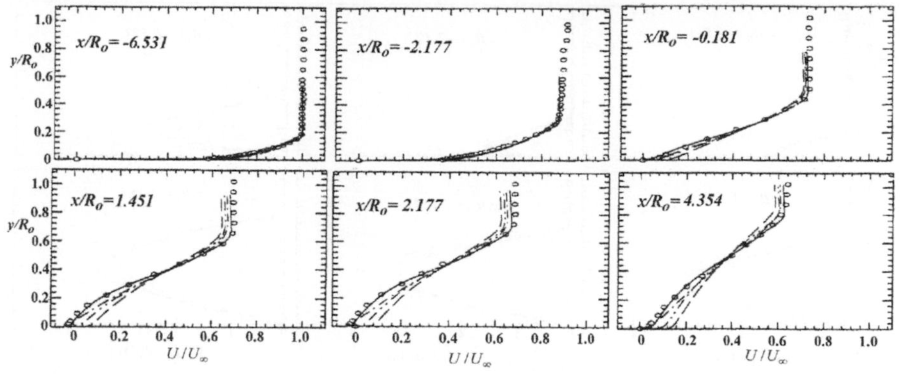

Fig. 9.8 Comparison of velocity profiles at different x/R_0-locations, after [2].

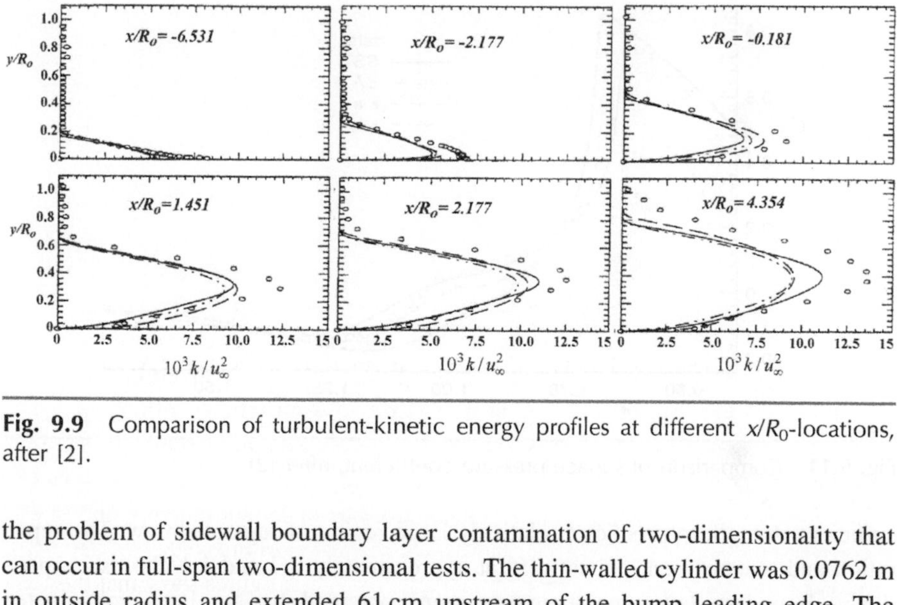

Fig. 9.9 Comparison of turbulent-kinetic energy profiles at different x/R_0-locations, after [2].

the problem of sidewall boundary layer contamination of two-dimensionality that can occur in full-span two-dimensional tests. The thin-walled cylinder was 0.0762 m in outside radius and extended 61 cm upstream of the bump leading edge. The straight section of the cylinder permitted natural transition and a turbulent boundary layer just ahead of the bump of sufficient thickness to allow accurate determination of boundary layer information. However, the boundary layer was not so thick, in comparison with the interaction on airfoils, that separation of greater severity would occur than is representative of full scale. The circular-arc bump had a 20.32 cm chord and a thickness of 1.905 cm. Its leading edge was joined to the cylinder by a smooth circular arc that was tangent to the cylinder and the bump at its two end points. Test conditions were a freestream Mach number of 0.875 and unit Reynolds number of 13.1×10^6/m. At this freestream Mach number, a shock wave was generated of

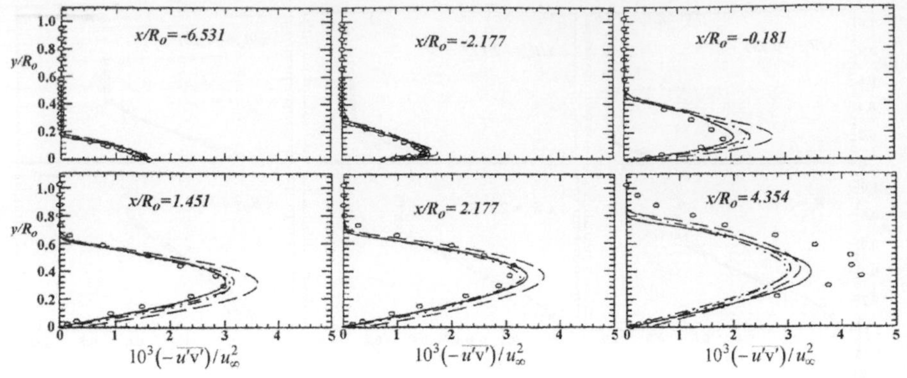

Fig. 9.10 Comparison of shear-stress profiles at different x/R_0-locations, after [2].

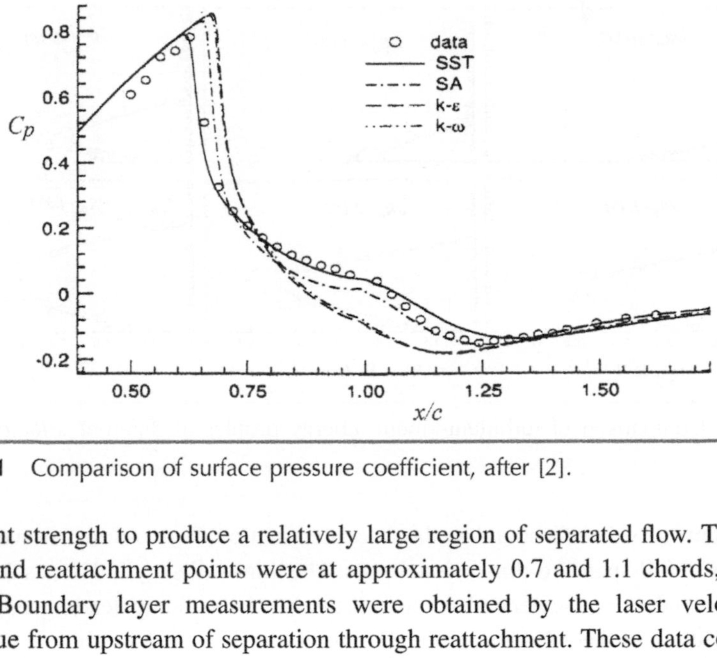

Fig. 9.11 Comparison of surface pressure coefficient, after [2].

sufficient strength to produce a relatively large region of separated flow. The separation and reattachment points were at approximately 0.7 and 1.1 chords, respectively. Boundary layer measurements were obtained by the laser velocimeter technique from upstream of separation through reattachment. These data consist of profiles of mean velocities, turbulence intensities, and shear stresses in the streamwise and normal direction. Separation and reattachment locations were determined from oil-flow visualizations, and local surface static pressures were obtained with conventional pressure instrumentation.

Figure 9.11 shows a comparison of the pressure coefficients along the surface of the axisymmetric "bump." Both k-ε and k-ω models predict a delay of the shock position and, therefore, underpredict the size of the flow separation. The *SST* model provides the best overall performance, and the *SA* model comes second.

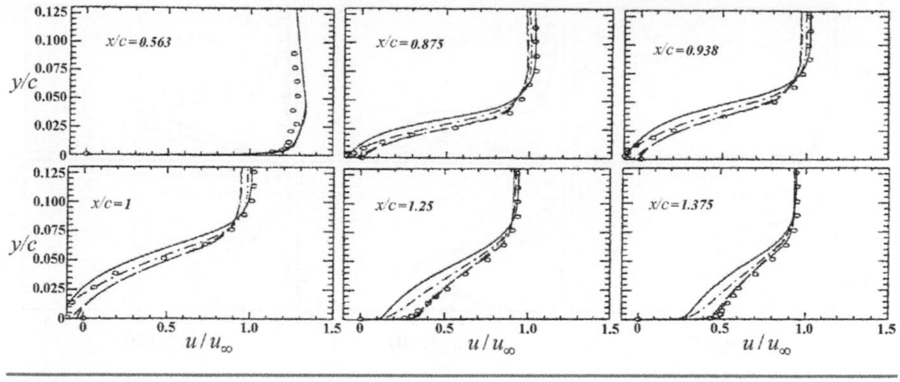

Fig. 9.12 Comparison of mean velocity profiles at different *x/c*-locations, after [2]. See symbols in Fig. 9.11.

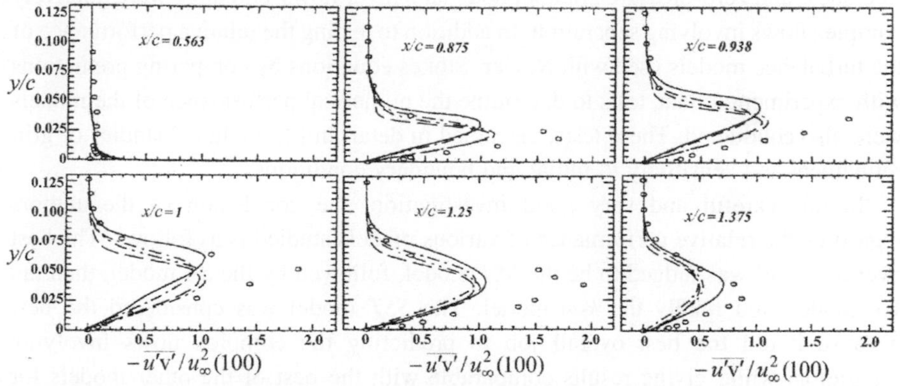

Fig. 9.13 Comparison of mean shear-stress profiles. See symbols in Fig. 9.11.

A sample of comparisons of the velocity, shear stress, and turbulent kinetic energy profiles at specified measured positions are shown in Fig. 9.12, 9.13, and 9.14, respectively. Since experimental data provide only two components of normal stresses, the turbulent kinetic energies shown in Fig. 9.14 were obtained by setting $w'^2 = (u'^2 + v'^2)/2$. The *SST* model gives the best agreement of the mean velocity profile with experiment, and closer agreement of mean shear-stress profiles.

9.5.3 Summary

The study conducted in [2] and briefly summarized in the previous two subsections investigated the relative performance of four turbulence models corresponding to *k-ε*, *k-ω*, *SA* and *SST* models. Of the ten flows tested, seven were relatively simple

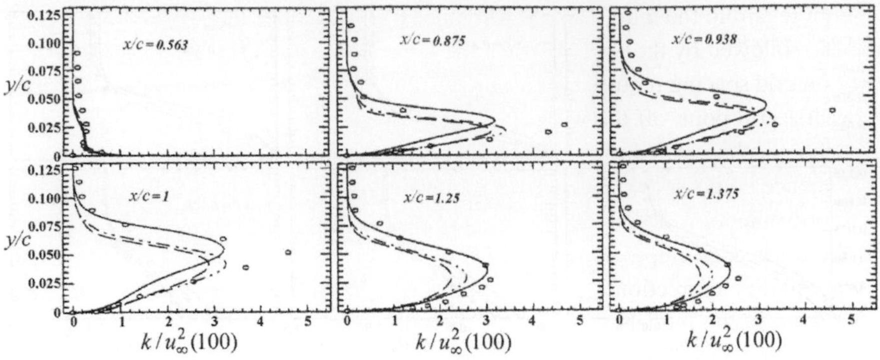

Fig. 9.14 Comparison of turbulent kinetic-energy profiles. See symbols in Fig. 9.11.

free-shear and zero-pressure gradient boundary-layer flows, and three were relatively complex flows involving separation. In addition to testing the relative performance of the turbulence models used with Navier–Stokes equations by comparing predictions with experimental data, tests to determine the numerical performance of the models were also conducted. These tests, discussed in detail in [2], included studies of grid refinement and sensitivity to initial and boundary conditions.

In this careful and very good investigation, the conclusion of the authors regarding the relative performance of various models studied is as follows. The best overall model was judged to be the *SST* model, followed by the *SA* model, then the k-ε model, and finally the k-ω model. The *SST* model was considered the best because it did the best overall job in predicting the complex flows involving separation, while giving results comparable with the best of the other models for the simple flows. For the simple free-shear flows, all of the models except the k-ω were about equal in their performance, with the *SA* giving best predictions of the mixing layer and plane wake flows, and the k-ε and *SST* models giving the best predictions of the jet flows. The performance of the k-ω model was judged to be poor for these flows because of its sensitivity to freestream conditions, with the resulting unreliability of solutions. None of the unmodified models did well in predicting the compressible mixing layer, although with compressibility modifications they did give improved predictions.

For the complex flows the best overall model was the *SST* because of its ability in predicting separation. The worst model in this regard was the k-ε, with the *SA* and k-ω models falling in between. The k-ω model did not appear to be as sensitive to freestream conditions for the complex flows (where a Navier-Stokes solver was used) as it was for the free-shear flows, although there was sensitivity. While there appear to be several possible explanations for this, the authors did not offer a definitive explanation at that time.

With regard to the numerical performance of the models, the *SA* was found to be the best, followed by the *SST*, and then the *k*-ε and *k*-ω models. This evaluation was based on grid spacing required for accurate solutions and the maximum y^+ allowable at the first grid point off the wall.

Although they stated that the *SST* and *SA* models were found to give superior performance compared with the other models needed, there was considerable room for improvement of these models. The *SST* needed improvement on the wake flow, and the *SA* needed improvement on the jet flows. All of the models needed better compressibility corrections for free-shear flows. Although not discussed in their study, none of the models appeared to do well on recovering flows downstream of reattachment. Corrections for rotation and curvature were still another area requiring attention.

An area that was not investigated in [2] is the ability of these turbulence models to predict flows with extensive regions of separation, i.e. airfoil flows near stall or post stall. Studies, either with Navier-Stokes or interactive boundary-layer methods, need to be conducted to explore the relative performance of these four transport equation turbulence models in predicting the accuracy of airfoil flows near stall or post stall. A study, for example, conducted in [17] with the *SA* and *CS* models showed that, while the predictions of the *SA* model were very good at low and moderate angles of attack, that was not the case at higher angles of attack.

Figures 9.15 and 9.16 show the results for the NACA 0012 airfoil at a Mach number of 0.3 and a chord Reynolds number of 3.9×10^6. Figure 9.15 shows a comparison between the calculated and measured lift coefficients. The calculated results with the *CS* model employing the modification of Cebeci-Chang discussed in

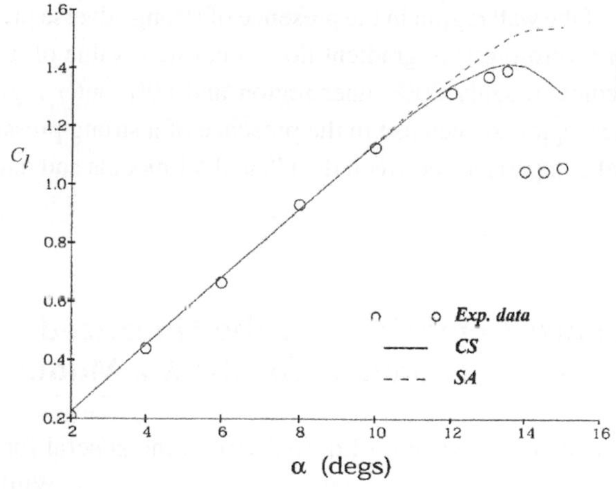

Fig. 9.15 Comparison of calculated lift coefficients with experimental data for the NACA 0012 airfoil at $M_\infty = 0.3$ and $R_c = 3.9 \times 10^6$.

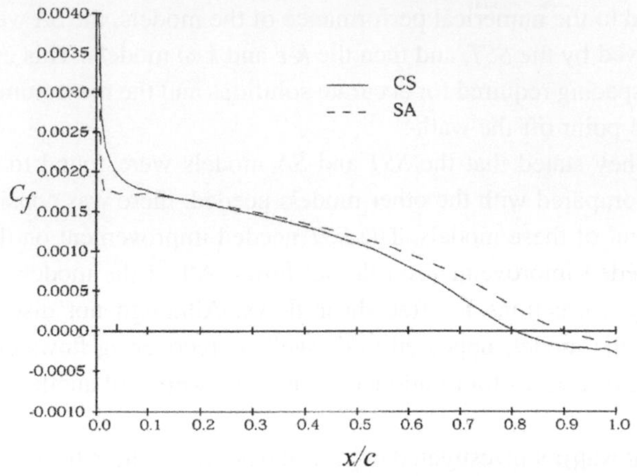

Fig. 9.16 Flow separation on the NACA 0012 airfoil at $M_\infty = 0.3$ and $R_c = 3.9 \times 10^6$ at $\alpha = 13.5°$.

subsection 5.4.2 indicate good agreement with data. The predictions of the *SA* model, while satisfactory at low and moderate angles of attack, are not satisfactory near and post stall. They resemble those obtained with the *CS* model with α in Eq. (5.4.14) as constant equal to 0.0168.

Figure 9.16 shows flow separation calculated with both turbulence models at $\alpha = 13.5°$, which is near the stall angle. As can be seen, while the *CS* model predicts separation at $\frac{x}{c} = 0.80$, the *SA* model predicts it at $\frac{x}{c} = 0.90$. Less flow separation predicted with the latter model is the reason why the calculated lift coefficient is higher than the experimental data. The reason for this may be the ability of the *SA* model not to decrease the law of the wall region in the presence of strong adverse pressure gradient. For example, for a zero-pressure gradient flow, a constant value of α ($\equiv$ 0.0168) in Eq. (5.4.14), predicts roughly 20% inner region and 80% outer region. A variable α allows the inner region to decrease in the presence of a strong pressure gradient.

For additional comparisons between the *CS* and *SA* models and data, the reader is referred to [13].

9A Appendix: Coefficients of the Linearized Finite-Difference Equations for the *k*-ε Model

We write the kinetic energy equation Eq. (6.2.19), in the general form

$$u\frac{\partial k}{\partial x} + v\frac{\partial k}{\partial y} = \frac{\partial}{\partial y}\left[\left(\nu + \frac{\varepsilon_m}{\sigma_k}\right)\frac{\partial k}{\partial y}\right] + \varepsilon_m\left(\frac{\partial u}{\partial y}\right)^2 - \left(\tilde{\varepsilon} + \Delta\right) + F \qquad (9A.1)$$

In terms of transformal variables, the above equation becomes

$$\left(b_2 k'\right)' + P - Q + F = 2\,mnk - m_1 f\,k' + x\left(u\frac{\partial k}{\partial x} - k'\frac{\partial f}{\partial x}\right) \tag{9A.2}$$

where $(b_2k')'$ denotes the diffusion term, P and Q defined by

$$P = \varepsilon_m^+ v^2, \quad Q = \tilde{\varepsilon} + \varDelta \tag{9A.3}$$

denote the production and dissipation terms, respectively. The right-hand side of Eq. (9.A.2) represents the convection term.

With $k' = s$, Eq. (9.A.2) becomes

$$\left(b_2 s\right)' + P - Q + F - 2mnk + m_1 fs = x\left(u\frac{\partial k}{\partial x} - s\frac{\partial f}{\partial x}\right) \tag{9A.4}$$

With x-wise derivation represented either by two- or three-point backward differences the finite-difference approximations to Eq. (9.A.4) are

$$\begin{aligned} &h_j^{-1}\left[(b_2 s)_j^n - (b_2 s)_{j-1}^n\right] + m_1^n (f\,s)_{j-1/2}^n - 2m^n (uk)_{j-1/2}^n + P_{j-1/2}^n \\ &-Q_{j-1/2}^n + F_{j-1/2}^n = x^n\left[u_{j-1/2}^n\left(\frac{\partial k}{\partial x}\right)_{j-1/2}^n - s_{j-1/2}^n\left(\frac{\partial f}{\partial x}\right)_{j-1/2}^n\right] \end{aligned} \tag{9A.5}$$

Linearizing, we get

$$\begin{aligned} &h_j^{-1}\left[(b_2)_j^n \delta s_j + s_j^n\left(\frac{\partial b_2}{\partial k}\right)_j^n \delta k_j + s_j^n\left(\frac{\partial b_2}{\partial \varepsilon}\right)_j^n \delta\varepsilon_j - (b_2)_{j-1}^n \delta s_{j-1}\right. \\ &\left.-s_{j-1}^n\left(\frac{\partial b_2}{\partial k}\right)_{j-1}^n \delta k_{j-1} - s_{j-1}^n\left(\frac{\partial b_2}{\partial \varepsilon}\right)_{j-1}^n \delta\varepsilon_{j-1}\right] \\ &+\frac{1}{2}\left[\left(\frac{\partial P}{\partial k}\right)_j^n \delta k_j + \left(\frac{\partial P}{\partial \varepsilon}\right)_j^n \delta\varepsilon_j + \left(\frac{\partial P}{\partial v}\right)_j^n \delta v_j\right. \\ &\left.+\left(\frac{\partial P}{\partial k}\right)_{j-1}^n \delta k_{j-1} + \left(\frac{\partial P}{\partial \varepsilon}\right)_{j-1}^n \delta\varepsilon_{j-1} + \left(\frac{\partial P}{\partial v}\right)_{j-1}^n \delta v_{j-1}\right] \\ &+\frac{m_1^n}{2}\left(f_j^n \delta s_j + s_j^n \delta f_j + f_{j-1}^n \delta s_{j-1} + s_{j-1}^n \delta f_{j-1}\right) \\ &-m^n\left[u_j^n \delta k_j + k_j^n \delta u_j + u_{j-1}^n \delta k_{j-1} + k_{j-1}^n \delta u_{j-1}\right] \end{aligned}$$

$$
\begin{aligned}
&-\frac{1}{2}\left[\left(\frac{\partial Q}{\partial k}\right)_j^n \delta k_j + \left(\frac{\partial Q}{\partial \varepsilon}\right)_j^n \delta\varepsilon_j + \left(\frac{\partial Q}{\partial k}\right)_{j-1}^n \delta k_{j-1}\right. \\
&\left. + \left(\frac{\partial Q}{\partial \varepsilon}\right)_{j-1}^n \delta\varepsilon_{j-1}\right] \\
&+\frac{1}{2}\left[\left(\frac{\partial F}{\partial k}\right)_j^n \delta k_j + \left(\frac{\partial F}{\partial \varepsilon}\right)_j^n \delta\varepsilon_j + \left(\frac{\partial F}{\partial k}\right)_{j-1}^n \delta k_{j-1}\right. \\
&\left. + \left(\frac{\partial F}{\partial \varepsilon}\right)_{j-1}^n \delta\varepsilon_{j-1}\right] \\
&= \frac{x^n}{2}\left\{\left(\delta u_j + \delta u_{j-1}\right)\left(\frac{\partial k}{\partial x}\right)_{j-1/2}^n + u_{j-1/2}^n\left[\frac{\partial}{\partial k}\left(\frac{\partial k}{\partial x}\right)_j^n \delta k_j\right.\right. \\
&\left. + \frac{\partial}{\partial k}\left(\frac{\partial k}{\partial x}\right)_{j-1}^n \delta k_{j-1}\right] - \left(\delta s_j + \delta s_{j-1}\right)\left(\frac{\partial f}{\partial x}\right)_{j-1/2}^n \\
&\left.- s_{j-1/2}^n\left[\frac{\partial}{\partial f}\left(\frac{\partial f}{\partial x}\right)_j^n \delta f_j + \frac{\partial}{\partial f}\left(\frac{\partial f}{\partial x}\right)_{j-1}^n \delta f_{j-1}\right]\right\} + (r_4)_j
\end{aligned}
\tag{9A.6}
$$

The coefficients of Eq. (9.2.35) can now be written as

$$(\alpha_1)_j = \frac{m_1^n}{2} s_j^n + \frac{x^n}{2} s_{j-1/2}^n \frac{\partial}{\partial f}\left(\frac{\partial f}{\partial x}\right)_j^n \tag{9A.7a}$$

$$(\alpha_2)_j = \frac{m_1^n}{2} s_{j-1}^n + \frac{x^n}{2} s_{j-1/2}^n \frac{\partial}{\partial f}\left(\frac{\partial f}{\partial x}\right)_{j-1}^n \tag{9A.7b}$$

$$(\alpha_3)_j = -m^n k_j^n - \frac{x^n}{2}\left(\frac{\partial k}{\partial x}\right)_{j-1/2}^n \tag{9A.7c}$$

$$(\alpha_4)_j = -m^n k_{j-1}^n - \frac{x^n}{2}\left(\frac{\partial k}{\partial x}\right)_{j-1/2} \tag{9A.7d}$$

$$(\alpha_5)_j = \frac{1}{2}\left(\frac{\partial P}{\partial v}\right)_j^n \tag{9A.7e}$$

$$(\alpha_6)_j = \frac{1}{2}\left(\frac{\partial P}{\partial v}\right)_{j-1}^n \tag{9A.7f}$$

$$
\begin{aligned}
(\alpha_7)_j = {} & h_j^{-1} s_j^n \left(\frac{\partial b_2}{\partial k}\right)_j^n + \frac{1}{2}\left(\frac{\partial P}{\partial k}\right)_j - m^n u_j^n - \frac{1}{2}\left(\frac{\partial Q}{\partial k}\right)_j^n \\
& + \frac{1}{2}\left(\frac{\partial F}{\partial k}\right)_j^n - \frac{x^n}{2} u_{j-1/2}^n \frac{\partial}{\partial k}\left(\frac{\partial k}{\partial x}\right)_j^n
\end{aligned}
\tag{9A.7g}
$$

$$(\alpha_8)_j = -h_j^{-1} s_{j-1}^n \left(\frac{\partial b_2}{\partial k}\right)_{j-1}^n + \frac{1}{2}\left(\frac{\partial P}{\partial k}\right)_{j-1} - m^n u_{j-1}^n$$

$$-\frac{1}{2}\left(\frac{\partial Q}{\partial k}\right)_{j-1}^n + \frac{1}{2}\left(\frac{\partial F}{\partial k}\right)_{j-1}^n - \frac{x^n}{2} u_{j-1/2}^n \frac{\partial}{\partial k}\left(\frac{\partial k}{\partial x}\right)_{j-1}^n \tag{9A.7h}$$

$$(\alpha_9)_j = h_j^{-1}(b_2)_j^n + \frac{m^n}{2} f_j^n + \frac{x^n}{2}\left(\frac{\partial f}{\partial x}\right)_{j-1/2}^n \tag{9A.7i}$$

$$(\alpha_{10})_j = -h_j^{-1}(b_2)_{j-1}^n + \frac{m^n}{2} f_{j-1}^n + \frac{x^n}{2}\left(\frac{\partial f}{\partial x}\right)_{j-1/2}^n \tag{9A.7j}$$

$$(\alpha_{11})_j = h_j^{-1} s_j^n \left(\frac{\partial b_2}{\partial \varepsilon}\right)_j^n + \frac{1}{2}\left[\left(\frac{\partial f}{\partial \varepsilon}\right)_j^n - \left(\frac{\partial Q}{\partial \varepsilon}\right)_j^n + \left(\frac{\partial F}{\partial \varepsilon}\right)_j^n\right] \tag{9A.7k}$$

$$(\alpha_{12})_j = -h_j^{-1} s_{j-1}^n \left(\frac{\partial b_2}{\partial \varepsilon}\right)_{j-1}^n$$

$$+\frac{1}{2}\left[\left(\frac{\partial f}{\partial \varepsilon}\right)_{j-1}^n - \left(\frac{\partial Q}{\partial \varepsilon}\right)_{j-1}^n + \left(\frac{\partial F}{\partial \varepsilon}\right)_{j-1}^n\right] \tag{9A.7l}$$

$$(r_4)_j = -\left[(b_2 s)_j^n - (b_2 s)_{j-1}^n\right]h_j^{-1}$$

$$-\left[P_{j-1/2}^n - Q_{j-1/2}^n + F_{j-1/2}^n - 2m^n (uk)_{j-1/2}^n\right.$$

$$\left. - m_1^n (fs)_{j-1/2}^n\right] \tag{9A.8a}$$

$$+x^n \left[u_{j-1/2}^n \left(\frac{\partial k}{\partial n}\right)_{j-1/2}^n - s_{j-1/2}^n \left(\frac{\partial f}{\partial x}\right)_{j-1/2}^n\right]$$

Remembering the definitions of diffusion, production, dissipation, convection and F terms, Eq. (9A.8a) can also be written as

$$(r_4)_j = -\left[\text{diffusion} + \text{production} - \text{dissipation} + F - \text{convection}\right] \tag{9A.8b}$$

The parameter P, Q and F are model dependent. As a result, the derivatives with respect to k, ε and v will be different for each model. The derivatives of $\frac{\partial k}{\partial x}$ and $\frac{\partial f}{\partial x}$ with respect to k and f are straightforward.

In term of transformed variables the parameters P, Q and F are (here ε is $\tilde{\varepsilon}$)

$$P = c_\mu f_\mu R_x \frac{k}{\varepsilon} v \tag{9A.9a}$$

$$Q = \varepsilon + \frac{1}{2}\frac{s^2}{k}, \quad \varDelta = \frac{1}{2}\frac{s^2}{k} \tag{9A.9b}$$

$$F = \left[\frac{1}{2Q}\left(ss' - \frac{1}{2}\frac{s^3}{k}\right)\right]' \tag{9A.9c}$$

The rate of dissipation equation is given by Eq. (6.2.20); in terms of transformed variables it can be written as ($\varepsilon = \tilde{\varepsilon}$)

$$(b_3\varepsilon')' + P_1 - Q_1 + E = x\left(u\frac{\partial \varepsilon}{\partial x} - \varepsilon'\frac{\partial f}{\partial x}\right) + (3m-1)u\varepsilon - m_1\varepsilon' f \qquad (9A.10)$$

where $(b_3\varepsilon')'$ denotes the diffusion term, P_1 and Q_1 defined by

$$P_1 = c_{\varepsilon 1} f_1 c_\mu f_\mu v^2 k \qquad (9A.11a)$$

$$Q_1 = c_{\varepsilon 2} f_2 \varepsilon^2 / k \qquad (9A.11b)$$

denote the generation and destruction terms, respectively. The right hand side of Eq. (9.A.10) represents the convective term.

With $\varepsilon' = q$, Eq. (9.A.10) becomes

$$(b_3 q)' + P_1 - Q_1 + E + m_1 f q - (3m-1)u\varepsilon = x\left(u\frac{\partial \varepsilon}{\partial x} - q\frac{\partial f}{\partial x}\right) \qquad (9A.12)$$

Following a procedure similar to the one used for the kinetic energy equation, the finite-difference approximations for the above equation are

$$\begin{aligned} &h_j^{-1}\left[(b_3 q)_j^n - (b_3 q)_{j-1}^n\right] + (P_1)_{j-1/2}^n - (Q_1)_{j-1/2}^n + E_{j-1/2}^n \\ &+ m_1^n (fq)_{j-1/2}^n - \left(3m^n - 1\right)(u\varepsilon)^n \\ &= x\left[u_{j-1/2}^n \left(\frac{\partial \varepsilon}{\partial x}\right)_{j-1/2}^n - q_{j-1/2}^n \left(\frac{\partial f}{\partial x}\right)_{j-1/2}^n\right] \end{aligned} \qquad (9A.13)$$

After linearization, the resulting expression can be expressed in the form given by Eq. (9.2.36),

$$(\beta_1)_j = \frac{m_1^n}{2} q_j^n + \frac{x^n}{2} q_{j-1/2}^n \frac{\partial}{\partial f}\left(\frac{\partial f}{\partial x}\right)_j^n \qquad (9A.14a)$$

$$(\beta_2)_j = \frac{m_1^n}{2} q_{j-1}^n + \frac{x^n}{2} q_{j-1/2}^n \frac{\partial}{\partial f}\left(\frac{\partial f}{\partial x}\right)_{j-1}^n \qquad (9A.14b)$$

$$(\beta_3)_j = -\frac{x^n}{2}\left(\frac{\partial \varepsilon}{\partial x}\right)_{j-1/2}^n - \frac{1}{2}(3m-1)\varepsilon_j^n \qquad (9A.14c)$$

$$(\beta_4)_j = -\frac{x^n}{2}\left(\frac{\partial \varepsilon}{\partial x}\right)_{j-1/2}^n - \frac{1}{2}(3m-1)\varepsilon_{j-1}^n \qquad (9A.14d)$$

$$(\beta_5)_j = \frac{1}{2}\left(\frac{\partial P_1}{\partial v}\right)_j^n \qquad (9A.14e)$$

$$(\beta_6)_j = \frac{1}{2}\left(\frac{\partial P_1}{\partial v}\right)_{j-1}^n \tag{9A.14f}$$

$$(\beta_7)_j = h_j^{-1} q_j^n \left(\frac{\partial b_3}{\partial k}\right)_j^n + \frac{1}{2}\left(\frac{\partial P_1}{\partial k}\right)_j^n - \frac{1}{2}\left(\frac{\partial Q_1}{\partial k}\right)_j^n \tag{9A.14g}$$

$$(\beta_8)_j = h_j^{-1} q_{j-1}^n \left(\frac{\partial b_3}{\partial k}\right)_{j-1}^n + \frac{1}{2}\left(\frac{\partial P_1}{\partial k}\right)_{j-1}^n - \frac{1}{2}\left(\frac{\partial Q_1}{\partial k}\right)_{j-1}^n \tag{9A.14h}$$

$$\begin{aligned}(\beta_{11})_j = {} & h_j^{-1} q_j^n \left(\frac{\partial b_3}{\partial \varepsilon}\right)_j^n - \frac{1}{2}\left(\frac{\partial Q_1}{\partial \varepsilon}\right)_j^n + \frac{1}{2}\left(\frac{\partial E}{\partial \varepsilon}\right)_j^n \\ & -\frac{1}{2}\left(3m^n - 1\right)u_j^n - \frac{x^n}{2} u_{j-1/2}^n \frac{\partial}{\partial \varepsilon}\left(\frac{\partial \varepsilon}{\partial x}\right)_j^n\end{aligned} \tag{9A.14i}$$

$$\begin{aligned}(\beta_{12})_j = {} & -h_j^{-1} q_{j-1}^n \left(\frac{\partial b_3}{\partial \varepsilon}\right)_{j-1}^n - \frac{1}{2}\left(\frac{\partial Q_1}{\partial \varepsilon}\right)_{j-1}^n + \frac{1}{2}\left(\frac{\partial E}{\partial \varepsilon}\right)_{j-1}^n \\ & -\frac{1}{2}\left(3m^n - 1\right)u_{j-1}^n - \frac{x^n}{2} u_{j-1/2}^n \frac{\partial}{\partial \varepsilon}\left(\frac{\partial \varepsilon}{\partial x}\right)_{j-1}^n\end{aligned} \tag{9A.14j}$$

$$(\beta_{13})_j = h_j^{-1}(b_3)_j^n + \frac{m_1^n}{2} f_j^n + \frac{x^n}{2}\left(\frac{\partial f}{\partial x}\right)_{j-1/2}^n \tag{9A.14k}$$

$$(\beta_{14})_j = -h_j^{-1}(b_3)_{j-1}^n + \frac{m_1^n}{2} f_{j-1}^n + \frac{x^n}{2}\left(\frac{\partial f}{\partial x}\right)_{j-1/2}^n \tag{9A.14l}$$

$$\begin{aligned}(r_5)_j = {} & -h_j^{-1}\left[(b_3 q)_j^n - (b_3 q)_{j-1}^n\right] \\ & +(P_1)_{j-1/2}^n - (Q_1)_{j-1/2}^n + E_{j-1/2}^n \\ & +m_1^n (fq)_{j-1/2}^n - \left(3m^n - 1\right)(u\varepsilon)_{j-1/2}^n] \\ & +x^n\left[u_{j-1/2}^n\left(\frac{\partial \varepsilon}{\partial x}\right)_{j-1/2}^n - q_{j-1/2}^n\left(\frac{\partial f}{\partial x}\right)_{j-1/2}^n\right]\end{aligned} \tag{9A.15a}$$

Equation (9A.15a) can also be written are

$$(r_5)j = -[\text{diffusion} + \text{generation} - \text{destruction} - \text{convection} + E] \tag{9A.15b}$$

Problems

9.1 Consider the SA model discussed in subsection 6.3.2. Using the Falkner-Skan transformation, Eq. (8.2.5), show that the transport equation for eddy viscosity, Eq. (6.3.10), can be written as

$$x\left(f'\frac{\partial \nu_t^+}{\partial x} - \nu_t^+\frac{\partial f}{\partial x}\right) - m_1(\nu_t^+)'f = c_{b_1}\left(1 - f_{t_2}\right)\tilde{s}^*\nu_t^*$$
$$-\left(c_{\omega_1}f_\omega - \frac{c_{b_1}}{k^2}f_{b_2}\right)\frac{(\nu_t^+)^2}{\eta^2} \tag{P9.1.1}$$
$$+\frac{1}{\sigma}\left[[(1+\nu_t^+)(\nu_t^+)']' + c_{b_2}[(\nu_t^+)']^2\right]$$

9.2 The solution of the transformed continuity and momentum equations, Eqs. (8.2.6) and (8.2.7) and Eq. (P9.1.1) can be obtained with the Box method discussed in Section 8.2.

(a) Show that Eqs. (8.2.6) and (P9.1.1) can be written as a system of five first-order equations by defining

$$g = (\nu_t^+)', \quad u = f', \quad v = u'$$
$$f' = u \tag{P9.2.1a}$$
$$u' = v \tag{P9.2.1b}$$
$$(\nu_t^+)' = g \tag{P9.2.1c}$$
$$(bv)' + m_1 f v + m(1 - u^2) = x\left(u\frac{\partial u}{\partial x} - v\frac{\partial f}{\partial x}\right) \tag{P9.2.1d}$$
$$\frac{1}{\sigma}\left[[(1+\nu_t^+)g]' + c_{b_2}g^2\right] + c_{b_1}(1 - f_{t_2})\tilde{s}^*\nu_t^+$$
$$-\left(c_{\omega_1}f_\omega - \frac{c_{b_1}}{\kappa^2}f_{t_2}\right)\frac{(\nu_t^+)^2}{\eta^2 + (\nu_w^*)^2} + m_1 g f \tag{P9.2.1e}$$
$$= x\left(u\frac{\partial \nu_t^+}{\partial x} - g\frac{\partial f}{\partial x}\right)$$

9.3 For the net rectangle shown in Fig. 8.1, write finite difference approximations to the equations in Problem 9.2 and check your answer with those given below. Use backward differences for the streamwise derivatives which are needed to avoid oscillations when initial conditions are specified for turbulent flow. Also, to keep the code simple, in Eq. (P9.2.1e) we have defined the diffusion term f_μ by

$$\frac{1}{\sigma}\left\{[(1+\nu_t^+)g]' + c_{b_2}g^2\right\} \tag{P9.3.1a}$$

production term $\tilde{p}_r$ by

$$\tilde{p}_r = c_{b_1}\left(1 - f_{t_2}\right)\left[\frac{\left(\nu^+\right)^2}{\kappa^2\eta^2} f_{v_2} + \sqrt{R_x}\left|\nu\right|\nu^+\right] \quad \text{(P9.3.1b)}$$

and dissipation term

$$\tilde{d}_e = \frac{\left(\nu^+\right)^2}{\eta^2}\left[c_{w_1} f_w - \frac{c_{b_1}}{\kappa^2} f_{t_2}\right] \quad \text{(P9.3.1c)}$$

9.4 Using Newton's method, linearize the algebraic system in Problem 9.3. In order to obtain quadratic convergence, differentiate variables $\tilde{f}_\mu, \tilde{p}_r, \tilde{d}_e$ with respect to ν^+. Show that the sesulting system of linear equations can be written in the form (note $\nu_t^+ \equiv \nu^+$ for convenience)

$$\delta f_j - \delta f_{j-1} - \frac{h_j}{2}\left(\delta u_j + \delta u_{j-1}\right) = \left(r_1\right)_j \quad \text{(P9.4.1a)}$$

$$\delta u_j - \delta u_{j-1} - \frac{h_j}{2}\left(\delta v_j + \delta v_{j-1}\right) = \left(r_4\right)_{j-1} \quad \text{(P9.4.1b)}$$

$$\delta \nu_j^+ - \delta \nu_{j-1}^+ - \frac{h_j}{2}\left(\delta g_j + \delta g_{j-1}\right) = \left(r_5\right)_{j-1} \quad \text{(P9.4.1c)}$$

$$\begin{aligned}&\left(s_1\right)_j \delta f_j + \left(s_2\right)_j \delta f_{j-1} + \left(s_3\right)_j \delta u_j + \left(s_4\right)_j \delta u_{j-1} + \left(s_5\right)_j \delta v_j \\ &+\left(s_6\right)_j \delta v_{j-1} + \left(s_7\right)_j \delta \nu_j^+ + \left(s_8\right)_j \delta \nu_{j-1}^+ = \left(r_2\right)_j\end{aligned} \quad \text{(P9.4.1d)}$$

$$\begin{aligned}&\left(e_1\right)_j \delta f_j + \left(e_2\right)_j \delta f_{j-1} + \left(e_3\right)_j \delta u_j + \left(e_4\right)_j \delta u_{j-1} + \left(e_5\right)_j \delta v_j \\ &+\left(e_6\right)_j \delta v_{j-1} + \left(e_7\right)_j \delta \nu_j^+ + \left(e_8\right)_j \delta \nu_{j-1}^+ + \left(e_9\right)_j \delta g_j \\ &+ \left(e_{10}\right)_j \delta g_{j-1} = \left(r_3\right)_j\end{aligned} \quad \text{(P9.4.1e)}$$

Here the coefficients of Eq. (P9.4.1) are

$$\left(s_1\right)_j = \tilde{\alpha} v_{j-1/2} + \frac{m_1}{2} v_j \quad \text{(P9.4.2a)}$$

$$\left(s_1\right)_j = \tilde{\alpha} v_{j-1/2} + \frac{m_1}{2} v_{j-1} \quad \text{(P9.4.2b)}$$

$$\left(s_3\right)_j = -\left(m + \tilde{\alpha}\right) u_j \quad \text{(P9.4.2c)}$$

$$\left(s_4\right)_j = -\left(m + \tilde{\alpha}\right) u_{j-1} \quad \text{(P9.4.2d)}$$

$$\left(s_5\right)_j = b_j h_j^{-1} + \frac{m_1}{2} f_j + 0.5x\left(\frac{\partial f}{\partial x}\right)_{j-1/2} \quad \text{(P9.4.2e)}$$

$$(s_6)_j = -b_{j-1}h_j^{-1} + \frac{m_1}{2}f_{j-1} + 0.5x\left(\frac{\partial f}{\partial x}\right)_{j-1/2} \quad \text{(P9.4.2f)}$$

$$(s_7)_j = \left(\frac{\partial b}{\partial v^+}\right)_j v_j h_j^{-1} \quad \text{(P9.4.2g)}$$

$$(s_8)_j = -\left(\frac{\partial b}{\partial v^+}\right)_{j-1} v_{j-1}h_j^{-1} \quad \text{(P9.4.2h)}$$

$$(r_2)_j = x\left(u\frac{\partial u}{\partial x} - \frac{\partial f}{\partial x}v\right)_{j-1/2} - \left[(bv)' + m_1 fv + m\left(1-u^2\right)\right]_{j-1/1} \quad \text{(P9.4.2i)}$$

The coefficients of Eq. (P9.4.1e) are

$$e_1 = \frac{m_1}{2}m_1 g_j + g_{j-1/2}x\frac{\delta}{\delta f}\left(\frac{\partial f}{\partial x}\right)_j \quad \text{(P9.4.3a)}$$

$$e_2 = \frac{m_1}{2}m_1 g_{j-1} + g_{j-1/2}x\frac{\delta}{\delta f}\left(\frac{\partial f}{\partial x}\right)_{j-1} \quad \text{(P9.4.3b)}$$

$$e_3 = -0.5x\frac{\partial v^+}{\partial x}, \quad e_4 = e_3 \quad \text{(P9.4.3c)}$$

$$e_5 = \frac{1}{2}\left[\left(\frac{\partial \tilde{p}_r}{\partial v}\right)_j - \left(\frac{\partial \tilde{d}_e}{\partial v}\right)_j\right] \quad \text{(P9.4.3d)}$$

$$e_6 = \frac{1}{2}\left[\left(\frac{\partial \tilde{p}_r}{\partial v}\right)_{j-1} - \left(\frac{\partial \tilde{d}_e}{\partial v}\right)_{j-1}\right] \quad \text{(P9.4.3e)}$$

$$e_7 = -x\frac{\partial}{\partial v^+}\left(\frac{\partial v^+}{\partial x}\right)_j + \left(\frac{\partial \tilde{f}_\mu}{\partial v^+}\right)_j + \frac{1}{2}\left[\left(\frac{\partial \tilde{p}_r}{\partial v^+}\right)_j - \left(\frac{\partial \tilde{d}_e}{\partial v^+}\right)_j\right] \quad \text{(P9.4.3f)}$$

$$e_8 = -x\frac{\partial}{\partial v+}\left(\frac{\partial v^+}{\partial x}\right)_{j-1} + \left(\frac{\partial \tilde{f}_\mu}{\partial v^+}\right)_{j-1} + \frac{1}{2}\left[\left(\frac{\partial \tilde{p}_r}{\partial v^+}\right)_{j-1} - \left(\frac{\partial \tilde{d}_e}{\partial v^+}\right)_{j-1}\right] \quad \text{(P9.4.3g)}$$

$$e_9 = \frac{1}{2}x\left(\frac{\partial f}{\partial x}\right)_{j-1/2} + \frac{m_1}{2}m_1 f_j + \left(\frac{\partial \tilde{f}_\mu}{\partial g}\right)_j \quad \text{(P9.4.3h)}$$

$$e_{10} = \frac{1}{2}x\left(\frac{\partial f}{\partial x}\right)_{j-1/2} + m_1 f_{j-1} + \left(\frac{\partial \tilde{f}_\mu}{\partial g}\right)_{j-1} \quad \text{(P9.4.3i)}$$

$$(r_3)_j = \left[u_{j-1/2}x\left(\frac{\partial v^+}{\partial x}\right)_{j-1/2} - g_{j-1/2}x\left(\frac{\partial f}{\partial x}\right)_{j-1/2} - m_1(fg)_{j-1/2}\right] - \left(\tilde{f}_\mu + \tilde{p}_r - \tilde{d}_e\right)_{j-1/2} \quad \text{(P9.4.3j)}$$

Here the diffusion, production and destruction terms, $\tilde{f}_\mu, \tilde{p}_r$ and $\tilde{d}_e$, respectively, are given by

$$\tilde{f}_\mu = \frac{(1+c_{b_2})}{\sigma}\left[(1+\nu_j^+)g_j - (1+\nu_{j-1}^+)g_{j-1}\right]h_j^{-1} - \frac{c_{b_2}}{\sigma}\nu_{j-1/2}^+(g_j - g_{j-1})h_j^{-1}$$

$$\tilde{p}_r = c_{b_1}(1-f_{t_2})\left[\frac{(\nu^+)^2}{\kappa^2(\eta^2+\nu_w^2)^2}f_{v_2} + \sqrt{R_x}|v|\nu^+\right]$$

$$\tilde{d}_e = \frac{(\nu^+)^2}{\eta^2}\left[c_{w_1}f_w - \frac{c_{b_1}}{\kappa^2}f_{t_2}\right]$$

and

$$\frac{\partial\tilde{p}_r}{\partial v} = c_{b_1}(1-f_{t_2})\sqrt{R_x}\left\{\begin{matrix}\nu^+ \\ -\nu^+\end{matrix}\right\}\begin{matrix}\text{if} & v>0 \\ \text{if} & v>0\end{matrix}$$

$$\frac{\partial\tilde{d}_e}{\partial_v} = \frac{(\nu^+)^2}{\eta^2}\left[c_{w_1}\frac{\partial f_w}{\partial v}\right]$$

$$\left(\frac{\partial\tilde{f}_\mu}{\partial\nu^+}\right)_j = \frac{1+c_{b_2}}{\sigma}\frac{g_j}{h_j} - \frac{1}{2}\frac{c_{b_2}}{\sigma}\frac{g_j - g_j - 1}{h_j}$$

$$\left(\frac{\partial\tilde{f}_\mu}{\partial\nu^+}\right)_{j-1} = -\frac{1+c_{b_2}}{\sigma}\frac{g_{j-1}}{h_j} - \frac{1}{2}\frac{c_{b_2}}{\sigma}\frac{g_j - g_{j-1}}{h_j}$$

$$\frac{\partial\tilde{f}_\mu}{\partial g} = \frac{1+c_{b_2}}{\sigma}\frac{1+\nu^+}{h_j} - \frac{c_{b_2}}{\sigma}\frac{\nu_{j-1/2}^+}{h_j}$$

$$\begin{aligned}\frac{\partial\tilde{p}_r}{\partial\nu^+} = &-c_{b_1}\left(\frac{\partial f_{t_2}}{\partial\nu^+}\right)\left[\frac{(\nu^+)^2}{\kappa^2\eta^2}f_{v^2} + \nu^+\sqrt{R_x}|v|\right] \\ &+c_{b_1}(1-f_{t_2})\left[\frac{(\nu^+)^2}{\kappa^2\eta^2}\frac{\partial f_{v_2}}{\partial\nu^+} + \frac{\partial f_{v_2}}{\partial\nu^+} + \sqrt{R_x}|v|\right. \\ &\left.+\frac{2\nu^+}{\kappa^2\eta^2}f_{v_2}\right]\end{aligned}$$

$$\frac{\partial d_e}{\partial\nu^+} = \frac{(\nu^+)^2}{\eta^2+(\nu_w^*)^2}\left[c_{w1}\frac{\partial f_w}{\partial\nu^+} - \frac{c_{b_1}}{\kappa^2}\frac{\partial f_{t_2}}{\partial\nu^+}\right] + \frac{2\nu^+}{\eta^2+(\nu_w^*)^2}\left(c_{w1}f_w - \frac{c_{b_1}}{\kappa^2}f_{t_2}\right)$$

In the above equations, $f_{v_1}, f_{v_2}, f_w, f_{t_2}$ and their variations with respect to ν and ν^+ are as follows

$$f_{t_2} = c_{t_3}\exp\left[-c_{t_4}(\nu^+)^2\right]$$

$$\frac{\partial f_{t_2}}{\partial \nu^+} = -2c_{t_4}\nu^+ f_{t_2}$$

$$f_{v_1} = \frac{(\nu^+)^3}{(\nu^+)^3 + c_{v_1}^3}$$

$$\frac{\partial f_{v_1}}{\partial \nu^+} = \frac{3c_{v_1}^3 (\nu^+)^2}{\left[(\nu^+)^3 + c_{v_1}^3\right]^2}$$

$$f_{v_2} = 1 - \frac{\nu^+}{1 + \nu^+ f_{v_1}}$$

$$\frac{\partial f_{v_2}}{\partial \nu^+} = \frac{(\nu^+)^2 \frac{\partial f_{v_1}}{\partial \nu^+} - 1}{(1 + \nu^+ f_{v_1})^2}$$

$$f_w = \left[\frac{\left(1 + c_{w_3}^6\right)}{\left(1 + c_{w_3}^6 g_1^{-6}\right)}\right]^{1/6}$$

$$g_1 = rr + c_{w_5}\left(rr^6 - rr\right)$$

$$rr = \frac{(\nu^+)^2}{\eta^2 \kappa^2}\left\{\nu^+ \sqrt{R_x}|v| + \frac{(\nu^+)^2}{\kappa^2 \eta^2} f_{v_2}\right\}^{-1}$$

$$\frac{\partial f_w}{\partial v} = \frac{\partial f_w}{\partial g_1}\frac{\partial g_1}{\partial (rr)}\frac{\partial (rr)}{\partial v}$$

$$\frac{\partial f_w}{\partial \nu^+} = \frac{\partial f_w}{\partial g_1}\frac{\partial g_1}{\partial (rr)}\frac{\partial (rr)}{\partial \nu^+}$$

where

$$\frac{\partial f_w}{\partial g_1} = \frac{c_{w_3}^6 g_1^{-6} f_w}{\left[g_1\left(1 + c_{w_3}^6 g_1^{-6}\right)\right]}$$

$$\frac{\partial g_1}{\partial (rr)} = 1 + c_{w_2}\left(6rr^5 - 1\right)$$

$$\frac{\partial (rr)}{\partial v} = -\frac{(\nu^+)^2}{\kappa^2\left(\eta^2 + \nu_w^{*2}\right)}\left[\nu^+ \sqrt{R_x}|v| + \frac{(\nu^+)^2}{\kappa^2 \eta^2} f_{v_2}\right]^{-2}$$

$$\times\left(\sqrt{R_x}\nu^+\right)\begin{Bmatrix} 1 \\ -1 \end{Bmatrix} \begin{matrix} \text{if } v > 0 \\ \text{if } v < 0 \end{matrix}$$

$$\frac{\partial(rr)}{\partial \nu^+} = -\frac{(\nu^+)^2}{\kappa^2(\eta^2+\nu_w^{*2})}\left[\nu^+\sqrt{\mathrm{R}_x}|v| + \frac{(\nu^+)^2}{\kappa^2(\eta^2+\nu_w^{*2})}f_{\upsilon_2}\right]^{-2}$$

$$\times\left[\frac{(\nu^+)^2}{\kappa^2(\eta^2+\nu_w^{*2})}\frac{\partial f_{\upsilon_2}}{\partial \nu^+} + \sqrt{\mathrm{R}_x}|v| + \frac{2\nu^+}{\kappa^2(\eta^2+\nu_w^{*2})}f_{\upsilon_2}\right]$$

$$+\frac{2\nu^+}{\kappa^2(\eta^2+\nu_w^{*2})}\left[\nu^+\sqrt{\mathrm{R}_x}|v| + \frac{(\nu^+)^2}{\kappa^2(\eta^2+\nu_w^{*2})}f_{\upsilon_2}\right]$$

9.5 Show that the linear system of equations given by Eqs. (P9.4) subject to the boundary conditions

$$\begin{aligned} &\eta = 0, \quad f = u = \nu^+ = 0 \\ &\eta = \eta_e, \quad u = 1, \quad \nu^+ = \nu_e^+ \end{aligned} \tag{P9.5.1}$$

which in linearized form

$$\begin{aligned} \delta u_0 &= \delta f_0 = \delta \nu_0^+ = 0 \\ \delta u_J &= \delta \nu_J^+ = 0 \end{aligned} \tag{P9.5.2}$$

can be written in matrix-vector form, given by Eq. (8.2.24) with five-dimensional vectors $\underset{\sim}{\delta}_j$ and $\underset{\sim}{r}_j$ for each value of j defined by

$$\underset{\sim}{\delta}_j = \begin{pmatrix} \delta f_j \\ \delta u_j \\ \delta v_j \\ \delta \nu_j^+ \\ \delta g_j \end{pmatrix}, \qquad \underset{\sim}{r}_j = \begin{pmatrix} (r_1)_j \\ (r_2)_j \\ (r_3)_j \\ (r_4)_j \\ (r_5)_j \end{pmatrix} \tag{P9.5.3}$$

and the 5×5 matrices A_j, B_j, C_j given by

$$A_0 = \begin{vmatrix} 1 & 0 & 0 & 0 & 0 \\ 0 & 1 & 0 & 0 & 0 \\ 0 & 0 & 0 & 1 & 0 \\ 0 & -1 & -\frac{h_1}{2} & 0 & 0 \\ 0 & 0 & 0 & -1 & -\frac{h_1}{2} \end{vmatrix} \tag{P9.5.4a}$$

$$B_j = \begin{vmatrix} 1- & -\frac{h_1}{2} & 0 & 0 & 0 \\ (s_2)_j & (s_4)_j & (s_6)_j & (s_8)_j & 0 \\ (e_2)_j & (e_4)_j & (e_6)_j & (e_8)_j & (e_{10})_j \\ 0 & 0 & 0 & 0 & 0 \\ 0 & 0 & 0 & 0 & 0 \end{vmatrix}, \quad 1 \le j \le J \tag{P9.5.4b}$$

$$A_j = \begin{vmatrix} 1 & -h_j/2 & 0 & 0 & 0 \\ (s_1)_j & (s_3)_j & (s_5)_j & (s_7)_j & 0 \\ (e_1)_j & (e_3)_j & (e_5)_j & (e_7)_j & (e_9)_j \\ 0 & -1 & -h_{j+1}/2 & 0 & 0 \\ 0 & 0 & 0 & -1 & -h_{j+1}/2 \end{vmatrix}, \quad 1 \le j \le J-1 \tag{P9.5.4c}$$

$$C_j = \begin{vmatrix} 0 & 0 & 0 & 0 & 0 \\ 0 & 0 & 0 & 0 & 0 \\ 0 & 0 & 0 & 0 & 0 \\ 0 & 1 & -h_{j+1}/2 & 0 & 0 \\ 0 & 0 & 0 & 1 & -h_{j+1}/2 \end{vmatrix}, \quad 0 \le j \le J-1 \tag{P9.5.4d}$$

$$A_j = \begin{vmatrix} 1 & -h_j/2 & 0 & 0 & 0 \\ (s_1)_j & (s_3)_j & (s_5)_j & (s_7)_j & 0 \\ (e_1)_j & (s_3)_j & (s_5)_j & (s_7)_j & (s_9)_j \\ 0 & 1 & 0 & 0 & 0 \\ 0 & 0 & 0 & 1 & 0 \end{vmatrix} \tag{P9.5.4e}$$

9.6 Using the matrix solver, MSA (subsection 10.7.3), write an algorithm for the linear system in Problem 9.5. Check your code with the one given on the companion site.

9.7 To develop a computer program to solve the equations using the *SA* model, it is necessary to specify initial profiles for f_j, u_j, v_j, υ_j^+ and g_j. It is also necessary to generate the boundary layer grid, account for the boundary layer growth, etc. A convenient procedure is to use the computer program described in Section 10.3 with initial velocity profiles incorporated in a new subroutine IVPT as described in Problem 8.9. The initial profiles for υ_j^+ and g_j can be calculated from those calculated in subroutine EDDY since $\upsilon_j^+ = (\varepsilon_m^+)_j$. Thus with initial profiles specified in this manner, we can replace subroutine COEF3 and SOLV3 in BLP2 with new subroutines, say COEF5 which contains the coefficients in Problem 9.4, and the algorithm discussed in Problem 9.6. Obviously, we need to incorporate other changes to the logic of the computations in order to extend the computer program of Problem 8.6 to solve the *SA* model equations.

9.8 The properties of a two-dimensional nonsimilar plane jet for a turbulent flow can also be calculated with an algebraic eddy viscosity formulation using the computer program described in Section 10.3. As before, we again use transformed variables.

(a) Show that, with the eddy viscosity concept and neglecting the pressure-gradient term, the continuity and momentum equations given by Eqs. (5.2.8) and (5.2.9) can be written as

$$(bf'')'+(f')^2+ff'' = 3\xi\left(f'\frac{\partial f'}{\partial \xi}-f''\frac{\partial f}{\partial \xi}\right) \tag{P9.8.1}$$

with the transformation defined by

$$\eta = \sqrt{\frac{u_0}{\nu L}}\frac{y}{3\xi^{2/3}}, \qquad \xi = \frac{x}{L}, \qquad \psi = \sqrt{u_0 \nu L}\,\xi^{1/3} f(\xi,\eta) \tag{P9.8.2}$$

Here u_0, L denote a reference velocity and length, respectively, and

$$b = 1+\varepsilon_m/\nu = 1+\varepsilon_m^+ \tag{P9.8.3}$$

(b) With

$$f' = u \tag{P9.8.4a}$$

$$u' = v \tag{P9.8.4b}$$

Eq. (P9.8.1) can be written as

$$(bv)'+u^2+fv = 3\xi\left(u\frac{\partial u}{\partial \xi}-v\frac{\partial f}{\partial \xi}\right) \tag{P9.8.4c}$$

Write finite difference approximations to the above equations and show that Eq. (P9.8.4c) can be expressed in the same form as Eq. (8.2.20c) and that the coefficients $(s_k)_j$ are identical to those given by Eqs. (8.2.22), provided that we take $m_1 = 1$ and $m_2 = -1$.

(c) With the boundary conditions given by

$$\eta = 0, \quad f = v = 0 \tag{P9.8.5a}$$

$$\eta = \eta_e, \quad u = 0 \tag{P9.8.5b}$$

or in linearized form

$$\delta f_0 = 0, \qquad \delta v_0 = 0, \qquad \delta u_J = 0 \tag{P9.8.6}$$

the linear system of equations can again be expressed in the form given by Eq. (8.2.24) and can be solved with minor changes to the computer program described in Section 10.3. One of the changes occurs in subroutine IVPL that defines the initial velocity profiles for laminar flows. For either laminar or turbulent plane jet, this subroutine requires changes. An initial velocity profile can be generated by assuming the profile to be of the form

$$\frac{u}{u_c} = \frac{1}{2}\left[1 - \tanh\beta\left(\frac{y}{h} - \varsigma_c\right)\right] \tag{P9.8.7}$$

Here β and ς_c are specified constants, and h is the half-width of the duct. This profile essentially corresponds to a uniform velocity at the exit of the duct. The fairing given by the above equation is to remove the discontinuity at $y/h = 1$. Plot the above equation for two values of β equal to 10 and 20 with $\varsigma_c = 1$ and show that in transformed variables, this equation can be written in the form

$$\frac{f'}{2\xi^{1/3}} = \frac{1}{2}\left\{1 - \tanh\beta\left[\frac{3\xi_0^{2/3}}{\sqrt{R_L}}(\eta - \eta_c)\right]\right\} \tag{P9.8.8}$$

Here R_L is a dimensionless Reynolds number, u_0L/ν and ξ_0 is the ξ-location at which the initial profiles are specified. The reference length L is usually taken to be equal to the half-width of the duct.

(d) In addition to subroutine IVPL, we need to make changes in subroutine EDDY and replace the eddy viscosity formulas in that subroutine with a new one. There are several formulations that can be used for this purpose. A simple one is

$$\varepsilon_m = 0.037u_c\delta \tag{P9.8.9}$$

where δ represents the half-width (taken as the point where $u/u_c = 0.5$ and u_c is the centerline viscosity.

Show that this equation can be written in transformed variables as

$$\varepsilon_m^+ = 0.037\sqrt{R_L}\xi^{1/3}\eta_{1/2}f'_c \tag{P9.8.10}$$

Here $\eta_{1/2}$ is the transformed η-distance where $u = \frac{1}{2}u_c$, and f'_c is the dimensionless centerline velocity.

(e) Compute the variation of the dimensionless centerline velocity for a turbulent flow with the revised computer program of Section 10.3. Take $R_L = 5300$, $\beta = 20$, $\varsigma_c = 1.0$ and $\xi_0 = 1$. Plot u/u_0 as a function of y/L at $\xi = 1.054$, 1.249, 1.581, 2.976, 4.484, 6.819, 10.

(f) Include the intermittency term γ

$$\gamma = \frac{1}{1 + 5.5(y/\delta)^6} \tag{P9.8.11}$$

in Eq. (P9.8.10) and repeat the calculations in (e) with this modification to ε_m. Compare your results with those in (e).

(g) Compare the results in (f) with experimental data given on the companion site.

(h) Repeat (g) with 0.037 in Eq. (P9.8.10) replaced with 0.035, 0.033 to study the effect of this constant on the solutions.

9.9 Repeat (e) in Problem P9.8 for $R_L = 100$.

References

[1] T. Cebeci, Turbulence Models and Their Applications, Horizons Publishing, Long Beach, CA, and Springer-Verlag, Heidelberg, Germany, 2003.

[2] J.E. Bardina, P.G. Huang, T.J. Coakley, Turbulence Modeling Validation, Testing and Development, NASA, Tech. Memo. 110446 (April 1997).

[3] H.W. Liepmann, J. Laufer, Investigations of Free Turbulent Mixing, NACA TN-1257 (1947).

[4] L.J.S. Bradbury, The Structure of a Self-Preserving Turbulent Plane Jet,, J. Fluid Mech. 23 (1) (1965) 31–64.

[5] S.J. Kline, B.J. Cantwell, G.M. Lilley, 1980-81 AFOSR-HTTM-Stanford Conference on Complex Turbulent Flows, Stanford University, Stanford, Calif, 1981.

[6] S. Birch, Planar Mixing Layer, In 1980-81 AFOSR-HTTM-Stanford Conference on Complex Turbulent Flows, in: S.J. Kline, B.J. Cantwell, G.M. Lilley (Eds.), Stanford University, Stanford, Calif., 1981. vol. 1, pp. 170–177.

[7] I. Wygnanski, H.E. Fiedler, Some Measurements in the Self-Preserving Jet, J. Fluid Mech. 38 (3) (1969) 577–612. Also in Boeing Scientific Research Labs, Flight Science Laboratory, Document D1-82-0712.

[8] A. Fage, V.M. Falkner, Note on Experiments on the Temperature and Velocity in the Wake of a Heated Cylindrical Obstacle. London, Proc. Roy. Soc. vol. A135 (1932) 702–705.

[9] G.S. Settles, L.J. Dodson, Hypersonic Turbulent Boundary-Layer and Free Shear Database, NASA CR-177610, Ames Research Center (Apr. 1993).

[10] S.F. Birch, J.M. Eggers, A Critical Review of the Experimental Data for Developed Free Turbulent Shear Layers. – Conference Proceedings, Langley Research Center, NASA, Hampton, Va., July 20–21, Free Turbulent Shear Flows vol. 1 (1972) 11–40.

[11] D.M. Driver, Reynolds Shear Stress Measurements in a Separated Boundary Layer Flow, AIAA Paper (1991) 91–1787.

[12] W.D. Bachalo, D.A. Johnson, Transonic Turbulent Boundary-Layer Separation Generated on an Axisymmetric Flow Model, AIAA J. 24 (1986) 437–443.

[13] T. Cebeci, K.C. Chang, An Improved Cebeci-Smith Turbulence Model for Boundary-Layer and Navier-Stokes Methods, paper No. ICAS-96-1.7.3, 20th Congress of the International Council of the Aeronautical Sciences, Sorrento, Italy, 1996.

Chapter 10

Companion Computer Programs

Chapter Outline Head

Analysis of Turbulent Flows with Computer Programs. http://dx.doi.org/10.1016/B978-0-08-098335-6.00010-0

10.1 Introduction

In this chapter we describe several computer programs for calculating two-dimensional laminar and turbulent incompressible flows. In Section 10.2 we first describe computer programs based on integral methods discussed in Chapter 7 and present sample calculations. The computer program in Section 10.3 is based on the differential method discussed in Chapter 8 and is applicable to both laminar and turbulent flows for a given external velocity distribution and transition location. In this section we also present sample calculations for an airfoil with the external velocity distribution obtained from the panel method discussed in Section 10.4. Sections 10.5 and 10.6 present computer programs for incompressible laminar and turbulent flows with heat transfer and for infinite-swept wing flows, respectively. Sections 10.7 to 10.10 present another differential method for two-dimensional incompressible turbulent flows with CS and k-ε models. The computer program with the CS model is essentially similar to the one in Section 10.3 except that the wall boundary conditions for the momentum and continuity equations are specified at some distance from the wall. The computer program for the k-ε model includes the zonal method with a combination of the CS model for the inner region and the k-ε model for the outer region as discussed in Section 9.2. It also includes the solution of the k-ε model equations with and without wall functions. Section 10.11 presents a differential method for the SA model using the numerical procedure discussed in Problems 9.1 to 9.7 and Section 10.12 for a plane jet discussed in Problem 9.8. Section 10.13 presents several subroutines discussed in Chapters 8 and 9. Section 10.14 presents the differential method for the inverse boundary-layer discussed in Section 8.8 and subsection 10.15.1 presents sample calculations for the panel method of Section 8.9 without viscous affects. Sample calculations for the inverse boundary-layer program is discussed in subsection 10.15.2 and those for the interactive boundary-layer program in subsection 10.15.3.

All programs, including the three programs in Chapter 5, can be found on the companion site, store.elsevier.com/companions/9780080983356.

10.2 Integral Methods

In Chapter 7 we discussed integral methods for calculating heat and momentum transfer in two-dimensional and axisymmetric laminar and turbulent flows. In this section we describe FORTRAN programs for them and present sample calculations.

10.2.1 Thwaites' Method

This method is applicable to both two-dimensional and axisymmetric laminar flows (see Problem 7.11) and has the following input requirements:

NXT	Total number of *x*-stations.
KASE	Flow index, 0 for two-dimensional flow, 1 for two-dimensional flow that starts as stagnation-point flow, and 2 for axisymmetric flow.
KDIS	Index for surface distance; 1 when surface distance is input, 0 when surface distance is calculated.
UREF	Reference velocity, u_{ref}, feet per second or meters per second.
BIGL	Reference length, L, feet or meters.
CNU	Kinematic viscosity, ν, square feet per second or square meters per second.
X	Dimensionless chordwise or axial distance, x/L. If KDIS = 1, then X is the surface distance, s.
UE	Dimensionless velocity, u_e/u_{ref}.
R	Dimensionless two-dimensional body ordinate or body of revolution radius; r/L.

Its output includes δ^*, θ, H, c_f and $R_\theta = u_e\theta/\nu$ together with Reynolds mumber, RS, based on surface distance and external velocity.

10.2.2 Smith-Spalding Method

This method is for laminar boundary-layer flows with variable u_e but *uniform* surface temperature. Its input is similar to Thwaites' method and consists of NXT, KASE, KDIS, UREF, BIGL, CNU and PR. We again read in X, R and UE.

The output includes X, S, UE and ST (Stanton number).

10.2.3 Head's Method

This method is applicable to only two-dimensional incompressible turbulent flows. Its input consists of the specification of the external velocity distribution, u_e/u_∞, UE(I), as a function of surface distance x/L, X(I), with u_∞ denoting the reference freestream velocity and L a reference length. The initial conditions consist of a dimensionless momentum thickness, θ/L, T(1), and shape factor H, H(1), at the first station. In addition, we specify a reference Reynolds number $R_L = u_\infty L/\nu$, RL and the total number of x-stations, NXT. In the code, the derivative of external velocity du_e/dx, DUEDX(I), is computed by using a three-point Lagrange-interpolation formula. The output includes X(I), UE(I), DUEDX(I), T(I), H(I), DELST(I), δ^*/L, CF(I) and DELTA(I), δ, defined by Eq. (7.3.2).

10.2.4 Ambrok's Method

This method is only for two-dimensional turbulent flows. Its input and output instructions are similar to Head's method.

10.3 Differential Method with CS Model: Two-Dimensional Laminar and Turbulent Flows

In this section we present the computer program discussed in Chapter 8 for two-dimensional incompressible laminar and turbulent flows. Its extension to flows with heat transfer is discussed in Section 10.5 and to infinite-swept wing flows in Section 10.6, to turbulent flows employing the SA model in Section 10.11 and to a plane jet in Section 10.12.

This computer program, called BLP2, and also described in [1], consists of a MAIN routine, which contains the logic of the computations, and seven subroutines: INPUT, IVPL, GROWTH, COEF3, SOLV3, EDDY and OUTPUT. The following subsections describe the function of each subroutine.

10.3.1 Main

BLP2 solves the linearized form of the equations. Thus an iteration procedure in which the solution of Eq. (8.2.24) is obtained for successive estimates of the velocity profiles is needed with a subsequent need to check the convergence of the solutions. A convergence criterion based on v_0 which corresponds to f''_w is usually used and the iterations, which are generally quadratic for laminar flows, are stopped when

$$|\delta v_0 (= \mathrm{DELV}(1))| < \varepsilon_1 \tag{10.3.1}$$

with ε_1 taken as 10^{-5}. For turbulent flows, due to the approximate linearization procedure used for the turbulent diffusion term, the rate of convergence is not quadratic and solutions are usually acceptable when the ratio of $|\delta v_0/v_0|$ is less than 0.02. With proper linearization, quadratic convergence of the solutions can be obtained as described in [1].

After the convergence of the solutions, the OUTPUT subroutine is called and the profiles F, U, V and B, which represent the variables f_j, u_j, v_j and b_j are shifted.

10.3.2 SUBROUTINE INPUT

This subroutine prepares data for boundary layer calculations. The data includes grid in ξ- and η-directions, dimensionless pressure gradient $m(\xi)$, mass transfer $f_w(\xi)$ parameter and calculation of the γ_{tr}-term in the CS model which is given by Eqs. (5.3.18) and (5.3.19).

The streamwise grid is generated by reading the values at ξ^i. In general, the ξ-grid distribution depends on the variation of u_e with ξ so that rapid variations in external velocity distribution and the approach to separation require small $\Delta\xi$-steps (k_n). For laminar flows, it is often sufficient to use a uniform grid in the η-direction. A choice of transformed boundary-layer thicknesses η_e equal to 8 often ensures that the dimensionless slope of the velocity profile at the edge, $f''(\eta_e)$, is sufficiently small ($< 10^{-3}$) and that approximately 61 grid points are adequate for most flows. For turbulent flows, however, a uniform grid is not satisfactory because the boundary-layer thickness η_e and the dimensionless wall shear parameter v_w ($\equiv f''_w$) are much larger in turbulent flows than laminar flows. Due to the rapid variation of the velocity profile close to the wall, it is necessary to take much smaller steps in η close to the wall.

The program uses a η-grid which has the property that the ratio of lengths of any two adjacent intervals is a constant, that is, $h_j = Kh_{j-1}$ and the distance to the j-th line is given by

$$\eta_j = h_1 \frac{K^J - 1}{K - 1} \quad j = 1, 2, \ldots, J \quad K > 1 \tag{10.3.2}$$

There are two parameters: h_1, the length of the first $\Delta\eta$-step, and K, the variable grid parameter. The total number of points, J, is calculated from

$$J = \frac{\ln[1 + (K - 1)(\eta_e/h_1)]}{\ln K} + 1 \tag{10.3.3}$$

In practice, it is common to choose h_1 [DETA(1)] and K (VGP) so that, for an assumed maximum value of η_e (ETAE), the number of j-points do not exceed the total number (NPT) specified in the code. For example, for $\eta_e = 50$ and $h_1 = 0.01$, the number of j-points depends on K. Figure 10.1 can help in the selection of K and shows that, for example for $J = 61$, the value of the variable grid parameter must be less than about 1.1.

With velocities known at each ξ-station, called NX-stations, the pressure gradient parameters m (P2) and m_1 (P1) are computed from their definitions for all ξ-stations, except the first one at which m is specified, after the derivative of $du_e/d\xi$ (DUDS)

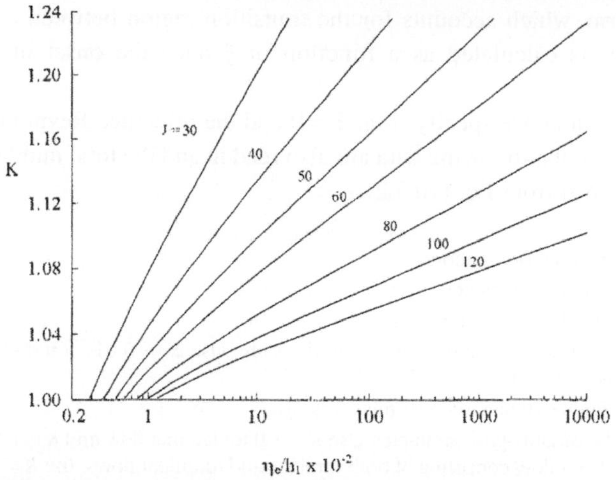

Fig. 10.1 Variation of K with h_1 for different η_e-values.

needed in the calculation of m is obtained by using three-point Lagrange interpolation formulas given by ($n < N$):

$$\left(\frac{du_e}{d\xi}\right)_n = -\frac{u_e^{n-1}}{A_1}(\xi_{n+1} - \xi_n) + \frac{u_e^n}{A_2}(\xi_{n+1} - 2\xi_n + \xi_{n-1}) + \frac{u_e^{n+1}}{A_3}(\xi_n - \xi_{n-1}) \tag{10.3.4}$$

Here N refers to the last ξ^n station and

$$\begin{aligned} A_1 &= (\xi_n - \xi_{n-1})(\xi_{n+1} - \xi_{n-1}) \\ A_2 &= (\xi_n - \xi_{n-1})(\xi_{n+1} - \xi_n) \\ A_3 &= (\xi_{n+1} - \xi_n)(\xi_{n+1} - \xi_{n-1}) \end{aligned} \tag{10.3.5}$$

The derivative of $du_e/d\xi$ at the end point $n = N$ is given by

$$\left(\frac{du_e}{d\xi}\right)_N = -\frac{u_e^{N-2}}{A_1}(\xi_N - \xi_{N-1}) + \frac{u_e^{N-1}}{A_2}(\xi_N - \xi_{N-2}) + \frac{u_e^N}{A_3}(2\xi_N - \xi_{N-2} - \xi_{N-1}) \tag{10.3.6}$$

where now

$$\begin{aligned} A_1 &= (\xi_{N-1} - \xi_{N-2})(\xi_N - \xi_{N-2}) \\ A_2 &= (\xi_{N-1} - \xi_{N-2})(\xi_N - \xi_{N-1}) \\ A_3 &= (\xi_N - \xi_{N-1})(\xi_N - \xi_{N-2}) \end{aligned} \tag{10.3.7}$$

The γ_{tr}-term, which accounts for the transition region between a laminar and turbulent flow, is calculated as a function of ξ once the onset of transition is specified.

In this subroutine we specify η_e at $\xi = 0$ and the reference Reynolds number R_L (RL). In addition, the following data are also read in and the total number of j-points J(NP) is computed from Eq. (10.3.2).

NXT	Total number of x-stations
NTR	NX-station for transition location x_{tr}
NPT	Total number of η-grid points.
DETA(I)	$\Delta\eta$-initial step size of the variable grid system. Use $\Delta\eta = 0.01$ for turbulent flows. If desired, it may be changed.
ETAE	Transformed boundary-layer thickness, η_e
VGP	K is the variable-grid parameter. Use $K = 1.0$ for laminar flow and $K = 1.12$ for turbulent flow. For a flow consisting of both laminar and turbulent flows, use $K = 1.12$.
RL	Reynolds number, $\frac{u_\infty L}{\nu}$
x	Surface distance, feet or meters, or dimensionless.
u_e	Velocity, feet per second or meter per second, or dimensionless.

10.3.3 Subroutine IVPL

At $x = 0$ with $b = 1$, Eq. (8.2.6) reduces to the Falkner-Skan similarity equation which can be solved subject to the boundary conditions of Eq. (8.2.7). Since the equations are solved in linearized form, initial estimates of f_j, u_j and v_j are needed in order to obtain the solutions of the nonlinear Falkner-Skan equation. Various expressions can be used for this purpose. Since Newton's method is used, however, it is useful to provide as good an estimate as is possible and an expression of the form.

$$u_j = \frac{3}{2}\frac{\eta_j}{\eta_e} - \frac{1}{2}\left(\frac{\eta_j}{\eta_e}\right)^3 \tag{10.3.8}$$

usually satisfies this requirement. The above equation is obtained by assuming a third-order polynomial of the form

$$f' = a + b\eta + c\eta^3$$

and by determining constants a, b, c from the boundary conditions given by Eq. (8.2.7) for the zero-mass transfer case and from one of the properties of momentum equation which requires that $f'' = 0$ at $\eta = \eta_e$.

The other profiles f_j, v_j follow from Eq. (10.3.4) and can be written as

$$f_j = \frac{\eta_e}{4}\left(\frac{\eta_j}{\eta_e}\right)^2\left[3 - \frac{1}{2}\left(\frac{\eta_j}{\eta_e}\right)^2\right] \tag{10.3.9}$$

$$v_j = \frac{3}{2}\frac{1}{\eta_e}\left[1 - \left(\frac{\eta_j}{\eta_e}\right)^2\right] \tag{10.3.10}$$

10.3.4 Subroutine GROWTH

For most laminar-boundary-layer flows the transformed boundary-layer thickness $\eta_e(x)$ is almost constant. A value of $\eta_e = 8$ is sufficient. However, for turbulent boundary-layers, $\eta_e(x)$ generally increases with increasing x. An estimate of $\eta_e(x)$ is determined by the following procedure.

We always require that $\eta_e(x^n) \geq \eta_e(x^{n-1})$, and in fact the calculations start with $\eta_e(0) = \eta_e(x_1)$. When the computations on $x = x^n$ (for any $n \geq 1$) have been completed, we test to see if $| v_J^n | \leq \varepsilon_v$ at $\eta_e(x^n)$ where, say $\varepsilon_v = 5 \times 10^{-4}$. This test is done in MAIN. If this test is satisfied, we set $\eta_e(x^{n+1}) = \eta_e(x^n)$. Otherwise, we call GROWTH and set $J_{\text{new}} = J_{\text{old}} + t$, where t is a number of points, say $t = 1$. In this case we also specify values of $(f_j^n, u_j^n, v_j^n, b_j^n)$ for the new n_j points. We take the values of $u_j = 1$, $v_j^n = 0$, $f_j^n = (\eta_j - \eta_e)u_J^n + f_J^n$, and $b_j^n = b_J^n$. This is also done for the values of f_j^{n-1}, v_j^{n-1}, and b_j^{n-1}.

10.3.5 Subroutine COEF3

This is one of the most important subroutines of BLP2. It defines the coefficients of the linearized momentum equation given by Eqs. (8.2.20) and (8.2.23).

10.3.6 Subroutine EDDY

This subroutine contains the CS algebraic eddy viscosity model in Section 5.8. For simplicity we do not include the low Reynolds number effect, roughness effect, mass transfer effect and strong pressure gradient effect (variable α) in this subroutine. Thcsc capabilities, if desired, can easily incorporated into the formula as defined in this subroutine. The formulas for the inner and outer eddy-viscosity expressions are given by Eqs. (5.2.11) with each side of equation multiplied by γ_{tr} given by Eqs. (5.3.18) and (5.3.19).

In terms of transformed variables $(\varepsilon_m^+)_\text{i}$ and $(\varepsilon_m^+)_\text{o}$ given by Eq. (5.2.11) can be written as

$$\left(\varepsilon_m^+\right)_i = 0.16R_x^{1/2}[1 - \exp(-y/A)]^2\eta^2 v\gamma_{\text{tr}} \tag{10.3.11a}$$

$$(\varepsilon_m^+)_0 = 0.0168 R_x^{1/2}\left(\eta_e - f_e\right)\gamma_{tr}\gamma \tag{10.3.11b}$$

$$\frac{y}{A} = \frac{N}{26} R_x^{1/4} v_w^{1/2} \eta, \quad R_x = \frac{u_e x}{\nu} \tag{10.3.12}$$

10.3.7 Subroutine SOLV3

This subroutine is used to obtain the solution of Eq. (8.2.24) with the block-elimination method discussed in subsection 8.2.3 and with the recursion formulas given in subsection 8.2.4.

10.3.8 Subroutine OUTPUT

This subroutine prints out the desired profiles such as f_j, u_j, v_j, and b_j as functions of η_j. It also computes the boundary-layer parameters, c_f, δ^*, θ and R_x.

10.4 Hess-Smith Panel Method with Viscous Effects

In this section we present a computer program for the panel method discussed in Section 8.9. This program can be used interactively with the boundary-layer program of Section 10.14 so that, as discussed in detail in [2,3] and briefly in Section 8.9, more accurate solutions of inviscid and viscous flow equations can be obtained by includingthe viscous effects in the panel method of Section 8.9.

The computer program of the panel method has five subroutines and MAIN, as described below.

10.4.1 MAIN

MAIN contains the input information which comprises (1) the number of panels alongthe surface of the airfoil, NODTOT, and the number of panels in the wake, NW. The code is arranged so that it can be used for inviscid flows with and without viscous effects. For inviscid flows, NW is equal to zero. (2) The next input data also comprises airfoil coordinates normalized with respect to its chord c, x/c, y/c, [≡X(I),Y(I)]. If NW $\neq$ 0, then it is necessary to specify the dimensionless displacement thickness δ^*/c (≡DLSP(I)), dimensionless blowingv elocity u_w/u_∞ (≡VNP(I)) distributions on the airfoil, as well as the wake coordinates XW(I), YW(I) of the dividing-streamline, the dimensionless displacement thickness distribution on the upper wake DELW(I,1) and lower wake DELW(I,2) and velocity jump QW(I). It should be noted that all input data for wake includes values at the trailing edge. The input also includes angle of attack α (≡ALPHA) and Mach number M_∞ (≡FMACH).

The panel slopes are calculated from Eq. (8.9.2). The subroutine COEF iscalled to compute A and $\vec{b}$ in Eq. (8.9.19) subroutine OBKUTA to calculate the off-body Kutta condition, subroutine GAUSS to compute $\vec{x}$, subroutine VPDIS to compute the velocity and pressure distributions, and subroutine CLCM to compute the airfoil characteristics correspondingto lift (CL) and pitchingmoment (CM) coefficients.

10.4.2 Subroutine COEF

This subroutine calculates the elements a_{ij} of the coefficient matrix A from Eqs. (8.9.21) and (8.9.23) and the elements of $\vec{b}$ from Eq. (8.9.24) We note that $N+1$ corresponds to KUTTA, and N to NODTOT

10.4.3 Subroutine OBKUTA

This subroutine is used to calculate the body-off Kutta condition.

10.4.4 Subroutine GAUSS

The solution of Eq. (8.9.19) is obtained with the Gauss elimination method described in Section 8.9.

10.4.5 Subroutine VPDIS

Once $\vec{x}$ is determined by subroutine GAUSS so that source strengths q_i $(i = 1, 2, \ldots, N)$ and vorticity τ on the airfoil surface are known, the tangential velocity component (V^t) at each control point can be calculated. Denoting q_i with Q(I) and τ with GAMMA, the tangential velocities $(V^t)_i$ are obtained with the help of Eq. (8.9.12b). This subroutine also determines the distributions of the dimensionless pressure coefficient C_p (≡ CP) defined by

$$C_p = \frac{p - p_\infty}{(1/2)_\varrho V_\infty^2} \tag{10.4.1a}$$

which in terms of velocities can be written as

$$C_p = 1 - \left(\frac{V^t}{V_\infty}\right)^2 \tag{10.4.1b}$$

It is common to use panel methods for low Mach number flows by introducingcompressibilit y corrections which depend upon the linearized form of the compressibility velocity potential equation and are based on the assumption of small perturbations and thin airfoils [4]. A simple correction formula for this purpose is the Karman-Tsien formula which uses the "tangent gas" approximation to simplify the compressible potential-flow equations. Accordingto this formula, the effect of Mach number on the pressure coefficient is estimated from

$$c_p = \frac{c_{pi}}{\beta + [M_\infty^2/(1+\beta)](c_{pi}/2)} \tag{10.4.2}$$

and the correspondingv elocities are computed from

$$V^2 = 1 + \frac{1}{c_6}\left[1 - (1 + c_8 c_p)^{1/c_7}\right] \tag{10.4.3}$$

Here c_{pi} denotes the incompressible pressure coefficient, M_∞ the freestream Mach number and

$$\beta = \sqrt{1 - M_\infty^2}, \quad c_6 = \frac{\gamma - 1}{2} M_\infty^2, \quad c_7 = \frac{\gamma}{\gamma - 1}, \quad c_8 = \frac{1}{2}\gamma M_\infty^2, \quad \gamma = 1.4 \tag{10.4.4}$$

In this subroutine we also include this capability in the HS panel method.

10.4.6 Subroutine CLCM

The dimensionless pressure in the appropriate directions is integrated to compute the aerodynamic force and the coefficients for lift (CL) and pitchingmoment (CM) about the leading edge of the airfoil.

10.4.7 Subroutine VPDWK

This subroutine calculates the total velocity and pressure coefficient at each control point alongthe upper and lower wakes separately. The normal and tangential components of the total velocities are computed from Eqs. (8.9.39a) and (8.9.39b).

10.5 Differential Method with CS Model: Two-Dimensional Flows with Heat Transfer

The program which for convenience we called BLP2H is the same computer program BLP2 which now includes the solution of the energy equation. Two subroutines, COEF2 and SOLV2, are added to BLP2 to calculate incompressible laminar and turbulent flows with heat transfer (see Problem 8.3). Sample calculations presented on the companion site, store.elsevier.com/components/9780080983356, represent the application of this code to Problems 8.4 and 8.5.

10.6 Differential Method with CS Model: Infinite Swept-Wing Flows

This program, called BLP2ISW, is also the extension of BLP2 to the calculation of infinite swept-wing equations for incompressible laminar and turbulent flows as discussed in Problem 8.6. Again two subroutines are added to BLP2 and changes are made to the eddy viscosity subroutines. Subroutine COEF2 includes the coefficients of the z-momentum equation and subroutine SOLV2 is the same solution algorithm used in BLP2H see subsection 10.13.1.

For three-dimensional turbulent flows, the eddy viscosity formulas require changes to those for two-dimensional flows. Here they are defined according to Eqs. (5.7.4) and (5.7.5).

Sample calculations for an infinite swept wing having the NACA 0012 airfoil cross section with a sweep angle of $\lambda = 30°$, an angle of attack of $\alpha = 2°$, chord Reynolds number $R_c = 5 \times 10^6$ and transition location at $x/c = 0.10$ are presented on the companion site, store.elsevier.com/components/9780080983356. See also Problem 8.7.

10.7 Differential Method with CS and *k*-ε Models: Components of the Computer Program Common to both Models

This section includes a MAIN routine which contains the logic of the computations and five subroutines, INPUT, IVPT, GROWTH, GRID and OUTPUT, described below.

10.7.1 MAIN

Here we first read in input data (subroutine INPUT) and generate the initial turbulent velocity profile (subroutine IVPT) and the eddy viscosity distribution for the CS model (subroutine EDDY), k-profile (subroutine KEINITK), ε-profile (subroutine KEINITG). Since linearized equations are being solved, we use an iteration procedure in which the solutions of the equations are obtained for successive estimates of velocity, kinetic energy, dissipation profiles with a subsequent need to check the convergence of the solutions. A convergence criterion based on $\frac{\delta v0}{v0} < 0.02$ is used and the iterations are stopped when

$$\left|\frac{\delta v0}{v0}\right| < 0.02$$

During this iteration procedure, we introduce an under-relaxation procedure for the iterations as described in MAIN. This is useful, especially with transport equation turbulence models.

When the solutions converge, we also check to see whether the boundary-layer thickness, η_e, used in the calculations for that x-station is large enough so that the asymptotic behavior of the solutions is reached. If this is not the case, we call subroutine GROWH.

After the convergence of the solutions. the OUTPUT subroutine is called and the profiles which represent the variables such as f_j, u_j, v_j, k_j, ε_j etc. are shifted.

10.7.2 Subroutine INPUT

In this subroutine we read in input data and set up the flow calculations according to the following turbulence models listed below.

Model	=	0	CS model
		1	Huang-Lin k-ε model
		2	Chien k-ε model
	=	−1	zonal method
	=	−2	high Re # k-ε model

In some problems, like airfoil flows, it is convenient to read in the dimensionless airfoil coordinates x/c, y/c rather than the surface distance required in the boundary-layer calculations. In all calculations, the external velocity $u_e(x)$ either dimensional or dimensionless, u_e/u_∞, and freestream or reference velocity, u_∞ (u_{ref}), kinematic viscosity ν (CNU), reference length c (chord), variable η-grid parameter K (VGP) discussed in subroutine GRID must be specified together with R_θ (RTHA) and c_f (CFA) needed to generate the initial turbulent velocity profile with subroutine IVPT.

The input also requires the specification of the first grid point needed in the η-grid generated by subroutine GRID. This is done by inputting y_0^+ (YPLUSW). defined by

$$\frac{y_0^+ u_\tau}{\nu}$$

where u_τ (UTAU) is the friction velocity, $u_e\sqrt{c_f/2}$ and $y0$ is the variable grid parameter $h1$, discussed in subroutine GRID. Its typical values for CS, zonal and high Reynolds number k-ε models are around 0.5 to 1.0. For low Reynolds number k-ε model, values of y_0^+ around 0.10 to 0.50 are typical. In the present program, K is set equal to 1.12, y_0^+ equal to 0.5 for low Reynolds number k-ε model and 1.0 for zonal and CS models.

Since equations use transformed variables where y is given by

$$y = \sqrt{\nu x/u_e}\eta$$

and since the location of x where the turbulent flow calculations are started, x_1, can be an arbitrary distance, in this subroutine we calculate x_0 in order to control y_0^+ better.

The calculation of the pressure gradient parameter $m(x)$ (P2) in the transformed momentum equation is achieved from the given external velocity $u_e(x)$ distribution and from the definition of m.

10.7.3 Subroutine IVPT

This subroutine is used to generate the initial turbulent velocity profile for both models by specifying a Reynolds number based on momentum thickness, $R_\theta = u_e\Theta/\nu$ and local skin-friction coefficient $c_f\left[\equiv \tau_w/\frac{1}{2}Qu_e^2\right]$. It makes use of Eq. (4.4.41) for $y^+ \leq 50$; and Eq. (4.4.35) for $y^+ \geq 50$. See problem 8.9. It is also given in subsection 10.13.1.

10.7.4 Subroutine GROWTH

This subroutine is similar to the one described in subsection 10.3.4. An estimate of $\eta_e(x)$ for turbulent flows is determined by the following procedure.

We always require that $\eta_e(x^n) \geq \eta_e(x^{n-1})$, and in fact the calculations start with $\eta_e(x^0) = \eta_e(x_1)$. When the computations on $x = x^n$ (for any $n \geq 1$) have been completed, we test to see if $|v_J^n| \leq \varepsilon_v$ at $\eta_e(x^n)$ where, say $\varepsilon_v = 5 \times 10^{-4}$. This test is done in MAIN. If this test is satisfied, we set $\eta_e(x^{n+1}) = \eta_e(x^n)$ Otherwise, we call GROWTH and set $J_{\text{new}} = J_{\text{old}} + t$, where t is a number of points, say $t = 1$. In this case we also specify values of $(f_j^n, u_j^n, v_j^n, b_j^n, k_j^n, \varepsilon_j^n$ etc.) for the new η_j points. We take the values of $u_j^n = 1$, $v_j^n = 0$, $f_j^n = (\eta_j - \eta_e)u_J^n + f_J^n, k_j^n = k_J^n, \varepsilon_j^n, s_j^n = 0, q_j^n = 0$.

10.7.5 Subroutine GRID

See subsection 10.3.2

10.7.6 Subroutine OUTPUT

This subroutine prints out the desired profiles of the momentum, kinetic energy and rate of dissipation equations, such as $f_j, u_j, v_j, k_j, \varepsilon_j$ as a function of η. It also computes the boundary-layer parameters, $c_f, \delta^*, \theta, R_\delta$ and R_θ.

10.8 Differential Method with CS and k-ε Models: CS Model

This part of the computer program which uses the CS model has five subroutines in addition to those described in Section 10.7. They include subroutines COEFTR, SOLV3, EDDY, GAMCAL, and CALFA and are briefly described in the following subsections.

10.8.1 Subroutine COEFTR

The solution of the momentum equation, Eq. (8.2.6), is much simpler than the solution of the k-ε model equations. Since this equation is third order, we have three first-order equations, the first two given by the first two equations in Eqs. (8.2.9a, b) (2.2.1) and the third by Eq. (8.2.9a). After writing the difference equations for Eqs. (8.2.9a, b) and linearizing them, we obtain Eqs. (8.2.20a, b) and (8.2.21a, b). The third equation is given by Eq. (8.2.20c) with (r_2) given by Eq. (8.2.21c)

The linearized boundary conditions correspond to Eqs. (8.2.23) at $\eta = 0$ and to $\delta u_J = 0$ at $\eta = \eta_j$. This system of equations is again written in matrix-vector form given by Eq. (8.2.24) with A_j, B_j and C_j matrices given by Eqs. (8.2.27) and $\vec{\delta}_j$ and $\vec{r}_j$ by Eq. (8.2.26).

The solution of Eq. (8.2.24) is again obtained with the block climination method described in subsection 8.2.3

This subroutine contains the coefficients of the linearized momentum equation given by Eqs. (8.2.20c), (8.2.20a,b), and (8.2.21c). Since the calculations are for turbulent flow only, these coefficients for the first two computed x-stations are slightly different due to the use of two-point backward difference formulas for the streamwise derivatives in the momentum equation. This is needed to avoid oscillations caused by the specified initial velocity profiles. At the third x-station, the calculations revert back to the central differences for the streamwise derivatives described in [1]. In this case the coefficients $(s_1)_j$ to $(s_6)_j$ and $(r_2)_j$ are given by Eqs. (8.2.22) and (8.2.21c).

10.8.2 Subroutine SOLV3

This subroutine is the same as the one described in subsection 10.3.7.

10.8.3 Subroutines EDDY, GAMCAL, CALFA

These subroutines use the CS algebraic eddy viscosity formulation discussed in Section 5.2. In terms of transformed variables, $(\varepsilon_m^+)_i$ and $(\varepsilon_m^+)_0$ are given by

$$(\varepsilon_m^+)_i = 0.16\eta^2\sqrt{R_x}v\Big\{1 - \exp\Big(-R_x^{1/4}v_w^{1/2}/26/c_n\Big)\Big\}\gamma_{tr} \qquad (10.8.1a)$$

$$(\varepsilon_m^+)_0 = -\alpha\sqrt{R_x}(\eta_J - f_J)\gamma_{tr}\gamma \qquad (10.8.1b)$$

where

$$c_n = \frac{m}{R_x^{1/4}v_w^{3/4}} \qquad (10.8.2)$$

Subroutine EDDY contains the expressions for the inner and outer regions. The intermittency expression used in the outer eddy viscosity formula is calculated in subroutine GAMCAL and the variable α in subroutine CALFA.

10.9 Differential Method with CS and *k*-ε Models: *k*-ε Model

The structure of the k-ε model, which includes the zonal method and the model for low and high Reynolds number flows, is similar to the CS model described above. It consists of the subroutines described below.

10.9.1 Subroutines KECOEF, KEPARM, KEDEF and KEDAMP

Again we need a subroutine for the coefficients of the linearized equations for momentum, turbutlent kinetic energy and rate of dissipation. We also need to generate initial profiles for the kinetic energy and rate of dissipation equations for both low and high Reynolds number flows. We do not need to generate the initial turbulent velocity profile for the momentum equation since it is already generated by subroutine IVPT discussed in subsection 10.7.3. Then we need an algorithm, like SOLV3, to solve the linear system of equations for the zonal method and k-ε model with and without wall functions for low and high Reynolds number flows.

To simplify the coding and discussion and the application of this computer program to other turbulence models, we use three additional subroutines to define the coefficients of the linearized equations for momentum, kinetic energy and rate of dissipation given in subroutine KECOEF. The first of these three subroutines is subroutine KEPARM, which calculates the parameters b_1, b_2, b_3 and production and dissipation terms and their linearized terms such as $\left(\frac{\partial b_2}{\partial k}\right)_j^n, \left(\frac{\partial b_3}{\partial \varepsilon}\right)_j^n, \left(\frac{\partial P}{\partial \varepsilon}\right)_j^n, \left(\frac{\partial Q}{\partial k}\right)_j^n, \left(\frac{\partial P}{\partial v}\right)_j^n$, etc. in the equations for kinetic energy and rate of dissipation.

The second of these three subroutines is subroutine KEDEF, which calculates D, E, F terms, (see subsection 6.2.1), and their linearized terms such as $\left(\frac{\partial E}{\partial \varepsilon}\right)_j^n, \left(\frac{\partial F}{\partial \varepsilon}\right)_j^n$, etc. in k-ε model associated with low Reynolds number effects, which in the present program correspond to the models of Huang-Lin and Chien discussed in Section 6.2.

The third of these subroutine is subroutine KEDAMP, which calculates near-wall damping terms $f_1, f_2, f_\mu, \sigma_k, \sigma_\varepsilon$ and their linearized terms which are for low Reynolds numbers and are model dependent.

The linearized coefficients of the momentum equation in subroutine KECOEF use both two and three point backward finite-difference approximations for the streamwise derivatives. For $j \le j_s$, the coefficients $(s_1)_j$ to $(s_6)_j$ are given by Eq. (8.2.22) for the CS model. At $j = j_s$.

$$(\varepsilon_m)_{CS} = (\varepsilon_m)_{k\text{-}\varepsilon}$$

and $(s_7)_j$ to $(s_{12})_j$ are given by the following equations,

$$(s_7)_j = h_j^{-1} v_j \left(\frac{\partial b}{\partial k}\right)_j \tag{10.9.1a}$$

$$(s_8)_j = -h_j^{-1} v_{j-1} \left(\frac{\partial b}{\partial k}\right)_{j-1} \tag{10.9.1b}$$

$$(s_9)_j = h_j^{-1} v_j \left(\frac{\partial b}{\partial s}\right)_j \tag{10.9.1c}$$

$$(s_{10})_j = -h_j^{-1} v_{j-1} \left(\frac{\partial b}{\partial s}\right)_{j-1} \tag{10.9.1d}$$

$$(s_{11})_j = h_j^{-1} v_j \left(\frac{\partial b}{\partial \varepsilon}\right)_j \tag{10.9.1e}$$

$$(s_{12})_j = -h_j^{-1} v_{j-1} \left(\frac{\partial b}{\partial \varepsilon}\right)_{j-1} \tag{10.9.1f}$$

for the k-ε model.

This subroutine also presents the coefficients of the kinetic energy equation, $(\alpha_1)_j$ to $(\alpha_{12})_j$ and $(r_4)_j$ in Eq. (9.2.35) and the coefficients of the rate of dissipation equation, $(\beta_1)_j$ to $(\beta_{14})_j$ $(\tau_5)_j$ in Eq. (9.2.36).

To discuss the procedure for obtaining the coefficients of the kinetic energy and rate of dissipation equations, consider Eq. (9.2.7). With x-wise derivatives represented either by two- or three-point backward differences, the finite difference approximations to Eq. (9.2.7) are given by Eq. (9A.5) and linearized Eqs. by (9A.6).

The coefficients of Eq. (9.2.35) are now given by Eqs. (9A.7a) to (9A.7l) and (9A.8a).

Remembering the definitions of the diffusion, production, dissipation, convention and F terms. Eq. (9A.8a) can also be written as Eq. (9A.8b)

The parameters P, Q and F are model-dependent. As a result, the derivatives with respect to k, ε and υ will be different for each model. The derivations of $\frac{\partial k}{\partial x}$ and $\frac{\partial f}{\partial x}$ with respect to k and f are straight forward.

In term of transformed variables the parameters, P, Q and F in Hung and Lin's model, for example, are (here ε is $\bar{\varepsilon}$) given by Eqs. (9A.9a) to (9A.9c)

We now consider the rate of dissipation equation given by Eq. (2.2.4). Following a procedure similar to the one used for the kinetic energy equation, the finite-difference approximation for this equation is given by Eq. (9A.13)

After linearization, Eq. (9A.13) can be expressed in the form given by Eq. (9.2.3b) with $(\beta_1)_j$ to $(\beta_{14})_j$ and $(r_5)_j$ given by Eqs. (9A.14a) to (9A.15a). The latter equation, (9A.15a), can also be written in the form given by Eq. (9A.15b).

10.9.2 Subroutine KEINITK

This subroutine generates the initial k-profile for low and high Reynolds numbers as well as the profile for the zonal method. For high Reynolds number flows, the kinetic energy profile k is determined by first calculating the shear stress τ from

$$\tau = (\varepsilon_m)_{CS} \frac{\partial u}{\partial y} \tag{10.9.2}$$

and using the relation between τ and k,

$$k = \frac{\tau}{a_1} \tag{10.9.3}$$

with $a_1 = 0.30$. The calculation of τ is easily accomplished in subroutine IVPT once the initial velocity profile is generated in that subroutine.

For low Reynolds number flows, we assume that the ratio of τ^+/k^+ is given by

$$\frac{\tau^+}{k^+} = \begin{cases} a(y^+)^2 b(y^+)^3 & y^+ \leq 4.0 \quad (10.9.4a) \\ c_1 + c_2 z + c_3 z^2 + c_4 z^3 & 60 \leq y^+ < 4.0 \quad (10.9.4b) \\ 0.30 & y^+ > 60 \quad (10.9.4c) \end{cases}$$

where $z = \ln y^+$. The constants in Eq. (11.9.4a) are determined by requiring that at $y^+ = 4$.

$$\frac{\tau^+}{k^+} = 0.054. \quad \left(\frac{\tau^+}{k^+}\right)' = 0.0145 \tag{10.9.5}$$

according to the data of [4]

The constants c_1 to c_4 in Eq. (10.9.4b) are taken as

$$\begin{aligned} c_1 &= 0.080015, \quad c_2 = -0.11169 \\ c_3 &= 0.07821, \quad c_4 = -0.0095665 \end{aligned} \tag{10.9.6}$$

10.9.3 Subroutine KEINITG

In this subroutine the rate of dissipation profile ε is determined by assuming

$$(\varepsilon_m)_{\text{CS}}(\varepsilon_m)_{k\text{-}\varepsilon} = f_\mu c_\mu \frac{k^2}{\varepsilon} \tag{10.9.7}$$

or from

$$\varepsilon = \frac{f_\mu c_\mu k^2}{(\varepsilon_m)_{\text{CS}}}$$

where $(\varepsilon_m)_{\text{CS}}$ is determined from the CS-eddy viscosity model in subroutine EDDY. Whereas f_μ is constant for high Reynolds number flows with a typical value of 1.0, it is not constant for low Reynolds number flows. Its variation differs according to different models developed close to the wall, say $y^+ \leq 60$ [4].

10.9.4 Subroutine KEWALL

This subroutine provides the wall boundary conditions for the k-ε model which includes low (with wall functions) and high Reynolds number (without wall functions) flows as well as the zonal method. For low Reynolds numbers, there are four physical wall boundary conditions and one "numerical" boundary condition. They are given by Eqs. (9.2.9)

For high Reynolds numbers, the "wall" boundary conditions are specified at a distance $y_0 = (\nu/u_\tau)y_0^+$. In this case we have a total of five boundary conditions.

10.9.5 Subroutine KESOLV

This subroutine performs both forward and backward sweeps for low and high Reynolds numbers, including the zonal method, by using the block elimination method. When the perturbation quantities *um* (1, *j*) to *um* (8, *j*) are calculated so that new values of f_j, u_j, v_j, etc., can be calculated, a relaxation parameter *rex* is used in order to stabilize the solutions.

In this subroutine, for the zonal method we also reset k, ε in the inner region only. Since the CS model is used for the inner region, there is no need for these quantities. For safety, they are arbitrarily defined in this region.

10.9.6 Test Cases for the CS and k-ε Models

There are five test cases for this computer program. They all use the notation employed in the Stanford Conference in 1968 [1]. For example, flow 1400 corresponds to a zero-pressure gradient glow. Flow 2100 has favorable, nearly-zero and adverse-pressure-gradient flow. All calculations are performed for Model $= 1, 2, -1$, and -2 (see subroutine 10.9.2). The predictions of four models with experimental data are given for c_f, δ^* and R_θ as a function of x in the companion site, store.elsevier.com/companions/9780080983356.

A summary of the freestream and initial conditions for each flow are summarized below.

1. Flow 1400: Zero-Pressure-Gradient Flow
   NXT $= 61$, $u_e/u_{ref} = 1.0$, $u_{ref} = 33$ ms^{-1},
   $c_f = 3.17 \times 10^{-3}$, $R_\theta = 3856$, $\nu = 1.5 \times 10^{-5}$ m^{2}s^{-1}, REF $= 1$
2. Flow 2100: Favorable, Zero and Adverse-Pressure-Gradient Flow
   NXT $= 81$, $u_{ref} = 100$ ft s^{-1}
   $c_f = 3.10 \times 10^{-3}$, $R_\theta = 3770$, $\nu = 1.6 \times 10^{-4}$ ft^{2}s^{-1}, REF $= 1$
3. Flow 1300: Accelerating Flow
   NXT $= 81$, $u_{ref} = 100$ ms^{-1},
   $c_f = 4.61 \times 10^{-3}$, $R_\theta = 1010$, $\nu = 1.54 \times 10^{-5}$ ft^{2}s^{-1}, REF $= 1$
4. Flow 2400: Relaxing Flow
   NXT $= 81$, $u_e/u_{ref} =$ tabulated values, $u_{ref} = 1$,
   $c_f = 1.42 \times 10^{-3}$, $R_\theta = 27{,}391$, $\nu = 1.55 \times 10^{-4}$ m^{2}s^{-1}, REF $= 1$
5. Flow 2900: Boundary Layor Flow in a Diverging Channel
   NXT $= 81$, $u_c/u_{ref} =$ tabulated values, $u_{ref} = 1$,
   $c_f = 1.77 \times 10^{-3}$, $R_\theta = 22.449.2$, $\nu = 1.57 \times 10^{-4}$ft^{2}s^{-1}, REF $= 1$

The input and output for each flow are given in tabular and graphical form and are included with the computer program. Figure 10.2 shows a comparison between the calculated results and experimental data for flow 1400. The calculations for Model $= 1, 2, -1$ and -2 correspond to low Reynolds number flows with Huang-Lin and Chien models, zonal method and high Reynolds number flows, respectively.

10.9.7 Solution Algorithm

When the system of first-order equations to be solved with the block elimination method becomes higher than, say 6, the preparation of the solution algorithm with the recursion formulas described in subroutine SOLV3 becomes tedious. A matrix-solver algorithm (MSA) discussed here can be used to perform the matrix operations required in the block elimination method. This algorithm consists of three

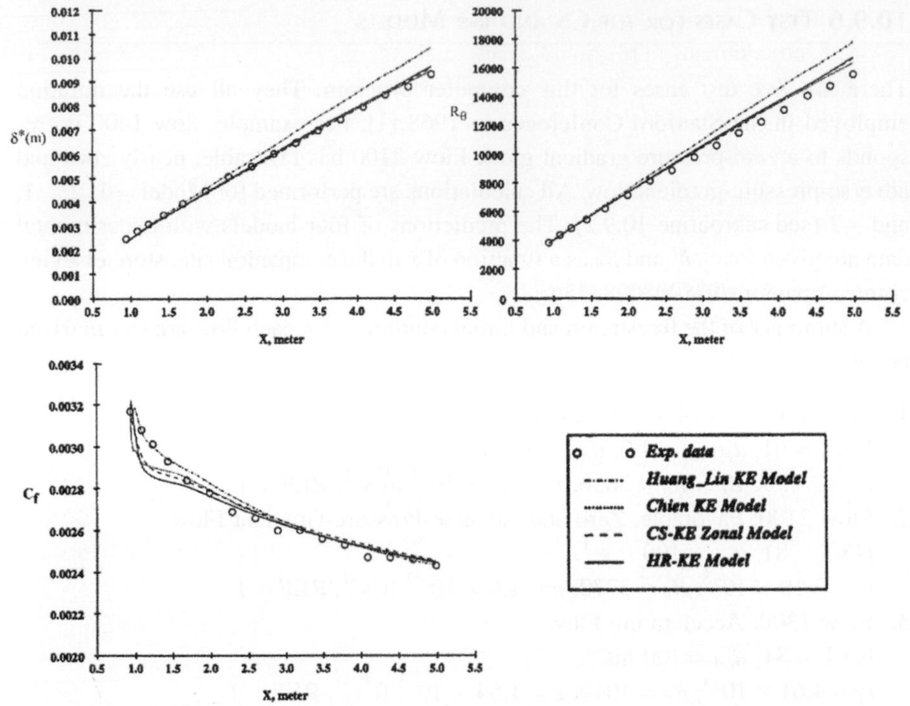

Fig. 10.2 Comparison of calculated results with the experimental data for flow 1400.

subroutines, namely, subroutines GAUSS, GAMSV and USOLV. To illustrate its use, we discuss the replacement of SOLV3 with MSA.

(1) Read in

```
DIMENSION DUMM(3), BB(2,3), YY(3,81),NROW(3,81),GAMJ(2,3,81).
          AA(3,3,,81),CC(2,3,81)
DATA IROW,ICOL,ISROW,INP/3,3,3,81/
```

Here IROW, ICOL correspond to number of maximum rows and columns respectively. ISROW denotes the number of "wall" boundary conditions and INP the total number of j-points in the η-direction, and

$$\text{BB} = B_j, \quad \text{YY} = \vec{w}_j, \quad \text{GAMJ} = \Gamma_j, \quad \text{AA} = A_j, \quad \text{CC} = C_j$$

The first and second numbers in the arguments of AA, BB, CC and GAMJ correspond to the number of nonzero rows and columns in A_j (or Δ_j), B_j, C_j and Γ_j matrices, respectively. Note that B_j and Γ_j have the same structure and the last row of B_j and the-first two rows of C_j are all zero. The index 81 in YY, NROW, GAMJ, AA and CC refers to INP.

(2) Set the elements of all matrices, A_j, B_j, C_j (and Δ_j) equal to zero.

(3) Define the matrices A_0 and C_0 by reading in their elements. Note that only those nonzero elements in the matrices are read in since in (2) we set all the elements equal to zero.
(4) Call subroutine GAUSS.
(5) Read in the elements of B_j and call subroutine GAMSV to compute Γ_1.
(6) Define A_j according to Eq. (9.2.33b), call GAUSS and read in the elements of C_j.
(7) Recall the elements of B_j and call GAMSV to compare Γ_2.
(8) Repeat (6) and (7) for $j < J$.
(9) At $j = J$. read in the last row of A_J which is also equal to the last row of Δ_J.
(10) Compute $\vec{w}_0$ according to Eq. (8.2.29a). Here $\vec{r}_0$ = RRR (1.81).
(11) Define the right-hand side of Eq. (8.2.29b) and compute $\vec{w}_j$ according to Eq. (8.2.29b)
(12) In the backward sweep, with δ_j corresponding to UM(I.J), compute $\vec{\delta}_J$ according to Eq. (8.2.30a) by calling USOLV at INP.
(13) Define the right-hand side of Eq. (8.2.30b) and solve for δ_j by calling USOLV for $j = J - 1, J - 2, \ldots, 0$.

This algorithm is very useful to solve the linear system for the k-ε model equations. With all A_j, B_j, C_j matrices and r_j nicely defined in subroutine KECOEF, the solution of Eq. (8.2.24) is relatively easy.

10.10 Differential Method with CS and *k*-ε Models: Basic Tools

The computer program also includes basic tools to perform smoothing, differentiation, integration, and interpolation. For example, subroutine DIFF-3 provides first. second and third derivatives of the input function at inputs. First derivatives use weighted angles, second and third derivatives, use cubic fits. Subroutine INTRP3 provides cubic interpolation. Given the values of a function (F1) and its derivatives at N1 values of the independent variable (X1), this subroutine determines the values of the function (F2) at N2 values of the independent variable (X2). Here X2 can be in arbitrary order.

Another subroutine used for interpolation is subroutine LNTP: it performs linear interpolation.

10.11 Differential Method with SA Model

This computer program called BLPSA is the extension of BLP2 with the procedure described in Problems 9.1 to 9.7. Many of the subroutines used in BLP2 remain the same except for some minor changes. Several additional subroutines are added. Subroutine COEF contains the coefficients of linearized continuity, momentum and eddy viscosity equations (Problem 9.4), subroutine MSA (subsection 10.10.3) is the solution algorithm to solve the linear system in Problem 9.4.

The companion site, store.elsevier.com/companions/9780080983356 has two test cases for this program and also includes another program for this model in which the continuity, momentum and eddy transport equations are all solved together. This program is referred to as 5×5 in contrast to the other one which is referred to as 2×2.

10.12 Differential Method for a Plane Jet

See Problem 9.8.

10.13 Useful Subroutines

In this section we present two subroutines that are useful to solve some of the problems in Chapters 8 and 9. They are briefly described below and are given in the companion site, store.elsevier.com/companions/9780080983356.

10.13.1 Subroutine IVPT

This subroutine is for generating initial velocity profiles turbulent flows with the method discussed in Problem 8.9. It requires the initial values of Reynolds number based on momentum thickness R_θ ($\equiv u_e\theta/\nu$) and local skin friction coefficient $c_f(\equiv 2\tau_w/\varrho u_e^2)$.

10.13.2 Subroutine SOLV2

This subroutines is similar to the solution algorithm, SOLV3, in subsection 8.2.4. It is designed to solve two first-order equations with the block-elimination method subject to the boundary conditions given by Eq. (P8.2.4). See also Problem 8.2.

10.14 Differential Method for Inverse Boundary-Layer Flows with CS Model

This computer program consists of a MAIN and 15 subroutines, INPUT, IVPL, HIC, EDDY, SWTCH, COEF, WAKEPR, DIFF1, LNTP, INTEG, AMEAN, SOLVA4,

EDGCHK, CALFA and GAMCAL MAIN, as before, is used to control the logic of the computations. Here the parameter g_i, in Eq. (8.8.7b) is also calculated with

$$\text{SUM1} = \sum_{j=1}^{i-1} C_{ij}\overline{D}_j \tag{10.14.1a}$$

and

$$\text{SUM2} = \sum_{j=i+1}^{N} C_{ij}\overline{D}_j \tag{10.14.1b}$$

The initial displacement thickness (δ^*) distribution needed in the calculation of $\overline{D}_j$ is computed in subroutine INPUT by assuming a δ^* distribution flat-plate flow and given by

$$\frac{\delta^*}{x} = 0.036HR_x^{-0.20} \tag{10.14.2}$$

with $H = 1.3$.

Of the 15 subroutines, subroutine WAKEPR is used to modify the profiles resulting from wall boundary layers for wake profiles. Except for this subroutine and except for subroutines INPUT, IVPL and HIC, the remaining subroutines are similar to those described in Sections 10.9.2 and 10.9.3. For this reason, only these three subroutines are described below.

10.14.1 Subroutine INPUT

This subroutine is used to generate the grid, calculate γtr in the eddy viscosity formulas, initial δ^* -distribution, and pressure gradient parameters m and m_1. The following data are read in and the number of j-points J(NP) is computed from Eq. (10.3.3)

NXT	Total number of x-stations
NXTE	Total number of x-stations on the body
NXS	NX-station after which inverse calculations begin
RL	Reynolds number, $u_\infty c/\nu$
XTR	x/c value for transition location
ETAE	Transformed boundary layer thickness η_e at $x = 0$, ETAE = 8.0 K is the variable-grid parameter. Take $K = 1.0$ for laminar flow and $K = 1.14$ for turbulent flow. For a flow consisting of both laminar and turbulent regions, take $K = 1.14$
DETA(1)	Δ_η/h_1-initial step size of the variable grid system. Take $h_1 = 0.01$ for turbulent flows
P2(1)	m at $x = 0$ (NX = 1)
x/c, y/c	Dimensionless airfoil coordinates
u_e/u_∞	Dimensionless external velocity

10.14.2 Subroutine HIC

This subroutine calculates the coefficients of the Hilbert integral denoted by C_{ij}. While they can be generated from any suitable integration procedure, we use the following procedure which is appropriate with the box method [5].

We calculate

$$H_i = \int_{\xi^i}^{\xi^L} G(\sigma) \frac{d\sigma}{\xi^i - \sigma} \tag{10.14.3}$$

where

$$G(\sigma) = \frac{dF}{d\sigma}$$

with F denoting any function, so that over each subinterval (ξ^{n-1}, ξ^n), except the two enclosing the point $\xi = \xi^i$, we replace $G(\sigma)$ by its midpoint value:

$$\int_{\xi^{n-1}}^{\xi^n} \frac{G(s)d\sigma}{\xi^i - \sigma} = G_{n-1/2} \int_{\xi^{n-1}}^{\xi^n} \frac{d\sigma}{\xi^i - \sigma} = G_{n-1/2} \ln\left|\frac{\xi^i - \xi^{n-1}}{\xi^i - \xi^n}\right| \tag{10.14.4}$$

Making the further approximation,

$$G_{n-1/2} - \frac{F_n - F_{n-1}}{\xi^n - \xi^{n-1}}$$

we can write

$$\int_{\xi^{n-1}}^{\xi^n} \frac{dF}{d\sigma} \frac{d\sigma}{\xi^i - \sigma} = E_n^i (F_n - F_{n-1}) \tag{10.14.5}$$

where for $n \neq i$ or $i + 1$

$$E_n^i = (\xi^n - \xi^{n-1}) \ln\left|\frac{\xi^i - \xi^{n-1}}{\xi^i - \xi^n}\right| \tag{10.14.6}$$

for the two subintervals ξ^{i-1} to ξ^i and ξ^i to ξ^{i+1}. Because of the cancellation with the constant term, account should be taken of the linear variation of G from one interval to the next. Thus, we take the linear interpolation

$$G = \frac{G_{i-1/2}(\xi^{i+1} - \xi^i) + G_{i+1/2}(\xi^i - \xi^{i-1}) + 2\left(G_{i+1/2} - G_{i-1/2}\right)(\sigma - \xi^i)}{(\xi^{i+1} - \xi^{i-1})}$$

so that

$$\int_{\xi^{i-1}}^{\xi^{i+1}} \frac{G\,d\sigma}{\xi^i - \sigma} = \frac{G_{i-1/2}\left(\xi^{i+1} - \xi^i\right) + G_{i+1/2}\left(\xi^i - \xi^{i-1}\right)}{\xi^{i+1} - \xi^{i-1}} \ln\left|\frac{\xi^i - \xi^{i-1}}{\xi^i - \xi^{i+1}}\right| -2\left(G_{i+1/2} - G_{i-1/2}\right) \tag{10.14.7}$$

Replacing the midpoint derivative values by difference quotients, we obtain

$$\int_{\xi^{i-1}}^{\xi^{i+1}} \frac{dF}{d\sigma}\frac{d\sigma}{\xi^i - \sigma} = E_i^i(F_i - F_{i-1}) + E_{i+1}^i(F_{i+1} - F_i) \tag{10.14.8}$$

where

$$E_i^i = \frac{\dfrac{\xi^{i+1} - \xi^i}{\xi^{i+1} - \xi^{i-1}}\ln\left|\dfrac{\xi^i - \xi^{i-1}}{\xi^i - \xi^{i+1}}\right| + 2}{\xi^i - \xi^{i-1}} \tag{10.14.9a}$$

$$E_{i+1}^i = \frac{\dfrac{\xi^i - \xi^{i-1}}{\xi^{i+1} - \xi^{i-1}}\ln\left|\dfrac{\xi^i - \xi^{i-1}}{\xi^i - \xi^{i+1}}\right| - 2}{\xi^{i+1} - \xi^i} \tag{10.14.9b}$$

Thus

$$\begin{aligned} H_i &= E_2^i(F_2 - F_1) + E_3^i(F_3 - F_2) + \ldots + E_{L-1}^i(F_{L-1} - F_{L-2}) + E_L^i(F_L - F) \\ &= -E_2^i F_1 + \left(E_2^i - E_3^i\right)F_2 + \ldots + \left(E_{L-1}^i - E_L^i\right)F_{L-1} + E_L^i F_L \end{aligned} \tag{10.14.10}$$

so, finally the C_{ij} of Eq. (4.0.6) are given by

$$C_{ij} = \frac{1}{\pi}\left(E_j^i - E_{j+1}^i\right) \tag{10.14.11}$$

and the E^i given by Eqs. (10.14.9a) and (10.14.9b) with $E_1^i = E_{L+1}^i = 0$.

10.15 Companion Computer Programs

10.15.1 Sample Calculations for the Panel Method Without Viscous Effects

This test case is for a NACA 0012 symmetrical airfoil, with a maximum thickness of 0.12c: the pressure and external velocity distributions on its upper and lower surfaces are computed and its section characteristics determined using the panel method. The airfoil coordinates are given on the companion site store.elsevier.com/companions/9780080983356 for 184 points in tabular form. This corresponds to NODTOT = 183. Note that the x/c and y/c values are read in starting on the lower surface trailing edge (TE), traversing clockwise around the nose of the airfoil to the upper surface TE. The calculations are performed for angles of attack of $\alpha = 0°$, $8°$ and $16°$. In identifying the upper and lower surfaces of the airfoil, it is necessary to determine the x/c locations where $\bar{u}_e(\equiv u_e/u_\infty) = 0$. This location, called the stagnation point, is easy to determine since the $\bar{u}_e$ values are positive for the upper surface and negative for the lower surface. In general it is sufficient to take the stagnation point to be the x/c location where the change of sign $\bar{u}_e$ occurs. For higher accuracy, if desired, the stagnation point can be determined by interpolation between the negative and positive values of $\bar{u}_e$ as a function of the surface distance along the airfoil.

Figures 10.3 and 10.4 show the variation of the pressure coefficient C_p and external velocity $\bar{u}_e$ on the lower and upper surfaces of the airfoil as a function of x/c at three angles of attack starting from 0°. As expected, the results show that the pressure and external velocity distributions on both surfaces are identical to each other at $\alpha = 0°$. With increasing incidence angle, the pressure peak moves upstream on the upper surface and downstream on the lower surface. In the former case, with the pressure peak increasing in magnitude with increasing α, the extent of the flow deceleration increases on the upper surface and, we shall see in the following section, increases the region of flow separation on the airfoil. On the lower surface, on the other hand, the region of accelerated flow increases with incidence angle which leads to regions of more laminar flow than turbulent flow.

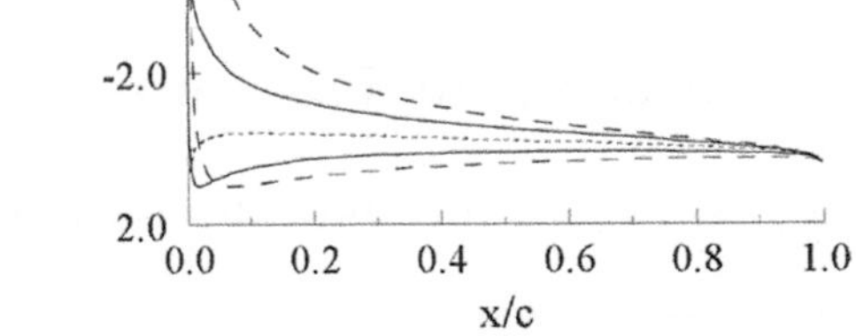

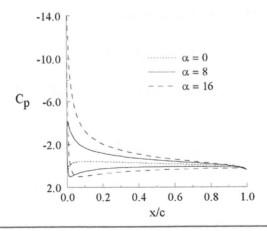

Fig. 10.3 Distribution of pressure coefficient on the NACA 0012 airfoil at $\alpha = 0°$, $8°$, and $16°$.

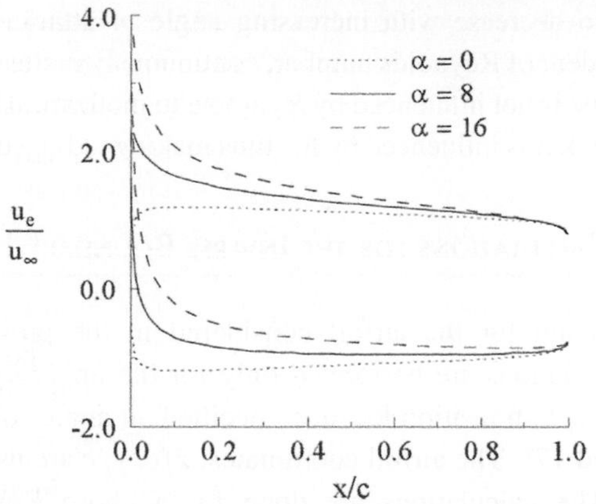

Fig. 10.4 Distribution of dimensionless external velocity on the NACA 0012 airfoil at $\alpha = 0°$, $8°$ and $16°$.

These results indicate that the use of inviscid flow theory becomes increasingly less accurate at higher angles of attack since, due to flow separation, the viscous effects neglected in the panel method become increasingly more important. This is indicated in Fig. 10.5, which shows the calculated inviscid lift coefficients for this airfoil together with the experimental data reported in [4] for chord Reynolds numbers, $R_c(\equiv u_\infty c/\nu)$, of 3×10^6 and 6×10^6. As can be seen, the calculated inviscid flow results agree reasonably well with the measured values at low and modest angles of attack. With increasing angle of attack, the lift coefficient reaches a maximum, called the maximum lift coefficient, $(c_\ell)_{max}$, at an angle of attack, α, called the stall angle. After this angle of attack, while the experimental lift

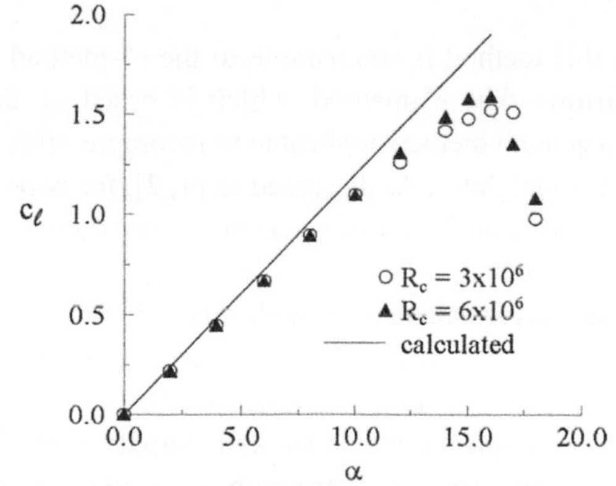

Fig. 10.5 Comparison of calculated (solid lines) and experimental (symbols) lift coefficients for the NACA 0012 airfoil.

coefficients begin to decrease with increasing angle of attack, the calculated lift coefficient, independent of Reynolds number, continuously increases with increasing α. The lift curve slope is not influenced by R_c at low to modest angles of attack, but at higher angles of attack it is influenced by R_c, thus making $(c_\ell)_{\max}$ dependent upon R_c.

10.15.2 Sample Calculations for the Inverse Boundary-Layer Program

This test case is again for the airfoil considered in the previous section. The boundary-layer calculations are performed only for the upper surface, for laminar and turbulent flows with transition location specified, at angles of attack of $\alpha = 4°$, $8°$, $12°$, $14°$, $16°$ and $17°$. The airfoil coordinates, x/c, y/c are used to calculate the surface distance. The calculations are done for a chord Reynolds number of 4×10^6.

In practice, it is also necessary to calculate the transition location. Two practical methods for this purpose are the Michel method and the e^n-method described, for example, in [1, 5]. The former is based on a empirical correlation between two Reynolds numbers based on momentum thickness, R_θ, and surface distance R_x. It is given by Eq. (10.15.1), also Eq. (5.3.22),

$$R_{\theta_{tr}} = 1.174\left(1 + \frac{22,400}{R_{x_{tr}}}\right) R_{x_{tr}}^{0.46} \tag{10.15.1}$$

where

$$R_\theta = \frac{u_e \theta}{\nu}, \quad R_x = \frac{u_e x}{\nu}$$

The accuracy of this method is comparable to the e^n-method at high Reynolds number flows on airfoils. The e^n-method, which is based on the linear stability theory, is, however, a general method applicable to incompressible and compressible two- and three-dimensional flows. As discussed in [1, 2], for two-dimensional flows at low Reynolds numbers, transition can occur inside separation bubble and can be predicted only by the e^n-method. For details, see [1, 2].

While the boundary-layer calculations with this program can be performed for standard and inverse problems, here they are performed for the standard problem, postponing the application of the inverse method to the following section.

Here we present a sample of the input and output of the calculations. The format of the inverse boundary-layer program is similar to the format of the interactive code and is discussed in the following section. Figure 10.6 shows the distribution of local skin friction coefficient, c_f, and dimensionless displacement

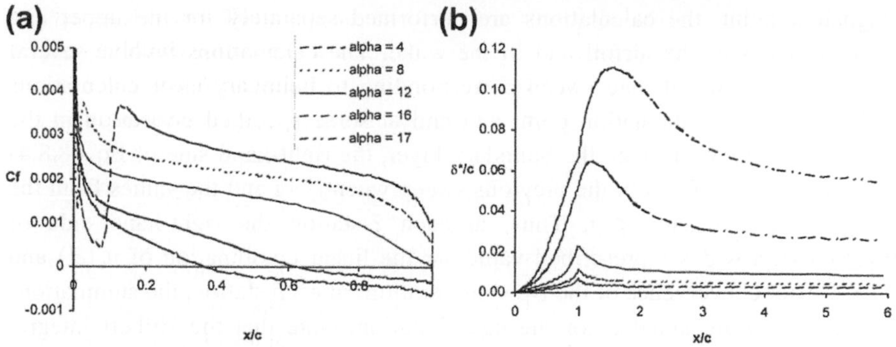

Fig. 10.6 Variation of (**a**) c_f and (**b**) δ^*/c on the NACA 0012 airfoil and its wake at several angles of attack for $R_c = 4 \times 10^6$.

thickness, δ^*/c, for several angles of attack. These results were obtained for the external velocity distribution provided by the panel method without viscous corrections. The boundary-layer calculations were performed in the inverse mode and several sweeps on the airfoil and in its wake were made. As can be seen, at low or medium angles of attack, there is no flow separation on the airfoil corresponding to the vanishing of c_f or v_w. At higher angles, however, as expected, the flow separates near the trailing edge and moves forward with increasing angle of attack. It is interesting to note that at $\alpha = 16°$, the flow separation occurs at $x/c = 0.6$ and at $\alpha = 17°$ at $x/c \cong 0.37$. As we shall see in the next section, interaction between inviscid and viscous results reduces the flow separation on the airfoil considerably. The results also show that, again as expected, transition location occurs very close to the stagnation point at higher angles of attack.

10.15.3 Sample Calculations with the Interactive Boundary-Layer Program

A combination of an inviscid method with a boundary-layer method allows the inviscid and viscous flow calculations to be performed in an interactive way. Using an inverse boundary-layer method allows similar calculations to be performed for flows including separation.

Before we present sample calculations with the interactive boundary-layer program, it is first useful to discuss the computational strategy in this program. For a specified angle of attack α and airfoil geometry $(x/c, y/c)$, the calculations are first initiated with the panel method in order to calculate the external velocity distribution and the lift coefficient. The external velocity distribution is then input to the inverse boundary-layer program in which, after identifying the airfoil stagnation point, the calculations are performed separately for the upper and lower surfaces of the airfoil and in the wake. The calculations involve several sweeps on the airfoil, one sweep corresponding to boundary-layer calculations which start at the stagnation point and end at some specified ξ-location in the wake. In sweeping through the boundary-layer, the right-hand side of Eq. (8.8.4) uses the values of δ^* from the previous sweep when $j > i$ and the values from the current sweep when $j < i$. Thus, at each ξ-station the right-hand side of Eq. (8.8.4) provides a prescribed value for the linear combination of $u_e(\xi^i)$ and $\delta^*(\xi^i)$. After convergence of the Newton iterations at each station, the summations of Eq. (8.8.4) are updated for the next ξ-station. Note that the Hilbert integral coefficients C_{ij} discussed in subsection 10.14.2 are computed and stored at the start of the boundary-layer calculations.

At the completion of the boundary-layer sweeps on the airfoil and in the wake, boundary-layer solutions are available on the airfoil and in the wake. The blowing velocity on the airfoil v_{iw} [see Eqs. (8.7.4) and (8.7.5)] and a jump in the normal velocity component Δv_i in the wake [see Eq. (8.7.6)], for which an incompressible flow are

$$v_{iw} = \frac{d}{dx}(v_{iw}\delta_A^*) \tag{10.15.2}$$

$$\Delta v_i = \frac{d}{dx}(u_{iu}\delta_u^*) + \frac{d}{dx}(u_{il}\delta_\ell^*) \tag{10.15.3}$$

are calculated and are used to obtain a new distribution of external velocity $u_{ei}(x)$ from the inviscid method. As before, the onset of transition location is determined from the laminar flow solutions and the boundary-layer calculations are performed on the upper and lower surfaces of the airfoil and in the wake by making several specified sweeps. This sequence of calculations is repeated for the whole flowfield until convergence is achieved.

The format of the input to this interactive boundary-layer (inviscid/viscous) program is similar to the input required for the inverse boundary-layer described in subsection 10.14.1. The code is arranged in such a way that it is only necessary to read in the airfoil geometry, the angles of attack to be calculated, Mach number and chord Reynolds number. The rest of the input is done internally.

We now present sample calculations for the NACA0012 airfoil for Reynolds numbers corresponding to 3×10^6. In this case, transition locations are calculated with Michel's formula. The calculations and the results are given on the companion site.

Lift, c_l, drag, c_d, pitching moment, C_m, coefficients for $R_c = 3 \times 10^6$ are shown in Table 10.1 for $\alpha = 2°$ to $16.5°$ and $M_\infty = 0.1$ together with lift coefficients calculated with the panel method. As can be seen, while at low and

TABLE 10.1 Results for the NACA 0012 airfoil ar Rc = 3× 106, M∞ = 0.1

α	Cl_{in}	Cl_{vi}	C_d	$C_{m_{ni}}$	$C_{m_{vi}}$
2.00000	0.24261	0.21099	0.00586	−0.06326	−0.04971
4.00000	0.48508	0.42567	0.00610	−0.12622	−0.10099
6.00000	0.72727	0.64337	0.00749	−0.18857	−0.15325
8.00000	0.96908	0.86241	0.00955	−0.25003	−0.20621
10.00000	1.21041	1.07109	0.01178	−0.31029	−0.25434
12.00000	1.45120	1.26253	0.01498	−0.36907	−0.29536
13.00000	1.57138	1.34396	0.01658	−0.39782	−0.31005
14.00000	1.69142	1.40836	0.01892	−0.42609	−0.31856
15.00000	1.81133	1.44754	0.02181	−0.45385	−0.31873
15.50000	1.93110	1.45653	0.02366	−0.48107	−0.31636
16.00000	1.99094	1.45811	0.02592	−0.49446	−0.31261
16.50000		1.44226	0.02837		−0.30540

modest angles of attack, the inviscid lift, $c_{l_{in}}$, and viscous lift, $c_{l_{vi}}$, coefficients agree reasonably well, at higher angles of attack, as expected, they differ from each other.

Figure 10.7 shows a comparison between the calculated and experimental values of lift and drag coefficients. The agreement is good and the stall angle

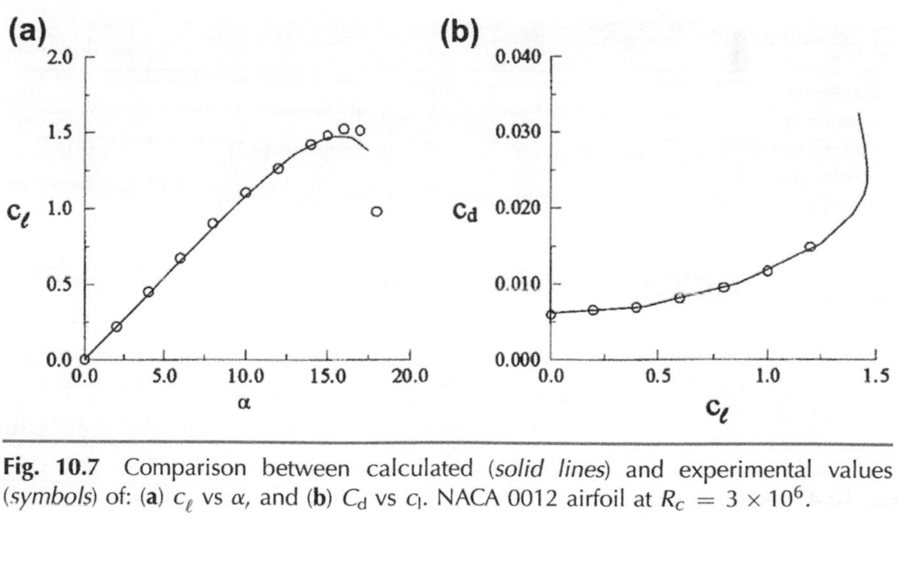

Fig. 10.7 Comparison between calculated (*solid lines*) and experimental values (*symbols*) of: (**a**) c_ℓ vs α, and (**b**) C_d vs c_l. NACA 0012 airfoil at $R_c = 3 \times 10^6$.

is reasonably well predicted. For additional comparisons with experimental data, see [2].

To describe the input and output of the computer program, we now present additional calculations for the same airfoil, this time for $R_c = 4 \times 10^6$.

The input to the IBL program (Fig. 10.8) includes airfoil geometry and/or the number of angles of attack (N), the freestream Mach number, M_∞, and the Reynolds number, R_c. The input file in the sample calculations contains the NACA 0012 airfoil coordinates which are specified by choosing either M1M4 or M1M4INP. The first choice contains only the airfoil geometry and does not contain either the angles of attack, Mach number or Reynolds number. The second choice contains airfoil geometry, angles of attack, Mach number and Reynolds number. If the first one is chosen, then it is necessary to specify N, M_∞ and R_c. For example if $N = 5$, then the angles of attack can be, say, 0°, 4°, 6°, 8° and 9°. Of course, these angles of attack as well as N can be changed.Then the calculations are started by specifying M_∞ and R_c. Figure 10.8 shows a sequence of the screens used for input.

Figure 10.9 shows the screen for starting the calculations and Fig. 10.10 shows the screen for the format of the output and the variation of lift coefficient with angle of attack. Other plots to include c_d vs α, c_m vs α and c_d vs c_l can also be obtained as shown in Fig. 10.11. Finally, the screen in Fig. 10.12 shows that one can copy the plot to the Microsoft Word file.

Figure 10.13 shows a comparison between the results of the previous section where the inviscid flow calculations did not include viscous effects, and the results of this section which include viscous effects in the panel method. Figures 10.13a and 10.13b show the strong influence of viscosity on c_l and c_d. Figure 10.13c shows that

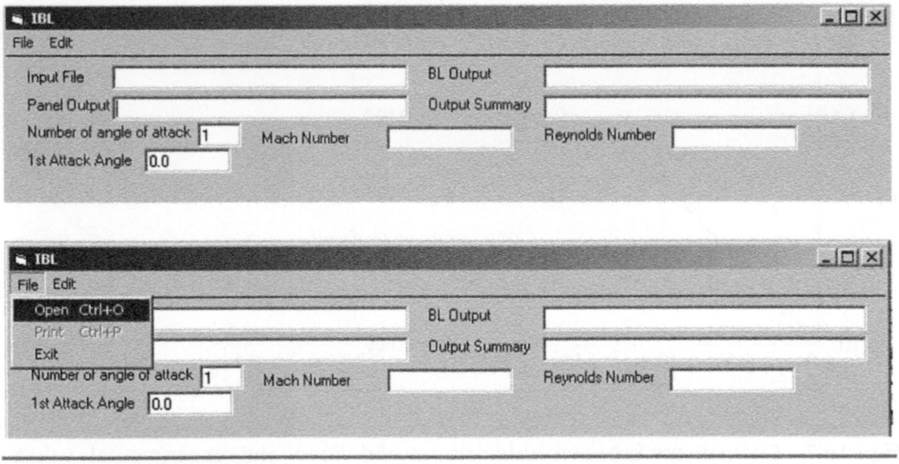

Fig. 10.8 Input format.

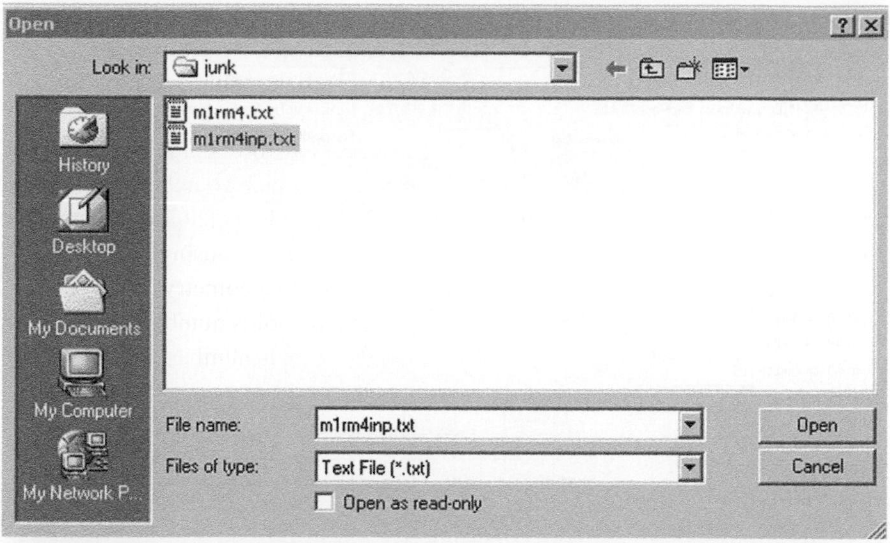

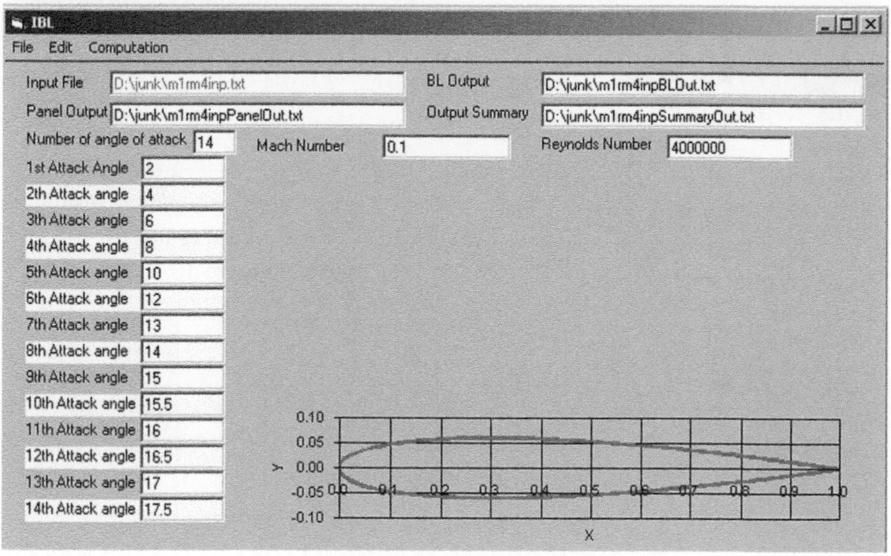

Fig. 10.8 *(Continued)*

with interaction, the extent of flow separation on the airfoil decreases. For example at $\alpha = 17^\circ$, without viscous effects in the panel method, the flow separation occurs around $x/c \cong 0.37$. With interaction, it occurs at $x/c \cong 0.62$. Similarly, with interaction, the peak in δ^*/c (Fig. 10.13d) decreases and is the reason for less flow separation on the airfoil.

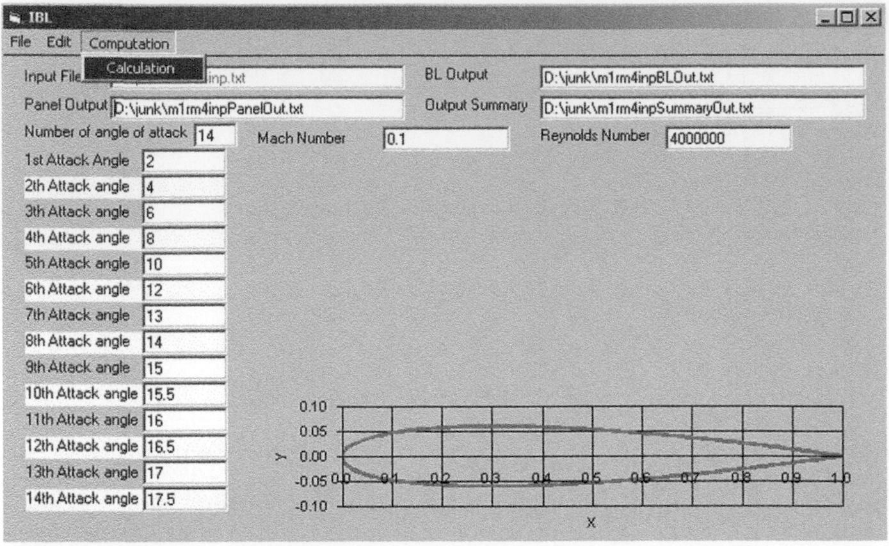

Fig. 10.9 Beginning of calculations.

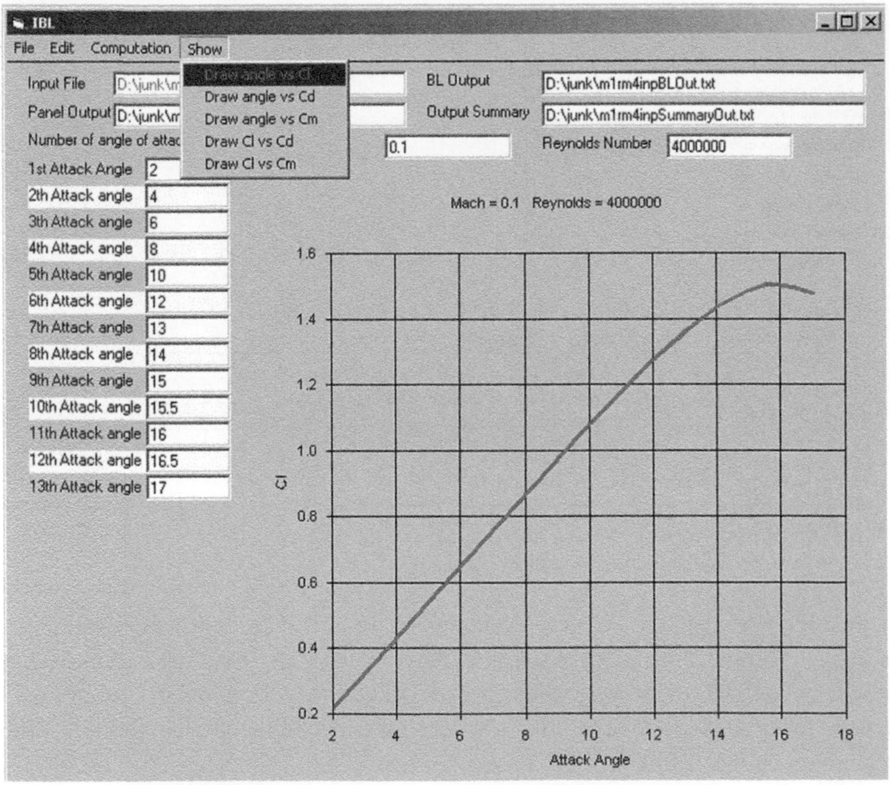

Fig. 10.10 Output format.

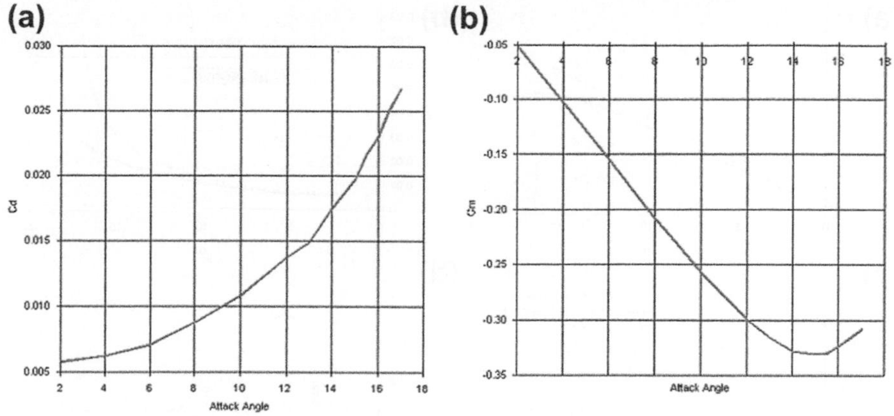

Fig. 10.11 Calculated results for the NACA0012 airfoil, $R_c = 4 \times 10^6$, $M_\infty = 0.1$. (**a**) c_d vs α, (**b**) c_m vs α.

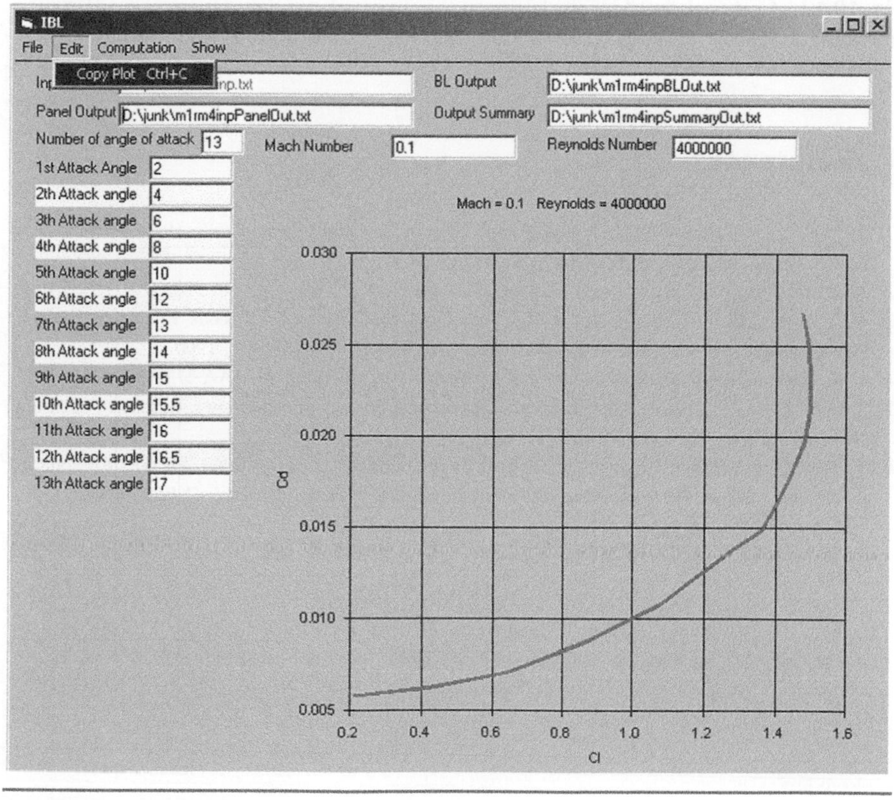

Fig. 10.12 Instruction for copying plots.

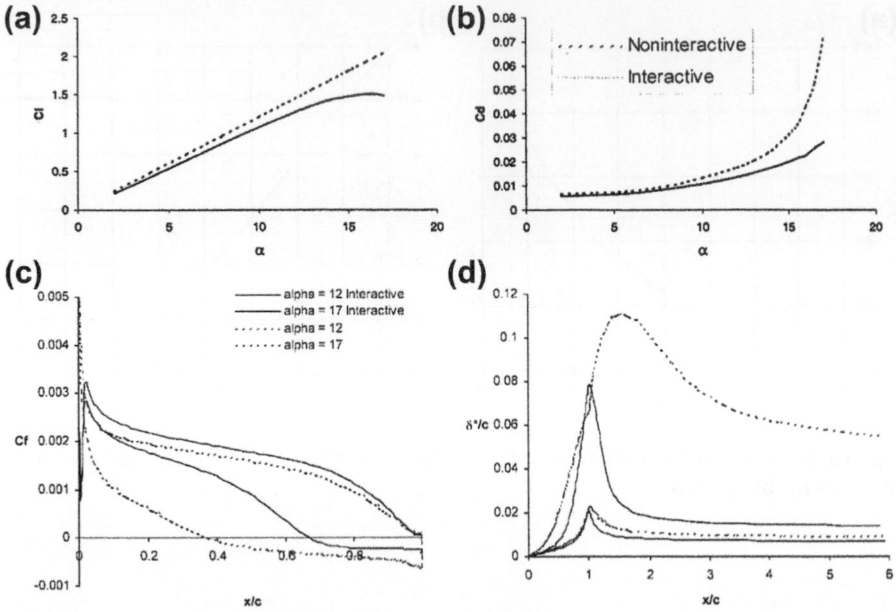

Fig. 10.13 Comparison of results between the inverse boundary-layer method and the interactive method. (**a**) c_l vs α, (**b**) c_d vs α, (**c**) c_f vs x/c, (**d**) δ^*/c vs x/c.

References

[1] T. Cebeci, J. Cousteix, Modeling and Computation of Boundary-Layer Flows, Horizons Pub. Long Beach, Calif. and Springer, Heidelberg, Germany, 1998.

[2] M.J. Lighthill, "On Displacement Thickness", J. Fluid Mech. vol. 4 (1958) 383.

[3] R.C. Lock, M.C.P. Firmin, "Survey of Techniques for Estimating Viscous Effects in External Aerodynamics", Royal Aircraft Establishment Tech Memo, AERO (1981) 1900.

[4] N.N. Mansour, J. Kim, P. Moin, Near-Wall k-ε Turbulence Modeling, J. AIAA 27 (1989) 1068–1073.

[5] T. Cebeci, An Engineering Approach to the Calculation of Aerodynamic Flows, Horizons Pub., Long Beach, Calif., and Springer, Heidelberg, Germany, 1999.

[6] D. Coles, E. A. Hirst, Computation of Turbulent Boundary Layers – 1968 AFOSR-IFP-Stanford Conference, vol. 2, Thermosciences Division, Stanford University, Stanford, Calif., 1969.

[7] J.H. Abbott, A.E. von Doenhoff, Theory of Wing Sections, Dover, 1959.

Index

Note: Page numbers followed by "f" and "t" indicate figures and tables respectively.

Printed and bound by CPI Group (UK) Ltd, Croydon, CR0 4YY
03/10/2024
01040416-0008